MICROSTRUCTURAL STABILITY OF CREEP RESISTANT ALLOYS FOR HIGH TEMPERATURE PLANT APPLICATIONS

MICROSTRUCTURAL STABILITY OF CREEP RESISTANT ALLOYS FOR HIGH TEMPERATURE PLANT APPLICATIONS

Edited by

A. Strang, J. Cawley & G.W. Greenwood

MICROSTRUCTURE OF HIGH TEMPERATURE MATERIALS
Number 2

Series Editor
A. Strang

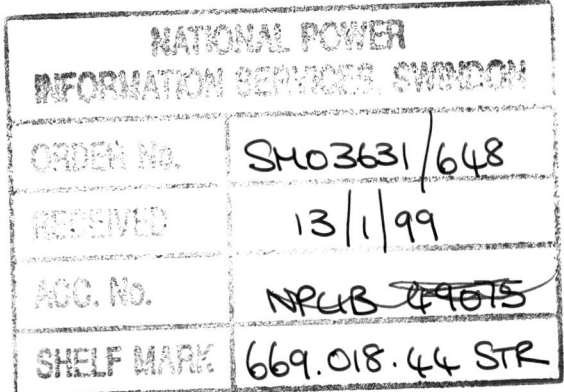

THE INSTITUTE OF MATERIALS

Book 682
Published in 1998 by
IOM Communications Ltd
1 Carlton House Terrace
London SW1Y 5DB

© IOM Communications 1998

IOM Communications Ltd
is a wholly-owned subsidiary of
The Institute of Materials

ISSN 1366–5510
ISBN 1 86125 045 2

Typeset by
Fakenham Photosetting Ltd
Fakenham, UK

Printed and bound by
The University Press
Cambridge, UK

Contents

Preface

Materials used for high temperature components in power generation, aerospace, chemical and process plant applications are known to undergo changes in their microstructure during operation in service which will effect their mechanical properties. These can take the form of particle coarsening, phase dissolution and recovery effects leading to material softening and reductions in tensile and creep strengths. In addition precipitation of new phases and diffusion of impurities to grain boundaries can also result in reduced creep ductility, impact strength and toughness as well as increased notch sensitivity. Microstructural degradation processes such as these, especially when coupled with any effects due to high temperature corrosion, can seriously compromise the integrity of critical high temperature plant components and lead to reduced operational life and possibly premature failure in service.

These microstructural changes are directly influenced by the operational conditions experienced by the material in service and are critically dependant upon the temperature and duration of exposure as well as any permanent deformation which occurs due to the imposed service stresses. Microstructural degradation also occurs at different rates in different materials and in this respect is strongly influenced by the material's chemical composition, heat treatment and initial microstructure. In consequence a clear understanding of the physical and metallurgical processes responsible for microstructural degradation during service and the mechanisms leading to subsequent changes in the mechanical properties of materials are of vital importance in determining the safe operating lives of critical high temperature plant components. Additionally, such an understanding may contribute to the design and development of new alloy systems with improved long term high temperature properties for applications in advanced high efficiency generating process plant.

These are the proceedings of the second in a series of specialist international conferences aimed at focusing attention on the microstructural changes occurring in high temperature materials during service exposure and identifying the processes and mechanisms leading to the observed degradation of their mechanical properties. Once more the papers presented at this conference, which was held at Sheffield Hallam University in March 1997 and attracted over 80 delegates from 15 countries, have highlighted the work currently in progress on the development of improved high temperature materials designed to be more resistant to microstructural degradation in service.

Series Editor

Andrew Strang
Chairman High Temperature Materials Performance Committee

Conferences Held in the Series

Microstructural Developement and Stability in High Chromium Ferritic Power Plant Steels
30 June 1995, Robinson College, Cambridge, UK

Microstructural Stability of Creep Resistant Alloys for High Temperature Plant Applications
24–26 March 1997, Sheffield Hallam University, Sheffield, UK

Microstructural Design and Stability in Engineering Ceramics
18 November 1997, UMIST, Manchester, UK

Future Conferences Planned in the Series

Microstructural Modelling and Prediction of High Temperature Properties of Creep Resistant Materials

Microstructure of Welded Materials and Structures for High Temperature Plant Applications

Microstructural Control and Stability of High Temperature Gas Turbine Blading Alloys

Microstructural Stability of Corrosion Resistant Coatings for High Temperature Gas Turbine Blading and Combustion Path Components

Microstructure of Advanced High Temperature Titanium Alloys

Microstructure of Advanced High Temperature Composites

Quantitative Evaluation of Microstructure for Modelling and Prediction of the High Temperature Properties of Creep Resistant Materials

The Microstructural Development of Austenitic Creep Resistant Steels for Elevated Temperature Service

D. Dulieu* and J.A. Whiteman†

*Avesta Sheffield Ltd (R&D), P.O. Box 161, Sheffield, S9 1TR
†Department of Engineering Materials, The University of Sheffield, P.O. Box 600, Sheffield S1 4DU

ABSTRACT

The historical development of the austenitic steels for high temperature service is outlined, with emphasis on the requirements of the aero engine, power generation and nuclear sectors as forcing the development of the original simple, solid solution strengthened compositions. The principal families of steels are defined, and the contribution of microstructural investigation methods, in understanding phase stability, precipitation sequences and high temperature deformation processes, is summarised.

INTRODUCTION

The austenitic stainless steels for elevated temperature service, in common with many other classes of steel product, were established as a class of materials before the advent of modern methods of microstructural characterisation. This reflects the great impetus to their development given by requirements of the aero engine, power generation and nuclear industries in the period 1937–1960. Later development at a slower pace has been driven by the needs of the high temperature reactor, advanced fossil fuel power and incineration technologies. To give this review a historical perspective, the evolution of the family of materials is traced briefly, illustrating some of the metallurgical methods used to optimise performance. The contribution of metallographic methods to the understanding of materials behaviour is summarised and the principal aspects of microstructural stability are outlined for the main classes of steels.

It is appropriate first to define metallurgically the austenitic stainless steels and to give an outline of the major classes of alloys. More extensive relevant reviews of their physical metallurgy and behaviour have been given by, among others, Gemmill,[1] (power generation applications), Pickering,[2] (physical metallurgy), Marshall,[3] (Structure and properties), Harries,[4] (physical metallurgy in the context of HTR applications) and Gladman,[5] (deformation and failure at elevated temperatures).

1

AUSTENITIC STAINLESS STEELS FOR ELEVATED TEMPERATURE SERVICE

The austenitic stainless steels in this context may be defined as:

> 'Steels containing more than 12% chromium, with austenite stabilising additions, principally nickel, and other alloying elements added such that the initial structure of the alloy is predominantly face centred cubic and iron is present at a level greater than any other addition, usually at 50% or more.'

The leanest standard austenitic steels, containing around 18% chromium and 10% nickel have predominantly austenitic structures in the standard, solution annealed heat treatment condition. However, depending upon their precise composition, they may contain traces of ferrite or may contain hcp, ϵ or bcc, α' martensite if cold deformed, or cooled to room temperature after prolonged ageing. Also, it is common to allow the controlled formation of a low level of ferrite in the weld regions of austenitic steels, to prevent liquidation cracking and hot tearing.

'Simple' steels may be defined as those with elevated temperature properties determined primarily by elements such as molybdenum in solid solution and whose principal strengthening precipitate during service is based on the $M_{23}C_6$ carbide. Sub-sets of the simple 18%Cr–10%Ni steels fall into four broad classes, and these classes also form the basis for steels using more potent strengthening mechanisms at high temperatures. The first comprises steels with increased nickel and reduced chromium contents, to improve stability with respect to sigma phase formation. The second comprises compositions with increased chromium (and compensating nickel) levels to improve oxidation resistance. These include steels with enhanced silicon additions and rare-earth element additions to improve oxide film stability.[6] The use of strengthening nitrogen additions gives the third sub-set and steels with higher levels of nickel, to confer resistance to carburisation and nitration, the fourth. Examples of 'simple' steels within these families numbered for reference in the text, are shown in Tables 1 and 2. (Note that reference is made to steel designations in two draft prEN standards, Table 1, this is to avoid confusion, where possible, between grade designations in common use derived from British, American and proprietary sources.)

More complex steels, numerically the most important of the creep resisting austenitic steels, contain alternative carbide-forming additions to chromium, principally niobium and titanium. Although the balance of matrix soluble niobium or titanium and the form of MC carbide precipitation are the principal factors in determining strength, compositional and kinetic factors may result in transient formation of chromium carbide. Examples of the compositions of these steels are given in Table 3.

Table 1 'Simple' and nitrogen Strengthened Steels

	Designation	%C	%Si	%Cr	%Mo	%Ni	%N	Others	Reference
1	1.4948	0.04-0.08	×1.0	17.0–19.0		8.0–11.0	×0.11		A*
2	316S51	0.04–0.10	×1.0	16.5–18.5	2.0–2.5	10.0–13.0			B†
3	1.4833	×0.15	×1.0	22.0–24.0		12.0–14.0	×0.11		C⁰
4	1.4845	×0.10	×1.5	23.0–26.0		19.0–22.0	×0.11		C
5	1.4315	×0.06	×1.0	18.0–20.0		8.0–11.0	0.12–0.22		A
6	1.4910	×0.04	×0.75	16.0–18.0	2.0–3.0	12.0–14.0	0.10–0.18	B 0.0015–0.005	A
7	16-25-6'	0.05	0.8	16.5	6.25	25	0.10		13
8	1.4872	0.2–0.3	×1.0	24.0–26.0		6.0–8.0	0.20–0.40	Mn 8.0-10.0	C

Footnote: Compositions given in Italics are as reported or typical
A* draft pr EN 10028–7
B† BS 1501 : pt 3: 1990
C⁰ draft pr EN 10095: 1997

Table 2   Oxidation Resistant Steels

		%C	%Si	%Cr	%Mo	%Ni	%N	Others	Ref
9	1.4828	×0.20	1.50–2.50	19.0–21.0		11.0–12.0	×0.11		C
10	1.4818	0.04–0.08	1.0–2.0	18.0–20.0		9.0–11.0	0.12–0.20	Ce 0.03–0.08	C
11	1.4835	0.05-0.12	1.40–2.50	20.0–22.0		10.0–12.0	0.12–0.20	Ce 0.03–0.08	C
12	1.4841	×0.20	1.5–20	24.0–26.0		19.0–22.0	×0.11		C
13	1.4864	×0.15	1.0–2.0	15.0–17.0		33.0–37.0	×0.11		C
14	1.4854	0.04–0.08	1.2–2.0	24.0–26.0		34.0–36.0	0.12–0.20	Cl 0.02–0.08	C

The third major family relies for strengthening on the formation of an intermetallic precipitate, based usually on gamma prime $Ni_3(Ti,Al)$ or upon copper. Carbide precipitation is less significant for strength, but can influence grain boundary properties. Steels within this family are given in Table 4.

An, as yet minor class of steels are those which use alternative methods to overcome the problems inherent in any solution-reprecipitation process to develop particle strengthening. Stable dispersions, principally of titanium nitride, have been obtained by combinations of external nitriding and internal nitriding via mechanical alloying.

An emerging category of austenitic steels for elevated temperature service, arises from the need to develop steels for fusion reactor applications, where the 'first wall' of the structure encounters very high neutron fluences. To minimise long term activation effects, and hence the disposal problems, there is interest in the concept of 'Low Activation' or LA alloys. These must be formulated without elements such as nickel, molybdenum or niobium which produce long-lived transmutation products on irradiation.

Table 3   Niobium/Titanium Strengthened Steels

	Designation	%C	%Si	%Cr	%Mo	%Ni	%Ti	%Nb	Other	Ref
15	1.4878	×0.10	×1.0	17.0–19.0		9.0–12.0	5×C ×0.80			C
16	1.4941	0.04–0.08	×1.0	17.0–19.0		9.0–12.0	5×C ×0.080		0.0015–0.005B	A
17	1.4961	0.04–0.08	0.30–0.60	15.0–17.00		12.0–14.0		10×C to 1.2		C
18	1.4571	×0.08	×1.0	16.5–18.5	2.0–2.5	10.5–13.6	5×C to 0.70			A
19	20–25–Nb	×0.02	0.8–1.0	19.0–21.0		24.0–26.0		8x(C+N)	<0.001	4
20	Esshete 1250	0.06–0.15		14.0–16.0	0.8–1.2	9.0–11.0		0.75–1.25	3.5–7.0Mn 0.15–0.4 V 0.003–0.009B	BS3059
21	12R72HV	0.10		15.0	1.2	15	0.4		0.006B	4
22	EV–548	0.06–0.09	×0.06	16.0–17.0	1.5	11.0–12.0	×0.015	×1.05	0.001–0.003	4
23	HR3C	0.06		25		20		0.45	0.2N	67
24	G18B	0.4		13	2.0	20		3.0	10Co 2.5W	10,17
25	1.4877	0.04–0.08	×0.3	26–28		31–33		0.6–1.0	Ce 0.05–0.10	C

Table 4   Intermetallic Precipitate Strengthened Steels

	Designation	%C	%Cr	%Mo	%Ni	%Cu	%Ti	%Al	Other	Ref
26	Tinidur	0.20	14.0	–	30.0	–	1.7			9, 13
27	Discalloy	0.05	13.0	3.0	25.0	–	1.8	0.2		9, 13
28	F-V Rex 78	0.08	19.0	4.0	18.0	4.0	0.6		0.25 V	9, 13
29	GT 45	0.08	17.0	3.0	14.0	3.0	0.30		0.45 Nb	13
30	A286	0.05	15.0	1.3	25	–	2.0	0.25	0.3 V	13
31	1.4958	0.03–0.08	19.0–22.0		30.0–34.0		0.25–0.65	0.25–0.65	Co < 0.05 (Ni+Co) 30.0–32.5	A
32	1.4959	0.05–0.10	19.0–22.0		30.0–34.0		0.25–0.65	0.25–0.65	Co < 0.5 (Ni+Co) 30.0–34.0	A
33	1.4876	×0.12	19.0–23.0		30.0–34.0		0.15–0.40	0.15–0.40		C
34	17 14 CuMo	×0.185	15.25–17.5	1.75–3.0	13.0–15.0	2.5–3.5	0.20–0.35	-	0.40–0.55Nb	13
35	Super 304H	0.07/0.13	17.0–19.0	–	7.5–10.5	2.5–3.5	–	–	0.30–0.60 Nb 0.05–0.12N	68

In terms of the temperature range of service for austenitic steels, pressure vessel applications invoke time-dependent properties at temperatures above around 540–580°C. For service at lower temperatures, factors governing the proof strength are important. To illustrate this, the lower carbon, nitrogen bearing variants of the simple steels, for example steel 6 compared with steel 2, (see Table 1) have higher allowed design strength values up to the temperatures given above. Pressure circuits, to which most elevated temperature performance data relate, usually involve relatively low stresses and long lives. For example, BS5500:1997 gives 100kh design strength values for pipes and tubes at 700°C of 33Nmm^{-2} for the controlled carbon content 18%Cr–10%Ni 'simple' steel 1 and 40Nmm^{-2} for the complex 'carbide strengthened' steel 20 (Table 3). Above this temperature, the austenitic steels also find widespread application, particularly as low pressure ductwork and radiation shields where there may be significant plastic strains, up to temperatures and durations limited by oxidation or self-weight collapse.

AN HISTORICAL INTRODUCTION

The nickel-chromium austenitic steels may be said to date from the Krupp patent of 1913,[7] covering steels containing 4–20% nickel, 15–40% chromium and less than 1% of carbon. Their strength and scaling resistance at high temperatures were recognised early. Further Krupp patents of 1923 covered steels containing 4–20% nickel, 18–30% chromium, 2–4% molybdenum, with 0.1–0.4% carbon[8] and later the addition of 2–6% of copper to their original composition range. A 25% nickel, 20% chromium–3% silicon steel was adopted early for scaling resistance. The oldest 'simple' creep resisting austenitic steel is thus a stable austenitic material with useful strain-induced precipitation of $M_{23}C_6$ carbide. This implies a minimum carbon level of around 0.04%, as recognised in current standards, Table 1. (Early stainless steelmakers would have had little difficulty in guaranteeing 'H' grades with a minimum carbon content above 0.04%.)

Molybdenum and tungsten were soon recognised as strengthening additions and Allen[9] and Oliver et al.[10] recognise in their reviews of the development of gas turbine materials that one of the first purpose-made austenitic creep resisting steels was a high carbon, tungsten-containing material, widely used for piston engine exhaust valves. The '16–25–6' steel, 7 (Table 1), was an early successful turbocharger grade.

Titanium and niobium were first used in steels for chemical plant service, to prevent intergranular corrosion. This results from local matrix depletion in chromium when chromium carbides form at slow cooling rates encountered in heat treatment of heavy section components, or in weld heat affected zones. The intention of making 'stabilising' additions was to form inert dispersions of TiC or Nb(C,N) and prevent $M_{23}C_6$ formation. By 1930 the benefits of these 'stabilising' additions for high temperature properties were being recognised, together with strengthening from the use of high levels of titanium.

In 1926, Moneypenny[11] described the French 'ATV' steel for elevated temperature service as containing 25–40% nickel, 10–15% chromium plus molybdenum, tungsten and vanadium. Hadfield both licensed manufacture of

Fig. 1   An early example an austenitic creep resisting steel in turbine applications. A 10 000kW steam turbine bladed throughout with the French 'ATV' steel. (Hadfield and Burnham, ref 12)

this steel in the UK, (Fig. 1) and patented 'a complex iron-silicon-chromium tungsten alloy which may also contain nickel, manganese, copper, cobalt vanadium, titanium and vanadium'.[12] Of these early complex alloys Moneypenny made the prescient comment:

> 'There is a fruitful field for research in investigating the properties of these and other complex alloys, particularly from the point of view of providing materials to withstand temperatures above atmospheric.'

Subsequently there was rapid progress in the development of austenitic steels for service at elevated temperatures. The period 1946–50 makes a convenient starting-point to review developments for three major reasons: By this time, most of the compositional families of austenitic steels were either anticipated or established. It saw the first published reports on the extensive development of materials for first the turbocharger and then the aircraft gas turbine during the second world war. It also represents a watershed between an era where microstructural studies were a minor and non-essential aspect of alloy development and one when they grew to make a significant contribution.

A review of the early development of applications and compositions from the American standpoint was given by Miller[13] in 1950. American work took as a starting-point turbocharger blade working temperatures of 540°C and then 815°C for gas turbine blades and 650°C for discs. Tables 3 and 4 include compositions from contemporary reviews. Steel 28, Firth Vickers Rex 78, is of historical interest as the material necessary for blading in the first of the British Whittle gas turbines to fly, replacing the Stayblade grade from the same company used in the experimental engine.[14]

The metallurgical design principles for these alloys were based initially on an empirical understanding of composition effects, in terms of solid solution strengthening, phase stability and the formation of carbides. Oliver[10] noted that there was an empirical understanding that coarse, optically visible inter-metallic particles contributed little to strengthening processes except perhaps at the highest operating temperatures. Techniques then available to follow microstructural processes on the sub-optical microscopical scale included measurements of resistivity, magnetic permeability and changes in matrix lattice parameter, as a function of solution treatment and ageing conditions. X-ray powder diffraction from electrolytically isolated precipitates was the principal means of phase identification. An example of this approach is the work of Monkman, Price and Grant.[15] They studied the effects of carbon and nitrogen levels in simple 18%Cr, 8%Ni steels on creep rupture properties. The benefits of the interstitials in increasing austenite stability and delaying the formation of sigma phase and ferrite were demonstrated.

However, the principal tool in the development of the more complex, high strength materials was experimental verification of the effects on short-time elevated temperature properties of compositional changes in matrices of experimental compositions. This is illustrated by the work of Harris and Child,[16] 'Examining the effects of different combinations of carbide forming elements in a 20% Nickel, 20% Chromium, 20% Cobalt base, they identified a 'complexity effect'. The high temperature strength properties were increased by increasing the number of carbide formers present, but 'the relative proportions of these carbide-formers must be adjusted to assure optimum properties'. (A task to be made easier when methods for the imaging and chemical analysis of individual strengthening carbide particles were developed.)

Understanding of the influence of processing variables on properties came from the thermomechanical working treatments required to shape the 'refractory alloys'. This is illustrated by the application of warm working techniques to increase the proof strength level in the 'G18B' complex austenitic gas turbine disc steel, 24 (Table 3), described by Harris and Bailey.[17]

THE IMPACT OF DEVELOPMENTS IN METHODS OF STRUCTURAL EVALUATION

The 1950s saw the development of carbon extraction electron microscopy techniques and the availability of microscopes capable of selected area electron diffraction. Nutting,[18, 19] amongst others, demonstrated the use of extraction replicas to reveal the morphology of fine-scale precipitates, focussing attention on processes such as strain-induced precipitation. They showed how replicas could be used not only in the microscope, but also, as a source of X-ray fluorescence and powder chemical and structural information.

With the addition of thin foil transmission techniques from the late 1950s onwards, it was possible to extend downwards the sizes of particles which could be imaged, to examine dislocation structures and to test models for verification of predictions of dislocation-particle interactions. By 1972 passing references were being made to in-situ microchemical analysis by wavelength dispersive analysis in the TEM. The rapid growth in in-situ X-ray techniques, and later methods for investigating particle and grain boundary chemistry, greatly enhanced the study of precipitation processes. Interest in the early stages of cavitation damage brought the application of both new techniques such as neutron small angle scattering and older ones such as precision density change, or pycnometric methods.

A parallel development to metallographic studies was an appreciation of the need to predict phase equilibria in complex systems, to help define and avoid compositional regions giving weakening or embrittling phases. Although developed mainly in the context of the nickel and cobalt-based alloys, the technique found application to the austenitic 'superalloys'. These early attempts applied electron-configuration hypotheses of phase stability to predict the terminal phases, after estimating the changes in matrix compositions as ageing processes took place. This approach is illustrated by the work of, amongst others, Woodyatt, et al.[20] It may be regarded as a step on the path to current thermodynamical data-bases such as Thermocalc and MTDATA used in alloy design.

MICROSTRUCTURAL OBSERVATIONS IN THE DEVELOPMENT OF AUSTENITIC STEELS

Simple Steels
The considerations for microstructural changes affecting time-dependent properties are:

1. The carbon supersaturation developed in the solution anneal and the levels of boron and nitrogen, determining the location and effect of chromium carbide, $M_{23}C_6$ precipitation.
2. Any dislocation structure induced by warm, cold work or residual stresses.
3. Ageing treatments prior to entering service.

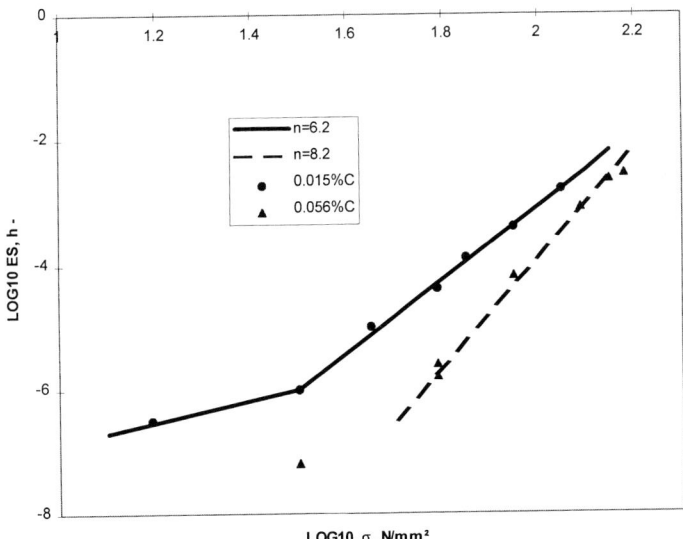

Fig. 2 The stress dependence of the minimum creep rates for 17.5% Cr–10% Ni steels with 0.015 and 0.056% carbon, showing the change in rate at lower stresses for the lower carbon steels. (Beckitt *et al.*[24])

4. The matrix stability in relation to service temperature, governing the initial and final presence of ferrite or martensite and the kinetics of formation of the sigma phase, intermediate alloy carbide or intermetallic precipitates.

The rate controlling creep deformation processes at a given temperature are stress-dependent; grain boundary sliding, solute drag and vacancy diffusional processes dominate at lower, and dislocation climb or glide processes at higher stresses. Chromium carbide is the principal precipitate phase to form first on ageing simple Cr–Ni steels. (The solubility of carbon reported in steel 4 is 0.107% at a typical solution annealing temperature of 1050°C and only 0.001% at 650°C in the 17%Cr 11%Ni 2.5%Mo steel 23.) Nitrogen may substitute for carbon in carbide phases,[4] although Alden and Aronsson[20] have pointed out the lattice structure limitations on substitution. Contributions to the strengthening effects of nitrogen come from its presence in solid solution and resultant influences on $M_{23}C_6$ carbide particle stability, through modification of the matrix lattice parameter and carbon diffusivity, (Horton *et. al.*,[22] Ceccarelli[23]).

Boron is added to austenitic steels to improve hot workability and rupture ductility. It has a significant role in grain boundary diffusion processes and the development of $M_{23}C_6$ particle distributions. Its presence increases the density of intragranular particles and raises the carbide lattice parameter,

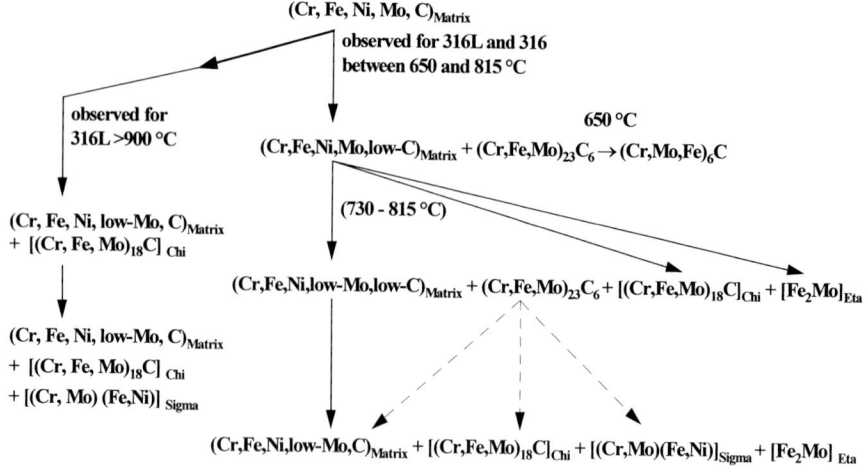

Fig. 3   Schematic diagram showing the possible sequence of phase reactions on ageing Types 316 and 316L 17%Cr–11%Ni–11%Ni–2.6%Mo over a range of temperatures. (Weiss and Stickler[27])

reducing the misfit with the austenite lattice and thus increasing particle stability.[1, 4]

Beckitt *et al.*[24] investigated the secondary creep characteristics and microstructural changes occurring in solution annealed 18%Cr–10%Ni steels at temperatures between 500 and 700°C over a range of stresses. They demonstrated that a low carbon (0.015% 304L) type stainless steel showed only grain and incoherent twin boundary nucleated $M_{23}C_6$ precipitation on creep testing. In contrast, in a steel containing 0.056%C, precipitation of the carbide on dislocations resulted in a reduction in creep rate. The stress dependence of the minimum creep rate for the two steels is shown in Fig. 2. The authors concluded that, at low stresses in the lower carbon content steel, the dependence was consistent with a Coble creep process.[5] They suggested that the lower strain rate at low stress for the higher carbon content steel was an indication of an effect of grain boundary carbides. The carbide is also significant for higher alloy simple steels. Keown[25] found significant $M_{23}C_6$ precipitation on dislocations during secondary creep in a 0.03%C 17.5%Cr, 26.4%Ni steel tested at 600°C.

At longer exposure times, sigma phase may be encountered in the simple steels. Horak *et al.*[26] described the structure and properties of both Types 304 and 316 steels after long periods of test and service exposure. The authors related the presence and location of ferrite to matrix solute depletion on ageing.

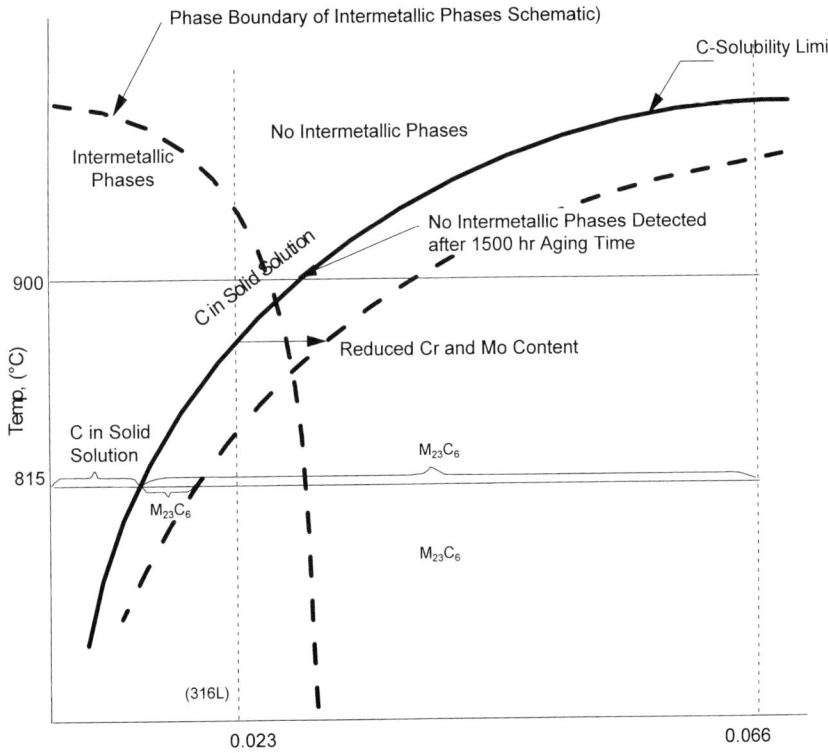

Fig. 4   Graphical presentation of phase stability effects; the dependence of the carbide-intermetallic phase balance on carbon content in the steels of Fig. 3.

Addition of molybdenum to the basic 18%Cr–10%Ni matrix adjusted to maintain austenite stability results in strengthening for 100kh service at temperatures below 650°C, delayed precipitation of the $M_{23}C_6$ phase and a more complex potential sequence of phases. This is well illustrated by the work of Weiss and Stickler,[27] who examined the time–temperature–precipitation sequences over the temperature range 480–950°C. The results are shown schematically in Fig. 3, demonstrating the metastable nature of the $M_{23}C_6$ carbide. These authors used their observations to predict the consequences of compositional adjustments on phase stability, as in Fig. 4. Horak *et al.*[26] indicated good agreement between the predictions of this work and their later observations.

Prior cold work, as may be encountered in fabrication operations, has been found to raise the short term creep strength by reduction in the creep rate of the simple steels, with a significant reduction in rupture ductility. Moen and Smith defined practical limits on cold working for the maintenance of standard properties.[28] Bernard *et al.*[29] attributed the effects to the formation of a

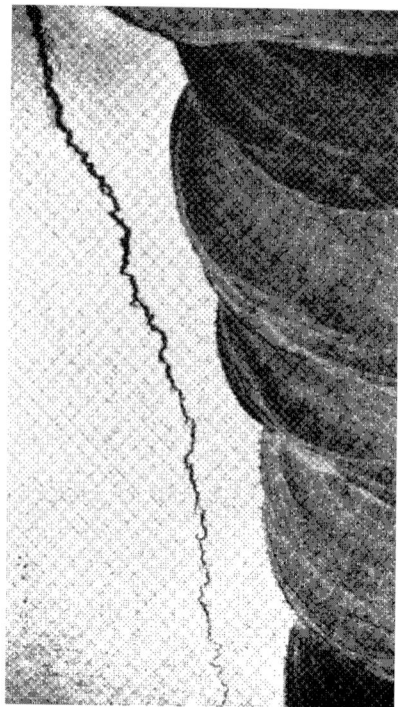

Fig. 5   Cracking in an 18%Cr,Ni,Nb steam pipe. (Gemmill[1])

finer and more homogeneous distribution of the $M_{23}C_6$ precipitates forming on the dislocation substructure. Guttmann and Timm[30] showed for a steel similar to 4, that the loss of ductility does not persist to longer durations, albeit at the relatively high temperature of 900°C.

The Niobium and Titanium (MC-Carbide) Strengthened Steels
The effects of niobium and titanium both depend critically upon the relationship between the addition level, carbon and nitrogen contents, the solubility product at the temperature of solution treatment and the presence of dislocation sites for nucleation. (At lower service temperatures, kinetic considerations may favour the initial formation of $M_{23}C_6$ in the precipitation sequences of both niobium and titanium stabilised steels.)

Early experience showed how properties, particularly creep rupture strength and ductility, were strongly influenced by prior processing and thermal history. White and Freeman,[31] studying the sensitivity to heat treatment conditions of the creep properties of the Type 321 grade, (steel 15) showed how the effects of prior cold work on rupture ductility seen in the simple steels were exacerbated with solution annealed titanium stabilised materials.

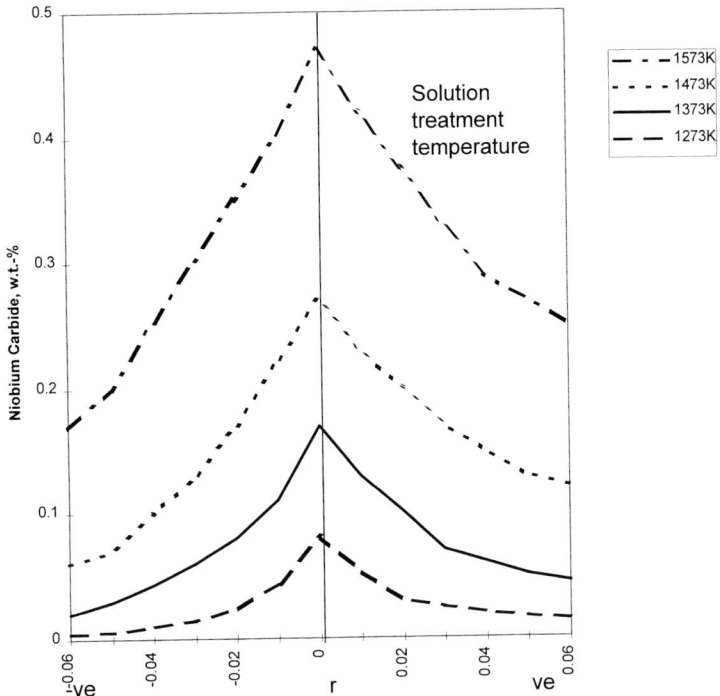

Fig. 6 Niobium Carbide available for Precipitation as a function of deviation from Stoichiometry. (Wadsworth *et al.*[34], as quoted by Keown and Pickering[35])

Problems in the 18%Cr–12%Ni–1%Nb steel were encountered when it had been used in the steam circuits of power stations in the USA and later in the UK, for headers, steampipes and valve bodies. Loss in ductility was found in material which had undergone a high temperature solution treatment, either in processing or as a result of welding. Strain-induced precipitation of niobium carbide, either as result of internal stresses or on initial loading, resulted in considerable strengthening within the grains, leading to low ductility intergranular creep failure. Kirkby and Truman[32] investigated the effects of variations in cooling rates from high solution treatment temperatures and of subsequent re-solution treatment at lower temperatures. Of these, slow cooling proved most effective, but variable, low rupture ductility levels remained. The problem was seen also in the form of stress relief cracking in restrained weldments.[1] Dissolution of NbC in regions of the HAZ reaching around 1300°C resulted in significant grain strengthening by strain-induced intragranular precipitation on reheating to temperatures in the range 700–850°C Fig. 5.

Understanding and control of properties came about as a result of establishing solubility data , reviewed by Keown and Pickering.[33] Figure 6, from the work of Wadsworth *et al.*[34] shows how the amount of niobium carbide available for precipitation is reduced at a given temperature of solution, as the ratio of niobium to carbon plus nitrogen deviates from the stochiometric value. Harries,[4] gives data showing the effect of nickel in increasing the solubility of niobium carbide in austenite.

Recognition of factors governing carbide solubility was supported by investigations, for example those by workers within CERL,[35-37] on the nucleation and growth of MC carbides. Mechanisms investigated by TEM methods included the generation of dislocations as a result of carbide particle growth and the role of larger, undissolved particles as a source of dislocations induced by thermal stress cycling. This 'internal warm-working' effect, and the possibility of solute drag effects concurrent with precipitation, complicated efforts to establish simple relationships between creep strength, rupture ductility and the basic microstructural parameters of grain size, and the initial niobium carbide supersaturation/particle dispersion. Keown and Pickering[38] studied 16–20%Cr, 13%Ni steels with varying niobium and carbon contents. They established the presence of $M_{23}C_6$ carbides where carbon was present above the stochiometric ratio and the Laves Fe_2Nb phase where there was a niobium excess. High stress tests at 700°C, indicated that rupture lives were controlled by the volume fraction of precipitated niobium carbide. Both life and ductility were at a maximum in steels with niobium and carbon contents close to the stoichiometric ratio. They attempted also to quantify the effect on strength of undissolved NbC.

Silcock and Willoughby[39] used 16%Cr, 12–30%Ni steels with varying niobium to carbon ratios, fully aged at 800°C to investigate structural effects on low stress creep behaviour in the range 710–780°C. They found that the creep rate depended upon the particle dispersion, particularly at lower alloy contents and larger grain sizes, creep rates decreasing with increasing particle density. At low stresses the stress dependence was consistent with grain boundary sliding, but the grain size dependence of creep rate was influenced by variations in the density of grain boundary particles.

A specific body of effort[40-42] was devoted to a higher alloy stabilised steel, the 20%Cr–25%Ni–Nb steel, (19 in Table 3) chosen for the fuel element cladding for the UK advanced gas-cooled reactors.

Carbide strengthened steels are susceptible to low ductility creep failure as a result of intergranular cavity formation. Gladman[5] has demonstrated that, in steels where there is carbide precipitation, the transition to low n values with reducing stress occurs at stress levels characteristic of the change from dislocation to Nabarro-Herring, volume diffusion controlled, rather than Coble, grain boundary diffusion, controlled creep. The presence of grain boundary precipitates influences both diffusion and the nucleation of grain boundary cavities. Needham and Gladman[43] studied quantitatively cavity

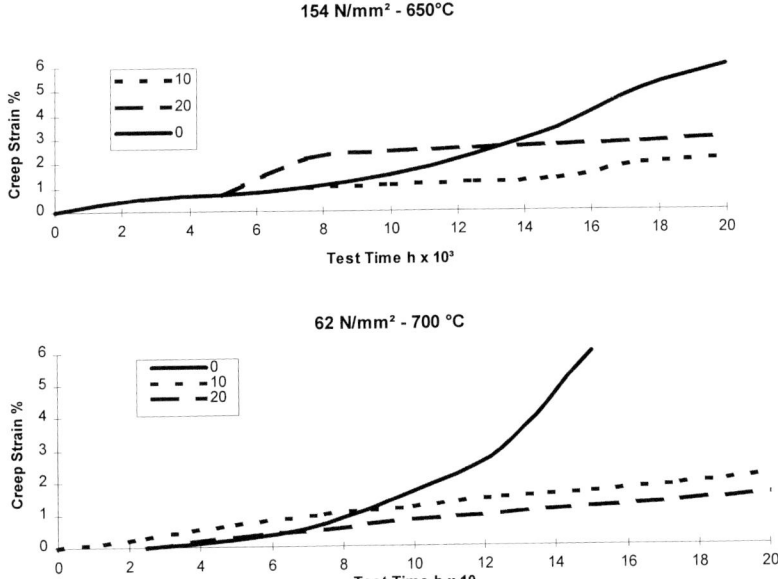

Fig. 7   The effects of prior cold straining by 10 or 20% on the creep behaviour of the complex, niobium-strengthened steel Esshete 1250. (Orr and Collins[46])

formation in the niobium stabilised steel, 17, in creep tests at 550 and 650°C. They found the number of cavities per unit volume to be linearly related to the time-dependent creep strain, being independent of stress or temperature at a given, time-dependent creep strain. They proposed a model to predict the total volume fraction of cavities under given conditions, and hence estimate rupture life. Many materials failing in this mode exhibit a common relationship between the area fraction of grain boundaries covered by cavities and the fraction life expended. Thus, quantitative measurement of the extent of cavitation provides a means of estimating component remanent life. Studies of cavitation kinetics have demonstrated the harmful effects of coarse grain sizes combined with conditions, as may be met in welding, where solution and grain boundary nucleated reprecipitation of manganese sulphides may occur.

The 16%Cr, 10%Ni, 6%Mn Mo,V,Nb,B Esshete 1250 steel (20 in Table 3) was compositionally optimised to give a combination of high creep strength and rupture ductility, primarily for superheater and reheater tube applications at temperatures above 570°C.[44] It remains one of the strongest of the austenitic steels for this application. Asbury and Willoughby[45] studied the structural changes occurring on ageing after a range of initial solution and ageing treatments and their effects on creep properties at 650°C. They showed that, at high levels of stress and initial supersaturation, the strength was influenced by precipitation of niobium carbide during the early stages of straining. The kinetics of the loss of strengthening with pre-ageing were consistent with those to be expected from niobium solute depletion. At low stresses the pre-existing niobium carbide and associated dislocations were the main factors affecting creep strength, with precipitation of sigma phase removing molybdenum and vanadium from solution.

Prior cold working may be unavoidable, as in the case of tube manipulation, or it may be deliberately used to increase the initial dislocation density, to improve properties under irradiation. Harries[4] has summarised the implications for recovery and primary recrystallisation for stabilised steels. Orr and Collins[46] studied microstructural changes occurring in steel 20 after initial cold working and creep testing, to identify permissible levels. They showed that cold straining by 10 or 20% in tension produced large increases in proof stress at ambient and elevated temperatures and a small increase in creep strength at 650 and 700°C. This behaviour was attributed to the formation of a recovered sub-grain structure, stabilised by precipitation of (NbV)C. The strength advantage from 10% initial strain was found to persist to beyond 30kh at 700°C, Fig. 7. The steel, like the earlier G18B composition, has been used in the warm worked condition, to improve the elevated temperatural properties for bolting applications.[47]

The general direction of development of the niobium strengthened steels was towards reducing the niobium content and optimising the niobium to carbon ratio to control the quantity of undissolved carbide. This had the advantages of minimising problems resulting from NbC/gamma eutectic regions at high temperatures and reducing heat treatment temperatures. (The reported normalising temperature of 1300°C for the G18B alloy, steel 18, was 20°C below the solidus[16].) The appreciation of solubility product effects allows niobium carbides to be used in both the grain refining and precipitation modes. This is illustrated by the work of Sawaragi et al.[48] They demonstrated how the grain size of the niobium-stabilised steel 17, as tubing, could be refined to improve resistance to steam side oxidation. A high temperature solution treatment is followed by cold drawing and re-solution at a lower temperature, the net niobiuim carbide precipitation being sufficient to give grain refinement.

Steels Strengthened by Intermetallic Precipitation
The early work of Hadfield, followed by the 'Tinidur' alloy developed in Germany and the American Discalloy for aircraft gas turbine components (steels 20 and 21 in Table 4) established the gamma prime hardened family of austenitic steels. The A286 steel, 24 in Table 4, is widely used, with a typical heat treatment involving solution at 980°C and ageing at 720°C. Studies of the compositional ranges, ageing response and precipitation sequences in steels with from around 1.0–4.0% titanium include those of Eichelberger,[49] Irvine

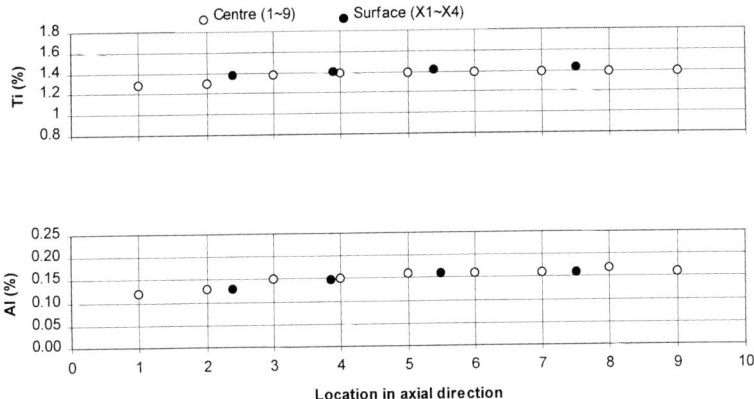

Fig. 8 Control over the titanium and aluminium levels achieved throughout a rotor forging made from a 40t ESR ingot of modified A286 steel. (Okamura et al.[53])

et al.[50] Wilson and Pickering[51] and Raynor and Silcock.[52] The evolution of these materials in outline has rested on determining the compositional balance to give the maximum volume fraction and range of stability of the gamma prime precipitate. This requires balancing the addition of aluminium, to stabilise the gamma prime precipitates and delaying or suppressing the formation of the equilibrium hexagonal Ni_3Ti phase, which grows in cellular colonies from grain boundaries. These steels tend to have low carbon contents, with limited carbide precipitation. Segregation effects can lead to the formation of titanium-based complex phases, for example the 'G'-phase nickel-titanium silicide.[53] Close compositional control is essential and Okamura et al.[54] have demonstrated the procedures necessary to produce a modification of the A286 alloy as 40t ingots for the production of components such as HP/IP rotors. Figure 8 shows the consistency of aluminium and titanium levels obtained the forging produced from the ESR ingot.

The alloy 800 composition (31–33 in Table 4) is of interest, in that the composition range allows for the selection of carbon, titanium and aluminium levels for properties to be controlled primarily by either carbide precipitation or the formation of gamma prime. The structure and properties have been reviewed by Orr,[55] who reported that a minimum of around 0.6%(Ti+Al) is required to develop gamma prime precipitation.

Kirman[56] has explored the strengthening possible from intermetallic precipitation of niobium in austenitic steels. However, alloys containing less than 25% nickel gave only Laves phase, Fe_2Nb precipitates. The most effective strengthening, from precipitation of the metastable DO_{22}-ordered γ^* Ni_3Nb, was found only at relatively high, around 50%, nickel contents.

Steels 28, 29 and 34 were established in the 1940s, combine MC carbide forming additions with copper, which precipitates on ageing. Tamura *et al.*[57] describe compositional modifications to the Armco 17%Cr, 14%Cr, Cu, Mo, Nb, Ti steel. Specifically they advocate reduction in chromium and an increase in copper levels, to promote the effect of copper precipitation, combined with rebalancing the carbon, niobium and titanium contents. These authors indicate that this modified composition equals the strength of steel 20 at temperatures around 680°C and exceeds it at 700°C.

Steels for Nuclear Applications
Nuclear applications, such as fuel element cladding, bring the additional complications of lattice damage and transmutation products from irradiation to the existing complexities of thermal and stress induced processes. This is particularly the case with materials for high temperature reactors. Cobalt levels are usually kept low to minimise activation. Boron can create helium bubbles, leading to intergranular failure, as a result of transmutation, hence the restriction for steel 19. Microstructural optimisation for nuclear processes in austenitic steels is concerned mainly with void swelling, irradiation creep, radiation induced solute segregation and precipitation. Ehrlich[58] has reviewed the mechanisms of irradiation creep in austenitic stainless steels and the evolution of theories of dislocation mechanisms in this area can be traced in the more recent review of Gelles.[59]

Lattice displacements caused by neutron bombardment result in the creation of vacancies and interstitials, which can be annihilated at suitable sites, including dislocations. Vacancies can alternatively condense to form voids, so that the presence of a stable dislocation structure induced by prior cold working is beneficial.

Wood[60] has given a comprehensive review of factors governing deformation and rupture of cold worked Type 316 (steel 2) in the specific application of fast reactor fuel pin cladding. The work shows how models of microstructural evolution and creep deformation available at that time could be applied to the performance of a specific component under complex conditions. The need for models combining mechanisms was noted, in his case, normal operating conditions required the simultaneous treatment of void swelling, irradiation and thermal creep.

Radiation induced segregation (RIS) is a mechanism which results from the flux of vacancies carrying with them to sinks atoms such as silicon and nickel, or creating depletion in molybdenum and chromium. The net result is the modification of the precipitation sequences found during thermal ageing, for example, the promotion of the formation of phases depleting the matrix in nickel and silicon, such as gamma prime Ni_3Si and the G phase (Ni,Nb,Ti) silicide. As discussed by Harries,[4] these depletions may promote local transformation of the austenitic matrix to martensite on cooling.

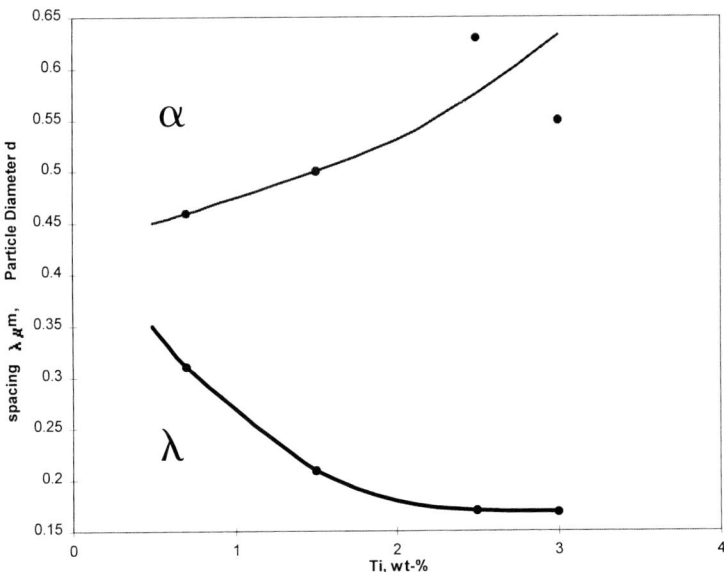

Fig. 9 Parameters of the dispersion of TiN particles, as a function of titanium content, obtained on nitriding a 20%Cr, 25%Ni–09%Si steel. (Roberts and Evans[62])

As an example of compositional and microstructural optimisation, Maziasz *et al.*[61] have described the development of the ORNL (Oak Ridge National Laboratory) modified 14%Cr–16%Ni void swelling resistant alloys. These depend upon the development of a fine and stable MC particle dispersion, nucleating early enough to give increased helium bubble nucleation which, in turn, reduces radiation induced segregation. They established the effects on the MC carbide distribution of varying additions of titanium, niobium and vanadium. Microanalytical techniques were used to determine precipitate compositions to identify the optimised steel composition of 16% nickel, 14–16%Cr, 2%Mn, 2.5%Mo, 0.3%Ti, 0.1–0.5%V with trace additions of P (0.02–0.07%) and B (0.005–0.007%).

The concept of the fusion reactor involves very high neutron fluences on the 'first wall' of the reactor structure. To minimise the long term activation of the components, 'Low Activation' or LA alloys are formulated without elements such as nickel, molybdenum or niobium which produce long-lived transmutation products on irradiation. Several groups have examined the routes to producing adequate elevated temperature properties, without sacrificing fabrication and room temperature properties requirements. These steels have base compositions ranging from around 20% chromium–11% manganese to 12% chromium–20% manganese, with additions of tungsten, titanium and vanadium, boron and phosphorus, often also with relatively high carbon contents, (0.25–0.8%), The ONRL LA steel is based on a 0.25%C, 12%Cr, and 20%Mn matrix, with 1%W, and additions of titanium and vanadium. It is described briefly by Maziasz et al.[61] and uses the same principles for control of the stability of the MC precipitate as the void swelling resistant alloys.

Steels Strengthened by Inert Particles
So far all the steels described have depended upon the balance between the initial solid solution, initial dislocation content, and precipitating phase reactions to provide the required through-life properties. This review will end with mention of a family of materials which sidesteps the requirement for a solution anneal/ageing treatment to develop the strengthening particle dispersion.

Nitriding was used to develop fine internal dispersions of TiN in the 20%Cr, 25%Ni, fuel cladding alloy matrix, containing 0.9%Si for oxidation resistance in the CO_2 atmosphere of the AGR. Roberts and Evans[62] showed the effect of increasing titanium content on both the diameter and volume fraction of the cruciform TiN particles. Figure 9 shows that the mean particle size, rather than the number of particles, increased with the increase in titanium content, Additionally, the nitriding process results in progressively coarser particles being formed with increasing depth into the strip. The addition of up to 3% titanium produced a thermally stable particle dispersion, giving a consistent increase in creep strength. The behaviour of this material in creep after cold deformation was examined by Wilson[63] and the particle dispersion was reported to be stable at temperatures up to 1000°C by Ferguson et al.[64]

Alternative processing methods were examined, to overcome the variability in particle size resulting from the diffusion front. Hamerton et al. described techniques of using nitrogen donor additions in powder metallurgy-based routes.[65] They presented initial mechanical properties data for the 20%Cr, 25%Ni matrix with TiN dispersion strengthening, developed by the 'Mechano-Fusion' route using a combination of matrix metal powders coated with chromium nitride as the donor.

FUTURE DEVELOPMENT OF THE AUSTENITIC STEELS
Despite the extensive optimisation which has taken place since the starting-point of this review, there will still be room for further developments. However, in respect of power generation applications, the austentic steels find themselves pressed increasingly between the performance of the ferritic steels and the range of adoption of nickel-based alloys. Thornton[66] has summarised the position in respect of bolting materials. He pointed out the relative design

limitiations of low proof strength, poor thermal conductivitiy and high thermal expansion characteristics of austenitic steels. to these must be added the ned for transition joints in mixed, austenitic-ferritic materials systems. However, subject to manufacturing capability, they remain candidates for turbine components for machines operating at 650°C, but are likely to displaced by nickel-based alloys at 700°C. However, the properties developed in the TiN strengthened steels indicate that some of the limitations set by solution-precipitation and cold/warm working strengthening mechanisms may be overcome by further developments. For service outside the specialised area of steam turbines, they will remain a widely used class of materials, offering a useful combination of strength, oxidation and environmental resistance and room temperature properties.

ACKNOWLEDGEMENTS
The authors wish to thank many colleagues for information and assistance in compiling this review. In particular, Professor T. Gladman made helpful comments on the manuscript. Christine Rawson, librarian at British Steel Swinden Technical Centre was of great assistance in tracking down reference material.

REFERENCES
1. M.G. Gemmill: *The Technology and Properties of Ferrous Alloys for High Temperature Use*, Newnes, London, 1966.
2. F.B. Pickering: *Physical Metallurgy and the Design of Steels*, Interscience, London & New York, 1978.
3. P. Marshall: *Austenitic Stainless Steels – Microstructural And Mechanical Properties*, Elservier Applied Science, Essex, 1984.
4. D.R. Harries: 'Physical Metallurgy of Fe-Cr-Ni austenitic steels', *Mechanical Behaviour and Nuclear Applications of Stainless Steels at Elevated Temperatures, Varese 1981*, The Metals Society, London, 1982, 1–21.
5. T. Gladman: 'High temperature aspects of stainless steels', *Stainless Steels '87*, The Institute of Materials, London 1988, 376–385.
6. M. Segerback, B. Ivarsson, R.E. Johansson and J.C. Kelly: 'Avesta Sheffield 353MA – a material for very high temperatures and harsh environments', *Proceedings of the Conference Corrosion '95*, paper 472, NACE, Houston.
7. *British Patent 13415*, 1913
8. *British Patent 201,915*, 1923
9. N.P. Allen: 'A survey of the development of Creep Resisting Alloys', *High Temperature Steels and alloys for Gas Turbines, Special Report 43*, The Iron and Steel Institute, London, 1952, 1–10.

10. D.A. Oliver, G.T. Harris and W.H. Bailey: 'Gas Turbine Steels and Alloys', *Steels in Modern Industry*, W.E. Benbow ed., Iliffe, London, 1951.

11. J.H.G. Moneypenny: *Stainless Iron and Steel*, Chapman & Hall, London, 1926, 258–259.

12. R.Hatfield and T.H. Burnham: *Special Steels*, Pitman, London, 1933, 141–143.

13. R.F.Miller: Appendix to, *The Properties of Metals at Elevated Temperatures*, G.V. Smith, McGraw-Hill, New York, 1950.

14. F. Whittle: 'Some Turbine Problems in the development of the Whittle Engine', *High Temperature Steels and alloys for Gas Turbines, Special Report 43*, The Iron and Steel Institute, London, 1952, 379–386.

15. F.C. Monkman, P.E. Price and N.J. Grant: *Trans. ASM.*, 1956, **48**, 418–445.

16. G.T. Harris and H.C. Child: 'Development of a high-temperature alloy for Gas-Turbine Rotor Blades', *High Temperature Steels and alloys for Gas Turbines, Special Report 43*, The Iron and Steel Institute, London, 1952, 67–80.

17. G.T. Harris and W.H. Bailey: 'Effect of warm-working on an Austenitic Steel (G18B)', *High Temperature Steels and alloys for Gas Turbines, Special Report 43*, The Iron and Steel Institute, London, 1952, 60–66.

18. J. Nutting: 'Some problems in the study of precipitations and phase transformation in metals', *Proceedings of the 4th International Conference on Electron Microscopy. Berlin 1958,Volume 1*, G. Mollenstedt, H. Niehrs, and E. Ruska, eds, Springer Verlag, Berlin, 1960, 574–579.

19. J. Nutting and J.M. Arrowsmith: 'The Metallography of Creep Resisting Alloys', *Structural Processes in Creep, SR 70*, The Institute of Iron, London, 1961, 147–165.

20. L.R. Woodyatt, C.T. Sims and H.J. Beattie : *Trans Met Soc AIME*, **236**, 1966, 519–527.

21. G. Alden and B. Aronsson: 'Some observations on the influence of carbon and nitrogen on the creep behaviours of AISI 316 type austenitic stainless steel', *Creep strength in steel and high temperature alloys*, The Metals Society, London, 1974, 67–71.

22. C.A.P. Horton, P. Marshall and R.G. Thomas: 'Time-dependant changes in microstructural and mechanical properties of type 316 steel and weld metal', *Mechanical Behaviour and Nuclear Applications of Stainless Steels at Elevated Temperatures, Varese 1981*, The Metals Society, London, 1982, 66–72.

23. M. Ceccarelli, R Santucci and A. Bennani: 'Hot mechanical properties of 316L stainless steel with boron and nitrogen additions', *Mechanical Behaviour and Nuclear Applications of Stainless Steels at Elevated Temperatures, Varese 1981*, The Metals Society, London, 1982, 39–46.

24. F.R. Beckitt, T.M. Banks and T. Gladman: 'Secondary creep deformation in 18%Cr–10%Ni steel', *Creep strength in steel and high temperature alloys*, The Metals Society, London, 1974, 71–78.

25. S.R. Keown: 'Microstructural changes occurring during the creep deformation of a simple austenitic steel at 600°C', *Creep strength in steel and high temperature alloys*, The Metals Society, London, 1974, 78–85.

26. J.A. Horak, V.K. Sikka and D.T. Raske: *Nuclear Power Plant ageing, Availability Factor and Reliability Analysis, San Diago 1985*, ASM, 1985, 301–313.

27. B. Weiss and R. Stickler: *Met. Trans.* 1972, **3**, 851–865.

28. R.A. Moen and G.V. Smith: *Trans ASME*, 1975 **79**, 162–171.

29. L. Bernard, E. Campo and S. Quaranta: 'Creep behaviour of AISI 304 and 316 stainless steel and influence of cold working', *Mechanical Behaviour and Nuclear Applications of Stainless Steels at Elevated Temperatures, Varese 1981*, The Metals Society, London, 1982, 88–93.

30. V. Guttman and J. Timm: 'Relationship between structure and creep properties of a predeformed austenitic steel', *Mechanical Behaviour and Nuclear Applications of Stainless Steels at Elevated Temperatures, Varese 1981*, The Metals Society, London, 1982, 106–112.

31. J.E. White and J.W. Freeman: *Trans ASME (J. Eng for Power)*, April 1963, 108–118.

32. H.W. Kirbyand R.J. Trueman: 'An investigation of the creep ductility of 18-12-1Nb and related steels', *Structural Processes in Creep, SR 70*, The Institute of Iron, London, 1961, 204–221.

33. S.R. Keown and F.B. Pickering: '*Niobium in Stainless steels*', *Niobium, Proceedings of a International Symposium 1981*, H. Stuart, ed, Met Soc. AIME, 1984, 1113–1142.

34. J. Wadsworth, J.H. Woodhead and S.R. Keown: *Met. Sci*, 1976, **10**, 347.

35. J.M. Silcock: *JISI*, 1963, **201**, 409–421.

36. J. Barfod and J. Myers: 'Structure related to hot ducitlity of these austenitic steels', *JISI*, 1963, **201**, 1025–1031.

37. J.M. Silcock and A. W. Denham: 'Precipitation of NaCl-Type Carbides in Austenite and Their Behaviour in the Neighbourhood of Grain Boundaries', 1968, 59–64.

38. S.R. Keown and F.B. Pickering: 'Effect of Niobium carbide on the creep-rupture properties of austenitic stainless steels', *Creep strength in steel and high temperature alloys*, The Metals Society, London, 1974, 134–144.

39. J. M Silcock and G. Willoughby: 'Ageing and creep behaviour of a Cr-Ni-Mn austenitic steel', *Creep strength in steel and high temperature alloys*, The Metals Society, London, 1974, 144–151.

40. R. Sumerling and J. Nutting: 'Precipitation ina 20%Cr–25%Ni steel stablized with Niobium', *JISI*, 1965, **203**, 398–405

41. J.D. Cook, D.R. Harries and A.C. Roberts: 'Some factors affecting the creep strength of 20%Cr-25%Ni-Nb austenitic steel at 750°C', *Creep strength in steel and high temperature alloys*, The Metals Society, London, 1974, 91–98.

42. B.D. Clay: 'Creep rupture properties of 20%Cr-25%Ni–Nb stainless steel in region 870-1320°C', *Mechanical Behaviour and Nuclear Applications of Stainless Steels at Elevated Temperatures, Varese 1981*, The Metals Society, London, 1982, 122–128.

43. N.G. Needham and T. Gladman: *Met. Sci.* 1980, **14**, 64–72.

44. J.D. Murray, J. Hacon and P.H. Wannell: *High Temperature Properties of Steels*, ISI, London. 1966, 403–425

45. F.E. Asbury and G. Willoughby: 'Aging and creep behaviour of a Cr-Ni-Mn austenitic steel', *Creep strength in steel and high temperature alloys*, The Metals Society, London, 1974, 144–151.

46. J.Orr and M.J. Collins: *Proceedings of the 2nd International Conference on Creep and Fracture of Engineering Materials and Structures*, vol 2, B. Wilshire and D.R.J. Owen, eds, Pineridge Press Ltd, Swansea, 1984, 775–788.

47. J. Orr: 'Warm worked Esshete 1250: A high strength bolting steel', Ironmaking steelmaking, 1994, **21** (5), 345–352.

48. Y. Sawaragi, H. Teranishi, M. Kubota and Y. Hayase: *Sumitomo Search*, 1989, **38**, 63–68.

49. T.W. Eichelberger: *Trans ASM*, 1958, **50**, 136–148.

50. K.J. Irvine, D.T. Llewellyn and F.B. Pickering: *JISI*, 1961, **199**, 153–175.

51. F.G. Wilson and F.B. Pickering: *JISI*, 1966, **204**, 628–637.

52. D. Raynor and J.M. Silcock: 'Strengthening Mechanisms in Gamma-Prime Precipitating alloys', *Metal Science Journal*, 1970, **4**, 121–130.

53. H.J. Beattie and W.C. Hagel: 'Intermatellic compounds in Ti-hardened Alloys', *Trans AIME*, 1957, **9**.

54. M. Okamura, M. Maeda and K. Hirose: *Proc. 6th Iron and Steel Congress, Nagoya*, Vol. 3, ISI, Japan. 1990, 715–722.

55. J. Orr: 'A review of the Strucutral Characteristics of Alloy 800', Alloy 800 (Proc. Conf.), Petten, 1978, 25–62.

56. I. Kirman: *JISI*, 1968, **207**, **1612**.

57. M. Tamura, N. Yamanouchi, A. Tohyama, T. Shiraishi, S. Murase and K. Matsuo: *Nippon Kokan Technical Report Overseas, 1987*, **49**, 29–39.

58. K.Erlich: 'Irradiation creep in austenitic stainless steel', *Mechanical Behaviour and Nuclear Applications of Stainless Steels at Elevated Temperatures, Varese 1981*, The Metals Society, London, 1982, 149–156.

59. D.S. Gelles: *J. Nuclear Materials*, 1993, **205**, 146–161.

60. M.H. Wood: *Nuclear Energy*, 1983, **22** (3), 159–176.

61. P.J. Maziasz, R.L. Klueh and A.F. Rowcliffe: *MRS Bulletin*, (July) 1989, 36–44.
62. A.C. Roberts and H.E. Evans: 'Effects of titantium and silicon additions on creep behaviour of TiN-dispersion-strengthened 20Cr-25Ni stainless steel', *Mechanical Behaviour and Nuclear Applications of Stainless Steels at Elevated Temperatures, Varese 1981*, The Metals Society, London, 1982, 51–57.
63. E.G. Wilson: 'Effect of cold working on creep behaviour of TiN-dispersion-strengthened 20%Cr-25% austenitic steel', *Mechanical Behaviour and Nuclear Applications of Stainless Steels at Elevated Temperatures, Varese 1981*, The Metals Society, London, 1982, 58–65.
64. P. Ferguson, J.H. Driver and A. Hendry: *J. Mat. Sci.*, 1983, **18**, 2951–2956.
65. R.G. Hamerton, D.M. Jaeger and A.R. Jones: 'Titanium Nitride Strengthened Steels for Bolting Applications', *Performance of Bolting Materials in High Temperature Plant Applications*, A. Strang ed., The Institute of Materials, London, 1995, 220–225.
66. D.V. Thorton: 'Bolting Requirements for Advanced Turbine Plant', *Performance of Bolting Materials in High Temperature Plant Applications*, A. Strang, ed., The Institute of Materials, London, 1995, 419–432.
67. M. Inoue, M. Yagi, H. Yuzawa, M. Yano, T. Masuda and N. Torii: 'Manufacturing System and Current Topics on Tubes and Pipe for Power Stations', *The Sumitomo Search*, 1984, **29**, 55–67.
68. Y. Sawaragi, K. Ogawa, S. Kato, A. Natori and S. Hirano: 'Development of the Economical 18-8 Stainless Steel (SUPER 304H) Having High Elevated Temperature Strength for Fossil Fired Boilers', *The Sumitomo Search*, 1992, **48**, 50–51.

Microstructure and Alloy Developments in Nickel-Based Superalloys

M.R. WINSTONE
Structural Materials Centre, DERA Farnborough, GU14 0LX

ABSTRACT

The history of nickel superalloys is the finest example of the parallel development of a material and its engineering application, both working in perfect harmony to achieve seemingly impossible targets. The application driver for nickel alloy development has been the gas turbine engine. This paper traces the history of nickel superalloys, from their humble beginnings nearly a century ago, to the high technology single crystal components of today. In the 50 years since the first flight of a gas turbine powered aircraft, the temperature capability of turbine blade alloys has been raised by over 400°C, so that blades can now operate at metal temperatures in excess of 1100°C. Advances have been achieved by a combination of alloy compositional modifications and innovative processing techniques. The review concentrates on alloys for turbine blades, but similar achievements have been made in derivation of nickel alloys for turbine discs and structural applications. These materials merit full reviews in their own right.

1. INTRODUCTION

The story of the evolution of nickel superalloys is the finest example of the parallel development of a material and its engineering application, both working in perfect harmony to achieve seemingly impossible targets. The application driver for nickel alloy development has been the gas turbine engine and, with it, the creation of the great aerospace industry of the second half of the twentieth century.

The origins of the material/application link are to be found in the late nineteenth century when the first gas turbines were conceived. The engines generated power by exploiting the Brayton thermodynamic cycle – inlet gas (air) is compressed, fuel injected, ignited and the hot gas allowed to expand through a turbine from which useful power is extracted. The remaining energy of the gas is used for direct jet thrust in aeropropulsion. It is a fundamental feature of this cycle that both the thermodynamic efficiency and the power output increase as the maximum cycle temperature is raised. Hence, there is a constant drive to create engines which run at higher temperatures but this is limited by the materials available for turbine construction. The turbine must operate in the high temperature gas stream and is subjected to high

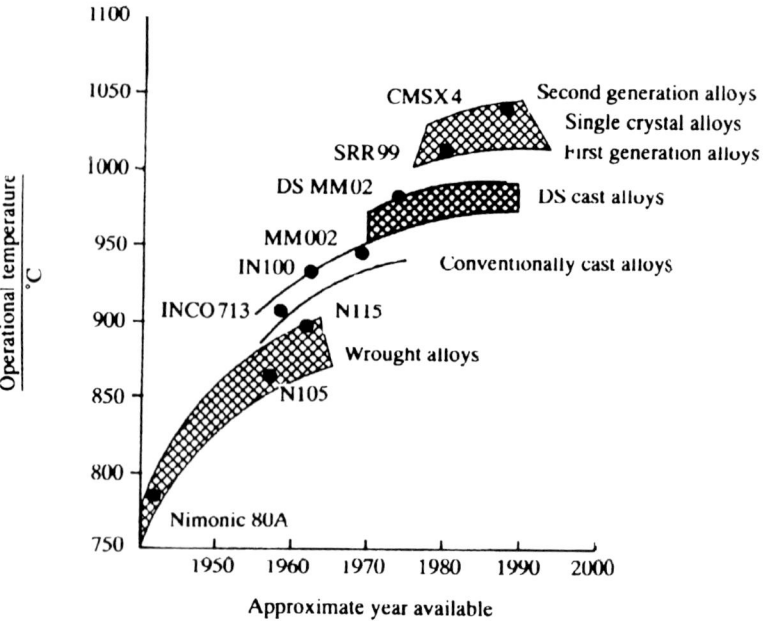

Fig. 1 Development of superalloys.

stresses due to centrifugal and thermal loads. Furthermore, the gas stream is highly oxidising and may contain corrodents such as oxides of vanadium and sulphur which often originate in the fuel. It is a triumph of materials engineering skill that commercial engines can now operate for thousands of hours with gas temperatures over 1600°C, all thanks to the advanced nickel superalloys developed specifically for this application. This paper will trace the history of nickel superalloys, from their humble beginnings nearly a century ago to the high technology single crystal components of today. Figure 1 shows schematically the temperature improvements which have been achieved for turbine blade alloys over the past 50 years,[1] highlighting the importance of the processing technique. This review will concentrate on the turbine blade alloys but that should not detract from the great achievements made in the development of alloys and processes for turbine discs and combustor applications. These materials merit full reviews of their own.

2. WROUGHT NICKEL ALLOYS

2.1 Historical Context
The origin of all nickel superalloys lies in the Ni–Cr system. Marsh's British Patent of 1906[2] highlighted the good oxidation resistance of Ni–Cr alloys and identified alloys that were particularly useful for electrical heating elements.

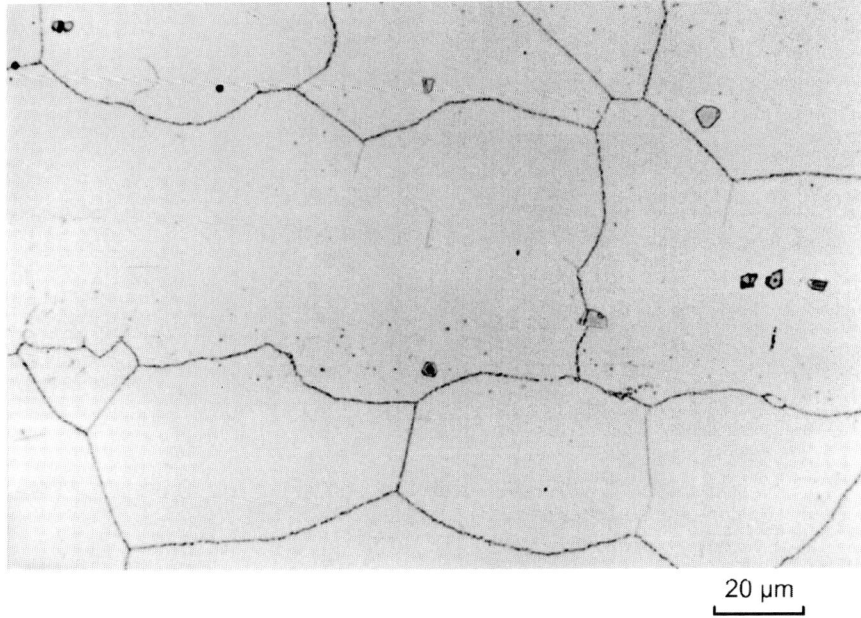

Fig. 2 Typical microstructure of wrought Nimonic 90 alloy.

Over the next 20 years it was established that the best balance of oxidation resistance and strength was achieved with 20%Cr,[3] the well known Nichrome alloy which is still in use today. Furthermore this alloy had better creep resistance than any other engineering material available in the 1920s.[4]

Nickel alloy research has always been a fertile bed for empirical research and within a few years it had been demonstrated that cobalt and titanium additions could further increase creep resistance and aluminium additions provided a high tensile strength at 800°C.[5] By coincidence these metallurgical advances occurred at precisely the time that Frank Whittle was building his first aircraft gas turbine engines. The new alloys found instant application to solve Whittle's major problem of creep failures in gas turbine blades. The alloys were commercialised by the Mond Nickel Co. as Nimonic 75 and Nimonic 80. The first flight of a Whittle gas turbine engine took place in 1941, with nickel alloy turbine blades operating at a gas temperature of 800°C.[6]

2.2 Structure and Properties
The microstructure of a typical wrought nickel superalloy is shown in Fig. 2 and the compositions of some Nimonic alloys are given in Table 1. These compositions were derived by purely empirical methods and it was not known why certain key elements were beneficial until many years later.

Table 1 Composition of typical nickel-based superalloys (weight %, balance Ni)

ALLOY	Co	Cr	Mo	W	Nb	Ta	Al	Ti	Zr	B	C	others
Nimonic 75	–	19.5	–	–	–	–	–	0.4	–	–	0.12	
Nimonic 80A	–	19.5	–	–	–	–	1.4	2.2	–	–	0.10	
Nimonic 90	18.0	19.5	–	–	–	–	1.5	2.5	–	–	0.13	
Nimonic 105	19.5	14.8	5.0	–	–	–	4.7	1.2	–	–	0.12	
Nimonic 115	14.2	15.0	4.0	–	–	–	5.0	4.0	0.10	0.015	0.16	
Udimet 700	18.5	15.0	5.2	–	–	–	4.3	3.5	0.05	0.03	0.08	
IN713C	–	12.5	4.2	–	××	××	6.1	0.8	0.10	0.012	0.12	Nb+Ta=2.2
G64	–	11.0	–	4.0	2.0	–	6.0	–	–	–	0.25	–
B1914	10.0	10.0	3.0	–	–	–	5.5	5.2	0.06	0.10	0.02	
B1925	8.5	12.0	1.7	4.5	–	4.0	3.5	4.0	0.10	0.10	0.02	
IN100	15.0	10.0	3.0	–	–	–	5.5	4.7	0.06	0.014	0.18	V, 0.90
IN738LC	8.5	16.0	1.7	2.6	0.9	1.7	3.4	3.4	0.05	0.011	0.11	
IN792+Hf	9.0	12.6	2.0	4.0	–	4.0	3.4	4.2	0.10	0.02	0.14	Hf, 1.0
IN939	19.0	22.4	–	2.0	1.0	1.4	1.9	3.7	0.10	0.01	0.15	
Mar-M200	10.0	9.0	–	12.5	1.0	–	5.0	2.0	0.05	0.015	0.15	
Mar-M200+Hf	9.0	8.0	–	12.0	1.0	–	5.0	1.9	0.03	0.015	0.13	Hf, 2.0
Mar-M246	10.0	9.0	2.5	10.0	–	1.5	5.5	1.5	0.05	0.015	0.15	
Mar-M002	10.0	9.0	–	10.0	–	2.5	5.5	1.5	0.06	0.015	0.15	Hf, 1.5
SRR99	5.0	8..5	–	9.5	–	2.8	5.5	2.2	–	–	–	
PWA1480	5.3	10.4	–	4.1	–	12	4.8	1.3	–	–	–	
Rene N4	8.0	10.0	2.0	6.0	0.5	5.0	4.2	3.5	–	–	–	Hf, 0.2
CMSX-4	9.0	6.5	0.6	6.0	–	6.5	5.6	1.0	–	–	–	Re, 3.0 Hf, 0.1
CMSX-10 (RR3000)	3.0	2.0	0.4	5.0	0.1	8.0	5.7	0.2	–	–	–	Re, 6.0 Hf, 0.1
Rene N6	12.0	4.0	1.0	6.0	–	7.0	5.8	–	–	–	–	Re, 5.0

Experience of ferrous alloys indicated that face-centred cubic crystal structures offered good creep resistance. This is now linked to the low stacking fault energy (SFE) which favours extended dislocation and inhibits dislocation climb, but in the 1930s dislocations were unknown although the twins observed in Nimonic alloys are further evidence of a low SFE.

The other feature clearly seen by optical metallography is the network of carbides around the grain boundaries. Grain boundary strength is a recurring theme in superalloy development which was appreciated from the very early days. The titanium additions provided stable carbides to control grain size and inhibit grain boundary sliding during creep. Figure 3 shows that the creep rate also decreases with increasing grain size.[7, 8]

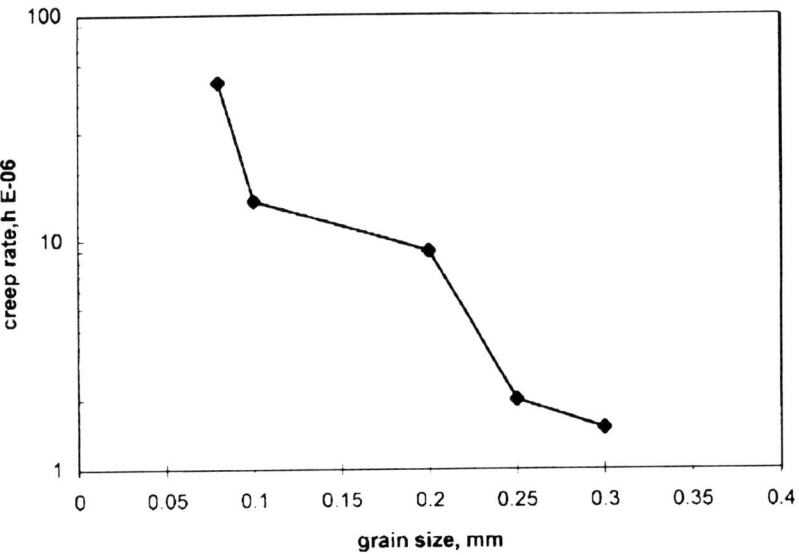

Fig. 3   Effect of grain size on the creep rate of a wrought superalloy (from Ref. 7).

The primary carbide, TiC, can be seen as angular particles in Fig. 2 but subsequent ageing allows reactions to occur, forming chromium rich particles at the grain boundaries. In Nimonic 75 the secondary carbides are based on Cr_7C_3 but the more common phase in higher Nimonics is $Cr_{23}C_6$. Optical examination of these early Nimonics reveals no detail within the grains. However, it was observed that aluminium additions produced significant age hardening, exploited in Nimonic 80A. Many years later the precipitate was identified as an ordered intermetallic phase which had a lattice parameter very similar to gamma nickel and formed extremely stable coherent particles. This phase is now known as gamma prime, based on Ni_3Al and has the ordered $L1_2$ structure. Cobalt was substituted for some of the nickel to create Nimonic 90. This raised the solubility temperature of gamma prime from 880°C to 960°C and thereby improved the hot workability. The cobalt also lowers the anti-phase boundary energy of the gamma prime which enhances the yield strength. It has been suggested that the cobalt lowers the stacking fault energy[9] which should enhance creep resistance, but this is not proven for complex alloys since the SFE is already very low and increases in cobalt also tend to raise the volume fraction of gamma prime.[10]

Thus, 60 years ago the key elements of the nickel superalloy system had been established; maintain a FCC structure, strengthen the grains by solid solution alloying, precipitate stable gamma prime and strengthen the grain boundaries with carbides.

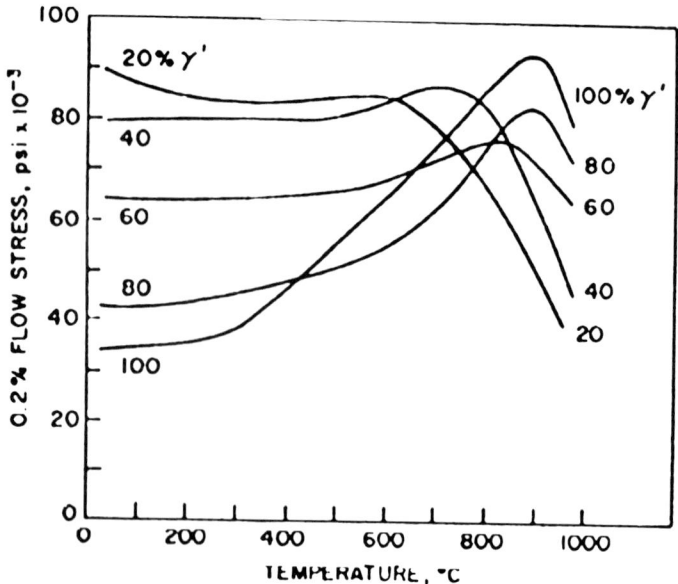

Fig. 4   Effect of gamma prime volume fraction on proof stress (from Ref. 11).

2.3   Development of High Strength Wrought Alloys

Further enhancement of the high temperature properties of wrought superalloys was achieved by increasing the volume fraction of gamma prime and adding solid solution strengthening elements. Beardmore et al.[11] investigated the temperature dependence of the flow stress in a series of ternary Ni–Cr–Al alloys in which the volume fraction of gamma prime was varied from 0 to 100%. The results are summarised in Fig. 4 and demonstrate the remarkable property of gamma prime – the flow stress *increases* with temperature up to about 850°C. Clearly, the low temperature properties of the 100% gamma prime alloy are too low for engine applications but a good balance of properties was achieved at 40–60% gamma prime which generates a focus for industrial alloy development. Under these conditions the dislocations are cutting the precipitate particles so solid solution strengthening of both the gamma and the gamma prime phases is required. This philosophy was applied very successfully to develop Nimonic 105 which contains 5% molybdenum to strengthen the matrix and 4.7% titanium to strengthen the precipitates. The high reactivity of aluminium and titanium resulted in a move to vacuum induction melting to ensure clean alloy production, free from oxide inclusions and to obtain good control of overall composition from strong inductive stirring. Furthermore, the alloy properties had become very sensitive to trace elements. The effects of trace elements has been reviewed by Holt and Wallace.[12] Only a few parts per million of elements like lead, bismuth, tel-

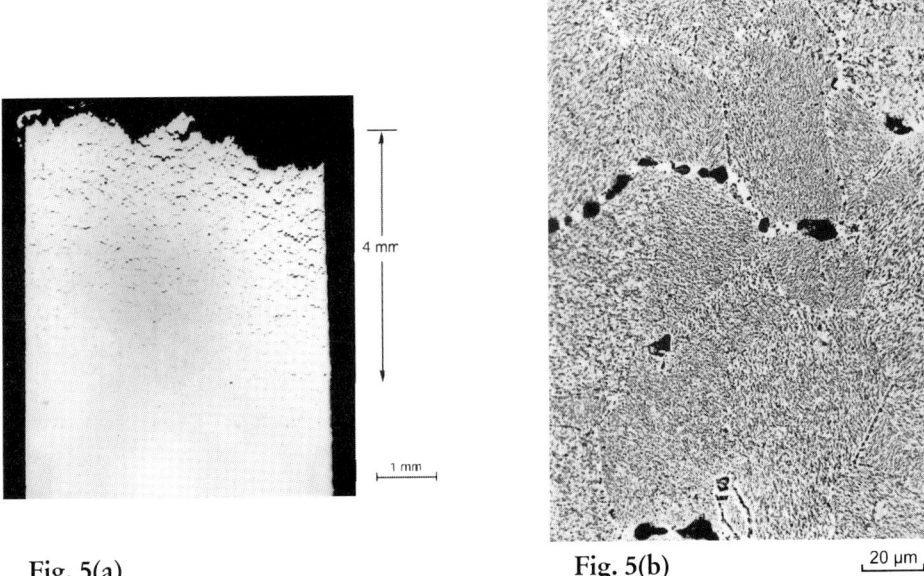

Fig. 5(a)   Fig. 5(b)

Grain boundary cracking in a Nimonic 90 turbine blade. (a) Section through failed blade. (b) Grain boundary voids.

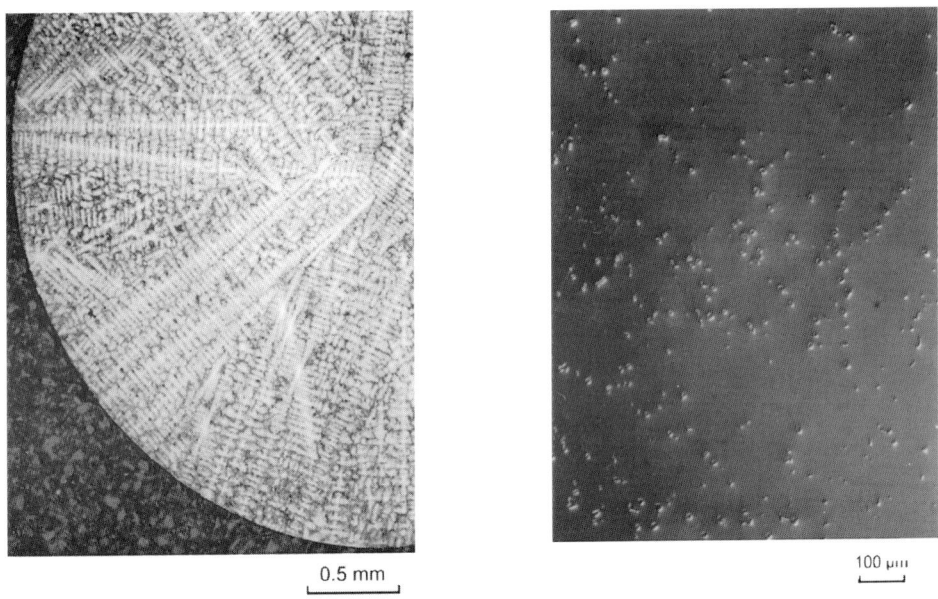

Fig. 6(a)   Fig. 6(b)

Structure of a cast superalloy. (a) dendrites. (b) blocky carbides.

lurium and thallium are highly damaging for the creep rupture properties.[13] The stirring action and the long melt time of vacuum induction melting bring impurities to the surface where they can be removed by the vacuum to lower the trace element concentration and enhance properties.[14]

The high temperature failure of wrought superalloy components is dominated by grain boundary cracking. Figure 5 shows a section through a failed Nimonic 90 turbine blade. Heat treatments to precipitate grain boundary carbides have always been an integral part of the processing. Nickel alloy studies in Russia identified boron as a creep strengthening element but the accidental contamination of an alloy in a crucible which contained both boron and zirconium identified a new grain boundary strengthening mechanism.[15] The addition of typically 0.015% boron and 0.1% zirconium is now applied universally to polycrystalline nickel superalloys. The action of zirconium is probably to reduce the concentration of grain boundary sulphur through the formation of zirconium carbosulphides.[16] The role of boron is less clear but may arise from the small atomic diameter which allows the boron to segregate to the grain boundaries, relax some of the grain boundary imperfections and reduce the concentration of vacancies available for dislocation climb.[17]

Further wrought alloy development beyond Nimonic 105 has been complicated by the problems of workability. It is very difficult to work alloys with more than 40% gamma prime and the alloys Nimonic 115 and Udimet 700 represent a limiting case where the high temperature strength of the alloys places an effective limit on the processing. Nimonic 115 can be extruded and is very difficult to forge but it is still used for blades in a few UK engines. From Fig. 1 it can be seen that the maximum blade metal temperature for a wrought superalloy had been raised from 750°C to about 900°C. On reaching this limit, alloy development needed to take a new turn. It was also necessary to recognise the need to produce hollow turbine blades since engines from 1960 onwards required channels up the centre of the blade for air cooling. Both the metal working and blade cooling problems could be solved by the investment casting of turbine blades.

3.   CAST NICKEL ALLOYS

3.1   Conventional Polycrystalline Alloys

With the requirements for hot working removed it was now possible to increase the gamma prime content of alloys above the 40% limit and approach the ideal requirement defined by Beardmore's work. This was achieved by increasing the aluminium content to about 6%. The commercial alloy IN713C also included 2% (niobium + tantalum) and gave about a 50° temperature improvement compared with the best wrought alloys.

Changing to the investment casting processing has two important effects on

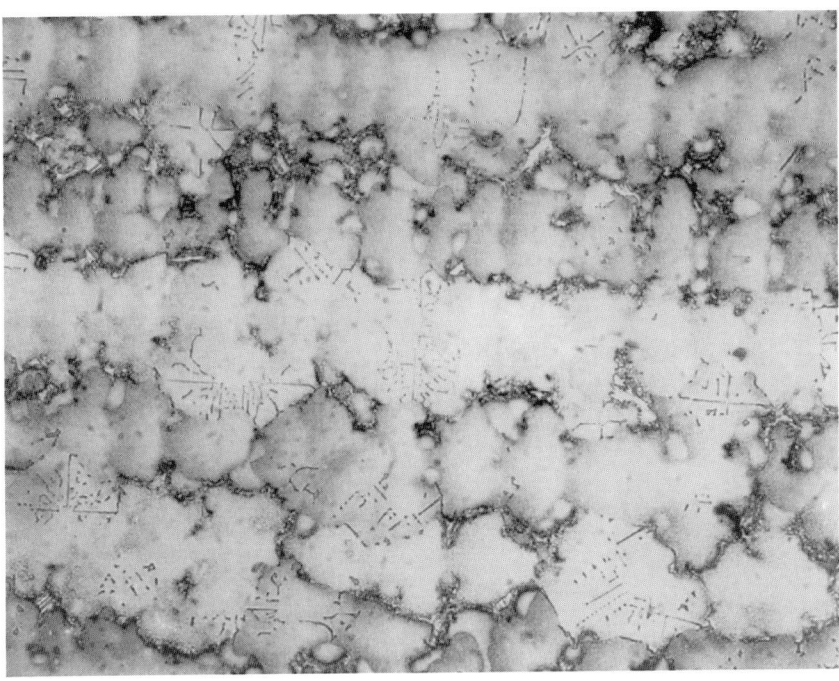

100 μm

Fig. 7 Script carbides in a cast superalloy.

the macrostructure. The castings have a heavily segregated, dendritic structure and the grain size is several millimetres, compared with a typical 50μm
for the wrought alloys (Fig. 6). The dendrite cores are nickel and cobalt rich.
Most of the alloying elements segregate to the interdendritic areas, with the
exceptions of tungsten and rhenium which segregate to the cores. Also within
the structure are coarse primary carbides with the general formula MC, where
the metal may be titanium, niobium or tantalum, depending on the alloy composition. These primary carbides are formed in the melt during solidification.
In Fig. 6 the carbides have a blocky morphology, but they may also take a
script morphology, dependent on the composition and the solidification conditions (Fig. 7).

The structure shown in Fig. 6 has the typical carbon level of about 0.1%
and 0.01% boron. It is perfectly feasible to produce alloys with much higher
boron levels which precipitate networks of borides with a very similar morphology to carbides. One of the first casting alloys was G64 produced by
Jessop Saville and used for turbine blades in the Bristol–Siddeley Proteus
engine which powered the Bristol Britannia aircraft in the late 1950s. This
alloy contained 0.2% boron and 0.13% carbon and provided good service in
the small Proteus blades. The high boron level gave the alloy good castability

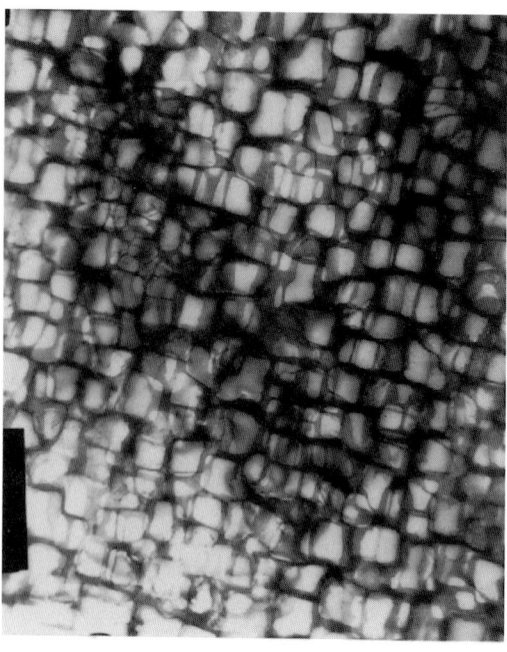

1 µm

Fig. 8   TEM micrograph of gamma prime precipitates in IN100.

but the excessive number of borides and carbides led to impact brittleness problems with larger castings. Twenty years later, Sorcery Alloys made a further attempt to market high boron alloys[18] but with low carbon levels, eg. Alloys B1914 and B1925. These alloys overcame the brittleness encountered in G64 and maintained the excellent castability, but the alloys enjoyed little commercial success; probably because the engine industry did not consider the benefits to be great enough to initiate a full scale development.

IN713 contains about 50% gamma prime and by 1960 the alloy IN100[19] was introduced with 60% gamma prime.[20] This signified the arrival of truly high strength, creep resistant cast superalloys. Fig. 8 is a dark field electron micrograph which shows the gamma prime precipitates in IN100. The regular array of cuboidal gamma prime precipitates is typical of nickel superalloys and the cube edge lengths are usually 0.25–0.5µm. IN100 has the remarkably low density of only 7.75g/cc which makes it ideal for high speed rotating turbine blades. It is still used today in many engines where the low density reduces the disc stresses.

The Achilles heel of the early cast alloys like IN713 and IN100 are their poor corrosion resistance (Fig. 9). Turbine blade life can be improved by applying corrosion resistant coatings[21] but ultimately the inherent corrosion behaviour of the substrate alloy remains important. The corrosion resistance

Fig. 9   Severely corroded turbine blade.

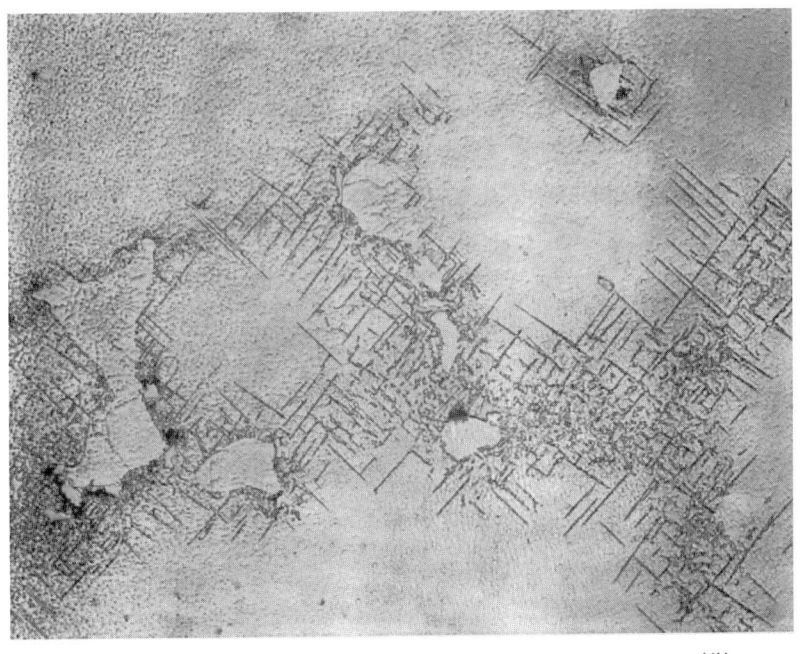

Fig. 10   Sigma precipitation in an experimental superalloy.

of conventional superalloys is proportional to the chromium content[22] but the solubility of chromium in gamma prime is only about 3%.[23] Consequently high gamma prime alloys have a maximum chromium content of about 10% to keep all the chromium in solution. If the solubility limit of the chromium is exceeded a complex sigma phase precipitates. In IN100 this phase is rich in Co, Cr, Mo and Ti and degrades the creep performance of the alloys.[24] Sigma forms most readily in alloys under stress at about 800–850°C and careful balancing of the composition is required to prevent in-service precipitation. Figure 10 shows sigma precipitation in the interdendritic areas of an experimental alloy after 1000h at 850°C. The more corrosion resistant alloys IN738 and IN939 were developed with higher chromium contents for long term industrial applications. Consequently they have lower gamma prime concentration and rely on heat treatments to optimise the gamma prime distribution. IN738 has about 45% gamma prime and a typical microstructure is shown in Fig. 11. A partial solution treatment at 1120°C has dissolved the gamma prime in the core of the dendrites but cubes of precipitate remain in the interdendritic areas. At higher magnification a bimodal distribution of gamma prime is observed (Fig. 11b) which ensures good properties over a wide range of conditions.

A significant step forward in the capability of cast superalloys was achieved by Carl Lund of Martin Metals when he introduced the first of the Mar–M series of alloys in 1965. These alloys contained a high concentration of tungsten to strengthen the matrix gamma phase and smaller additions of tantalum or niobium to strengthen the gamma prime. The first alloys in the series, typified by Mar–M200, had poor ductility in the 700–800°C temperature range but this was improved with the addition of 1.5% hafnium.[25] These hafnium modified alloys are still in widespread usage today and most engines from Rolls-Royce and Pratt and Whitney contain turbine blades cast from Mar–M alloys. The microstructural and constitutional effects of hafnium additions are complex. Hafnium is a strong sulphide former and would have a chemical interaction with residual sulphur [26, 27] (Fig. 12) but it also enters the carbides and modifies the gamma prime morphology.[27, 28] Gamma/gamma prime rosettes and blocky particles are formed at grain boundaries (Fig. 13) which appear to retard grain boundary sliding.[29, 30]

Investment casting had allowed the turbine blade metal temperature to be raised to over 950°C but grain boundary weakness was a life limiting problem.

3.2 Directionally Solidified Alloys

The maximum stresses in a turbine blade are along the leading and trailing edges. It follows that the grain boundaries perpendicular to the blade edge are usually the initiation points for blade cracking. If a process can be conceived which removes these boundaries then the blade life should be increased. In

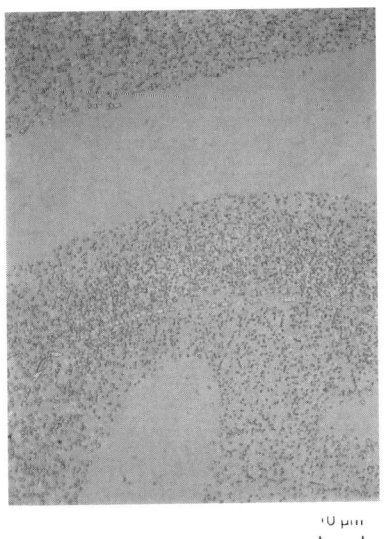

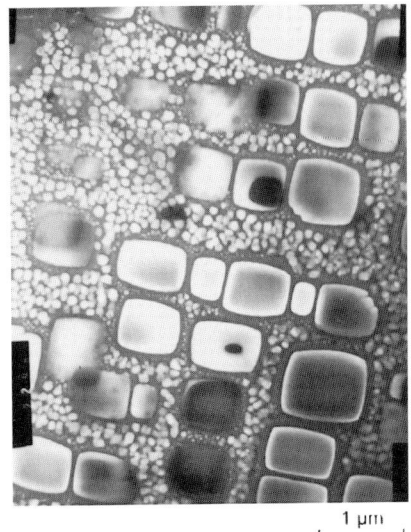

Fig. 11(a)   **Fig 11(b)**

Microstructure of IN738, heat treated 1120°C, 2h + 845°C, 24h. (a) Optical micrograph. (b) TEM micrograph.

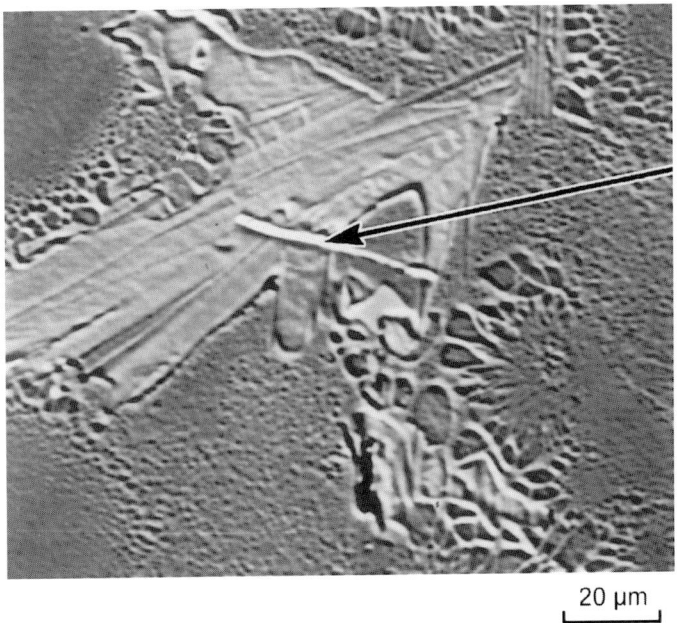

Fig. 12 Interdendritic carbosulphide particle in a hafnium containing alloy (IN792 + Hf).

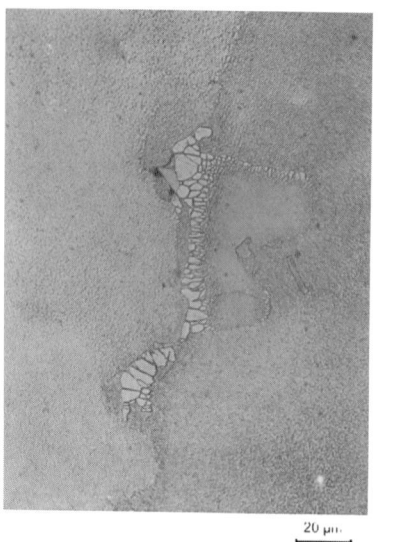

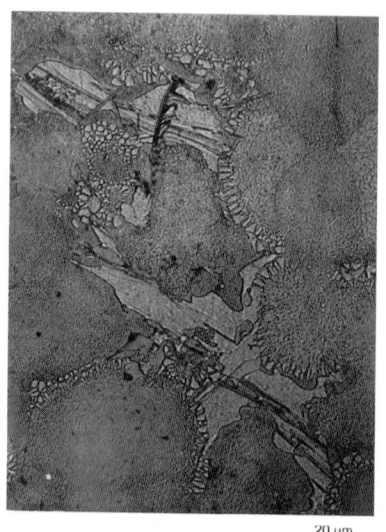

Fig. 13(a)   **Fig. 13(b)**

Effect of hafnium additions on microstructure of IN792. (a) no hafnium. (b) 1% hafnium.

        (a)   (b)   (c)

Fig. 14   Macrostructures of cast turbine blades.
              (a) Conventionally cast
              (b) Directionally solidified
              (c) Single crystal.

1960, VerSnyder and Guard[31] described a casting technique for the directional solidification (DS) of nickel alloy ingots and showed that testpieces cut parallel to the solidification direction had higher creep lives and ductilities. This principle was developed rapidly to produce investment cast turbine blades[32] and a minor change in mould design was used to produce single crystal superalloy blades.[33] Figure 14 shows the macrostructure of turbine blades cast by conventional and directional solidification techniques. The conventionally cast blade has equiaxed crystals a few millimetres in diameter whereas the DS blade has columnar grains running the full length of the component. At higher magnification the segregated dendritic nature of the DS casting is apparent, with a network of MC script carbides (Fig. 7). The early work on directional solidification was reviewed by VerSnyder and Shank in 1970,[34] but what is not generally realised is that the potential benefits of single crystal turbine blades had been highlighted 30 years earlier by Taylor.[35] The Frank Whittle engines suffered from grain boundary failures, coarse grain structures were better[6] and a clear extrapolation led to the deduction that single crystals were the ideal.

The introduction of polycrystalline DS castings provided a substantial increase in turbine blade life due to the improved creep resistance and much increased resistance to thermal fatigue cracking,[35] using existing superalloys. DS castings also provided a major improvement in the properties of thin section castings, typical of a cooled aerofoil. Coarse-grained conventional castings have a significant loss of creep strength when the section thickness is comparable with the grain size, but this is virtually eliminated in DS castings.[36, 37]

3.3   Single Crystal Casting

Casting conventional alloys in single crystal form does not provide any further advantage over the polycrystalline DS. It is not possible to homogenise and solution treat conventional cast high strength superalloys due to low melting point phases at the grain boundaries. These phases arise from the grain boundary strengtheners (C, B Zr and Hf) which reduce the incipient melting point to below the gamma prime solvus. However, single crystal applications do not need these additions and this has been exploited in alloy developments specifically for the single crystal castings.

The first generation of single crystal alloys were relatively simple; SRR99 is a typical example. Carbon has been reduced to a minimum so that few carbides are formed and the as-cast structure shows the strong segregation typical of superalloys. This structure can be solution treated to dissolve the gamma/gamma prime islands and decrease the dendritic segregation (Fig. 15). Subsequent ageing then precipitates the gamma prime in the optimum size and distribution. This process provided a 30°C increase in creep temperature capability compared with the equivalent DS alloy.

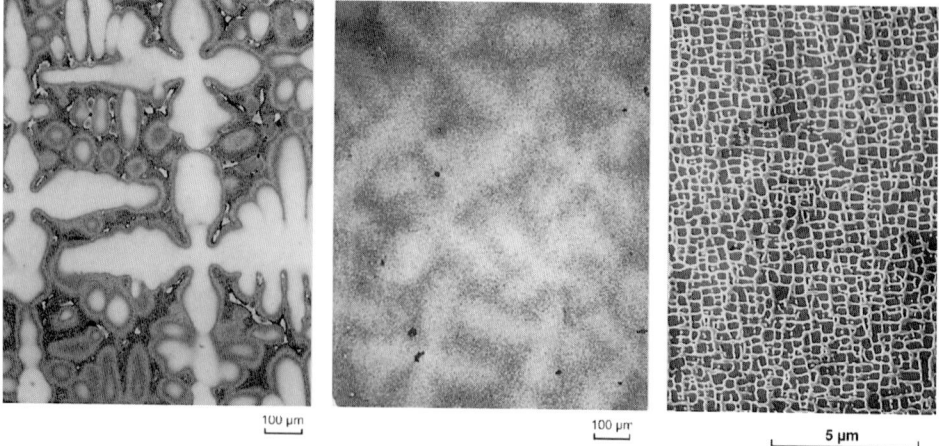

Fig. 15 (a, b, c) Single crystal alloy microstructure. (a) as cast. (b) solution treated – optical. (c) SEM.

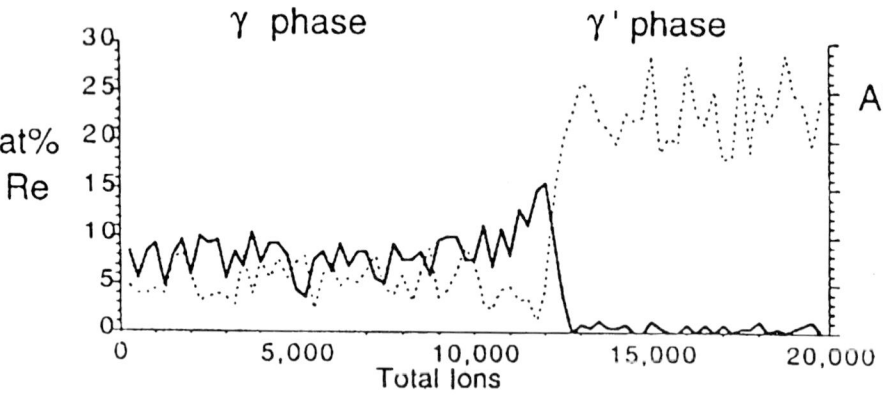

Fig. 16 Atom probe trace of rhenium concentration across gamma/gamma prime interface in RR3000 alloy.

3.4   Rhenium-Containing Alloys

While the first generation of single crystal alloys was being introduced into service during the early 1980s it was recognised that a further increment in temperature capability could be achieved by reducing the tungsten and molybdenum content and adding rhenium. Both DS alloys and single crystal alloys[38] have been developed with rhenium additions. Probably the best known is CMSX–4 which contains 3% rhenium. This second generation single crystal alloy has a 25° increase in temperature capability over alloys like

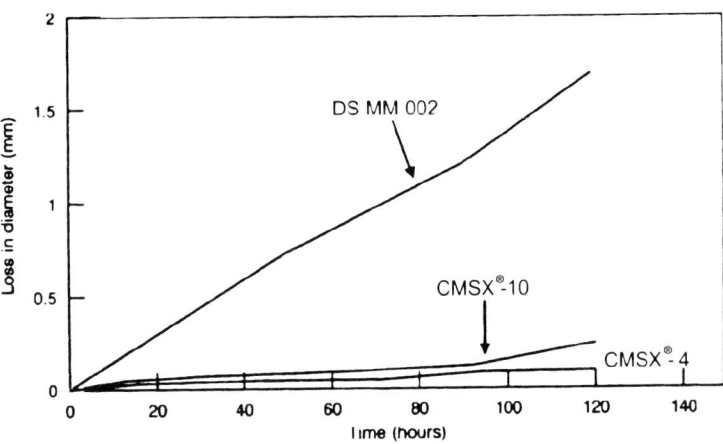

Fig. 17   Burner rig test results for rhenium containing alloys.

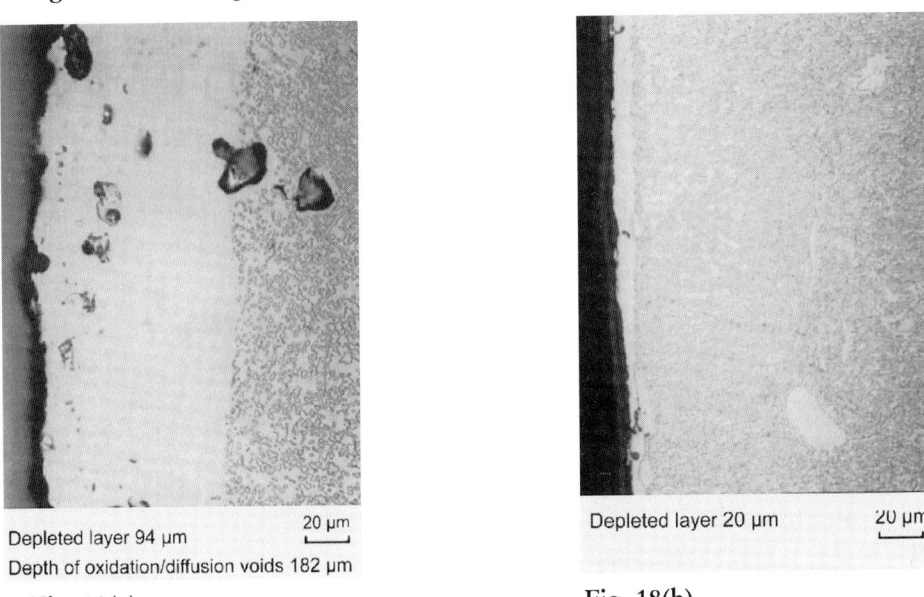

Depleted layer 94 μm

Depth of oxidation/diffusion voids 182 μm

Depleted layer 20 μm

Fig. 18(a)   **Fig. 18(b)**

Optical micrograph of cyclic oxidation testpiece after 450h at 1177°C (2150°F).
(a) CMSX–4 (<2ppm S). (b) CMSX–4 (<2ppm S + 15ppm Y).

SRR99. More recently a third generation of alloys has been announced which contain about 6% rhenium and provide a further 30° increment of temperature capability.[39,40] Rhenium is a large atom which segregates to the gamma phase and the dendrite cores. It has a very slow diffusion rate which makes homogenisation difficult but is also responsible for excellent microstructural stability at high temperatures. Atom probe work has identified rhenium clusters in the gamma phase[41] which may also contribute to the high creep resist-

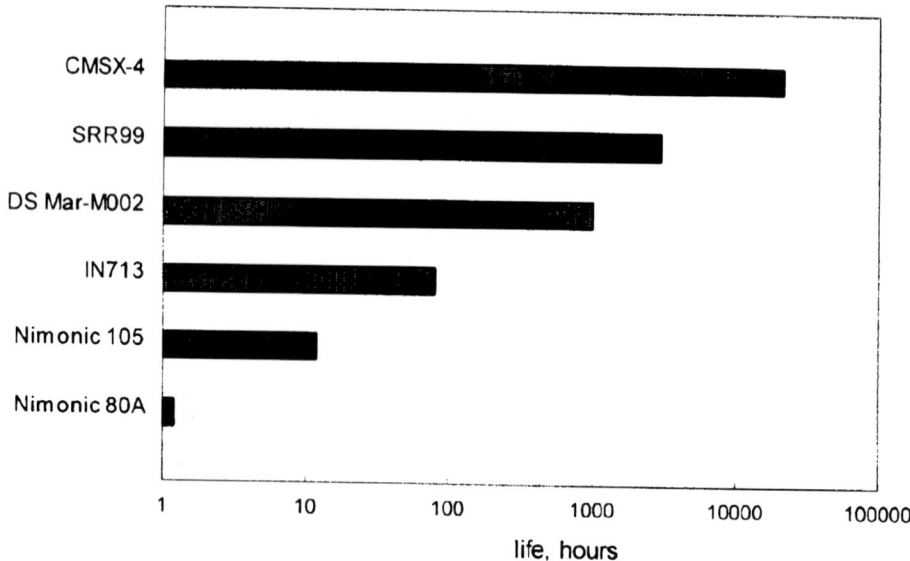

Fig. 19 Creep rupture lives of typical superalloys at 850°C, 350MPa.

ance. A similar study at Oxford University[42] has not confirmed the presence of clusters, but it has identified a rhenium rich region in the gamma phase immediately adjacent to the gamma prime cuboids (Fig. 16). It is reasonable to assume that this rhenium is rejected by the growing particle but it will also be extremely slow to diffuse away, which would inhibit further precipitate growth. This could account for the very slow gamma prime coarsening kinetics observed in rhenium-containing alloys.

Rhenium additions have made it possible to raise the turbine blade metal temperature to over 1100°C, but an unexpected property of the rhenium alloys is good oxidation and corrosion resistance. Burner rig tests[42] show that the second and third generation single crystal alloys have better oxidation and corrosion resistance than conventional alloys like Mar–M002 (Fig. 17). This would not have been predicted in alloys which contain only 2–4% chromium. It has also been shown that small additions of lanthanum or yttrium are advantageous in reducing the oxidation rate, especially in combination with an ultra-low sulphur levels (Fig. 18). This is now being considered for third generation single crystal alloys.[42]

4. CONCLUDING REMARKS

In their 50 year history superalloys have made great advances, such that alloys have now been demonstrated in engines at metal temperatures of 1130°C.[42] This is about 87% of the melting point and a far higher homologous temperature than any other structural material. Research continues to derive new superalloys with even greater creep resistance. Developments in ceramics and refractory metal alloys have failed to produce materials which are useful in the engineering context of highly loaded turbine components and the superiority of the nickel-based system is unchallenged.

The emphasis on temperature capability has been driven by the need for greater engine thrust and improved thermal efficiency. However, the trend in the second half of the 1990s is to concentrate on total life cycle cost. This shift in emphasis places great importance on component life, and less on the temperature of operation. A recalculation of data for various superalloys can be plotted to assess how successful superalloy development has been in terms of life rather than performance. Figure 19 plots the creep rupture lives of a selection of alloys, assuming a creep stress of 350MPa at 850°C.

The early Nimonic alloy achieved a life of only 1 hour, compared with 21 000 hours for single crystal CMSX–4. Third generation alloys should increase this to nearer 100 000 hours. This 10^5 increase in life-time is a dramatic indication of how successful the nickel superalloy development programmes have been. While creep resistance has been the primary driver for turbine blade alloy development, other issues, such as fatigue and oxidation/corrosion resistance, can dominate when service lives are very long. It has not been possible to include these in this review but they are a vitally important part of the lifing of components. The study of the three damage mechanisms, creep-fatigue-environment, is now becoming an integrated activity, and all are driving future composition and microstructure developments.

ACKNOWLEDGEMENTS

The author gratefully acknowledges the assistance of many colleagues in the preparation of this paper and for the permission to use some of the figures. In particular thanks are due to John Sutton of European Gas Turbines (Figs. 2, 5), Bob Broomfield of Rolls-Royce plc (Fig. 17), Ken Harris of Cannon-Muskegan (Fig. 18) and Prof. George Smith of Oxford University (Fig. 16).

REFERENCES

1. D.A. Ford: *Proc. Instn. Mech. Engrs*, 1996, **210**, 147–155.
2. A.L. Marsh: *British Patent No. 2129*, 1906.
3. C.J. Smithells, S. V. Williams and J. W. Avery: *J. Inst. Met.*, 1928, **40**, 269–290.
4. H.J. Tapsell and J. Bradley: *Engineering*, 1925, **120**, 614–615.

5. A.L. Sanford and O.E. Harder: *Trans. Amer. Soc. Met.*, 1939, **27**, 538–559.
6. T.A. Taylor: *Proc. Inst. Mech. Eng.*, 1945, **153**, 505–512.
7. T.B. Gibbons and B.E. Hopkins: *Met. Sci. J.*, 1971, **5**, 233–240.
8. T.B. Gibbons and B.E. Hopkins: *Met. Sci. J.*, 1984, **18**, 273–280.
9. B.E.P. Beeston and L.K. France: *J. Inst. Met.*, 1968, **98**, 105.
10. R.N. Jarrett and J.K. Tien: *Met. Trans.*, 1982, **13A**, 1021–1032.
11. P. Beardsmore, R.G. Davies and T.L. Johnson: *Trans. Met. Soc. AIME*, 1969, **245**, 1537.
12. R.T. Holt and W. Wallace: *Int. Met. Rev.*, 1976, **203**, 1–24.
13. D.R. Wood and R.M. Cook: *Metallurgia*, 1963, **67**, 109–117.
14. P.P. Turillon: *Trans. 6th Int. Vacuum Metallurgy Conf.*, American Vacuum Society, 1963.
15. R.F. Decker, J.P. Rowe and J.W. Freeman: *NACA Report 1392*, 1958.
16. J.H. Schneibel, C.L. White and M.H. Loo: *Met. Trans.*, 1985, **16A**, 651–660.
17. R.S. Polvani, A.W. Ruff and P.R. Strutt: *J. Mat. Sci Let.*, 1983, **3**, 287–290.
18. D.H. Maxwell, J.F. Baldwin and J.F. Radavich: *Metallurgia*, 1975, 332–337.
19. C.G. Bieber: *US Pat 3,061,426*, 1960.
20. O.H. Kriege and J.M. Baris: *Trans ASM*, 1969, **62**, 195–200.
21. D.S. Rickerby and M.R. Winstone: *Materials & Manufacturing Processes*, 1992, **7**, 495–526.
22. H.-J. Rätzer-Scheibe and M.R. Winstone: *Mat. Sci. Tech.*, 1993, **9**, 253–258.
23. S. Chakravort, S. Sadiq and D.R.F. West: *J. Mat. Sci.*, 1989, **24**, 577–583.
24. J.R. Mihalisin, C.G. Bieber and R.T. Grant: *Trans. Met. Soc. AIME*, 1968, **242**, 2399–2414.
25. D.N. Duhl and C.P. Sullivan: *J. Metals*, 1971, **23**, 88.
26. J.E. Doherty, A.F. Giamei and B.H. Kear: *Canadian Metals Quarterly*, 1974, **13**, 229.
27. C. Lund and J.F. Radavich: *Superalloys*, 1980, **80**, 85–98.
28. J.M. Dahl, W.F. Danesi and R.G. Dunn: *Met. Trans.*, 1973, **4**, 1087–1096.
29. P.S. Kotval, J.D. Venables and R.W. Calder: *Met. Trans.*, 1972, **3**, 453–458.
30. F.L. VerSnyder and R.W. Guard: *Trans ASM*, 1960, **52**, 485–493.
31. B.J. Pearcey and F.L. VerSnyder: *J. of Aircraft*, 1966, **3**, 390–397.
32. B.H. Kear and B.J. Pearcey: *Trans. Met. Soc. AIME*, 1967, **239**, 1209–1215.
33. F.L. VerSnyder and M.E. Shank: *Mat. Sci. Eng.*, 1970, **6**, 213–247.
34. T.A. Taylor: 'Production of metals with high temperature creep resisting properties', *Report No. 3*, Power Jets Ltd, August 1940.
35. J.E. Northwood: *Metallurgia*, 1979, **46**, 437–442.

36. M.R. Winstone and J.E. Northwood: *Solidification Technology in the Foundry and Cast House*, The Metals Society, London, 1980, 298–303.
37. K. Harris, G.L. Erickson, S.L. Sikkenga, W.D. Brentall, J.M. Aurrecoechea and K.G. Kuarych: *Superalloys 1992*, 1992, 297–306.
38. W.S. Walston, K.S. O'Hara, E.W. Ross, T.M. Pollock and W.H. Murphy: *Superalloys 1996*, 1996, 27–34.
39. G.L. Erickson: *Superalloys 1996*, 1996, 35–44.
40. D. Blavette, P. Caron and T. Khan: 'Superalloys 1988', *TMS*, 1988, 305–314.
41. R.W. Broomfield, D.A. Ford, H.K. Bhangu, M.C. Thomas, D.J. Frasier, P.S. Burkholder, K. Harris, G.L. Erickson and J.B. Wahl: *ASME Turbo Expo '97*, Orlando, Florida, June 1997.

Principles of Microstructural Stability in Creep Resistant Alloys

T. GLADMAN
University of Leeds, Leeds LS2 9JT, UK

ABSTRACT
Attention is focused on the stability of grain size and the factors controlling particle coarsening in creep resistant alloys. In microalloyed steel particle coarsening is important to the control of austenite grain size at elevated temperatures, and examples are discussed where the solubilities of the carbides and nitrides are important in controlling the rate of bulk diffusion controlled particle coarsening.

Consideration is given to the stability of the microstructure of creep resisting steels where, because of the higher alloy contents and lower temperatures, other mechanisms of particle coarsening may operate, e.g. grain boundary diffusion controlled coarsening, and dislocation pipe diffusion control. The use of these mechanisms requires a considerable data base, particularly in the area of particle solubility. An example is given of the data usage in the case of chromium carbides in the Fe–Cr–C ternary system.

At present, there is an extensive data base, but the complexity of creep resistant steel chemistries may yet prove that the data base is incomplete.

1. INTRODUCTION

The mechanisms of both creep deformation and creep rupture are numerous, and are dependent on the stress, stress state, and temperature, for a given material. However, virtually all of these mechanisms are strongly dependent upon the microstructure. Dislocation creep rates are reduced by finely distributed second phase particles, the by-passing of particles by dislocations being highly dependent upon dislocation climb at the creep exposure temperature. Grain boundary diffusion creep and grain boundary sliding can be reduced by second phase particles precipitated on the grain boundaries. In creep rupture, the nucleation of r-type cavities may be influenced by second phase particles at the boundaries; manganese sulphide and the larger M_6C or $M_{23}C_6$ carbides are capable of enhancing the cavity nucleation rates. The long term exposure of metallic materials at elevated temperatures permits considerable changes of the microstructure to occur, and these changes in turn affect the creep response of the material.

The changes in microstructure that may occur as a result of long term creep exposure include carbide or nitride precipitation, which depletes the solute content of the matrix, carbide transformation, grain growth and particle

Table 1 Grain Growth Forces

	Driving	Pinning
Zener	$-2\gamma/R$	$3f\gamma/2r$
Gladman	$+2\gamma/R - 3\gamma/2R0$	$3f\gamma/2r$
Senogles	$+2\gamma/R - 3\gamma/2R0$	$\Sigma 3f_i\gamma/2r_i$
Doherty	$+2\gamma/R - 3\gamma/2R0$	$f\gamma R_0/2r^2$

coarsening. Several of the papers in the present conference consider precipitation and carbide transformation, and this paper is aimed at describing the principles involved in the grain growth and particle coarsening phenomena that can occur during creep exposure. It should be pointed out that the author's experience of these principles was mainly derived from studies of microalloyed steels at much higher temperatures than are involved in creep applications, but the principles remain the same for both sets of conditions. The differences between microalloyed and creep resistant steels are mainly associated with the alloying elements and the concentrations that are used. Some consideration is given to these differences, and the attendant modification of the constitutive equations that may be required. Further, the creep exposure itself may bring other considerations into play. Whilst the thermally induced changes in microstructure may cause changes in the creep resistance of the material, the creep process itself, e.g. dislocation creep, may affect the rate of change of the microstructure.

The paper considers first the principles involved in grain size stability, which has some application in creep resisting steels, but which illustrates many of the principles involved in particle size stability. Secondly, the principles of particle coarsening are broadened to illustrate the complexity imposed by the range of creep conditions, and the attendant multiplicity of particle coarsening mechanisms that may be invoked. Finally, some consideration is given to the potential data sources that are necessary to the successful prediction of particle coarsening rates.

2. GRAIN COARSENING

The principles and practices of grain growth inhibition by second phase particles are well established. Following the initial work of Zener,[1] who considered the grain boundary pinning force of a random array of fine particles, and an estimate of the driving force for grain growth, based on a shrinking spherical grain, several more refined models have been proposed with more realistic estimates of the driving force for grain growth.[2,3,4] Models which cover grain refinement by the simultaneous addition of several species of particle, each of very different size,[3] and which consider grain boundary precipitates rather than randomly distributed particles,[4] are available.

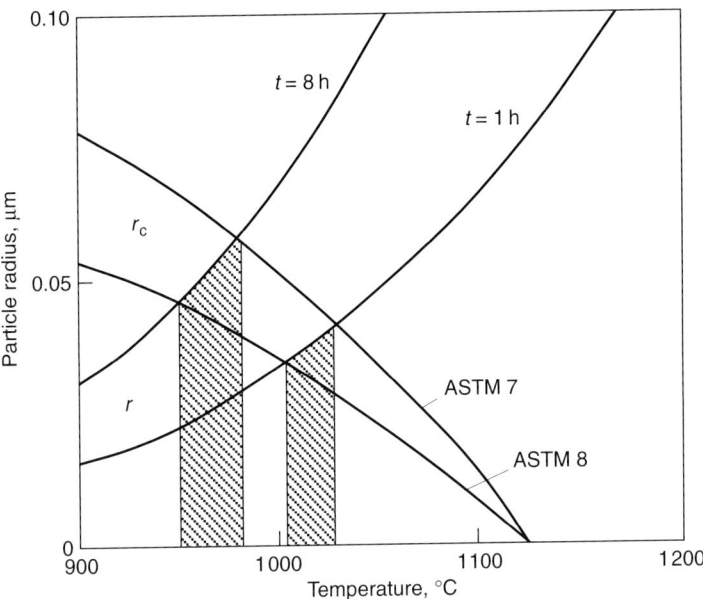

Fig. 1 Changes in the critical particle size for grain growth and the actual particle sizes of A1N as a function of temperature and time.

The rates of change of energy accompanying the growth of a grain of radius R in a matrix of grains of radius R_0 are shown in Table 1. The term $2\gamma/R$ represents a growth resisting force due to expansion of the grain interface; the term $-3\ \gamma/2R_0$ represents a growth encouraging force due to disappearance of neighbouring grain boundaries, and the particle pinning forces resist grain growth. If the summed rates of change of energy (all of which are expressed as per unit area of the growing grain) are negative, grain growth will occur. If the summed rates are positive, growth will be inhibited. There is obviously a boundary between these domains where the summed rates of change of energy are equal to zero. In the case of uniform randomly distributed particles,

$$2\ \gamma/R - 3\ \gamma/2R_0 + 3f\ \gamma/2r = 0$$

or
$$(1 - 4/3Z) = R_0 f/r \tag{1}$$

where $Z = R/R_0$ and represents the size advantage of the large grain of radius R, and usually has a value between $2^{1/2}$ and 2. Equation 1 shows the maximum particle radius that will inhibit grain growth for any given level of matrix grain size, R_0, and volume fraction of particles, f. It can also be used to express the minimum volume fraction, f, required to inhibit growth, given r and R_0, or the minimum grain size, R_0, that can be stabilised given f and r.

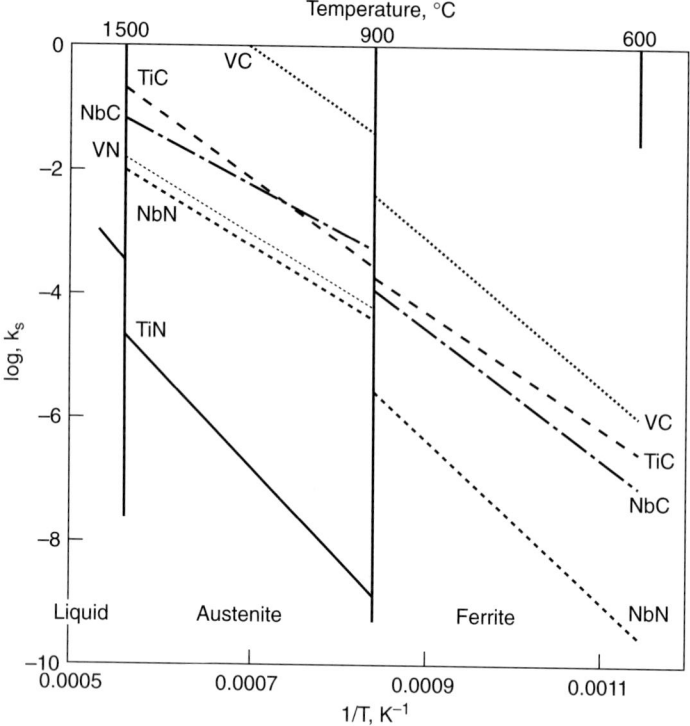

Fig. 2 The solubilities of microalloy carbides in liquid, austenite and ferrite.

A similar treatment of particles exclusively formed on grain boundaries yields the following equation:

$$(3 - 4/Z) = R_0^2 f/r_c^2 \qquad (2)$$

The use of these equations is illustrated in Fig. 1. The critical particle size can be defined by the volume fraction of particles and the matrix grain size from eqn 1. As the volume fraction of sparingly soluble carbide/nitride decreases with increasing temperature, so the minimum critical particle size, r_c, decreases. A knowledge of *solubility* of the particulate material in the matrix is clearly required. Meanwhile, exposure at progressively higher temperatures causes increased Ostwald ripening of the particles, i.e. r increases. When the particle size exceeds the critical particle size, then grain growth commences. The importance of *Ostwald ripening* is thus established.

2.1 *The Solubility of Microalloy Carbo-Nitrides*
The microalloy carbides, nitrides or carbo-nitrides are reasonably well documented. In a simple ternary system, Fe–M–X, where *M* is the microalloying

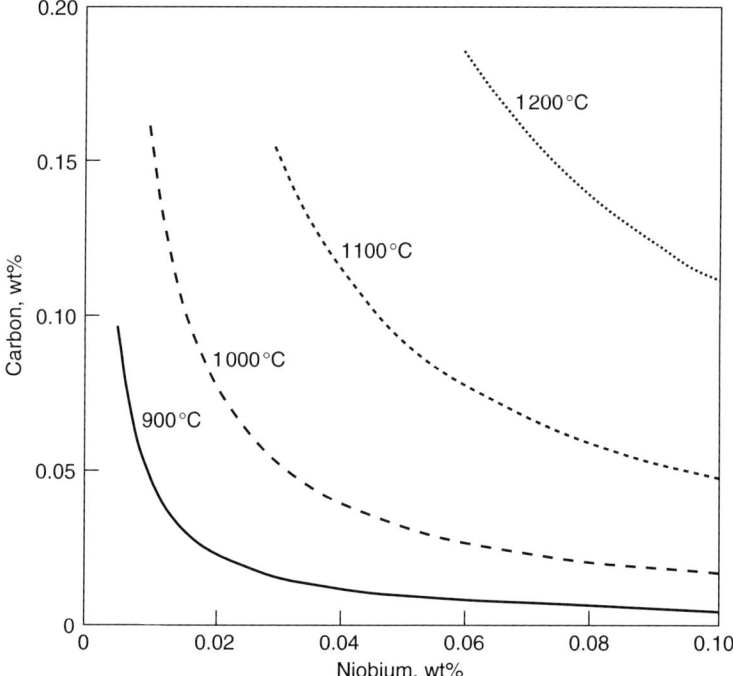

Fig. 3   The effect of temperature on the solubility of niobium carbide in austenite.

element, and X is either carbon or nitrogen, the solubility product, k_s, for the compound MX is given by:

$$k_s = [M][X] \tag{3}$$

where $[M]$ is the dissolved microalloying concentration in equilibrium with the compound MX, and $[X]$ is the corresponding dissolved carbon or nitrogen concentration.

The solubility product in microalloyed steels is usually based on wt % concentrations for convenience. The activity of MX is assumed to be unity, and the activity coefficients of M and X in solution are also assumed to be unity. The solubility product increases with increasing temperature according to the Arrhenius type relationship, usually expressed in the form:

$$\log_{10}k_s = -A/T + B \tag{4}$$

The published data for microalloy carbide and nitride solubilities are summarised[5] in Fig. 2. The solubilities of the various carbides and nitrides in austenite are very similar with two notable exceptions:

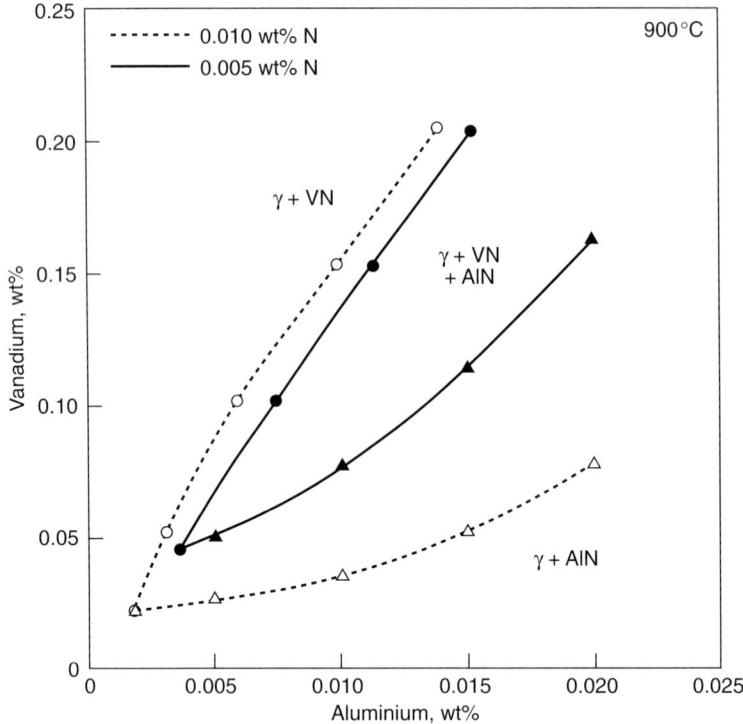

Fig. 4 Composition ranges for the presence of both aluminium and vanadium nitrides.

(i) The solubility of vanadium carbide is about three orders of magnitude more than most.

(ii) The solubility of titanium nitride in austenite is about three orders of mag nitude less than most of the other carbides and nitrides.

The nitrides are more stable than the carbides. The solubility of microalloy carbides and nitrides can be represented as isothermal intersections with the solvus surface of the ternary system, as shown for niobium carbide in austenite in Fig. 3. The high solubility of carbon in austenite prevents any interference from the formation of cementite in microalloyed steels, but this is not necessarily true for microalloyed steels in the ferrite range, where the limited solubility of carbon in ferrite, Fig. 2, may result in the formation of cementite.

The formation of more than one carbide or nitride phase due to the addition of a fourth element to the system has been considered[5] as for example in the Fe–V–N–Al system. The solubility products of VN and AlN in austenite can result in the formation of both hexagonal AlN and the fcc VN over certain composition ranges, Fig. 4. The nitrogen remaining in solution must comply with both solubility products.

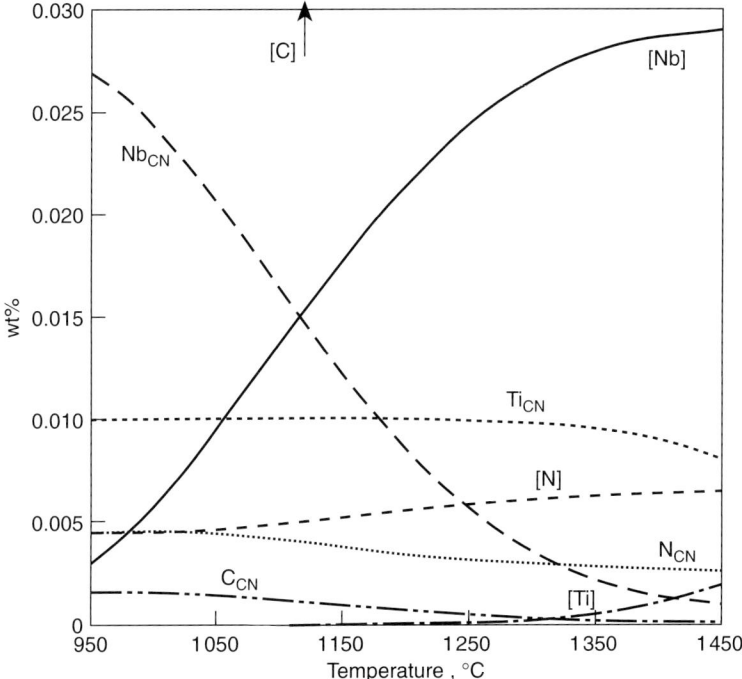

Fig. 5 Soluble and insoluble elemental contents in an Fe–Nb–Ti–C–N quinary alloy.

The addition of the fourth element may result in the formation of two compounds which show complete mutual solubility. The carbides and nitrides of niobium, titanium and vanadium all have fcc structures and are all mutually soluble. The solubility product of the complex carbide or carbonitride is modified as shown by Hudd et al.,[6] but remains a complex function of the individual solubility products. For example in the Fe–Nb–C–N quaternary system, the compound $NbC_xN_{(1-x)}$ is formed. The value of x is dependent upon the solubilities of the individual carbide and nitride phases, the temperature, and the alloy composition, including the carbon, nitrogen and niobium contents. Nevertheless, it is possible to calculate the soluble and insoluble elemental concentrations and the value of x from the analytical solution given by Hudd et al. An important aspect of this treatment is that the activity of NbC in the $NbC_xN_{(1-x)}$ is assumed to be x, i.e. the solid solution of NbC and NbN is an ideal solution, and the activity of NbN is therefore $(1-x)$.

The use of these principles in more complex alloys, i.e. quinary and higher systems, has been demonstrated by Adrian,[7] who has presented results for the Fe–Nb–V–Ti–C–N–Al system. Also included in this model are the interac-

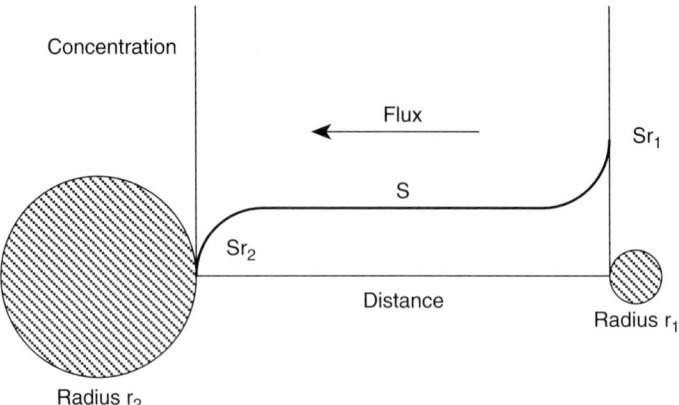

Fig. 6 Composition gradients developed as a result of the Gibbs–Thomson relationship.

tion coefficients for the effects of, for example, manganese on the activity of nitrogen. Such results for quinary and higher systems are obtained by computer based iterative procedures. An example solution for the quinary Fe–Ti–Nb–C–N system is shown in Fig. 5. Interaction coefficients were not used in this calculation. The dominance of the more stable titanium nitride at the elevated temperatures is clearly illustrated.

A feature of the formation of these mutually soluble compounds is that they can enhance the phase stability and can give more extensive precipitation at any given temperature than would be expected of the individual compounds, although the magnitude of the effect is secondary with respect to that of the most stable compound.

The important point here is that it is possible to calculate the phase proportions from the published solubility products. Whilst this has a direct bearing on the critical particle size and grain growth inhibition in microalloyed steels, such principles will also be relevant to the phase proportions in creep resistant steels.

2.2 Ostwald Ripening

In any array of precipitated particles having a distribution of particle sizes, extended exposure to elevated temperatures (but below the solvus temperature) causes the large particles to grow and the small particles to shrink. Given that there is no substantive change in the volume fraction of particles, this process is known as Ostwald ripening. The increase in size and the decrease in number of particles effectively reduces the total interfacial energy in the system, as is the case with grain boundary energy and grain growth. However, grain growth is a tactile process in that a large grain is influenced by contact angles and/or boundary curvature deriving from the presence of adjacent smaller grains. This is not so for distributed particles. The question arising for each particle is whether it should shrink because of a large particle lurking somewhere in the vicinity but beyond the immediate matrix, or whether it should grow because of a small particle lurking in the vicinity but beyond the immediate matrix. The answer to the problem is found in the Gibbs–Thomson equation where the solute content in equilibrium with an elemental second phase particle is inversely related to the particle radius.

$$\ln (S_r/S) = 2\sigma V/RTr \tag{5}$$

where S_r is the solute content in equilibrium with a particle of radius r,
    S is the equilibrium solubility,
    σ is the interfacial energy,
and V is the molar volume.

   This gives rise to concentration gradients in the matrix as indicated in Fig. 6. Diffusion of solute down the concentration gradients results in depletion of solute near the small particles, with dissolution of the small particle to re-establish the Gibbs–Thomson relationship, and accretion of solute near the large particles, with attendant growth of the large particles.
   In the case of a compound MX (or any other compound – M_2X, M_7X_3 etc.), eqn 5 may be rewritten as:

$$\ln(k_r/k_s) = 2\sigma V/RTr \tag{6}$$

where k_r and k_s are the relevant solubility products.
   There are several solutions to the rate of change in particle size resulting from the composition gradients, depending upon the rate limiting step in the growth process. For microalloyed steels, with their low volume fraction of randomly distributed particles, the volume diffusion of the microalloy solute is much slower than the volume diffusion of the interstitial solutes, i.e. carbon and nitrogen. The solution to the change in mean particle size under these circumstances has been derived by Lifshitz and Slyozov,[8] Wagner[9] and Greenwood.[10]

$$r^3 - r_0^3 = 8D\sigma Vst/9RT \tag{7}$$

where r is the mean particle radius at time t
    r_0 is the initial mean particle radius
    D is the diffusion coefficient of the rate limiting specie
    s is the equilibrium solute content of the rate limiting specie
    σ is the interfacial energy
and V is the molar volume (of the carbide/nitride per mole of the rate limiting specie).

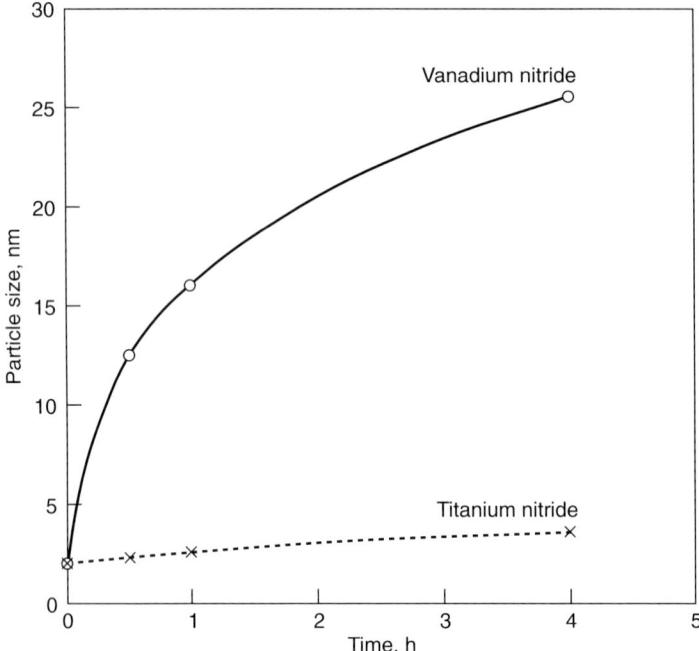

Fig. 7 Calculated coarsening rates for vanadium nitride and titanium nitride in austenite.

This equation, originally developed for an elemental precipitate, e.g. the precipitation of chromium in cobalt, appears to work well for the simpler carbides or nitrides, e.g. VN, but there has been little rigorous testing of the more complex carbo-nitrides where changes in the solubility product with particle size may imply changes in the equilibrium chemistry of the precipitate. There has been little examination of this aspect.

Calculated coarsening rates of vanadium nitride and titanium nitride are shown in Fig. 7. The diffusion rates of the microalloying elements are all very similar at the elevated temperatures of the austenite temperature range, and the difference between the coarsening rates of vanadium and titanium nitrides stem mainly from the differences in their solubilities shown in Fig. 2. The interfacial energies were assumed to be 1 Jm^{-2}, a value which might be expected for an incoherent particle.

2.3 Stoichiometry and Titanium Nitride Technology

Given the significance of the solubility of the rate limiting specie on the particle coarsening rate, attention was given[11] to the optimisation of steel composition with respect to titanium and nitrogen contents. The solubility curves for titanium nitride are shown in Fig. 8. The stoichiometric ratio for titanium

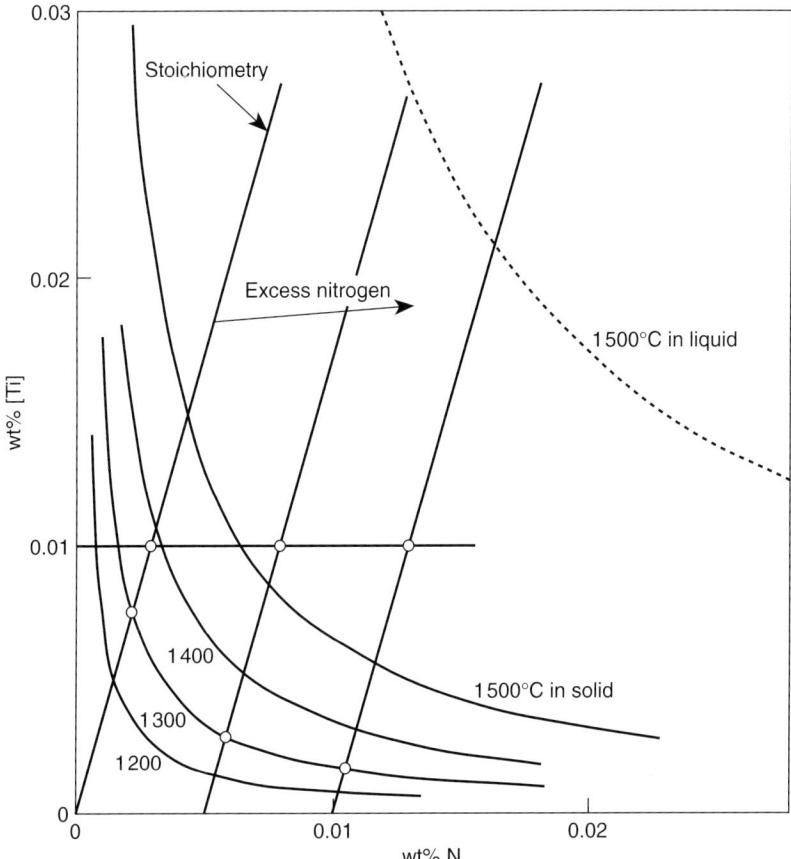

Fig. 8 Solubility isotherms for titanium nitride in austenite and liquid.

nitride is also shown for two sub-stoichiometric titanium contents and a sto-ichiometric titanium content. It can be seen that the titanium in solution is much lower for the substoichiometric steels and that the dissolved titanium decreases as the excess nitrogen in the steel increases. The coarsening rate of the nitride changes correspondingly. The coarsening rate for titanium nitride, shown in Fig. 7, was in fact based on a substoichiometric titanium content. It can also be seen that the amount of titanium and nitrogen that can be added to a steel may be limited by precipitation in the liquid, before or during sol-idification, when large cuboids are formed with edge lengths in the range of 1 μm–10 μm. These are far too coarse to influence grain growth at conventional microalloying levels.

Optimisation of the composition in this way results in a limited volume fraction of very fine and relatively stable particles that resist Ostwald ripen-ing. The effect on particle coarsening and on the attendant grain growth cri-

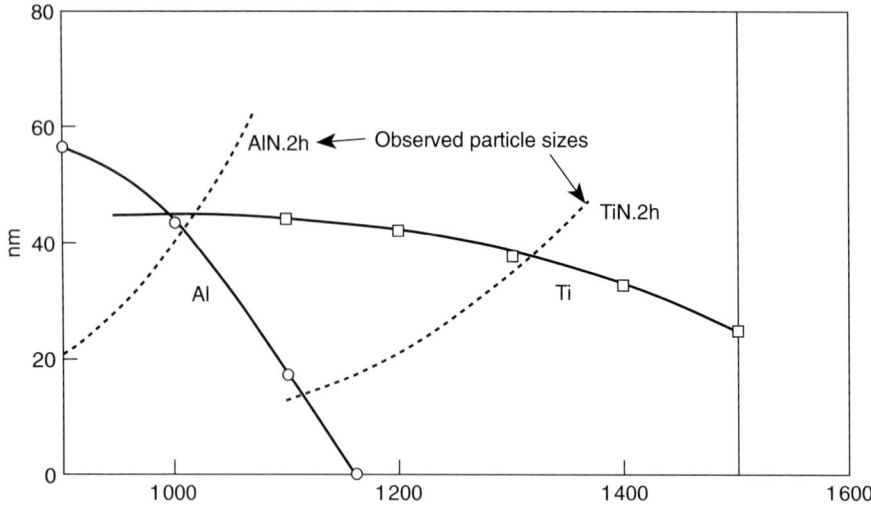

Fig. 9   Comparison of particle coarsening and grain coarsening in aluminium and titanium steels.

terion is illustrated in Fig. 9. It can be seen that the more stable titanium nitride gives a smaller particle size than does aluminium nitride (or would niobium carbide or vanadium nitride) for a given thermal exposure, and the austenite grain coarsening temperature is raised to about 1300°C. This allows steels to be heated for forging without losing their fine grained characteristics and forms the basis of the 'titanium nitride technology'.

The important point being made for the purpose of the present paper, however, is that the rate limiting solute content of the matrix is governed not only by the temperature and solubility considerations, but also by the chemical composition of the steel with regard to the stoichiometric ratios.

3. CREEP RESISTANT STEELS

The principles outlined for microalloyed steels in the preceding sections should also be applicable to creep resistant steels, but some consideration should be given to important differences in the steel chemistry and constitution. Firstly, the volume fractions of second phases are much higher in creep resistant steels due to the relative effects of particles on creep resistance and grain coarsening. Secondly, the nature of the precipitation is far more complex in steels containing chromium, molybdenum and vanadium. Chromium may substitute for some of the iron in the orthorhombic Fe_3C, or, depending upon the steel composition, may form the trigonal Cr_7C_3 or the complex cubic $Cr_{23}C_6$. Even in ternary alloys, kinetic effects may result in the transitory formation of Cr_7C_3 when $Cr_{23}C_6$ is the ultimately stable composition. In all of these carbides, there is also a substantial level of iron solubility. In the case of molybdenum, the hexagonal Mo_2C and the complex cubic Mo_6C are known stable carbides, again showing some solubility for iron. In steels containing both chromium and molybdenum, the carbides show extended solubility for both chromium and molybdenum, and the Mo_6C may be replaced as the more stable carbide by $M_{23}C_6$ of high molybdenum and chromium contents. The addition of vanadium or niobium to this list may add the f.c.c. VC (or more usually the f.c.c. V_4C_3 which is a vacancy modified form of VC).

In the case of steels in the austenite temperature range, the solubility of cementite is high in relation to the carbon content of most steels, and can be disregarded, as is the case for austenitic creep resistant steels. However, creep resistant ferritic and martensitic steels have a carbon content significantly in excess of that in equilibrium with cementite. Consequently, the usual treatment of alloy carbide solubility product has to be modified because of the limited soluble carbon contents in the ferrite.

A further complication may arise as for example when the tetragonal Z phase is formed in Cr–Mo–Nb steels. This has a composition $(Cr_xMo_yNb_{1-x-y})C$, and is not formed unless all elements are present.

Despite these complications it is possible, in principle, to define the solubility products of the stable carbides and the solute contents as a function of alloy content and temperature.

3.1 Bulk Diffusion Controlled Ostwald Ripening

3.1.1 The constitutional equation. The Wagner solution for bulk diffusion controlled Ostwald ripening, eqn 7, is based on extremely low volume fractions of precipitate, such that the composition gradients associated with each particle do not overlap. Ardell[12] has considered the situation where the precipitate fraction is large, and offers the equation:

$$r^3 - r_0^3 = 8k_mD\sigma VSt/9RT \tag{8}$$

This differs from eqn 7 only in the introduction of the proportionality constant k_m, which has a value of unity at very low volume fractions, 2 at 1% by volume, 5 at 7%, and 10 at 25%. Thus the rate of coarsening by bulk diffusion is expected to increase with increasing volume fraction of the second phase.

3.1.2 Diffusion data. Data are already available for the diffusion of chromium, molybdenum, vanadium and niobium in austenite and for chromium, molybdenum and vanadium in ferrite. The diffusion coefficient, D, varies with temperature according to:

$$D = D_0 \exp(-Q/RT)$$

Table 2   Diffusion Data

		D_0 cm²/s	Q KJ/mole	D cm²/s
α-iron at 500°C	Cr	8.5	240	2×10^{-14}
	Mo	10	240	2×10^{-14}
	V	10	240	2×10^{-14}
γ-iron at 600°C	Cr	11	280	8×10^{-12}
	Mo	0.07	236	2×10^{-15}
	V	0.25	252	6×10^{-15}
	Nb	530	330	7×10^{-17}

Values of D_0 and Q taken from Smithells[13] are given in Table 2, and values of D are also shown for the above elements at 500°C in ferrite, and at 600°C in austenite. Due consideration must be given to the method of determination of the diffusion coefficient as some are based on dilute solution tracer methods and others on interdiffusion in highly alloyed materials. Possible variations in diffusion coefficients with concentration are possible, and the appropriate diffusion coefficient must be used. In less dilute alloys, coupled diffusion coefficients may be desired.

The diffusion data for these elements in ferrite are based on the non-magnetic α-iron. Diffusion rates in magnetic alpha iron at temperatures relevant to creep exposure are generally slightly less than those in non-magnetic α-iron.

3.1.3 Interfacial energy. The interfacial energy (f.c.c. carbo-nitride with austenite) in microalloyed steels may conveniently be assumed to have a value of 1 Jm⁻². The austenite is most often transformed from a ferrite-pearlite structure after the f.c.c. carbo-nitride has precipitated. In ferritic creep resisting steels, the carbides or nitrides may be precipitated within the ferritic structure destined for service, and the carbides, particularly Mo_2C and Cr_7C_3, often have a coherent interface. Coherent precipitates may have low interfacial energies, say as low as 0.05 Jm⁻². However continued growth of the precipitates may lead to a loss of coherency with an attendant increase in the interfacial energy, although there has been a suggestion that further growth may be inhibited by the energy requirements involved in loss of coherency. Such considerations may lead to changes in particle morphology, but, irrespective of this, the possibility of significant variations in interfacial energy must be considered. Data on specific interfacial energies would be of direct importance in any assessment of particle coarsening rates.

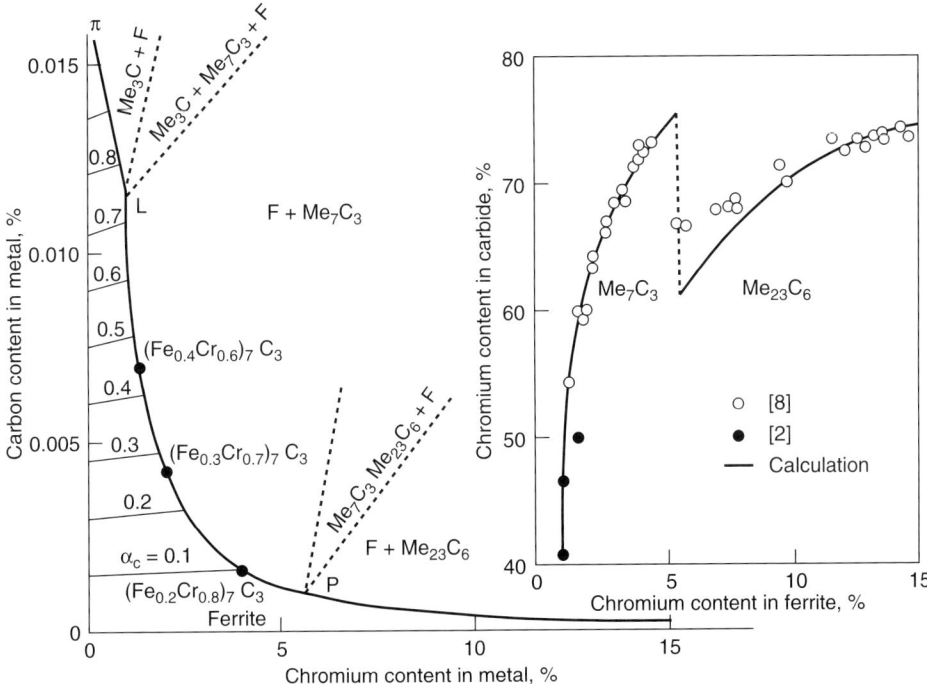

Fig. 10   Fe–Cr–C system – the 700°C isotherm – after Leonovich *et al.*[14]

3.1.4 Solubility data. As pointed out for the microalloyed steels, solubility data is vital to the consideration of Ostwald ripening, especially when coupled with stoichiometry considerations and alloy design. A complete list of solubility equations for the carbides (and nitrides?) of chromium, molybdenum and vanadium in both austenite and ferrite, corresponding to the list for microalloy carbides and nitrides shown in Fig. 2, would be extremely helpful. Although the thermodynamic data for the various systems and their interactions is undoubtedly available, it has not yet been collated in the form illustrated in Fig. 2.

Consider for example the Fe–Cr–C ternary system. A recent study of the phase equilibrium diagram has bee carried out using the thermodynamic data from Calphad. An isothermal section calculated by Leonovich *et al.*[14,15] at 700°C is shown in Fig. 10. The expected transitions from M_3C to M_7C_3 and then to $M_{23}C_6$ are clearly shown, together with the areas in which two carbides are in equilibrium with the ferrites. In this respect the system is not dissimilar to the Fe–Al–V–N system referred to earlier. The main difference of course arises from the extended solubility of iron in the chromium carbide phases. This isotherm shows the extent of the single phase region of ferrite. The line KL shows the limit of solubility of carbon and chromium in equilibrium with M_3C, whilst the line LP shows the limit of solubility of carbon and chromium in ferrite in equilibrium with M_7C_3. Also shown in Fig. 10 are the iron and chromium fractions in the carbide at the corresponding tied compositions of the ferrite. Any steel lying in the α + M_7C_3 two phase region at 700°C will thus fall on a tie line indicating the specific ferrite and M_7C_3 composition. The simple concept of stoichiometry so easily used in the microalloyed steel is clearly modified by the extended solubility of iron (in the range 25% to 60% in the Cr_7C_3 structure.

The implications of this diagram are quite dramatic. The solubilities in ferrite are calculated directly from the thermodynamic data, and the tie lines between the matrix and precipitate compositions are also predicted. In the three phase areas, the ferrite composition is unique and alloy composition variations merely change the proportions of the two carbides each of which has a specific composition in accord with the metallurgical version of the phase rule.

$$F = C - P + 1$$

where the number of components C = 3 (Fe, C, Cr) and the number of phases P = 3 (α, M_3C, M_7C_3), leaving the number of degrees of freedom F = 1.

The act of selecting the temperature (T = 700°C) uses this degree of freedom and the phase compositions are fixed.

This information, coupled with the diffusion data, molar volumes of the carbides, and the interfacial energies, would allow calculation of the coarsening rates of the carbides in a ternary system. If the coarsening rate were to be limited by the bulk diffusion of chromium, for example, due account should be given to the fraction of chromium in the carbide when estimating the molar volume, e.g. the arrival of 7 gm atoms of Cr would create 2 gm moles of M_7C_3, if the iron content was 50%.

This analysis of the Fe–Cr–C system is quite rigorous in that the activities of the solutes are used and also the solute interaction terms for the effects of chromium on the carbon activity, as shown in Fig. 10. This has obvious importance in alloys containing several percent of the substitutional solutes.

The study is quite revealing in that the Gibbs energies for Fe_3C, Cr_7C_3 and $Cr_{23}C_6$ are matched by information on the Gibbs energies for Cr_3C, Fe_7C_3 and $Fe_{23}C_6$, thus obviating the need for formulation of their respective solubility products and combination for the mutual solubility of the carbide phases.

The extension of this approach to the quaternary system Fe–Mo–Cr–C, given that there is thermodynamic data for Fe_2C, Cr_2C, Mo_2C etc. is quite obvious. The invariant compositions in the isothermal section of a three-phase area in a ternary system are not invariant in a quaternary system, due to the extra degree of freedom. The further specification of the content of one element in one of the phases is required to define the three-phase region of

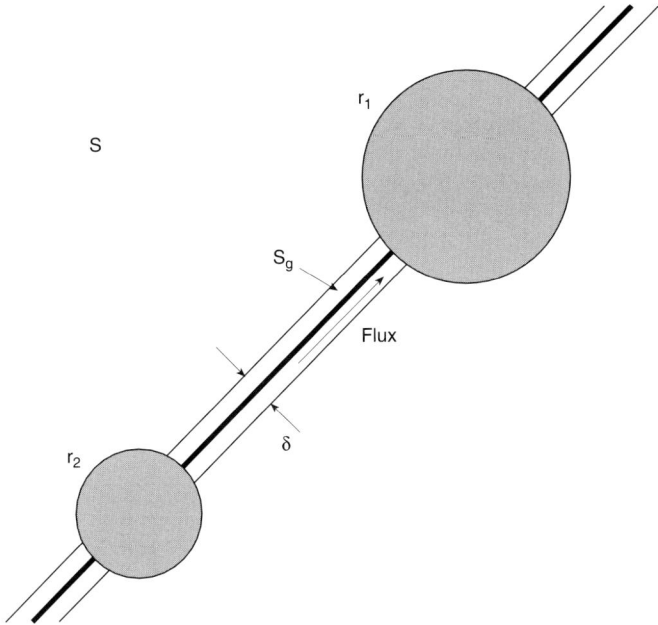

Fig. 11 The Grain Boundary Coarsening Model.

the quaternary. Given the availability of the thermodynamic data and an adequate level of computing power, there is clearly no obstacle to evaluating the more complex Fe–Cr–Mo–V–C–N system, as suggested by the work of Adrian.[7]

Although the solubility product data is available for an assessment of vanadium carbide coarsening based on simple assumptions of virtual insolubility of vanadium in Fe_3C and iron in VC (or V_4C_3), the thermodynamic approach would include the partial solubilities and the solute activities and interaction terms.

3.2   Grain Boundary Diffusion

A striking feature of the microstructure of creep resisting steels is the distribution of the carbide phases. Small intragranular particles of, say, M_2C may co-exist with much larger intergranular particles of, say, $M_{23}C_6$. The fine particles may be important to dislocation creep resistance, but the coarser grain boundary carbides may be equally important to the Coble creep mechanism. This raises the question of the coarsening of grain boundary carbides. In part, the grain boundary carbides may be coarse because of nucleation difficulties, but the further coarsening of these particles must be considered in the light of a different rate controlling process, i.e. one of grain boundary diffusion control. The grain boundary offers an easy diffusion path for solute elements so that the flux of solute atoms is greatly enhanced. The situation with regard to intergranular particles is illustrated in Fig. 11. The flux of atoms is assumed to occur along the boundary from the small particle to the large particle. The width of the grain boundary, δ, is generally assumed to be three atom spacings. The constitutive equation for grain boundary diffusion controlled ripening derived by Kirchner[16] is:

$$r^4 - r_0^4 = 9\delta D_g \sigma S_g Vt/32ABRT \tag{9}$$

where D_g is the grain boundary diffusion coefficient
S_g is the solute concentration on the grain boundary

and

$$A = 2/3 - \sigma_b/2\sigma + (\sigma_b/2)^3/3 \tag{10}$$

where σ_b is the grain boundary energy

and

$$B = [\ln(1/f)]/2 \tag{11}$$

where f is the fraction of grain boundary covered by precipitated particles.

The grain boundary diffusion coefficients may be related to the bulk diffusion coefficients in that, for f.c.c. metals, the activation energy for grain boundary diffusion is approximately one half of that for bulk diffusion, whilst for b.c.c. metals, the activation energy for grain boundary diffusion is some 70% of the value for bulk diffusion. Grain boundary diffusion will be the dominant diffusion path when

$$\delta D_g >> dD \tag{12}$$

where d is the grain size. This condition is satisfied for both f.c.c. and b.c.c. iron when the temperature (K) is less than half of the melting temperature,[17] i.e. in the creep range.

The solute content S_g given in eqn 9 is the solute content at the grain boundary, and this can be very different from that in the lattice, depending upon the specific binding energy between the solute and the grain boundary. This area has been studied intensively in connection with reversible temper embrittlement and although, for a single element, the relationship is relatively simple, i.e.

$$\ln(S_g/S) = \Delta H/RT \tag{13}$$

there is competition between elements for grain boundary sites. Phosphorus segregation is well established but may ultimately be displaced by the segregation of antimony at longer times, the kinetics of segregation being dependent upon the initial concentrations and diffusion rates. There are also significant interactions between elements, such that in chromium steels, phosphorus segregation may be more intense due to Cr–P coupling in the grain boundary. Thus the segregation of molybdenum, a known horophile element, to grain boundaries may be radically affected by the impurity levels in the steel, and by elements which enhance coupling. However, some data are available that would enable a reasonable estimate of grain boundary concentrations on the basis of known lattice solute contents.

3.3.   Other Coarsening Mechanisms

Bulk diffusion and grain boundary diffusion controlled particle growth are not the only particle coarsening mechanisms. The rate limiting step in the growth process may involve the flux of atoms across the interface from matrix to particle, giving the condition of interface control where the time dependence of growth is expressed by:

$$r^2 - r_0^2 = 64K\sigma VSt/81RT \tag{14}$$

Here, K describes the flux across the interface and replaces the diffusion coefficient.

Also of importance is the dislocation pipe diffusion mechanism, where the diffusion of solute along the core of dislocations is very much enhanced compared with lattice diffusion. For relatively low dislocation densities a r^5 versus t relationship is observed whilst for well recovered subgrains, and low angle boundaries, the relationship reverts to a relationship very similar to that for grain boundary diffusion. The significance of dislocations is an important consideration for creep resisting steels, where dislocations and low angle boundaries are a persistent feature of creep exposure.

Even in a bulk diffusion controlled system, the rate controlling feature may not be obvious. The often cited case of cementite coarsening during the tempering of plain carbon martensites provides an example where the rate controlling step is the diffusion of vacancies required by the local volume changes accompanying particle growth and particle dissolution.

SUMMARY

Particle coarsening and grain growth are important to creep responses during long term exposure at elevated temperatures. The principles of both processes have been considered in relation to the carbides and nitrides in microalloyed steels. A simple solubility product approach to particle solubility is used extensively in microalloyed steels, where small volume fractions of particles at elevated temperature usually result in a bulk diffusion controlled coarsening process.

Consideration of creep resisting steels, with their much higher alloy contents, suggests that a more sophisticated thermodynamic approach may be more useful, making use of known thermodynamic data, activities, and solute interactions. There is a considerable data base for many of the constitutional diagrams required in order to provide input to the particle coarsening equations, but this needs to be reviewed and tested.

The mechanisms of particle coarsening have been considered, and it is clear that, under creep exposure conditions, grain boundary diffusion and dislocation pipe diffusion may play an important part in controlling the rate of particle coarsening and the ensuing changes in creep resistance.

REFERENCES

1. C. Zener: 'Private communication to C.S. Smith', *Trans. AIMME*, 1948, **175**, 15.
2. T. Gladman: *Scripta Met et Mat.*, 1992, **27**, 1569.
3. T. Gladman and D.J. Senogles: 'Grain Refinement of Low Carbon Structural Steels by Titanium Oxide Particles', *Titanium Technology in Microalloyed Steels*, T.N. Baker ed., The Institute of Materials, London, 1997, 83–97.
4. R.D. Doherty, D.J. Srolovitz, A.D. Rollett and M.P. Anderson: *Scripta Met.*, 1987, **21** 675.
5. T. Gladman: *The Physical Metallurgy of Microalloyed Steels*, The Institute of Materials, London, 1997.
6. R.C. Hudd, A. Jones and M.N. Kale: *J. Iron Steel Inst.*, 1971, **209**, 121.
7. H. Adrian: *Microalloying '95*, ISS, Pittsburgh, 1995, 285.
8. I.M. Lifshitz and V.V. Slyozov: *J. Phys. Chem. Solids*, 1961, **19**, 35.
9. C. Wagner: *Z. Elektrochem.*, 1961, **65**, 581.
10. G.W. Greenwood: *The Mechanism of Phase Transformations in Crystalline Solids*, The Institute of Metals, London, 1969, 103.
11. T. George and J.J. Irani: *J. Australian Inst. Metals*, 1968, **13**, 94.
12. A.J. Ardell: *Acta Met.*, 1972, **20**, 61.
13. *Smithells Metals Reference Book (6th Edn)*, E.A. Brandes, ed. Butterworths, London, 1983.
14. B.I. Leonovich, V.E. Serebryakov, N.R. Frage and A.P. Kartavsev: *Steel in the USSR*, 1989, **19**, 12.
15. B.I. Leonovich, O.I. Kachurina, G.G. Mikhailov and T.D. Kozyreva: *Steel in the USSR*, 1991, **21**, 107.
16. H.P.K. Kirchner: *Met. Trans.*, 1971, **2**, 2861.
17. J.W. Martin and R.D. Doherty: *Stability of Microstructure in Metallic Systems*, Cambridge University Press, Cambridge, 1980.

The Effects of Microstructural Evolution on the Creep Rupture Behaviour of CrNi(Mo)N Austenitic Steels

V. VODÁREK, M. SOBOTKOVÁ AND J. SOBOTKA

R & D Division, VÍTKOVICE, Ostrava, Czech Republic

ABSTRACT

The creep resistance of austenitic CrNi(Mo) steels strongly depends on the microstructural evolution prior and during long-term creep exposure. Small changes in chemical composition of steels result in significant differences in creep behaviour. The contents of carbon, nitrogen, molybdenum and boron in CrNi(Mo) austenitic steels are of decisive importance. A better understanding of the effect of individual alloying and residual elements on creep rupture behaviour requires long-term creep rupture testing of casts with intentional additions, followed by detailed microstructural investigations and surveys on the failure mechanisms of creep testpieces.

This paper deals with results of microstructure-creep property studies on wrought AISI 304LN and 316LN steels with niobium and/or boron additions. It has been proved that an appropriate addition of boron to AISI type 316LN steels results in an increase of both creep rupture strength and creep rupture ductility. On the other hand, the creep rupture ductility in boron-treated AISI type 304LN steels is rather low. A small addition of niobium to AISI type 316LN steels results in a significant reduction of minimum creep rate and shortening of the tertiary creep stage. The observed changes in creep rupture behaviour of individual casts are related to differences in solid solution hardening, precipitation reactions, coarsening of minor phases and the failure mechanisms. The results demonstrate the decisive role of the microstructural development on the high temperature creep behaviour of austenitic steels.

INTRODUCTION

Research has been in progress since the 1960s on ways of improving high temperature properties of CrNi(Mo) austenitic steels. The most frequently used non-stabilised creep resistant austenitic steels are the AISI 304 and 316 grades.[1] It is very well known that the creep properties of steels depend very significantly on the microstructural evolution prior and during creep exposure.[2] Since many factors can influence microstructure, it is not surprising that dramatic changes in creep rupture properties of the above mentioned grades have been observed.[3] The most important factors are chemical composition, heat treatment, cold or hot working prior to creep exposure and the testing temperature and environment.[1] Furthermore, extrapolation procedures are expected to be partly responsible for the scatter in the predicted creep properties.

It is expected that revised chemical specifications enable us to produce steels with more consistent creep properties. The emphasis in recent years has been placed on the role of elements which strengthen and stabilise austenitic steels. As the importance of recycling of steel scrap grows, it also becomes crucial to understand the role of residual elements. Furthermore it has been proved that the contents of carbon, nitrogen, molybdenum and boron in CrNi(Mo) austenitic steels are of decisive importance.[2] For example, the creep rupture strength of AISI type 304 and 316 steels at 600°C has been expressed by the linear regressive equation:

$$R_{mT}/10^5/600 = 46 + 22.6 \text{ Mo} + 63.4 \text{ N} + 6343 \text{ B} + 381 \text{ C} \qquad (1)$$

with the correlation coefficient value of 0.912, all element concentrations in eqn (1) are in per cent by weight.[2]

Austenitic steels of AISI type 304 and 316 steels may contain up to 0.1 wt.% of carbon. Carbon acts as an interstitial hardener. However, at typical temperatures of creep exposure carbon precipitates as chromium rich $M_{23}C_6$ type carbides which preferentially nucleate at grain boundaries. The local depletion of matrix in chromium results in deterioration of corrosion resistance of austenitic steels. It has been suggested that nitrogen reduces the diffusion rate of carbon and thereby delays the coalescence of carbides. Nitrogen additions to commercial austenitic steels have been used for a number of years to improve mechanical properties and have led to the development and production of high nitrogen grades of 304 and 316 steels.[1, 4–6] Nitrogen suppresses the formation of δ-ferrite and intermetallic phases both of which may be associated with reduced creep rupture life. The maximum nitrogen content used in commercial AISI 304 and 316 steels is about 0.25 wt.%.[7] Nitrogen levels above 0.3 wt.% result in a discontinuous precipitation reaction where Cr_2N lamellae precipitate in the austenitic matrix.[8] This reaction causes a reduction in creep rupture strength.

Molybdenum improves high temperature creep rupture properties of CrNi austenitic steels.[2] The reason for the beneficial role of molybdenum arises because the element acts as a substitutional hardener. However, the prolonged ageing of molybdenum-bearing steels at temperatures close to 650°C promotes the formation of molybdenum rich phases, such as η-Laves, $M_6(C,N)$, which reduce the positive effect of this element.[9]

Boron was initially added to austenitic steels to improve hot workability. Nevertheless, it was subsequently found that small additions of boron increased creep rupture strength and in some grades ductility too.[10] It would

Table 1 Chemical Composition of Steels (wt.%)

Type	Cast	C	N	Mn	Si	P	S	Cr	Ni	Mo	B	Nb
304LN	A	0.027	0.177	1.46	0.72	0.012	0.012	18.8	10.5	0.06	0.0023	–
304LN	B	0.020	0.209	1.36	0.36	0.009	0.015	17.7	9.8	0.07	0.0027	–
316LN	C	0.023	0.145	2.50	0.27	0.015	0.017	17.8	14.6	2.68	0.0037	–
316LN	D	0.021	0.250	0.44	0.28	0.012	0.019	17.7	12.1	2.20	0.0030	–
316LN+Nb	E	0.023	0.161	1.34	0.48	0.014	0.013	18.1	12.5	2.82	0.0012	0.106
316L+Nb	F	0.021	0.158	1.11	0.42	0.025	0.009	17.8	12.6	2.64	0.0020	0.300

appear that boron increases creep rupture life by delaying the onset of the tertiary stage due to suppression of grain boundary cavitation and wedge cracking rather than increasing the creep strength by decreasing creep rate.[4, 11, 12] Boron has low solubility in austenitic steels and segregates to grain boundaries, precipitate/matrix interfaces and can partly replace carbon in $M_{23}C_6$ phase. Due to low solubility of boron several types of borides have been observed in CrNi(Mo) austenitic steels.[12] On the other hand, the strengthening effect of boron is preserved in nitrogen-bearing austenitic steels in spite of the high affinity of boron to nitrogen. Shimada[13] has reported that boron segregation to grain boundaries in austenitic steels containing more than 0.2 wt.% of nitrogen is not accompanied by precipitation of BN phase.

One of the most successful methods of improving the long-term creep resistance of austenitic steels is based on increasing the extent of precipitation strengthening during creep exposure.[14] This can be achieved by alloying of these steels with small amounts of strongly carbide or nitride forming elements such as niobium, vanadium and titanium.

A better understanding of the causes of scatter in creep properties of CrNi(Mo) austenitic steels requires controlled creep experiments involving casts with intentional additions followed by detailed microstructural investigations and surveys of the failure behaviour of creep rupture testpieces. It is very important to evaluate the creep properties of austenitic steels by long-term laboratory testing because empirical methods based on extrapolation of relatively short-term data are likely to cross creep mechanism boundaries and tend the prediction of to unreliable values of creep strength.

In this paper the results of microstructure – creep property studies on several casts of CrNi(Mo)N austenitic steels are reported.

EXPERIMENTAL

The evaluation of long-term creep properties was carried out on wrought AISI 304LN and 316LN steels with additions of niobium and/or boron. The chemical compositions of steels investigated are listed in Table 1. The steels

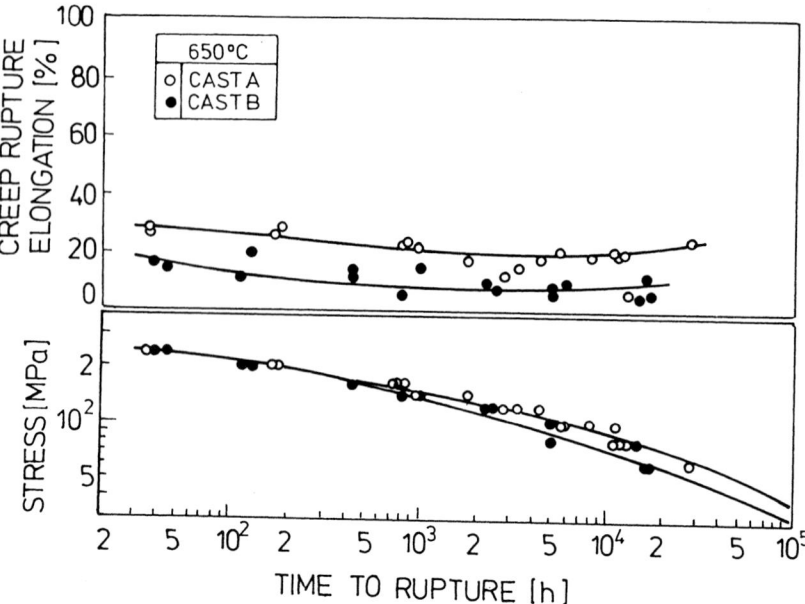

Fig. 1 Creep rupture strength and creep ductility of AISI type 304LN steels.

underwent solution annealing at temperatures in the range 1050–1120°C. All the casts then had fully austenitic microstructures.

Long-term creep rupture tests with a constant tensile load were carried out in air in the temperature range 575–650°C. The stress-temperature dependences of the time to rupture were described by the Seifert parametric equation.[15] The creep ductility evaluation of individual casts was performed using isothermal dependences of creep rupture elongation versus time to rupture. The creep properties of steels investigated were also judged by the comparison of the shape of creep curves plotted at identical experimental parameters.

Detailed microstructural studies were conducted on ruptured testpieces of individual casts in order to understand the factors and processes responsible for the observed differences in the creep behaviour. Both transverse and longitudinal sections were taken from ruptured testpieces. Microstructural examinations were conducted by means of optical metallography, transmission electron microscopy on a JEM-200CX microscope fitted with a Kevex energy dispersive spectrometer and X-ray microanalysis on a JCXA-733 microprobe. Furthermore, the fracture surfaces of the testpieces were investigated in the secondary electron mode using a TESLA BS300 scanning electron microscope. TEM studies were performed on carbon extraction replicas and on thin foils. The thin foils were prepared by spark cutting, grinding and finally jet polishing to perforation using a Struers Tenupol set at 70 V with a 5% perchloric acid in glacial acetic acid electrolyte at room temperature.

RESULTS AND DISCUSSION

1   AISI 304LN Steels

As evident in Table 1 the nitrogen additions in casts A and B were close 0.2 wt.%. Combined additions of nitrogen and boron to AISI type 304 steels resulted in an increase of creep rupture strength. As the testing temperature and time to rupture increased the creep rupture elongation decreased.[16] Within the temperature range of 600 to 650°C the long-term creep ductility, despite boron additions, was rather low and was comparable with boron-free AISI 316H type steels.[2] The creep rupture strength and creep ductility results at 650°C on the casts investigated are shown in Fig. 1. Fractographic exami-nations revealed that up to 30 000 hours of creep testing at 600 and 650°C the dominant failure mechanism of testpieces was intergranular fracture. This failure mode was accompanied by a high concentration of creep voids beneath the fracture surface.

In common non-stabilised CrNi austenitic steels we may, depending on the testing temperature and exposure time, encounter precipitation of $M_{23}C_6$ car-bides and σ-phase.[1] The solubility of nitrogen in the austenitic matrix is high and nitrogen is known to retard the formation of σ-phase. In steels contain-ing more than ~0.15 wt.% of nitrogen, Cr_2N particles can form.[17] Microstructural investigations on Casts A and B after short term exposure at 600 and 650°C revealed the presence of Cr_2N and $M_{23}C_6$ phases. Most pre-cipitated particles occurred on grain boundaries. Prolonging the time to rup-ture resulted in the formation of a minor phase which had a face centered cubic unit cell with a lattice parameter of a = 1.08 nm.[18] Furthermore, dif-fraction patterns along <001> directions of the examined precipitates revealed an absence of {hk0} reflections for which h + k ≠ 4n, where n is an integer. This indicates that the space group of this phase contains a diamond glide plane.[20] The lattice parameter established here is close to the figures published for $M_{23}C_6$, G-phase and M_6C (η-carbide). However, only the crystal struc-ture of η-carbide (Fd3m) is in conformity with above discussed results of dif-fraction analysis. Figure 2 shows the [001] zone axis pattern of the η-carbide.

Table 2 lists the chemical composition of η-carbides, as determined by EDS. Qualitative analyses of the very light elements in this phase have confirmed the presence of carbon as well as of nitrogen. The results of diffraction and X-ray microanalysis studies entitle us to interpret this phase as $M_6(C,N)$ phase with the following stoichiometric formula: $(Si_1Cr_{2.7}Fe_{0.6}Ni_{1.7})(C,N)$.[18]

The term η-carbide is generally applied to double carbides which are stable

Fig. 2   Diffraction pattern along the [001] direction of η-carbide.

Table 2   Chemical Composition of Minor Phases (wt.%)

Phase	Si	Cr	Fe	Ni	V
η-carbide	9	48	9	33	1
π-phase	6	60	23	11	–

Note: The data are normalized to yield 100 per cent.

only in the presence of two metals whose locations in the Periodic Table conform to the diagonal rule.[19] The phase under investigation in AISI 304LN steels fails to conform to conventional ideas on the composition of η-carbides as it contains no transition metal belonging to groups IVA, VA or VIA of the Periodic Table.

Comparing the precipitation reactions in the AISI 304L with those in the AISI 304LN steels we have to conclude that the precipitation of $M_6(C,N)$ is closely related to the nitrogen additions. At 650°C and times to rupture exceeding 10 000 hours, $M_{23}C_6$ and $M_6(C,N)$ were the dominant intergranular minor phases in both casts investigated. The rate of coarsening of $M_6(C,N)$ particles was high as evident in Fig. 3. The character of precipitation on grain boundaries resulted in dominant intergranular failure mode and corresponding low creep ductility.

Furthermore, in testpiece regions which had suffered severe cavitation damage during creep exposure coarse particles were found. The regions with these particles also contained oxide particles, evidently formed by an internal oxidation mechanism. The diffraction analysis indicated that this phase had a cubic unit cell with a lattice parameter of a = 0.623 nm and it was interpreted

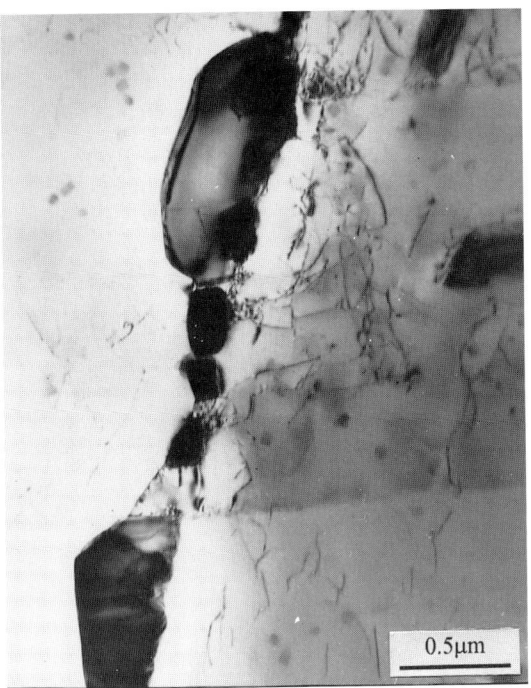

Fig. 3   Coarse intergranular particles of $M_6(C,N)$ and $M_{23}C_6$ particles in the testpiece of Cast A after creep exposure 650°C/28,840 hours.

as π-phase.[21] π-phase is a general designation for phases isomorphous with β-Mn. Atoms of nitrogen and carbon are liable to dissolve in this phase which has led to its sometimes being called π-nitride or π-carbide.[19] In high-alloy austenitic CrNi(Mo) steels which contain high nitrogen additions the π-nitrides are known to form as an equilibrium nitride precipitate.[28] Table 2 presents the composition of π-phase particles in AISI 304LN steels, as established by EDS. It is evident that the composition of the π-phase is similar to that of η-carbides. Qualitative X-ray microanalysis for very light elements revealed that π-phase particles contained some carbon, nitrogen and oxygen too. Nevertheless, the evidence of oxygen might only be the effect of local surface oxidation of π-phase.

It has to be stressed that π-phase particles were detected only in the extensive cavitation-damaged regions, Fig. 4. We suppose that it is the atmospheric penetration in regions of extensive grain boundary damage that provides the source of nitrogen for the formation of π-phase and the source of oxygen for an internal oxidation. Microstructural investigations suggested that the nitrogen content of the AISI 304LN steels was too low for π-phase to precipitate in those specimen areas which remained substantionally undamaged by cavitation processes. In those regions where cavitation-induced damage was

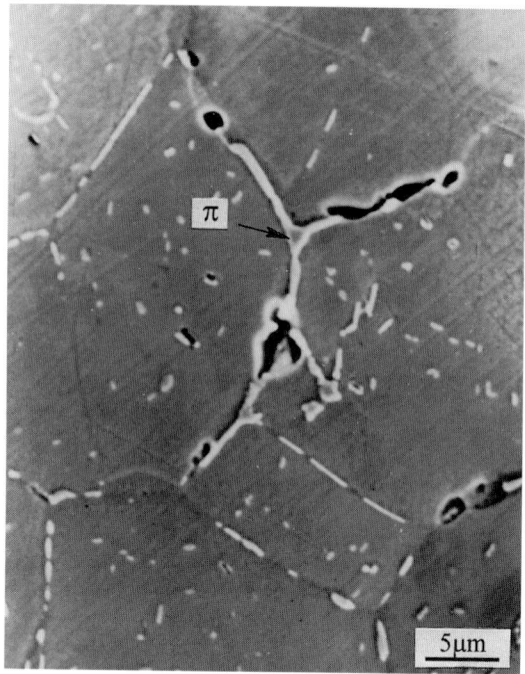

Fig. 4   π-phase particles in proximity to cavities in the testpiece of Cast A after creep exposure 650°C/28 840 hours.

extensive the increase in nitrogen concentration might be attributable to atmospheric penetration along cavities and/or enhanced grain boundary or surface diffusion at damaged boundaries.

No σ-phase particles were detected in either of the casts investigated after creep exposure for up to 30 000 hours at 650°C. This demonstrates that nitrogen retards the formation of σ-phase in CrNi austenitic steels.

2   AISI 316LN Steels

Combined additions of nitrogen and boron to AISI type 316 steels result in a significant increase of both creep rupture strength and creep rupture ductility. For instance, a nitrogen addition of 0.1 to 0.2 wt.% to AISI 316L increases the creep strength on level characteristic of AISI 316H steel.[2] The stress dependence of time to rupture in Cast C at 650°C is shown in Fig. 5. The creep elongation dependence on time to rupture for Casts C and D in the temperature range 575–650°C is shown in Fig. 6. Fractographic examinations on testpieces from the region of the highest applied stresses revealed the transgranular ductile fracture mode. At test temperatures up to 650°C a decline of applied stress to values corresponding to time to rupture about 1 000 hours was accompanied by a change of failure mode to intergranular

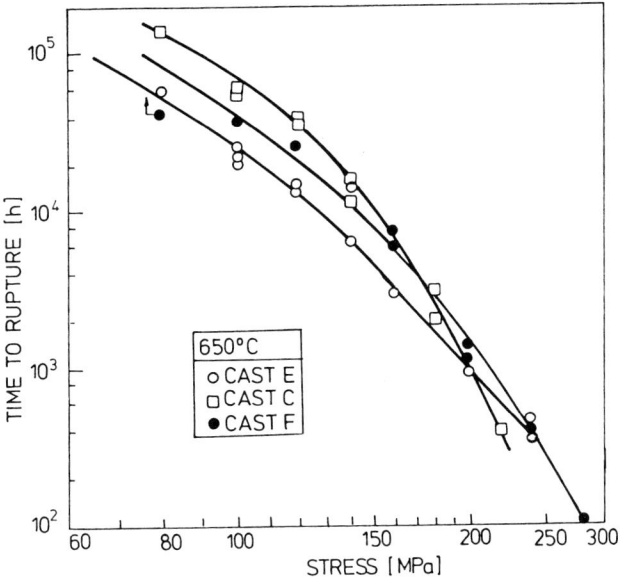

Fig. 5   Stress dependence of time to rupture in Casts C, E and F.

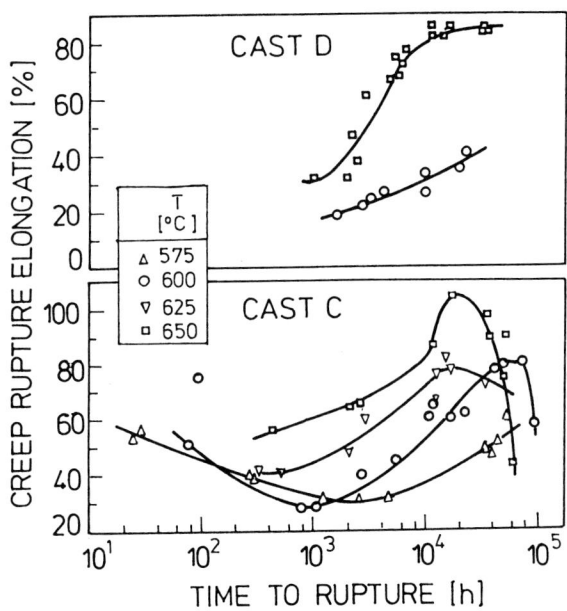

Fig. 6   Time-temperature dependence of relative creep rupture elongation of AISI type 316LN steels.

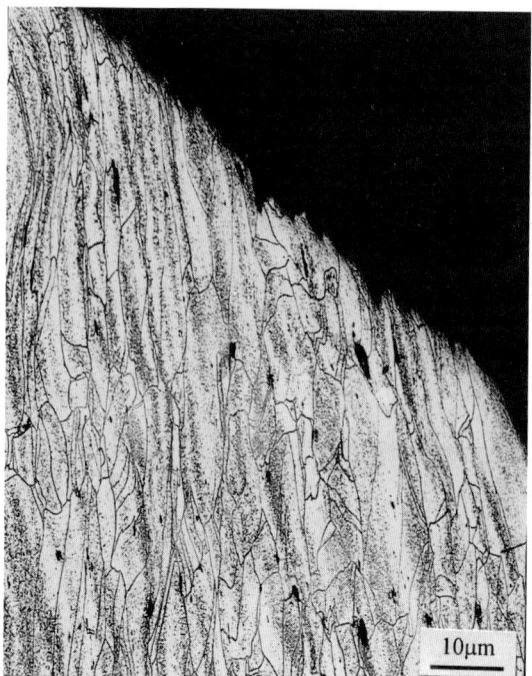

Fig. 7 Transgranular shear fracture of the testpiece of Cast C after creep exposure 650°C/16 336 hours.

ductile fractures.[22] This change resulted in a reduction in the creep rupture ductility. At longer creep rupture durations a gradual increase of creep elongation up to a maximum was observed. The test durations at which the maximum values of creep rupture ductility occurred were found to be temperature dependent.[22] As the creep ductility increased a gradual transition from intergranular to transgranular ductile shear fractures was observed. Transgranular shear fractures were accompanied by a relatively slight concentration of creep voids beneath fracture surface, Fig. 7. The most probable mechanism accounting for the shear fractures is thought to be transgranular failure of the bridge-like partitions between voids which have been initiated at inclusions, minor phase particles or at grain boundary triple points.[23] The negligible amount of surface oxidation found on the fracture surfaces suggests a fracture process which in this sense is not time dependent. A decline of creep ductility at the longest times to rupture has been ascribed to a predominantly intergranular ductile failure mode.[4] The many coarse cavities were largely interconnected, even in regions distant from the fracture surface. The failure mechanism involved comprises the nucleation, growth and interlinking of cavities, till in the final phase a main crack develops and spreads.

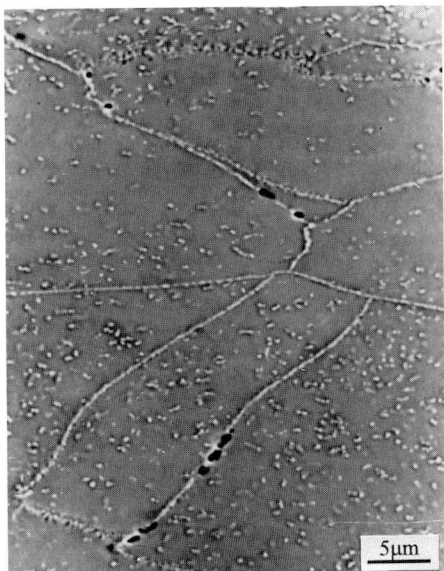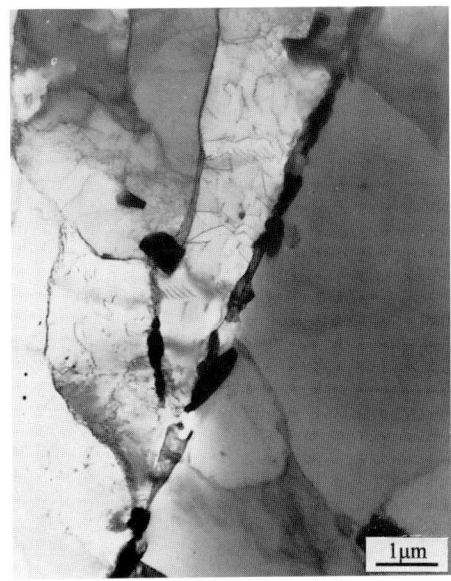

Fig. 8   Examples of fine precipitate particles along grain boundaries in Cast C after creep exposure, a. 650°C/16 366 hours, b. 650°C/59 496 hours, most particles were identified as σ-phase.

The observed differences in creep ductility are decisively affected by failure mechanism which depends on the temperature, applied stress and microstructural evolution during creep exposure. At low applied stresses where the fracture is a consequence of nucleation and growth of intergranular voids, it is necessary to pay attention to the character of intergranular precipitation. Particles on grain boundaries can prevent local grain-boundary sliding. Restriction of boundary sliding results in local stress concentrations which are necessary for nucleation of cavities. Many authors attributed the low level of creep ductility observed in some CrNi(Mo) austenitic steels to intergranular precipitation of σ-phase.[24, 25] Howewer, the microstructural investigations on testpieces of Casts C and D revealed that in spite of the nearly continuous network of fine σ-phase particles along grain boundaries the observed level of creep ductility was high. The character of precipitation on grain boundaries of Cast C after creep at 650°C is shown in Fig. 8. An important role in this process is probably played by the strength of the σ-phase/matrix interface, which is enhanced by the presence of boron and nitrogen.[11] As the temperature and exposure time increase, the particles of σ-phase and other minor phases coarsen, Fig. 9 and thus facilitate the nucleation of cavities. This is accompanied by a gradual transition to predominantly intergranular ductile fracture. It has been observed that in this region of intergranular fracture mode a very detrimental role is played by grain boundary coarse particles,

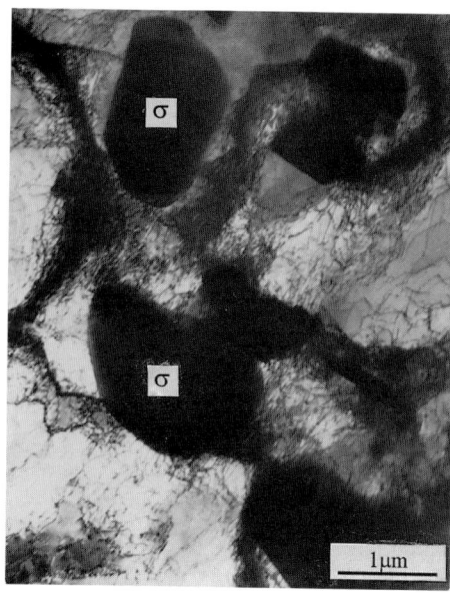

Fig. 9   Intragranular particles of σ-phase in Cast C after creep 650°C/132 524 hours.

regardless of phase type. Coarse particles are surrounded by denuded zones which can deform more easily than surrounding matrix.

Coarse intergranular particles give rise to local stress concentrations and consequently to nucleation of a number of cavities at the particle/matrix interfaces. The microhardness measurements inside of austenitic grains as well as quantitative evaluation of intragranular precipitate density[23] indicated that intensive precipitation strengthening of the austenitic matrix was not necessary for realisation of intergranular failure mode.

During long-term high temperature exposure a number of minor phases can form in AISI type 316 steels.[1] Precipitation processes depend on chemical composition of steel, steelmaking procedures, thermomechanical history of material, applied stress, temperature of exposure, etc. Precipitation reactions in the Casts C and D at 600 and 650°C started with the formation of intergranular Cr_2N particles. In accordance with theoretical expectations only a small amount of Cr_2N phase was found in testpieces of Cast C containing ~0.15 wt.% of nitrogen.[17] However, in testpieces of Cast D very intensive precipitation of Cr_2N occurred. Increasing nitrogen content retarded the formation of $M_{23}C_6$ carbides. $M_{23}C_6$ particles nucleated preferentially on grain boundaries and Cr_2N particles continued to form inside the austenitic grains, Fig. 10. Prolonging the creep exposure was accompanied by a gradual replacement of Cr_2N particles on grain boundaries by η-Laves and finaly by σ-phase particles. After creep exposure for ~30 000 hours at 650°C

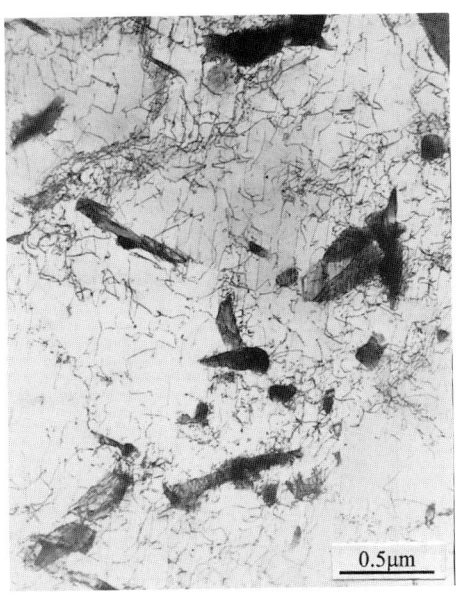

Fig. 10 Intragranular Cr$_2$N particles in Cast D after creep exposure 650°C/18 842 hours.

most intergranular particles were identified as σ-phase while particles of other minor phases were found predominantly inside the austenitic grains. Due to the chemical inhomogeneity of the casts investigated precipitation processes gave rise to distinct precipitate bands parallel to the original rolling direction of the material. Furthermore, M$_{23}$C$_6$ particles were gradually replaced by M$_6$(C,N) phase. In the course of creep exposure up to 650°C only very small amounts of intermetallic χ-phase were formed in the casts investigated.

In the extensive cavitation-damaged regions of the long-term testpieces particles of π-phase were observed. The nitrogen content of the AISI 316LN steels was too low for π-phase to precipitate in testpiece regions which had not been substantially damaged by cavitation processes. It was the atmospheric penetration in regions of extensive grain boundary damage that provided the source of nitrogen for the formation of π-phase.

The results of microstructural studies indicate that precipitation sequences in AISI type 316LN steels are the same as reported for the AISI 316L steel,[1] except for the Cr$_2$N phase. Nevertheless, a nitrogen addition to AISI 316L steel retards the precipitation of M$_{23}$C$_6$ as well as the intermetallic phases and slows down the rate of coarsening of those phases. On the other hand, nitrogen accelerates the precipitation of M$_6$(C,N) phase.

3   AISI 316LN Steels with Niobium Additions

The long-term creep resistance of austenitic CrNi(Mo) steels can be improved

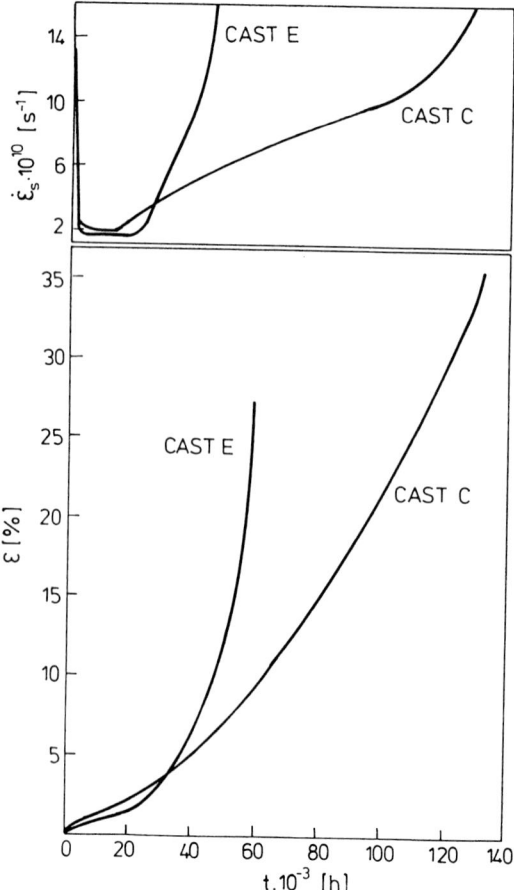

Fig. 11   The comparison of creep curve shapes at 650°C/80 MPa for Casts C and E.

due to small additions of strongly carbide or nitride forming elements.[14] Investigations of the way niobium affects the creep strength of austenitic CrNi(Mo) steels soon revealed that the Nb/(C + N) ratio which yielded the best creep resistance was much lower than the stoichiometric ratio for the formation of Nb(C,N) phase.[26] Furthermore, in nitrogen alloyed austenitic steels with niobium additions a nitride phase forms which is called Z-phase.[27] This phase can form only in steels containing both chromium and niobium atoms and its ideal composition is CrNbN.[28] Particles of this phase can precipitate over a wide range of temperatures, up to 1250°C.[29] This creates favourable conditions for austenitic grain refinement and for improvement of yield strength.[31]

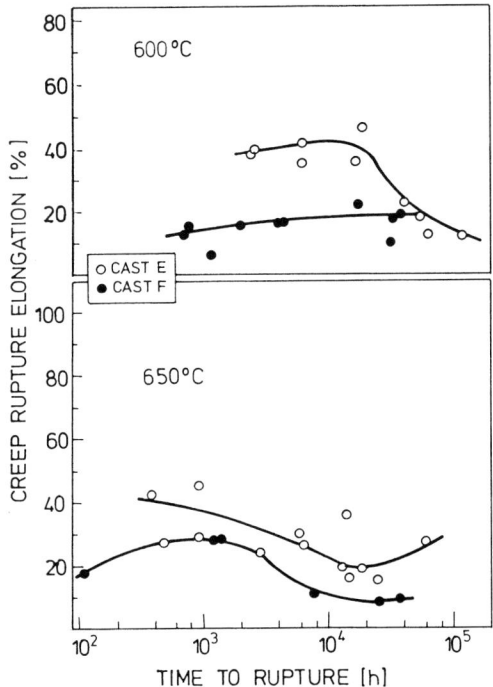

Fig. 12   Time dependence of creep rupture ductility for Casts E and F at 600 and 650°C.

Table 3   Selected Parameters of Creep Curves at 650°C/100 MPa

Cast	$\epsilon_s.10^{10}$ [s^{-1}]	t_3 [h]	t_r [h]	t_3/t_r
C	7.0	6,230	51,921	0.12
E	3.5	7,600	23,616	0.32
F	0.86	21,500	37,890	0.57

The most obvious effect of niobium upon the creep life of AISI 316LN steels is apparent from the profound change of the creep curves. Figure 11 illustrates the shape of creep curves of Casts C and E plotted at 650°C/80 MPa. The results of analysis of creep curves at 650°C/100 MPa for niobium-free Cast C and for Casts E and F containing 0.106 and 0.300 wt.% of niobium respectively are listed in Table 3.

It is evident that increasing niobium contents strongly reduces the minimum creep rate, ϵ_s and prolongs the time to the onset of the tertiary stage of creep, t_3. The experimentally established dependence of the time to rupture, t_r on the applied stress and temperature is shown in Fig. 5. At shorter times and lower temperature niobium additions produced no demonstrable effect

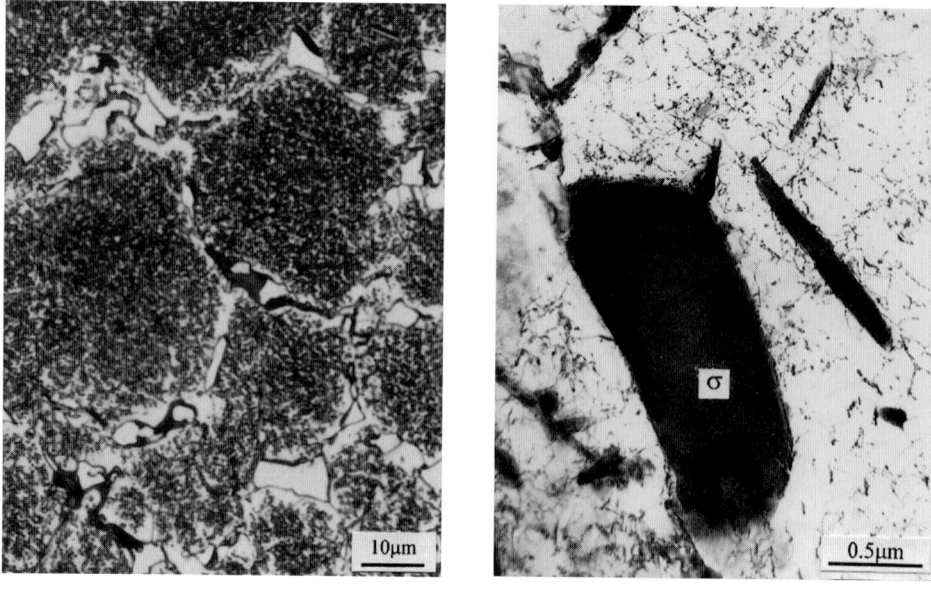

Fig. 13 Coarse particles of σ-phase at grain boundaries in Cast E after creep exposure: a. 650°C/58,936 hours, b. 650°C/14,864 hours.

on the creep rupture strength. At 650°C and times to rupture exceeding 10 000 hours, additions of 0.1 to 0.3 wt.% of niobium seemed actually to impair the creep strength. Figure 12 shows how the creep rupture elongation of Casts E and F varied with the time to rupture at the various test temperatures. On the contrary to the Casts C and D there is no well-defined trend at times to rupture shorter than 10 000 hours, but a pronounced decline of creep ductility at longer times. In conclusion, it is virtually impossible to predict creep life of the casts investigated from an experimentally ascertained minimum creep rate. The enhanced creep resistance of niobium-bearing AISI 316LN steels in the primary and secondary stages of creep is not accompanied by the longer creep life that might have been expected.[30]

After solution treatment, niobium-bearing Casts E and F contained both intragranular and intergranular Z-phase particles and some undissolved NbX particles. Z-phase particles continued to precipitate during creep exposure. These particles nucleated both at the grain boundaries and on dislocations inside grains. Z-phase particles formed during creep exposure were considerably dimensionally stable. The average equivalent diameter of Z-phase particles in Cast F after creep exposure at 650°C for 37 890 hours reached only 12 nm.[30] The niobium additions strongly enhance the formation of η-Laves and M_6(C,N) phases. Particles of η-Laves were identified in Cast F after a few tens of hours exposure at 650°C. M_6(C,N) particles gradually replaced less

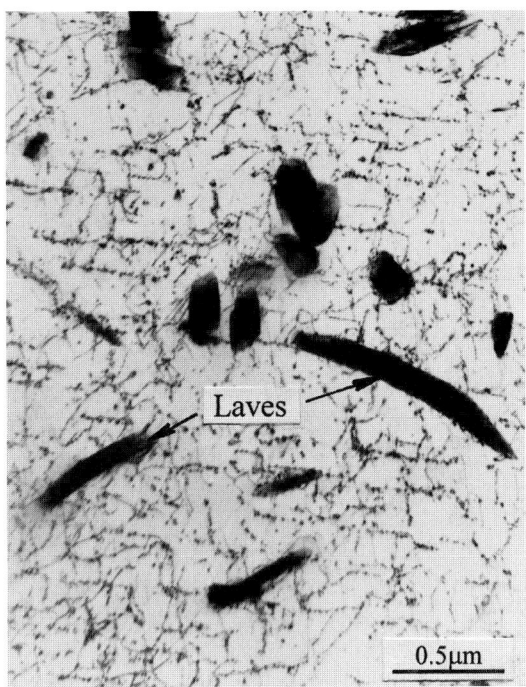

Fig. 14 The character of intragranular precipitation of Z-phase and η-Laves particles in Cast F after creep exposure 650°C/26,505 hours.

stable $M_{23}C_6$ phase. No Cr_2N and/or NbX particles were found to precipitate during creep exposure.

Microstructural changes at grain boundaries included the Z-phase formation, followed by the precipitation of $M_{23}C_6$, η-Laves, $M_6(C,N)$ and σ-phase. At longer times of creep exposure at 650°C the intergranular precipitates were identified as Z-phase, $M_6(C,N)$ and σ-phase particles. Coarse isolated intergranular σ-phase particles were surrounded by precipitate free zones. The results of microstructural investigations indicate that niobium enhances the rate of coarsening of σ-phase, Fig. 13. The coarse intergranular particles form potential sites for initiation of intergranular failure. Most intragranular precipitates were found to be Z-phase and η-Laves particles. The intragranular precipitation strengthening in niobium-bearing AISI 316LN steels is very intensive as can be seen in Fig. 14. In conclusion, the relatively poor creep ductility of Casts E and F corresponds to microstructural evolution of material in the course of creep exposure. Fractographic studies on ruptured testpieces revealed the dominant intergranular failure mode. The coarse intergranular particles facilitated the formation of cavities, their growth and interlinking which resulted in intergranular fracture.

CONCLUSIONS

Creep resistance of CrNi(Mo) austenitic steels strongly depends on the character of the microstructural evolution during long-term high temperature exposure under stress. Small changes in chemical composition of austenitic steels can result in significant differences in creep behaviour. In order to understand the effect of individual alloying and residual elements on creep resistance of CrNi(Mo) austenitic steels long-term creep testing of casts with intentional additions is required.

The microstructure-creep property relationship in the steels investigated can be summed up as follows:

1. Nitrogen and boron additions to AISI type 304L steel increase creep rupture strength. Creep ductility, however, is rather low. Nitrogen promotes the formation of Cr_2N and $M_6(C,N)$ phases and retards the precipitation of intermetallic σ-phase. The occurrence of coarse $M_6(C,N)$ particles on grain boundaries contributed to dominant intergranular fractures.

2. Boron and nitrogen additions to AISI type 316L steel increase both creep rupture strength and ductility. Additions of these elements retard the formation of $M_{23}C_6$ and intermetallic phases and slow down the rate of coarsening of those phases. Furthermore, nitrogen promotes the precipitation of Cr_2N and $M_6(C,N)$ phases. High creep rupture ductility in boron-treated AISI type 316LN steels is preserved even at times to rupture exceeding 10,000 hours. A nearly continuous network of fine intergranular particles was observed in testpieces which exhibited high creep ductility. The coarsening of intergranular precipitates was accompanied by a change to predominantly intergranular fracture and by a corresponding decline of creep rupture ductility. Coarse particles on grain boundaries were found to be harmful to creep ductility, regardless of phase type.

3. Small additions of niobium to AISI 316LN steel resulted in a significant reduction of steady creep rate and shortening of the tertiary creep stage. On the other hand, the enhanced creep resistance in primary and secondary stages of creep were not accompanied by a longer creep life. Creep rupture ductility of these steels was rather poor. Microstructural investigations revealed that niobium provoked the formation of Z-phase particles during solution treatment and/or creep exposure. These particles were considerably dimensionally stable. Furthermore, niobium accelerated both the formation and coarsening of η-Laves, $M_6(C,N)$ and σ-phase. The coarse intergranular particles facilitated the intergranular failure mode and contributed to relatively poor creep ductility.

ACKNOWLEDGEMENTS
The authors wish to acknowledge the financial support from the Grant Agency of the Czech Republic under the contract number 106/95/1129.

REFERENCES
 1. P. Marshall: *Austenitic Stainless Steels*, Elsevier, London–New York, 1984.
 2. J. Sobotka, V. Vodárek and M. Sobotková: *Metallurgical Journal*, 1992, **47**, (12), 31(in Czech).
 3. C. Horton, P. Marshall and R. Thomas: *Int. Conf. Mech. Behaviour and Nuclear Applications of Stainless Steel at Elevated Temperatures*, 1981, Metals Society, London, 1981.
 4. J. Sobotka, M. Liška, M. Sobotková, M. Tomášová, V. Vodárek and V. Bína: *Int. Conf. High Nitrogen Steels 88*, The Institute of Metals, London 1989, 363.
 5. K. Lorenz, H. Fabritius and E. Kranz: *Stahl und Eisen*, 1972, **92**, 393.
 6. Y. Kawabe, R. Nakagawa and T. Mukoyama: *Trans. ISIJ*, 1968, **8**, 353.
 7. British Steel Corporation, *Ref. No. SSD–7276*, 1973.
 8. M. Kikuchi, M. Kajihara and Si Kyung Choi: *Materials Sci. and Eng. A*, 1991, **146**, 131.
 9. J.K.L. Lai: *Scripta Met.* 1991 **25**, 2371.
10. S.R. Keown and F. B. Pickering: *Met.Sci.*, 1997, **7**, 225.
11. J.K.L. Lai and A. Wickens: *Acta Met.*, 1979, **27**, 217.
12. L. Karlsson and H. Norden: *Acta Metall.*, 1988, **36**, 35.
13. M. Shimada: *ISIJ International*, 1990, **30**, 579.
14. P.J. Masiasz: *JOM*, 1989, July 14.
15. W. Seifert: *Warmfeste metallische Werkstoffe*, Zittau, 1987, 89.
16. M. Liška, V. Vodárek, M. Sobotková and J. Sobotka: *Proc. High Nitrogen Steels 90*, Verlag Stahleisen, Düsseldorf 1990, 78.
17. A. Mercier, R. Levegue and G. Remy: *Rev. Met.*, 1967, **64**, 1085.
18. V. Vodárek, M. Sobotková, J. Sobotka: *Proc. High Nitrogen Steels '93*, Kiev 1993, 284.
19. H.J. Goldschmidt: *Interstitial Alloys*, Butterworth, London, 1967.
20. J. Gjonnes and A.F. Moodie: *Acta Cryst.*, 1954, **19**, 65.
21. V. Vodárek, M. Sobotková and J. Sobotka: *Metallic Materials*, 1993, **2**, 122.
22. M. Liška, M. Sobotková and J. Sobotka: *Czech Journal of Phys.*, 1985, **35**, 329.
23. V. Vodárek, J. Sobotka and M. Sobotková: *Metallic Materials*, 1989, **27**, 23 (in Czech).
24. L.T.M. Hopkin and L.H. Taylor: *JISI*, 1967, **205**, 17.
25. T.M. Williams, D.R. Harries and J. Furnival: *JISI*, 1972, **210**, 351.

26. T. Ishii, T. Shinada and R. Tanaka: *Trans ISIJ*, 1973, **13**, 125.
27. V. Vodárek: *Scripta Met.*, 1991, **25**, 549.
28. D.H. Jack, K.H. Jack: *JISI*, 1972, **209**, 790.
29. H. Gerlach and E. Schmidtmann: *Archiv fúr Eisenhútt.*, 1968, **39**, 189.
30. V. Vodárek, J. Sobotka, M. Sobotková and J. Sojka: *Proc. Applications of Stainless Steel '92*, Stockholm 1992, 123.
31. N.C. Mataya, C.A. Perkins, S.W. Thompson and D.K. Matlock: *Met. Trans.*, 1996, **27A**, 1251.

Evolution of Microstructure in P91-Type Steel in High Temperature Creep

A. ORLOVÁ, J. BURŠÍK, K. KUCHAŘOVÁ AND
V. SKLENIČKA

Institute of Physics of Materials, Academy of Sciences of the Czech Republic, Žižkova 22, 616 62 Brno, Czech Republic

ABSTRACT

The modified 9%Cr P91 steel is one of the materials actually employed in power plant pipework components. The detailed microstructural analysis of a trial melt produced by Vitkovice Steel, Ostrava is reported in the present work. Microstructure evolution during creep at 873 K was investigated by means of transmission electron microscopy (TEM) and computer image analysis on specimens subjects to creep tests conducted up to several predetermined creep life fractions. Two main microstructural elements, namely subgrains and secondary phase particles, were studied quantitatively. The separate contribution of stress free ageing and stress under creep conditions on particle coarsening and subgrain growth is determined.

The microstructure evolution consists in a growth of subgrain size and shape changes. This process is retarded by the influence of the dispersed phases that are also subject to further evolution in the course of the creep and annealing process. The smooth increase of subgrain size resulting from stress free ageing is strongly accelerated by the applied stress. Contrary to the subgrain size, the evolution of particle mean size is not monotonic: the initial increase is followed by a final decrease. Due to this fact, care should be taken over the evaluation of true particle growth rate values.

1. INTRODUCTION

The tempered martensite ferritic steels (TMFS) are widely used in today's power plant pipework components for their considerable high temperature creep strength, high corrosion cracking resistance, low oxidation rate and excellent weldability.[1-4] An increasing demand for power plants with higher efficiencies finds the most economical solution in the use of steam plants operating at higher temperatures and higher pressures. This is the reason why TMFS, namely P91-type, have been a subject of extensive investigations within projects of European co-operation (COST) as well as national projects of the member countries and a lot of new results on their heat resistance and mechanical behaviour at elevated and high temperatures have been gathered.[5-8]

89

Table 1.   Basic creep test data of investigated specimens

specimen	location	T[K]	σ[MPa]	t[h]	$\dot{\varepsilon}_s[s^{-1}]$	ε[%]
SP00	unimportant	–	–	–	–	
SP209	head	873	–	77.2	–	–
SP209	gauge length	873	175	77.2	3.3×10^{-8}	2.34
SP207	head	873	–	193.0	–	–
SP207	gauge length	873	175	193.0	4.1×10^{-8}	4.37
SP211	head	873	–	270.2	–	–
SP211	gauge length	873	175	270.2	2.8×10^{-8}	4.59
SP206	head	873	–	386.7	–	–
SP206	gauge length	873	175	386.7	2.9×10^{-8}	14.8
SP212	head	873	–	1822	–	–
SP212	gauge length	873	125	1822	8.0×10^{-10}	1.60
SP214	head	873	–	4555	–	–
SP214	gauge length	873	125	4555	7.5×10^{-10}	2.44
SP208	head	873	–	6472	–	–
SP208	gauge length	873	125	6472	7.5×10^{-10}	3.72
SP64	head	873	–	9111	–	–
SP64	gauge length	873	125	9111	7.0×10^{-10}	7.1
SP220	head	873	–	3306	–	–
SP220	gauge length	873	110	3306	1.0×10^{-10}	0.81
SP86	head	873	–	8347	–	–
SP86	gauge length	873	110	8347	1.0×10^{-10}	0.81
SP120	head	873	–	11658	–	–
SP120	gauge length	873	110	11658	2.5×10^{-10}	2.0
AX41	head	873	–	16650	–	–
AX41	gauge length	873	110	16650	n.a.	n.a.

The experimental material of this work was the trial melt of a P91-type steel produced by Vitkovice Steel, CR. The work tries to contribute to the stock of available data a systematic study of a microstructural development in the course of high temperature creep. While the major part of investigations are confined to fractured specimens, the present authors investigated specimens subjected to creep exposures to several fractions of the creep life under the chosen applied stress. This gives a chance to judge the real kinetics of the processes under certain creep conditions.

2.   MATERIAL AND EXPERIMENTAL TECHNIQUE

The chemical composition of the steel was as follows (in mass %): 0.09C, 0.56Mn, 0.20Si, 0.021P, 0.009S, 0.470Ni, 8.36Cr, 0.86Mo, 0.20V, 0.06Nb, 0.007Al, 0.065N, 0.05Cu, bal.Fe. The initial two-stage heat treatment (1060°C/1 h/air + 750°C/2 h/air) resulted in mechanical properties at room temperature: $R_{0.2} = 556$MPa, $R_m = 706$MPa, $A_5 = 22.9\%$. Prior to creep test-ing, the material was subjected to further annealing 600°C/6 h. Creep speci-mens 5.0 × 3.2 mm² in cross section and 50 mm in gauge length were subjected to tensile creep tests under the constant applied stress in the pro-tective atmosphere. Three levels of applied stress were chosen for the present investigation. The creep tests were conducted up to predetermined fractions of the creep life t_F, namely $t/t_F = 0.2, 0.5, 0.7$ and 1.0, on different specimens to generate the corresponding microstructural stages. The experimental data of all investigated specimens are listed in Table 1 including the steady state creep rate $\dot{\epsilon}_s$ and the total plastic deformation ϵ.

After the creep exposures, the Vickers hardness HV10 was measured and then a detailed investigation of the microstructure by means of TEM was per-formed. Two basic microstructural elements, namely the subgrain structure and the dispersed particles of secondary phases, were investigated. The qual-ity of minor phase particles was determined by the electron diffraction analy-sis using the selected area (SAD) technique. In some cases also the EDAX microanalysis in the STEM mode was applied. The metric parameters of both structure elements were evaluated by the image analysis using the computer software package DIPS 3.0.[9, 10] The microstructure both in the gauge length and in the specimen's head was investigated to provide data on the influence of the ageing under unstressed conditions in a comparison with the influence of ageing under the stress in creep.

3.   RESULTS

3.1   Creep Curves

The material exhibits a normal three-stage creep behaviour, following a rela-tively small instantaneous elongation (Fig. 1a). The primary stage of a decreasing creep rate is relatively short, approximately $0.2t_F$ only. The very short steady-stage stage is followed by a relatively long tertiary one, repre-senting the major part of the test duration and the major part of total creep strain. Creep curves repeated on several specimens show the following differ-ences: (i) different initial elongation causes a shift of the curve along the axis of strain, and (ii) an earlier, or later, start of the tertiary behaviour and its sub-sequent development by different rates lead probably to different times to fracture. However, a redrawing of the creep curves to the physically more informative $\dot{\epsilon}(t)$ representation illustrates that the differences are of minor

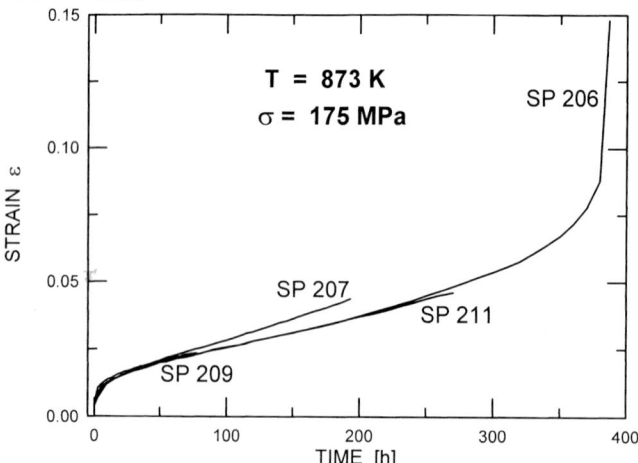

Fig. 1(a) Creep curves conducted to various predetermined t/t_F ratios at 175 MPa.

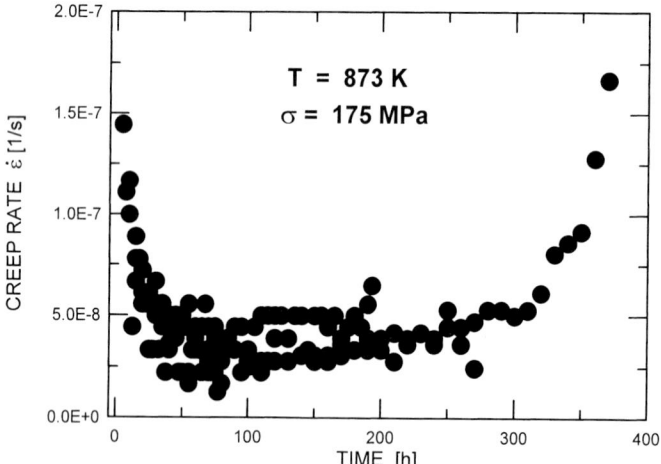

Fig. 1(b) Creep rate vs time plot for all specimens crept at 175 Mpa.

importance and the reproducibility of the creep behaviour is quite good (see Fig. 1b).

3.2 *Specimen Hardness*

Hardness HV10 being 241 in the initial material decreased to 237 in reaching the testing temperature 873 K (i.e., in 5 h heating + 3 h delay at the temperature). The hardness measured along the crept specimens in 1 mm intervals has shown a further decrease of the mean value, more pronounced in the gauge

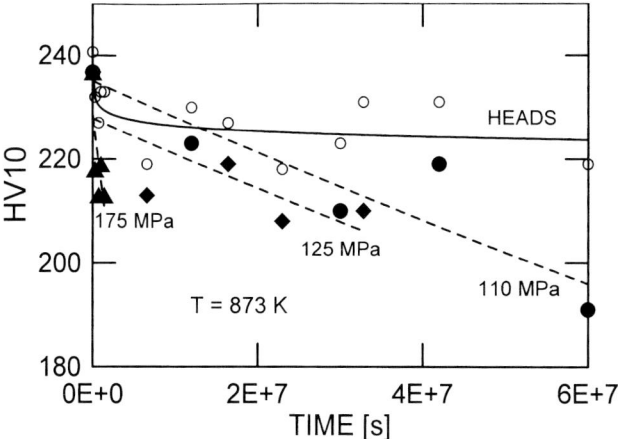

Fig. 2(a) Time and stress dependence of the hardness.

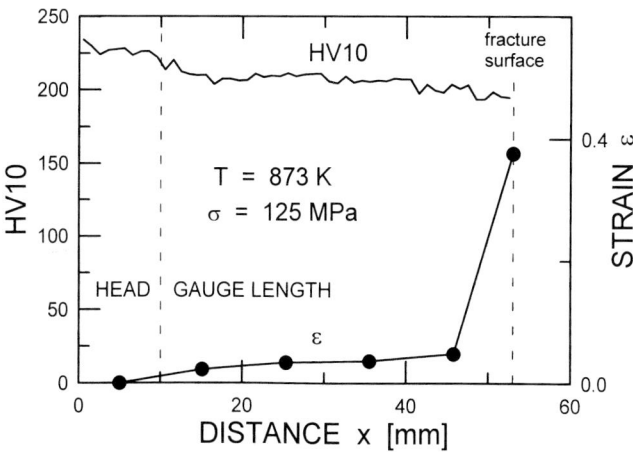

Fig. 2(b) Hardness and local strain changes along the specimen.

length than in the specimens' heads and growing with the time of testing (Fig. 2a). In the non-fractured specimens the hardness slightly fluctuates about a constant value along the gauge length. In the specimens deformed to fracture, a decrease of hardness, accompanied by an increase of the local strain, is observed along the whole specimen length on the approach of the fracture surface (Fig. 2b). A slight gradient of local strain coinciding with a gradient of hardness can be seen also at the non-fractured specimens deformed to more than about $0.5t_F$.

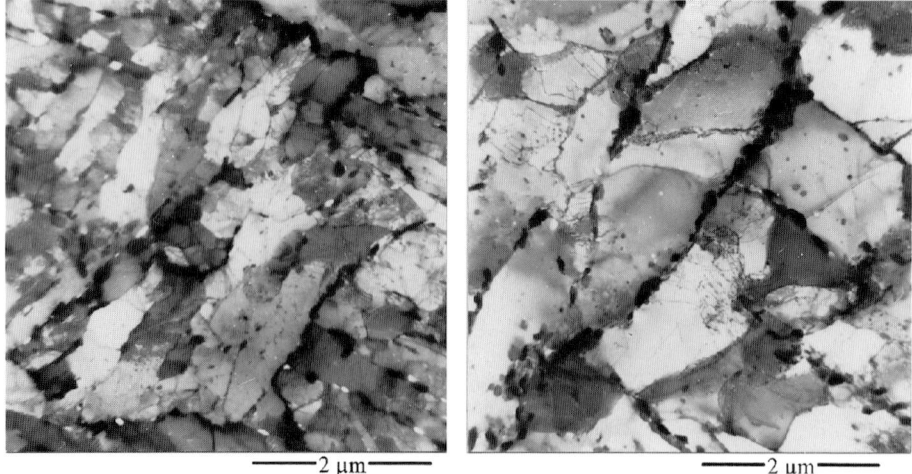

———2 µm———   ———2 µm———

Fig. 3   Typical microstructures: (a) after stress-free ageing (6472 h at 600°C, specimen SP208) (b) after creep (125 MPa, 0.7t$_F$, specimen SP208).

3.3   Microstructure

The initial microstructure as well as the stress free aged one evidently relates to the original martensite (Fig. 3a). It consists of elongated subgrains formed by fragmenting the original martensite laths by transversal sub-boundaries. The present subgrain boundaries are sites of numerous minor phase precipitates, first of all the carbides $M_{23}C_6$, that result mostly from the initial heat treatment. Particles decorating the subgrain boundaries represent the upper end of the particle size interval observed in the specimens. Particles of the carbide and other minor phases appear also in the subgrain interior. They are finer and their different morphologies indicate different phases. Electron diffraction patterns correspond to phases suggested by Foldyna et al.[11] whose chemical composition was also signalled by the EDAX microanalysis in the previous work,[8] i.e., carbides or carbonitrides NbC, Nb(C,N), Nb_2C, MoC and Mo_2C.

Microstructures observed after the creep exposures were qualitatively similar to the initial microstructure. The influence of the initial tempered martensitic structure remains to be well apparent in the two shorter creep tests (σ = 125 and 175 MPa) and seems to disappear in the third creep test of the longest duration (Fig. 3b). As to the minor phases indicated by the SAD analysis, carbides V_4C_3 and Cr_2C appear, while Nb(C,N) seem to be less frequent. The dislocation pinning effect of particles in subgrain interiors is frequently observed (Fig. 4a). In several specimens (the fractured specimen crept at 175 MPa, the specimens crept for 0.2t$_F$, 0.5t$_F$ and 1.0t$_F$ at 110 MPa) SAD indi-

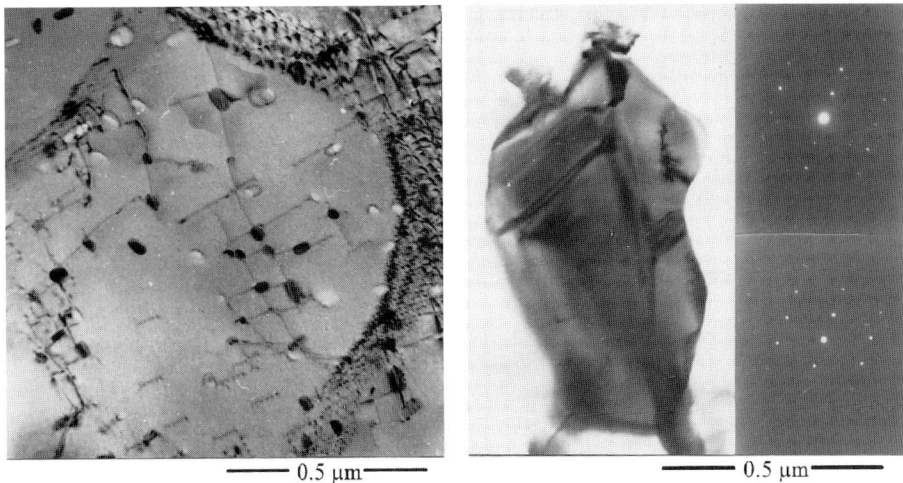

Fig. 4 (a) Minor phase particles distribution inside subgrain, dislocation pinning (SP208, gauge length). (b) Particle of Laves phase, SAD patterns of [$\bar{4}$11] and [$\bar{3}$11] zone axes (SP206, gauge length).

cated also bigger particles of a hexagonal Laves phase of Fe_2Mo type (Fig. 4b). The ratio of metallic elements 56Fe:36Mo:8Cr indicates the stoichiometry $(Fe,Cr)_2Mo$.

Besides the homogeneous structure of the tempered martensite, regions (of the area of about $10\,\mu m^2$) of a homogeneous ferritic matrix were locally observed in the creep tested specimens. They seem to fill a space between two original martensitic 'plates' or in a 'corner' of grain at a triple point. These regions contain a relatively high dislocation density and fine minor phase particles ($M_{23}C_6$ or V_4C_3).

3.4 Subgrain Structure Evolution

The image analysis of the TEM micrographs was aimed to a determination of the parameters of shape and size and evaluation of their mean values and cumulative frequency distributions. As the subgrain size, the equivalent subgrain diameter determined from the subgrain area measured in the foil plane was taken. The shape of the elongated subgrain was expressed by the shape factor obtained as a ratio of the minimum and the maximum subgrain dimensions d_{MIN}/d_{MAX}. The limitations and weak points of the procedures were as follows: (1) data samples of about 200 subgrain data per specimen were probably small, (2) the image of an anisotropic substructure following the structure of the original martensite is dependent on the crystallographic plane of the investigated thin foil, and (3) the existence of regions without any substructure cannot be classified by the procedures.

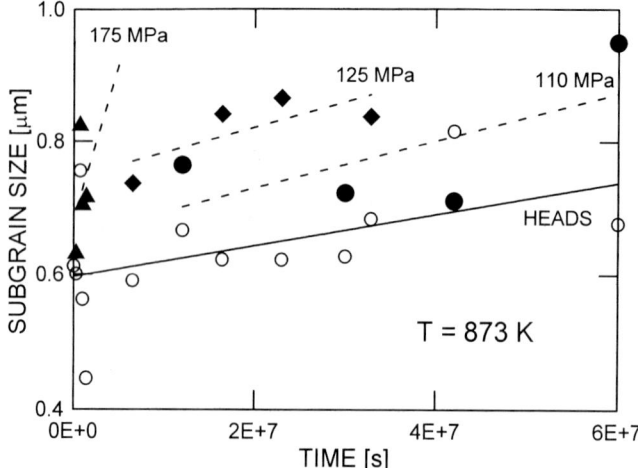

Fig. 5(a) The stress and time dependence of the mean subgrain size.

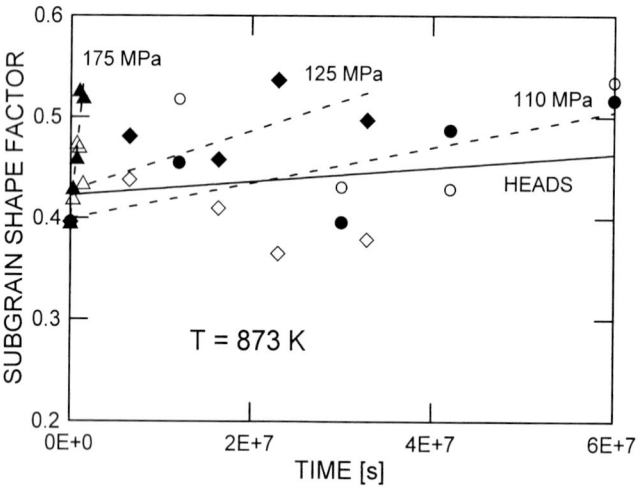

Fig. 5(b) The stress and time dependence of the mean subgrain shape factor.

The subgrain structure evolution shows the growth of the mean subgrain size in the creeping gauge length, which is supported by the influence of the applied stress (Fig. 5a) and is more pronounced than the growth in the heads subjected only to the influence of temperature. Simultaneously with the subgrain size growth we can observe an increase of the subgrain shape factor (Fig. 5b), indicating that the subgrains tend to approach the equiaxed shape. The change is again strongly accelerated by the influence of the applied stress in creep.

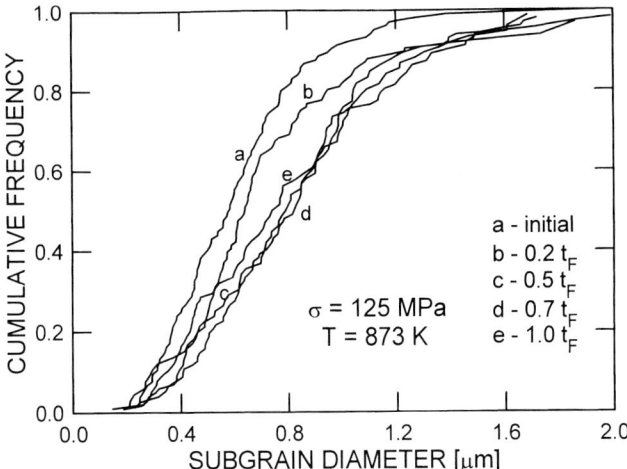

Fig. 6(a)   Cumulative frequency curves of equivalent subgrain diameter distribution in crept specimens.

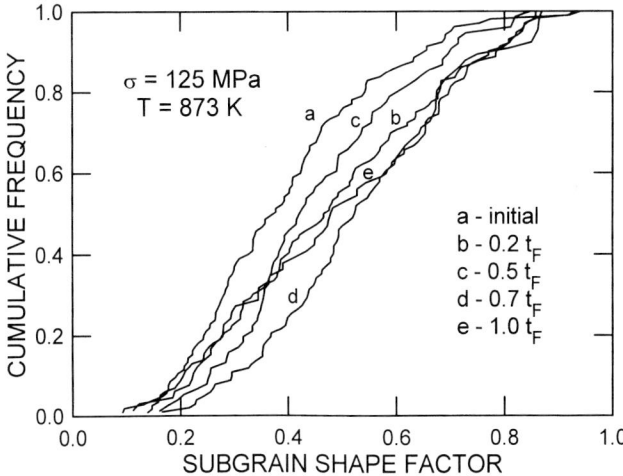

Fig. 6(b)   Cumulative frequency curves of subgrain shape factor distribution in crept specimens.

A view of the statistics of the parameters is illustrated in Fig. 6. Cumulative distribution of the subgrain size widens to higher values of the subgrain size in the creeping material (Fig. 6a), indicating that some of the subgrains have grown. This tendency increases with the time of testing. The simultaneous shift of the subgrain shape factor curves to the right indicates changes towards more equiaxed subgrains (Fig. 6b). The changes of the mentioned parameters in the unstressed heads are less pronounced. The annealing seems to result in steeper cumulative frequency functions of the subgrain size with respect to the initial state, indicating that the smallest subgrains have grown.

The evolution of the substructure in creep follows a tendency to reach a subgrain size value characteristic of the applied stress. According to the suggestion of Raj and Pharr[12] deduced from an analysis of a large collection of data, the characteristic subgrain size d can be expressed as

$$d = K \cdot b \cdot \left(\frac{G}{\sigma}\right)^m, \tag{1}$$

where K is a dimensionless constant, b is the Burgers vector length, G is the shear modulus, σ is the applied stress and m is an exponent, $0 < m < 2$; with $m = 1$ corresponds $K = 23$. Figure 7 shows the dependence of the minimum and maximum subgrain dimensions d_{MIN} and d_{MAX} on the applied stress. The data of Sklenička et al.[8] measured on the fractured specimens are included for comparison. The stress dependencies of both data samples are relatively weak and in rather good accord, the exponent m being 0.153 and 0.174 for d_{MIN} and only slightly higher, i.e., 0.317 and 0.202 for d_{MAX}.

3.5   Minor Phase Dispersion Parameters

The image analysis of the TEM micrographs was oriented to the characteristics of size and shape of the dispersed particles and to their quantity in the volume. The data sets represent *all* particles resolved in the microstructures, i.e., those situated both in the subgrain boundaries and in the subgrain interiors. The weak point of this type of sampling is (i) non-distinguishing between different minor phases in the structure (ii) effect of anisotropy related to the anisotropy of the matrix: particles preferentially oriented in certain crystallographic directions or planes, and (iii) regions with different dislocation and phase structure (regions without substructure) contain different spectra of particles.

The changes of particle characteristics are shown in Fig. 8. While the number of particles in the non-stressed heads shows a monotonic increase with the time of ageing, in the creep exposed region its behaviour is uneven – an increase in the initial period of creep up to $0.2t_F$ is followed by a decrease in the latter stages below the values reached in heads and even below the value of the initial state (Fig. 8a). A non-monotonic behaviour in creep is exhibited also by the particle mean size in the stressed part (it was again evaluated as the equivalent particle diameter of particle projection) (Fig. 8b). The particle shape factor indicates an increasing elongation of the particles in the stressed conditions (with an exception of the fractured specimen at $\sigma = 110$ MPa) and probably no systematic change in annealing without the influence of stress. The complicated behaviour of dispersoid populations is indicated also by sta-

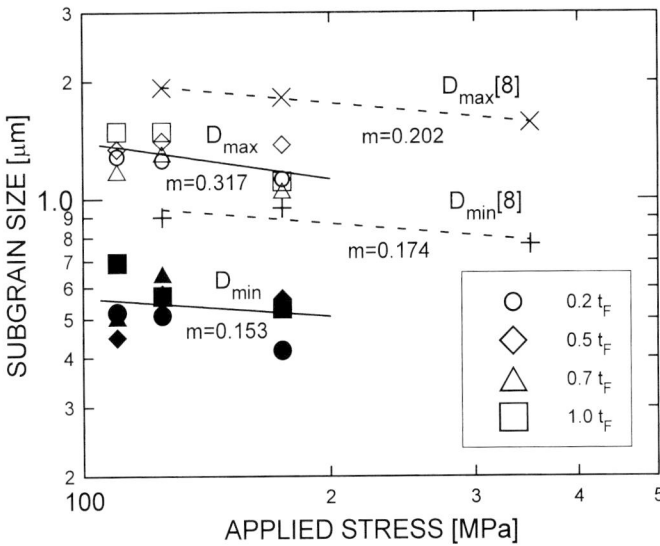

Fig. 7   The applied stress dependence of the minimum and maximum subgrain dimensions.

tistical data. An example in Fig. 8c shows an increasing spread of particle sizes to higher values in the course of creep (curves *b*, *c*, and *d*) and a reduction of the width of spectra in the fractured specimen (curve *e*). The simultaneous changes in the annealed heads are less pronounced.

4.   DISCUSSION

The major part of the creep curves is the tertiary stage of a continuously increasing creep rate. This behaviour means a continual degradation of a creep strength resulting from the microstructure evolution. The process is connected also with a development of a gradient of local strain along the specimen on the approach of the future fracture, accompanied by a gradient of hardness (Fig. 2). The microfractographic analysis in[8] indicated a gradient of cavity concentration in the close neighbourhood of the fracture, giving an evidence of a role of cavitation at least in the late tertiary stage.

The subgrain structure resulting from tempering of the initial martensite is not stable. The subgrain size is growing with the time of testing, the growth being evidently enhanced by the influence of the applied stress. Simultaneously there is a tendency to get the subgrain shape more equiaxed. The driving force for this evolution is a non-equilibrium state of the initial substructure. According to the general semi-empirical eqn (1) with $m = 1$ and $K = 23$ the steady-state subgrain sizes formed at given values of the normalised stress σ/G^{12} should be 3.3, 2.9 and 2.1 µm at applied stresses 110, 125

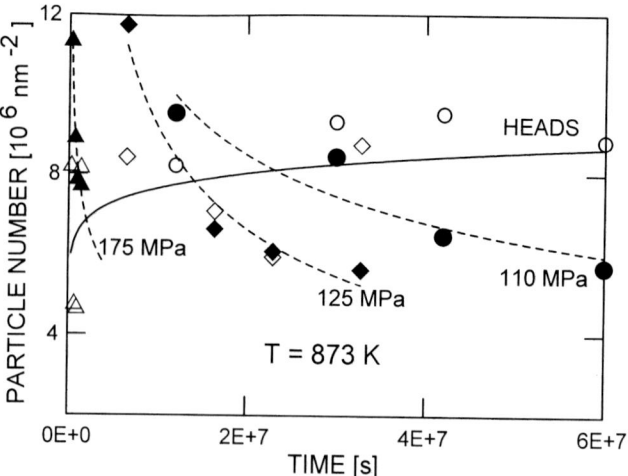

Fig. 8(a) Time dependence of particle number per unit area.

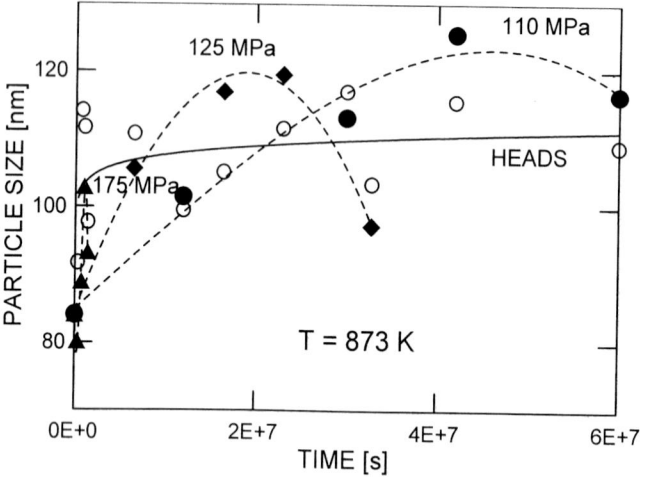

Fig. 8(b) Time dependence of mean particle size.

and 175 MPa, respectively. Recent analysis of Straub et al.[13] suggests to expect steady-state values about 2.3 times smaller in accord with K = 10 in eqn (1) which was proved experimentally for this type of steels. Finer subgrain size than that in the steady-state substructure corresponding to the actual applied stress is a source of substructure hardening as suggested by Vastava and Langdon[14] and the growth of the subgrain size is a process of material's softening or recovery. The coarsening of subgrains follows the kinetics suggested by Straub et al.,[15] based on a model of subgrain growth due to recombination

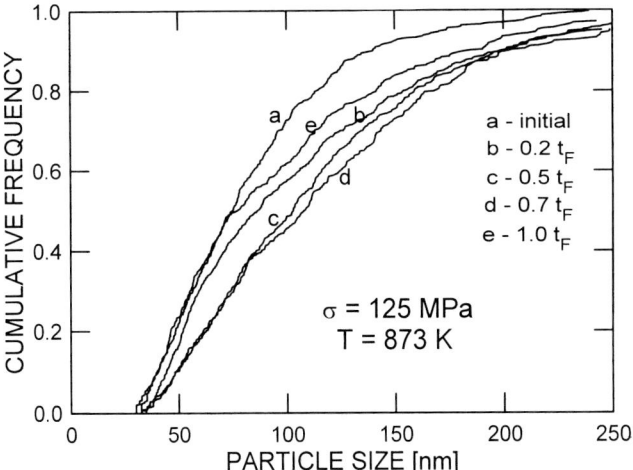

Fig. 8 (c)   Cumulative frequency distribution of particle size.

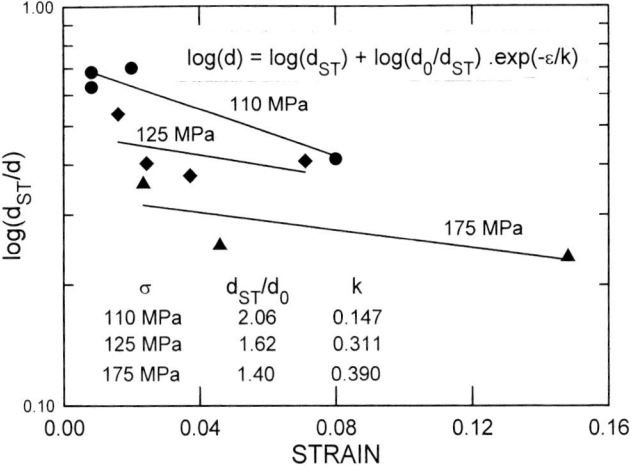

Fig. 9   The kinetics of subgrain coarsening.

and migration of sub-boundaries. Explicitly, this model can work without taking into account minor phase dispersed particles. The mean subgrain size evolution can be expressed according to the model by the relation

$$\log(d/d_{ST}) = \log(d_0/d_{ST}) \exp(-\epsilon/k), \qquad (2)$$

where d_{ST} is stationary subgrain size and d_0 is the initial subgrain size at zero strain. The fitted values for our experimental data are indicated in Fig. 9, they are comparable with those obtained by Straub[13] on X20 steel.

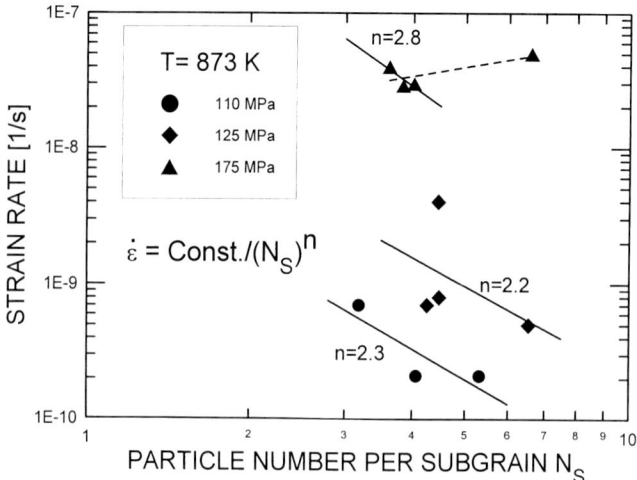

Fig. 10   Correlation of strain rate with particle number per subgrain.

The element determining the original subgrain size and shape is the initial lath martensitic microstructure. The subsequent tempering generates a complex structure of dispersed particles that help to conserve the initial subgrain structure in the next annealing and creep exposures by fixing the sub-boundaries' positions and thus retarding the process of subgrain growth. According to recent work by Dlouhý et al.,[16] a strong correlation of the creep rate to the density of carbide particles at subgrain boundaries indicates a decisive role of blocking of sub-boundaries by particles in the creep strength. The last fact can be documented by results of the present work, too: the number of particles per subgrain (N_s) in the given section can be correlated with the creep rate reached in the chosen points of the creep curve (Fig. 10). The course of the creep curve thus might reflect the changes of particle blocking of the sub-boundaries.

The particle structure also develops in the course of annealing and creep. While the particle number and size is growing in the non-stressed conditions (annealing), the evolution in the creep exposed material is non-monotonic. The start of the creep curve up to $0.2t_F$ is characterised by an increase of the particle number, indicating a precipitation of new particles. The prerequisite of this jump in particle number in the stressed material may be the newly generated fresh dislocations that serve as nucleation sites for new particles. This precipitation gives rise to a further role of dispersed particles – opposing the glide motion of dislocations inside subgrains (e.g. the Orowan interaction).

In the creep process the mean particle size grows and particle number per unit area is decreasing. This behaviour indicates particle coarsening which can reduce the strengthening effect of the dispersed phases. With the exception of

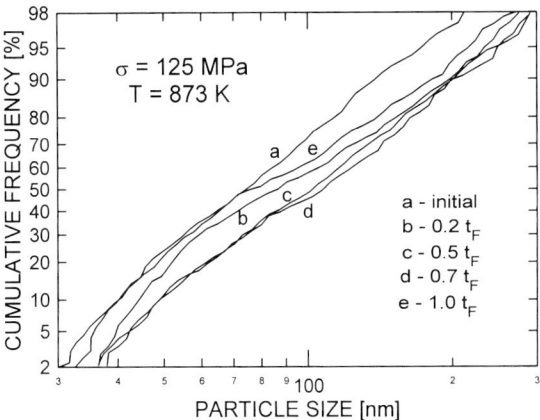

Fig. 11 Cumulative particle size distributions (y-axis subdivided according to the inverse error function).

the fractured specimens, the evolution of particle mean size d can be described by the Ostwald ripening law

$$d^3 = d^3_0 + k{\cdot}t, \tag{3}$$

where d_0 is the initial particle size at $t = 0$ and k is the particle growth rate. The measured data give $d_0 = 90$ nm, $k = 0.010$ nm^3 s^{-1} in heads, $k = 0.031$, 0.048 and 0.498 nm^3 s^{-1} for $\sigma = 110$, 125 and 175 MPa. While the first two k values can be accepted, the last one is not understood in view of other published data.[5, 15, 17]

The change of the shape factor indicates an increase of elongation, corresponding to the growth of particles along the subgrain boundaries. The decrease of the mean particle size with simultaneous decrease of particle number per unit area is observed in the creep exposed specimens for the time higher than $0.7t_F$. This fact can be explained by particle size redistribution caused by both particle coarsening and precipitation of new ones. In this respect the mean particle size values are not descriptive enough and the overall shape of the particle size distribution curves must be considered.

The cumulative particle size distributions from Fig. 8c are replotted in Fig. 11, where the y-axis is subdivided according to the inverse error function. This yields a straight line for normally distributed x-values. As can be seen from Fig. 11 the distribution of particle projected diameter is approximately lognormal in the initial state. Curves (b) and (e) in Fig. 11 show a deviation indicating an increased frequency of particle diameters in the range from 30 to 90 nm. This fact suggests the idea of precipitation of new particles at the beginning as well as in the final stage of creep life. At the initial stage most probably the new MX and M_2X fine particles are precipitating in subgrain interiors. A rapid increase of deformation at the final stage can stimulate a nucleation of Laves phase.[18]

The complexity of the particle structure, consisting of different phases with different kinetics of their evolution appears in the complex behaviour of the statistical data. The most important as subgrain stabiliser appear the $M_{23}C_6$ carbides that represent the major part of the dispersed phases situated in the subgrain boundaries. Fine particles of these carbides and also of other phases can act as obstacles for the free dislocation motion in the subgrains. Minor phases embedded in the ferritic matrix, e.g., the Laves phase, may be a weak point in the microstructure, inclined to failure.

5. CONCLUSIONS

The investigation of a microstructural evolution of P91 tempered martensite ferritic steel during creep at 600°C has shown that the high creep strength of the material is due the coexistence of subgrains and precipitates, acting as a subgrain boundary stabiliser. Characteristic features of both particle and subgrain subsystems were discussed on a quantitative base.

The microstructure evolution consists of a growth of subgrain size and a change of subgrain shape towards a more equiaxed one. This process, connected with a removal of substructure hardening of the material, is retarded by the influence of the dispersed phases that originate from the process of tempering and are subject of further evolution in the course of the creep and annealing process.

The smooth increase of subgrain size resulting from stress free ageing is strongly accelerated by the applied stress. The changes in mean subgrain size during creep are monotonic.

The particle coarsening upon ageing is accelerated under creep conditions. Contrary to the subgrain size, the evolution of particle mean size is not monotonic: the initial increase is followed by final decrease. A care should be taken over the evaluation of true growth rate values.

ACKNOWLEDGEMENTS

The work was supported by the Grant Agency of the Czech Republic, Grant No. 106/95/1530 and the Grant Agency of the Academy of Sciences of the Czech Republic, Grant No. 241403. The authors greatly appreciate valuable discussions with Dr. V. Foldyna from Vitkovice Steel, C.R.

REFERENCES

1. F.B. Pickering: *Physical Metallurgy and the Design of Steels*, Applied Science, London, 1978.
2. B. Walser, P. Brezina and T. Geiger: *Arch. Eisenhütt.*, 1979, **50**, 249.

3. W. Blum and S. Straub: *Steel Research*, 1991, **62**, 72.

4. W.B. Jones, C.R. Hills and D.H. Polonis: *Metall. Trans.*, 1991, **22A**, 1049.

5. G. Eggeler: *Acta Met.*, 1989, **37**, 3225.

6. F. Brühl, H. Cerjak, P. Schwaab and H. Weber: *Steel Research*, 1991, **62**, 75.

7. W. Bendick, K. Haarmann, G. Wellnitz and M. Zschau: *VGB-Kraftwerkstechnik*, 1993, **73**, 77.

8. V Sklenička, K. Kuchařová, A. Dlouhý and J. Krejčí: *Proc. Fifth Conf. on Materials for Advanced Power Engng.*, Oct. 3–6, D. Coutsouradis *et al.* eds, Liége, Belgium, 1994, 435–444.

9. P. Heriban and M. Druckmüller: *DIPS 3.0 User's Manual*, SOFO, Brno, 1992.

10. J. Němec, A. Dlouhý, K. Kuchařová and M. Cans: *Prakt. Metallographie*, 1996, **33**, 198.

11. V. Foldyna, A. Jakobová, V. Vodárek and Z. Kuboň: *see [8]*, 453–464.

12. S.V. Raj and G.M. Pharr: *Mater. Sci. Engng.*, 1986, **81**, 217.

13. S. Straub: *Verformungsverhalten und Mikrostruktur warmfester martensitischen 12%-Chromstähle*, Fortschritt-Berichte VDI, Reihe 5, Nr. 405, VDI Verlag, Düsseldorf, 1995.

14. R.B. Vastava and T.G. Langdon: *Advances in Mater. Technology in the Americas – 1980*, Vol. 2, I. LeMay, ed. ASM Publ., New York, 1980.

15. S. Straub, M. Meier, J. Osterman and W. Blum: *VGB Kraftstechnik*, 1993, **73**, 744.

16. A. Dlouhý, K. Kuchařová and J. Němec: *Research Report to Project No. 106/93/0965*, GACR, Brno, 1995.

17. V. Foldyna and A. Jakobová: *Kovove Mater.*, 1984, **22**, 402.

18. V. Foldyna: *private communication.*

Microstructural Evolution of P91 Steel After Long Term Creep Tests

P. BIANCHI[*], P. BONTEMPI[*], A. BENVENUTI[**] AND
N. RICCI[**]
*ENEL, CRAM, Milan, Italy
**CISE, Milan, Italy

ABSTRACT
A microstructural investigation on P91 steel after creep tests is presented; creep test duration ranged from few thousands hours up to more than 20 000 hours at temperatures from 550 to 650°C and stresses from 60 to 190 MPa. Material softening was the main physical phenomenon observed in the crept material. Morphological measurements performed by image analysis indicated both cell/subgrain and carbide coarsening, particularly marked in the gauge length of the investigated specimens. In all creep specimens enrichment of Cr in $M_{23}C_6$ carbides in comparison with the as received material was also measured and related to the time-temperature Larson Miller parameter; presence of Laves phase was detected; its composition was homogeneous among the investigated specimens.

INTRODUCTION
High Cr (9–12%Cr) ferritic steels represent a valuable alternative to austenitic stainless steel for high temperature applications up to 600°C both in power and petrochemical plant, as they show good strength and toughness, good creep strength, as well as good resistance to oxidation and corrosion. Their long term microstructural evolution may be of concern owing to its influence on mechanical properties.[1, 2, 3]

An investigation on creep-tested P91 steel has been performed to clarify the main mechanisms of microstructural evolution in this material and point out microstructural parameters indicative of material ageing during steel service in plant.

MATERIALS AND METHODS
The chemical analysis of the steel under investigation is shown in Table 1. The creep specimens were machined from a commercially manufactured header pipe (outer diameter 343 mm, thickness 74 mm), normalised and tempered according to the ASTM 335 specification (4 hours at 1040°C air cooling + 4 hours at 760°C air cooling). The fully tempered martensitic structure grain size was n° 6.5 ASTM, while the Vickers hardness (HV10) values fell in the

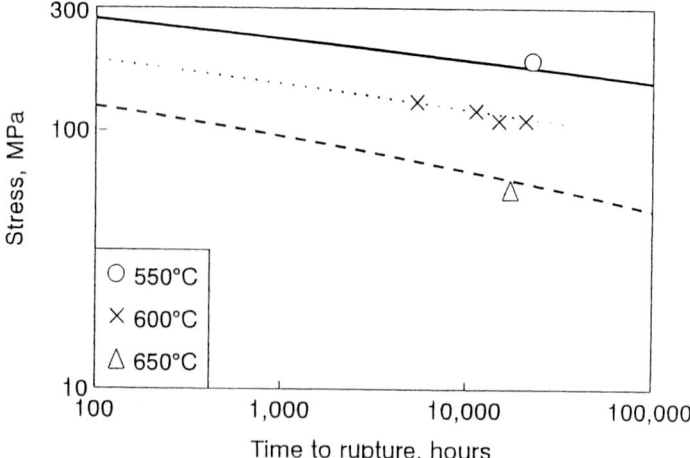

Fig. 1 Stress vs time to rupture: comparison to reference curves derived from.[4]

Table 1 Chemical composition of the P91 steel examined

C	Si	Mn	P	S	Al*	Cr	Mo	Ni	V	Nb	Cu	N
0.09	0.44	0.45	0.018	0.001	0.005	9.08	0.96	0.08	0.19	0.08	0.02	0.041

(*) Soluble Al.

Table 2 Creep test parameters and results for the P91 specimens

Specimen	T (°C)	σ (MPa)	t_r (hours)	Rupture elongation (%)	Reduction of area (%)
A	550	190	22239	21	88
B	600	130	5420	33	89
C	600	120	11237	22	84
D	600	110	14858	30	90
E	600	110	20497	22	88
F	650	60	16981	28	89

range 225 ± 5 kgmm^{-2}. As indicated in Table 2, the creep test duration ranged from few thousand hours up to more than 20 000 hours at temperatures from 550 to 650°C and stresses from 60 to 190 MPa. The rupture times are compared to reference curves derived from the Larson-Miller relation of Yang *et al.*[4] in Fig. 1.

The stress rupture master curve for steel P91 used here is the following one:

$$P(s) = T_k/1000 * (CLM + \log t_f)$$

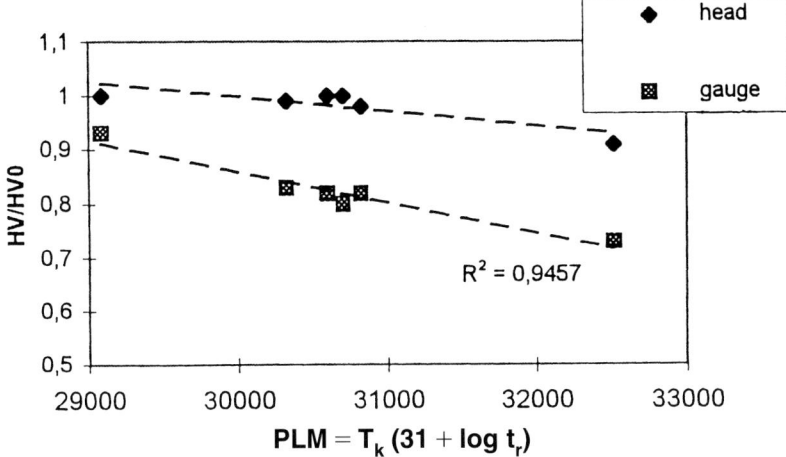

Fig. 2   Material softening in the investigated creep specimens vs Larson Miller time-temperature parameter.

s = stress (MPa), CLM = 31, t_f = rupture time, hours

$P(s) = a + b \log s + c(\log s)^2 + d(\log s)^3 + e(\log s)^4$

$a = 50\,681.03$, $b = 27\,974.33$, $c = 20\,321.24$, $d = 7169.06$, $e = 785$.

In order to evaluate the influence of deformation as well as pure thermal ageing on microstructural evolution, foils and extraction replica from both the gauge length and the head of the creep specimens were examined and compared to the as-received condition of the material.

Thin sections were first examined by Transmission Electron Microscopy (Jeol 200 CX TEM) to provide a qualitative description of the microstructure. Quantitative measurements of the most relevant microstructural parameters (concerning the size and shape of carbides and cells/subgrains, as set out in the framework of the COST 501/3WP11 study for advanced 9–12%Cr creep resisting steels[5]) were then performed by image analysis (Q500 Leica Image Analysis System) on digitised TEM micrographs.

The composition of secondary phases (Cr carbides and Laves phase) was measured on the extraction replicas by an EDS microanalysis system (Noran TN 5500, SiLi detector with Be window). X-ray spectra. The acquisition was performed at 200 kV with the highest available spatial resolution (electron beam diameter ≈10 nm). Quantitative microanalysis was based on experimentally determined Cliff-Lorimer conversion factors between X-rays intensities and element concentrations, with Fe as reference element.

Table 3(a)   Cell/subgrain (a) and carbides (b) morphological parameters.

Specimen	L_{min} (nm)	L_{max} (nm)	Aspect ratio (L_{min}/L_{max})
as received	691	1553	0.49
E – head	716	1339	0.53
E – gauge length	905	1807	0.54
F – head	723	1461	0.52
F – gauge length	1075	1979	0.55

Table 3(b)

Specimen	$D_{eq.}$ (nm)	D_{max}/D_{min}	Surface density (n°/μm²)	Area fraction (%)	Volume density (n°/μm³)	Interparticle spacing (nm)
as received	104	2.1	3.8	3.8	22.6	240
E – head	135	2.1	2.9	4.8	12.6	273
E – gauge length	152	1.9	3.1	6.9	14.2	216
F – head	156	1.9	3.3	7.0	15.1	198
F – gauge length	202	1.7	2.0	8.1	7.7	236

RESULTS

Morphological Evolution

Softening was the main physical phenomenon observed in the crept material. All the creep specimens showed pronounced strain softening in the gauge length with a decrease in Vickers hardness (HV10) of up to 30% in comparison to the as-received material. Thermal softening in the head of creep specimens was lower, with a maximum hardness decrease less than or equal to 10%. The amount of softening (both thermal and strain-induced), expressed as HV/HV0 (HV0 = as-received material hardness), could be linearly correlated with the Larson Miller parameter ($LPM = T_K (31 + \log t_r)$) as shown in Fig. 2.

TEM observations indicated that dislocation recovery and carbide coarsening phenomena were the primary causes of material softening. In all the creep specimens, and particularly in the gauge length, the original martensite laths became coarser and round-shaped, with formation of dislocation cells and subgrains, while analysis of the precipitates showed carbide coarsening and Laves phase formation (Fig. 3).

In order to quantify the above observations, morphological measurements using image analysis were performed on two of the creep specimens (E and F, which had the longest rupture times at 600°C and 650°C respectively) and the as-received material. The main results obtained are summarised in Table 3. An

Fig. 3   (a) Comparison between microstructures in the as received material (b)
and in crept material (gauge length of specimen C)

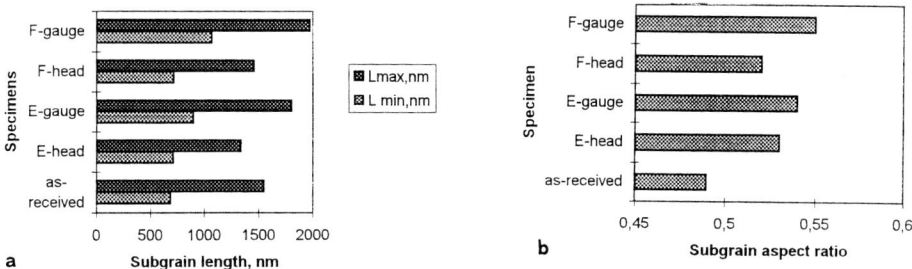

Fig. 4   Subgrain lengths (a) and aspect ratios (b) in specimens E and F and comparison with as received material

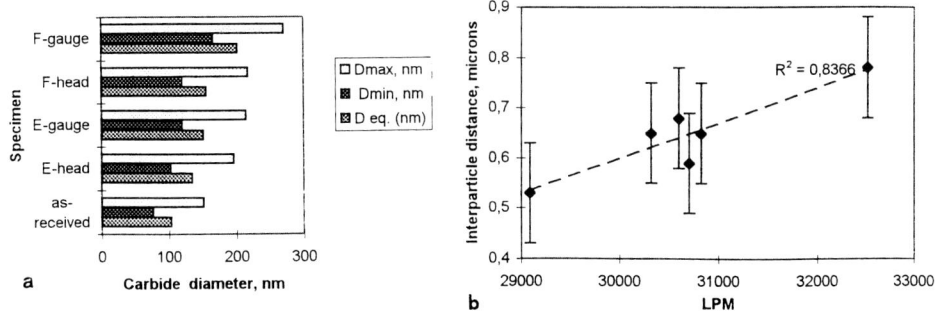

Fig. 5   (a) Carbide diameters in specimens E, F and in the as received material,
(b) interparticle distance of carbides (gauge length) vs LPM parameter

increase in cell-subgrain size (as expressed by measurements of maximum and minimum length, L_{max} and L_{min}) is evident in the gauge length of creep specimens in comparison to the as-received material (see also the bar chart in Fig. 4a). The subgrain aspect ratio, L_{min}/L_{max}, followed the same trend (Fig. 4b).

As far as the size of secondary phases ($M_{23}C_6$ carbides, M_2X, MX and Laves phases) is concerned, diameter measurements (D_{max}, D_{min} and $D_{equiv.}$) showed a general increase both in the head and, most markedly, in the gauge length of the creep specimens (Table 3b and Fig. 5a). Additionally, the particle shape showed a tendency towards spheroidisation. A reduction in area density N_A ($n°/\mu m^2$) and volume density $N_V(n°/\mu m^3)$ of particle was also measured in the creep specimens (Table 3). This was attributed to particle coalescence and agglomeration. The inter-carbide spacing, l, as determined by:

$$l = 1/2 \, (N_V \cdot d)^{-1/2} - d \cdot 2/3^{1/2}$$

where d is the particle equivalent diameter, failed to show an increase, implying that the reduction in volume density was evidently counterbalanced by carbide coarsening. This effect has already been observed for aged 9–12%Cr steels by Cerjak et al.[6] A moderate increase in carbide interparticle spacing in the gauge length of creep specimens as a function of the Larson Miller parameter (Fig. 5b) could be established, where l was simply evaluated as $l = N_A^{-1/2}$ on secondary electron images obtained in STEM mode, following a method already set up for low alloy steels.[7]

Compositional Evolution in Secondary Phases

The evolution of microstructure morphology in the creep specimens was accompanied by a significant change in composition of secondary phases. The following two main aspects were detected:

a) Cr enrichment in $M_{23}C_6$ carbides;
b) Laves phase precipitation.

As far as intragranular M_2X and MX precipitation is concerned, EDS spectra for these phases both in as-received material and the creep specimens indicated that they contained V and Nb (in variable amounts) as well as traces of Cr; no quantitative composition measurements on these very fine intragranular phases was performed, since getting a statistically good EDS spectra is very time consuming.

(a) Attention was focused on the measurement of the $M_{23}C_6$ carbide composition in extraction replicas. All the creep specimens showed Cr enrichment in this respect relative to the as-received material, Table 4. The Mo content in $M_{23}C_6$ carbides was found to be quite stable at about 10 wt%. These findings are in agreement with thermodynamic predictions for $M_{23}C_6$ carbide compo-

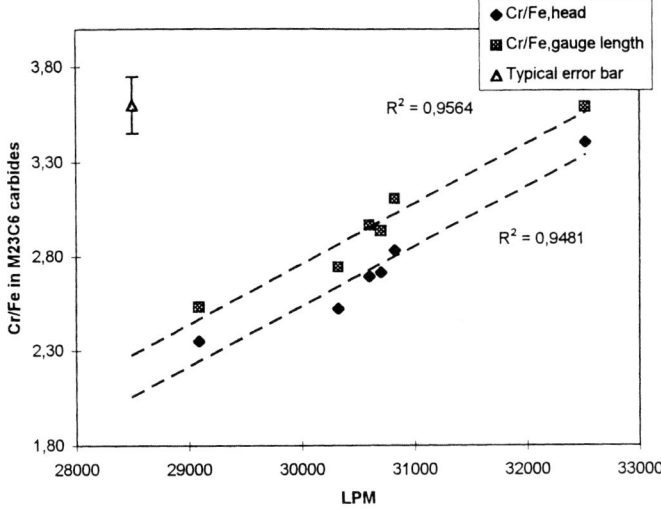

Fig. 6 Cr/Fe ratio in $M_{23}C_6$ carbides (measured both in head and gauge length of creep specimens) vs LPM parameter

Table 4 Chemical composition (wt%) of $M_{23}C_6$ carbides in the examined specimens (measurements on specimen heads).

Element	as received	A	B	C	D	E	F
Fe	28.0	26.3	24.9	24.0	24.0	23.3	20.2
Cr	61.3	61.7	62.8	64.5	65.1	66.0	68.6
Mn	0.7	0.7	1.1	0.8	0.4	0.7	0.5
Mo	8.8	9.8	10.0	9.6	9.5	9.3	10.1
V	1.1	1.2	1.0	0.9	0.8	0.6	0.7
Nb	0.2	0.36	0.2	0.2	0.2	0.1	0.1

sition as a function of temperature. The equilibrium content of Cr in carbides is in fact higher at creep test temperatures than at the steel tempering temperature.[8] The Cr enrichment was more pronounced in the gauge lengths than in the specimens heads, indicating that creep strain may have contributed to accelerating the diffusion of Cr into carbides.

The Cr enrichment in $M_{23}C_6$ carbide yields a contribution to depletion of Cr into the matrix; a decreasing Cr content vs time of ageing at elevated temperature have already been detected on a 12CrMoVNb[9] even if there it was assumed that Cr depletion were caused by the MC $\rightarrow M_{23}C_6$ transformation.

A linear correlation could be established between the Cr/Fe ratio and the Larson Miller time-temperature parameter (Fig. 6). This provides a potential tool for quantifying material ageing during service, so extending to high Cr steels the approach already used for low alloy steels.[10] Further aspects which however need to be investigated for P91 steels include:

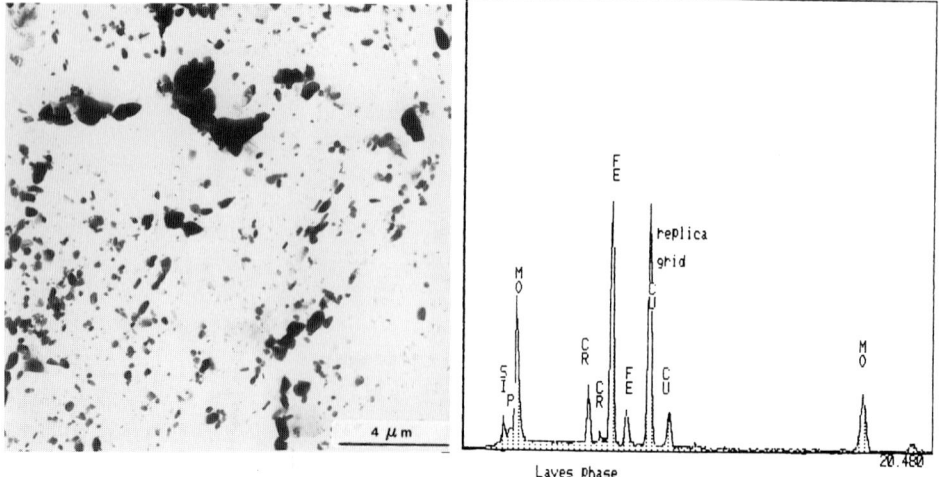

Fig. 7   Intergranular coarse Laves phase in crept material (a) and EDS spectrum (b)

– batch to batch variability of $M_{23}C_6$ carbide composition;
– stability limits of Cr carbides after long term ageing;
– influence of strain on Cr carbide composition and its relevance to service exposed materials.

(b) The precipitation of the intermetallic Laves phase was observed both in the gauge length and the heads of all the creep specimens. The amounts however varied from specimen to specimen. Laves phase was very abundant in specimen F (tested at 650°C), while in specimen A (tested at 550°C) its presence was less evident. This is in agreement with the time temperature precipitation diagram for Laves phase in P91 steel with 0.41% Si (Si content in the investigated material batch is 0.44 wt%), that predicts a maximum precipitation rate around 650°C, with a C shape kinetics.[11]

Laves phase was mainly detected at grain and subgrain boundaries, Fig. 7a, in form of coarse highly faulted particles, in close contact with $M_{23}C_6$ carbides; its composition was homogeneous among the creep specimens investigated and between specimen head and gauge length on each specimen, indicating no strain influence on its kinetics.

Beside Fe (≈40 wt%), Mo (≈44 wt%) and Cr (≈8 wt%), Laves phase always contained a significant amount of Si (≈5 wt%) and smaller amounts of P, Mn, Nb and V (see also EDS spectrum of Laves Phase in Fig. 7b).

The relatively high Si content of the material batch may have accelerated the formation of this phase: from the comparison of long term ageing of steels having different content of Si it has shown that this element promotes Laves phase formation;[1, 2] also P should accelerate the precipitation of this phase, as it segregates at $M_{23}C_6$ carbide/matrix interfaces, where Lave phase nucleates and grows.

CONCLUSIONS

- Material softening in P91 steel after creep test is caused by dislocation re-organisation and carbide coarsening: the tempered martensite structure of the material in the as received condition undergoes dislocation recovery phenomena with dislocation cell and subgrains formation; reduction of both precipitation and solid solution strengthening mechanisms occurring during creep test accelerate dislocation recovery.
- Carbide coarsening and decrease in carbide surface and volume density yield a reduction of precipitation strengthening mechanism, as well as Laves Phase precipitation and Cr enrichment in carbide $M_{23}C_6$ weakens the contribution of solid solution strengthening owing to alloying element matrix depletion.
- The linear correlation here established between Cr/Fe ratio in $M_{23}C_6$ and the Larson Miller time-temperature parameter may provide a potential tool to non destructively evaluate material ageing during steel service.

REFERENCES

1. C.R. Brinkman, B. Gieseke and P.J. Maziasz: 'The influence of long thermal aging on the microstructure and mechanical properties of modified 9Cr-1Mo steel', *Int. Conf. on Microstructures and Mechanical Properties of Ageing Materials*, Liaw *et al.* eds, The Minerals, Metals & Materials Society, 1993, 107–114.
2. B.V. Gieseke, P.J. Maziasz and C.R. Brinkman: 'The influence of thermal ageing on the microstructure and fatigue properties of modified 9Cr 1Mo steel', *Int. Conf. on Microstructures and Mechanical Properties of Ageing Materials*, Liaw, *et al.* eds, The Minerals, Metal & Materials Society, 1993, 197–205.
3. D.J. Alexander, P.J. Maziasz and C.R. Brinkman: 'The effect of long term aging on the impact properties of modified 9Cr 1Mo steel', *Int. Conf. on Microstructures and Mechanical Properties of Ageing Materials*, Liaw, *et al.* eds, The Minerals, Metal & Materials Society, 1993, 343–348.
4. Z. Yang, M.A. Fong and T.B. Gibbons: 'Steels for Thick Section Parts: Comparison of Economics of Usage in a Typical Design', *EPRI/National Power Conference New Steels for Advanced Plant up to 620°C*, May 1995.
5. R.W. Vanstone, H. Cerjak, V. Foldyna, J. Hald and K. Spiradek: 'Microstructural development in advanced 9–12%Cr creep resisting steels – a collaborative investigation in COST 501/3 WP11', *Micro-struc-*

tural Development and Stability in High Chromium Ferritic Power Plant Steels, A. Strang and D.H. Gooch eds, The Institute of Materials, London 1997.

6. H. Cerjak, P. Hofer and P. Warbichler: 'Microstructural evaluation of aged 9–12%Cr steels containing W', *Int. Symp. Mat. Ageing & Component Life Extension*, Milano, 10–13 Ottobre, 1995.

7. A.B. Ponti, P. Battaini, P. Bianchi and V. Regis: 'New Method for Interparticle Distance Measurements in 2 1/4 Cr 1Mo Ferritic Steels', *Electron Microscopy*, 1986, **35**, 1641–1643

8. R.C. Thomson and H.K.D.H. Bhadesia: Carbide Precipitation in 12Cr1MoV Power Plant Steel, *Met. Trans.*, 1992, **23A**, 1171.

9. H. Chikwanda, A. Strang and M. McLean: 'The Evolution of Carbide Chemistry and Morphology of a 12CrMoVNb Steel and its Influence on Long Term Mechanical Performance', *Proc. of Materials for Advanced Power Engineering, Vol. a, Part 1*, Kluver Academic Publisher, 1994, 291–297

10. A. Benvenuti, P. Bontempi, S. Corti and N. Ricci: 'Assessment of Material Thermal History in Elevated Temperature Components', *Materials Characterisation*, 1996, **36**, 271–278.

11. V. Foldyna, Z. Kubon, A. Jakobová and V. Vodárek: 'Structural Stability of 9–12%Cr Steels for Advanced Power Plant', *Central European & World Connection Electric Power Industry Forum 1995*, R.H. Kozlowski ed., Kraków, October 1995, 156.

Microstructural Stability of Creep Resistant Martensitic 12%Cr Steels

A. STRANG* AND V. VODÁREK†

* GEC ALSTHOM Large Steam Turbines, Rugby, UK
† Materials Research Institute, Vitkovice, Ostrava, Czech Republic

ABSTRACT

Detailed microstructural studies have been carried out on a series of high temperature martensitic 12CrMoVNb steels which exhibited sigmoidal creep rupture behaviour when tested at temperatures of 550°C and above out to durations beyond 100 000 hours. This behaviour has been shown to be associated with marked softening of the materials due to microstructural degradation effects occurring during the creep process. Electron microscopy studies on creep exposed materials have shown that these effects take the form of dissolution of the fine M_2X and MX matrix precipitates, precipitation of M_6X, Z and Laves phases and progressive coarsening of the $M_{23}C_6$ carbides at prior austenite, interlath and subgrain boundaries. These processes result in the creep rupture strength of the steels progressively changing during creep exposure from being initially controlled by precipitation strengthening to being dependent in the long term on solid solution hardening, thus accounting for the observed sigmoidal inflexion in their high temperature creep rupture characteristics.

INTRODUCTION

Long term creep rupture studies on a series of commercial casts of creep resistant 12CrMoVNb steels used for high temperature components in large steam turbine power generation plant have shown that these materials can be microstructurally unstable particularly when exposed at temperatures of 550°C and greater. This behaviour manifests itself in the form of an inflexion in their creep rupture characteristics where a rapid reduction in rupture strength is generally accompanied by a corresponding increase in creep rupture ductility. This phenomenon, which is generally referred to as *sigmoidal* behaviour is well known and thought to be associated with progressive softening and microstructural degradation of the material during long term creep exposure at high temperatures.[1–5] Since these changes can lead to a significant loss in creep rupture strength in service, a clear understanding of the microstructural processes responsible for metallurgical instability in these materials is necessary. This is especially important in the case of alloys which may be particularly prone to this type of behaviour when exposed to modern

117

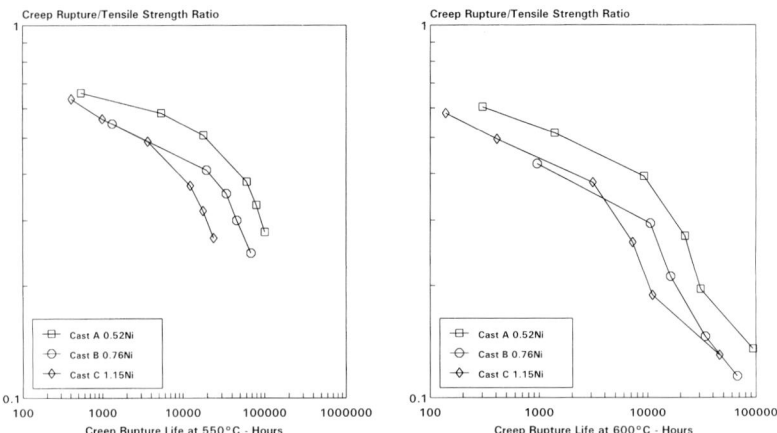

Fig. 1   Creep rupture properties of 12CrMoVNb steels at 550°C and 600°C

steam plant operating conditions, as well as to ensure the safe operation of critical plant components designed for service lives of 250 000 hours or more.

In this paper, the results of detailed transmission and analytical electron microscopy studies on a series of tempered martensitic 12CrMoVNb steels in both the as-received and creep exposed condition are reported. The effects of composition, quality heat treatment and creep exposure on the microstructural degradation and phase evolution kinetics leading to the observed changes in the long-term creep rupture strengths of these alloys are discussed. In addition, the effects of these observations on the possible long-term properties of modern high strength creep resistant 9–12%Cr steels are also considered.

MATERIALS

Three casts of 12CrMoVNb steel were investigated in this study. These were selected from a series of different commercial casts of this alloy, all of which exhibited sigmoidal behaviour in creep rupture tests extending to test durations of up to 100 000 hours at test temperatures of 550°C and 600°C. The three casts chosen had the highest, mean and lowest creep rupture strengths,

Table 1   Chemical compositions of the 12CrMoVNb steels (wt%)

Cast Identity	C	Si	Ni	Cr	Mo	V	Nb	N
A	0.16	0.28	0.52	11.20	0.61	0.28	0.29	0.074
B	0.14	0.37	0.76	11.10	0.57	0.36	0.32	0.062
C	0.14	0.13	1.15	11.74	0.50	0.29	0.30	0.064

Table 2 Commercial heat treatments for the 12CrMoVNb steels

Cast Identity	Heat Treatments	Hardness Hv30
A	Solution Treated 1150°C AC Tempered 6 hours at 650°C AC	346
B	Solution Treated 1150/1160°C AC Tempered 4 hours at 700°C AC	320
C	Solution Treated 1165°C AC Tempered 4 hours at 675°C AC	346

as well as respectively containing the lowest, intermediate and highest nickel contents, viz., 0.52 wt%, 0.76 wt% and 1.15 wt%, Fig. 1.

The compositions, heat treatments, hardnesses and tensile strengths of the steels investigated are shown respectively in Tables 1–3.

Table 3 Room temperature tensile properties of the 12CrMoVNb steels

Cast Identity	$Rp_{0.2}$ MPa	R_m MPa	Elongation %	R in A %
A	831	990	15	43
B	848	967	17	56
C	914	1084	14	46

EXPERIMENTAL PROCEDURES

Optical and transmission electron microscopy studies were carried out on all three casts of the 12CrMoVNb steels in the as-received condition as well as on longitudinal and transverse microsections taken from each of the fractured creep rupture testpieces. Longitudinal Vickers diamond hardness surveys were also performed at a load of 10 kg on microsections taken through the central axis of each of the fractured creep rupture testpieces. The results of these studies were compared with similar hardness measurements on each cast of material in the as-received condition.

Transmission and analytical electron microscopy studies were carried out on carbon extraction replicas and thin foils prepared from both as-received and creep tested materials using a Philips CM20 transmission electron micro-scope fitted with an ultrathin window EDAX 9900 energy dispersive X-ray microanalyser. Quantitative analyses of minor phase particles, extracted on the carbon extraction replicas, were carried out using PMTHIN software and the results normalised to 100% with respect to the elements detected. The carbon extraction replicas were prepared using the standard Smith and Nutting technique,[6] and the thin foils using a Struers Tenupol 3 twin jet electropolishing unit, operated at 60 volts with an electrolyte consisting of 5% $HClO_4$ in glacial acetic acid which was maintained at room temperature.

The microstructural parameters measured for each material in both the as-received and creep tested conditions are summarised in Table 4. Measurements of the size and distribution of minor phases and the sizes of subgrains were carried out on each of the steels using image analyses techniques. Transparencies showing the particle and subgrain structures observed were drawn from montages of thin foil micrographs prepared from each of the samples examined. These were digitised using a Hewlett Packard HP-4C scanner and the resulting TIF images analysed using the feature analysis software of a PGT IMIX EDX microanalyser. Dislocation densities on materials in both the as-received and creep tested conditions were determined using the method proposed by Ham and Sharpe.[7] The total area of micrographs used for image analyses of individual samples varied between 0.9×10^{-4} and $1.8 \times 10^{-4} \, mm^{-2}$.

Table 4 Microstructural parameters measured in the 12CrMoVNb steels in their as-received and creep tested conditions

Minor Phases	Grains / Subgrains
Area equivalent diameter (D_{eqv}) Minimum dimension (D_{min}) Maximum dimension (D_{max}) Aspect ratio (D_{min}/D_{max}) Number of particles/unit area (N_A) Area fraction of particles (A_A)	Minimum width (L_{min}) Maximum length (L_{max}) Aspect ratio (L_{min}/L_{max}) Dislocation density

RESULTS

Microstructure of 12CrMoVNb Steels in the As-Received Condition
Optical microscopy indicated that in the as-received condition all of the 12CrMoVNb steels had fully tempered martensitic microstructures with no evidence of the presence of any δ-ferrite. The average hardnesses of the three casts were found to be in the range 320–346 Hv30, Table 2.

TEM studies of the 12CrMoVNb steels in the as-received heat treated condition revealed extensive precipitation of coarse $M_{23}C_6$ particles at prior austenite and martensite lath boundaries in each cast of material. Large spherodised primary NbX particles were also present randomly dispersed throughout the matrix of each cast. Intragranular precipitation of finer particles of M_2X and MX phases was also found to be present in varying degrees depending on the initial tempering treatment received by each cast of material. The effects of tempering temperature on the type and presence of these additional finely dispersed intragranular phases present in these steels in the as-received heat treated condition are shown in Table 5.

In Cast A, which was tempered at 650°C, only M_2X was observed, while in Cast B, tempered at 700°C, the predominant phase was MX together with a small amount of M_2X. However in Cast C, which was tempered at 675°C, the predominant phase was M_2X together with a small amount of MX. EDX and electron diffraction studies on the M_2X phase indicated that it was rich in Cr and had a crystal lattice which was isomorphous with Cr_2N. Furthermore, EDX analyses on the MX precipitates showed that they in turn were rich in V and Nb and had the general form $(V,Nb)X$, Table 6.[8] The compositions of the $M_{23}C_6$ and M_2X precipitates were also influenced by the tempering treatment, Table 6. In the case of the $M_{23}C_6$ precipitates the Cr and Mo contents decreased and the Fe content increased with increasing tempering temperature. These results are in good agreement with MTDATA studies carried out to predict the equilibrium composition of $M_{23}C_6$ precipitates in 12CrMoVNb and 12CrMoV steels.[9,10] In the M_2X precipitates present in these steels, increases in the V and Nb contents and a decrease in Cr content were observed with increasing tempering temperature. The effects of tempering temperature on the composition of the MX phase could not be determined since this phase was only present in sufficient quantities in Cast B to permit satisfactory EDX analysis. Finally since the NbX particles constituted a primary phase in these steels their composition was not systematically influenced by the tempering treatments.

The compositions of all of the phases present in the 12CrMoVNb steels in the as-received heat treated condition are summarised in Table 6.

EFFECTS OF CREEP EXPOSURE

Hardness Studies on Creep Exposed 12CrMoVNb Steels

The results of the hardness studies on the 12CrMoVNb steels in the as-received condition and following creep exposure at 550°C and 600°C are shown in Fig. 2. These indicated that progressive softening of the materials

Table 5   Minor phases present in the 12CrMoVNb steels in their as-received heat treated condition

Cast Identity	Tempering Temperature °C	Nickel Wt%	Phases			
			NbX	$M_{23}C_6$	M_2X	$(V,Nb)X$
A	650	0.52	√	√	√	–
B	700	0.76	√	√	√*	√
C	675	1.15	√	√	√	√*

*Only small amounts present

Table 6 Compositions of NbX, $M_{23}C_6$, M_2X and MX phases in the 12CrMoVNb steels in their as-received heat treated condition

Cast Identity	Nickel wt %	Phase	Normalised Composition in wt%					
			Cr	V	Mo	Nb	Fe	Ni
A	0.52	NbX	2.4	3.4	0.8	93.4	0	*
B	0.76		1.8	5.0	0.6	91.5	1.10	*
C	1.15		2.7	2.6	0.7	94.5	0.5	*
A	0.52	$M_{23}C_6$	67.8	1.0	8.8	0.6	21.9	0.7
B	0.76		64.0	1.9	7.1	0.6	25.7	0.7
C	1.15		65.6	1.1	7.5	0.6	24.1	1.1
A	0.52	M_2X	71.9	7.1	9.5	4.5	6.7	0.3
B	0.76		65.5	12.9	9.1	6.2	5.3	1.1
C	1.15		69.5	9.7	10.1	6.0	3.6	1.1
A	0.52	MX	*	*	*	*	*	*
B	0.76		12.7	59.3	3.1	18.2	6.2	0.5
C	1.15		*	*	*	*	*	*

*Not analysed

had occurred with increasing creep exposure duration at each of these temperatures. As previously reported for other creep resistant martensitic 9–12%Cr steels the extent of softening in the testpiece heads was consistently found to be significantly less than that found in the creep strained gauge length regions.[5,11] For example, in Cast A containing the lowest nickel content the hardness following creep exposure for 94 000 hours at 600°C, was reduced by 32% in the testpiece head and 48% in the creep strained gauge length compared with that of the material in the as-received condition. Similar results were found for Casts B and C, although in these higher nickel containing casts softening occurred more rapidly and to a greater degree compared with the lower nickel containing Cast A. These results clearly demonstrate that the hardnesses and hence the tensile strengths of all three casts of the 12CrMoVNb steel, are significantly reduced due to the effects of thermal exposure and plastic strain accumulation during creep exposure at these temperatures, with softening occurring more rapidly with increasing nickel content. These changes were also observed to closely follow a sigmoidal pattern similar to that for the creep rupture behaviour already observed in each of these materials, Fig. 2.

Microstructure of Creep Exposed 12CrMoVNb Steels
Electron microscopy studies of Casts A, B and C indicated that whilst evolution of the phases was similar in each material after creep exposure in the range 475°C to 600°C, the kinetics varied significantly from one cast to another. In Cast A, creep exposure at 600°C, resulted in progressive coarsen-

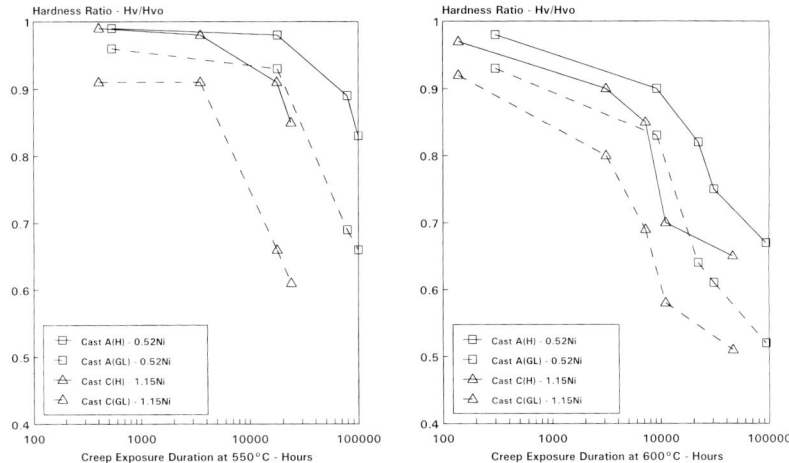

Fig. 2   Hardness changes in 12CrMoVNb steels due to creep exposure at 550°C and 600°C

ing of the $M_{23}C_6$ precipitate together with dissolution of the fine background M_2X phase, accompanied by the simultaneous precipitation of fine platelets of secondary MX of the form (V,Nb)X. This phase also proved to be thermodynamically unstable, gradually redissolving following creep exposure for about 20,000 hours at 600°C. This was due to the simultaneous precipitation of large platelets of a complex nitride having the general formula Cr(V,Nb)N which has recently been shown to be a modified form of Z–phase; a nitride of the form CrNbN, previously only reported in CrNi austenitic steels containing Nb and N.[12–14] Coarsening of the $M_{23}C_6$ and Z–phase particles continued with further creep exposure at 600°C, together with the precipitation of a small amount of Fe$_2$Mo Laves phase, which was present after 94 000 hours at this temperature. X-ray diffractometer studies also indicated that the volume fraction of the primary NbX phase was reduced with increasing exposure at 600°C, eventually reaching approximately 50% of that present in the as-received condition after 94 000 hours at this temperature.[15,16]

Primary NbX, $M_{23}C_6$ and M_2X were also found to be present in Cast A after creep exposure at 550°C at all test durations up to 79 000 hours. At test durations beyond this, whilst primary NbX and $M_{23}C_6$ continued to be present in the microstructure, together with Z–phase and Fe$_2$Mo Laves phase, dissolution of the fine M_2X phase had occurred. At 550°C, Laves phase precipitated after creep exposure at this temperature for approximately 18 000 hours while Z–phase was present after 79 000 hours under the same exposure conditions. In the single testpiece of Cast A examined following creep exposure for 84 000 hours at 475°C, primary NbX, $M_{23}C_6$, M_2X and Fe$_2$Mo

Laves phase were found to be present. No evidence of Z–phase was found after creep exposure at this temperature.

Although the precipitation sequences during creep exposure at 600°C and 550°C, in Cast B containing the higher nickel content, were very similar to those observed in Cast A, the coarsening rate of the $M_{23}C_6$ precipitate appeared to be greater. Furthermore, the creep exposure durations at which the M_2X and MX phases redissolved and the Z–phase was precipitated in Cast B, were much shorter than those observed in Cast A containing the lower nickel content. In addition, whilst Fe_2Mo Laves phase was found to be present in Cast A after long term creep exposure at 475°C, 550°C and 600°C, none of this phase was observed in Cast B following creep exposure at any of these test temperatures. However, M_6X was found to have precipitated in Cast C after creep exposure for approximately 18 000 hours at 550°C and 39 000 hours at 475°C. Furthermore, M_6X was found in Cast B after creep exposure for approximately 68 000 hours at 550°C. No evidence of M_6X was however found to be present in Casts B and C after creep exposure at 600°C, or in Cast A containing the lower nickel content at any of the test temperatures investigated. Comparisons of the phases detected and the precipitation sequences at 600°C and 550°C in each of the 12CrMoVNb steels studied are shown in Tables 7 and 8.

Table 7 Minor phase evolution in the 12CrMoVNb steels after creep exposure at 600°C

Materials	Exposure Conditions Hours	Minor Phases Present					
		NbX	$M_{23}C_6$	M_2X	MX (V,Nb)X	Z- Phase Cr(V,Nb)N	Laves Phase Fe_2Mo
Cast A 0.52 wt% Ni	As received	√	√	√			
	302	√	√	√			
	9 203	√	√	√	√		
	22 052	√	√	√*	√	√	
	30 770	√	√			√	
	94 021	√	√			√	√
Cast B 0.76 wt% Ni	As received	√	√	√*	√		
	960	√	√		√		
	10 558	√	√		√		
	16 237	√	√		√*	√	
	34 204	√	√			√	
	67 260	√	√			√	
Cast C 1.15 wt% Ni	As received	√	√	√	√*		
	3 113	√	√		√		
	7 247	√	√			√	
	11 081	√	√			√	
	46 157	√	√			√	
	74 795	√	√			√	

*Only small amounts present

Table 8 Minor phase evolution in the 12CrMoVNb steels after creep exposure at 550°C

Materials	Exposure Conditions Hours	Minor Phases Present						
		NbX	$M_{23}C_6$	M_2X	MX (V,Nb)X	Z- Phase Cr(V,Nb)N	Laves Phase Fe$_2$Mo	M_6X
Cast A 0.52 wt% Ni	As received	√	√	√				
	541	√	√	√				
	18 042	√	√	√			√	
	79 171	√	√	√		√	√	
	100 538	√	√			√	√	
Cast B 0.76 wt% Ni	As received	√	√	√*	√			
	68 234	√	√		√		√	√
Cast C 1.15 wt% Ni	As received	√	√	√	√*			
	410	√	√	√	√*			
	3 567	√	√	√*	√			
	17 875	√	√			√		√
	24 024	√	√			√		√

*Only small amounts present

Finally, with increasing creep exposure a considerable amount of recovery of the microstructure was observed, particularly in the strained regions of the creep exposed testpieces. For example in Cast A the dislocation density was found to decrease in the gauge length of the creep rupture testpieces by more than an order of magnitude following creep exposure for 94 000 hours at 600°C, viz. from $9.86 \times 10^{-14}\,m^{22}$ to $0.76 \times 10^{-14}\,m^{-2}$.

Equilibrium Studies
The equilibrium minor phases expected as a result of MTDATA studies on the 12CrMoVNb steels in the as-received condition and following long term exposure at 550°C and 600°C are shown in Tables 9 and 10.

For each of the three 12CrMoVNb steels in the as-received heat-treated con-

Table 9 Predicted and observed minor phases in the 12CrMoVNb steels in their as-received heat treated condition.

Cast Identity	Tempering Temperature °C	Nickel Content Wt%	Minor Phases in as-received Condition	
			Predicted	Observed
A	650	0.52	NbC, $M_{23}C_6$, VN	NbX, $M_{23}C_6$, M_2X
B	700	0.76	NbC, $M_{23}C_6$, VN	NbX, $M_{23}C_6$, M_2X, MX
C	675	1.15	NbC, $M_{23}C_6$, VN	NbX, $M_{23}C_6$, M_2X, MX

Table 10   Predicted and observed minor phases in each of the 12CrMoVNb steels following their longest creep exposure durations at 550°C and 600°C.

Cast Identity	Nickel Content wt%	Predicted Equilibrium Minor Phases		Observed Minor Phases	
		550°C	600°C	550°C	600°C
A	0.52	NbC, $M_{23}C_6$ VN	NbC, $M_{23}C_6$ VN	NbX, $M_{23}C_6$ Fe_2Mo, Z–phase	NbX, $M_{23}C_6$ Fe_2Mo, Z–phase
B	0.76	NbC, $M_{23}C_6$ VN	NbC, $M_{23}C_6$ VN	NbX, $M_{23}C_6$, Fe_2Mo M_6X, MX	NbX, $M_{23}C_6$ Z–phase
C	1.15	NbC, $M_{23}C_6$, VN	NbC, $M_{23}C_6$ VN	NbX, $M_{23}C_6$ M_6X, Z–phase	NbX, $M_{23}C_6$ Z–phase

dition, the minor equilibrium phases predicted are primary NbC, $M_{23}C_6$ and VN. However the electron metallographic studies indicate that whilst NbX and $M_{23}C_6$ are present in all of these steels in the as received condition, this is accompanied by varying amounts of MX and M_2X. This appears to be dependent on the tempering temperature used during the quality heat-treatment of these materials, with the lower and higher tempering temperatures favouring the precipitation of M_2X and MX respectively. In the case of Cast A, which was tempered at 650°C, M_2X was the only additional phase present, while in Cast B tempered at 700°C the predominant phase was MX. Finally in Cast C, which was tempered at 675°C, M_2X was initially present together with a small amount of MX.

Primary NbC, $M_{23}C_6$ and VN are also the equilibrium phases predicted for all three casts of the 12CrMoVNb steel following exposure at 550°C and 600°C. Electron microscopy studies indicate that primary NbC and $M_{23}C_6$ are present in all three casts of this steel following long-term creep exposure at 550°C and 600°C. However in tests conducted at 550°C these phases are accompanied by M_2X, Z–phase and Fe_2Mo Laves phase in Cast A, MX, Fe_2Mo Laves phase and M_6X in Cast B and finally by Z–phase and M_6X in Cast C. Similarly, following long-term creep exposure at 600°C, the accompanying phases found in Cast A are Z–phase and Fe_2Mo Laves phase, while in Casts B and C only Z–phase is present.

These results indicate that none of the steels contain all of the predicted equilibrium phases in their quality heat-treated condition or following long-term creep exposure at 550°C and 600°C. Whilst MTDATA would not be expected to predict the precipitation of the newly reported presence of Z–phase in these steels, its failure to predict Fe_2Mo Laves phase and M_6X is surprising, unless of course that these too are also transient and not equilibrium phases. Also these coupled with other electron microscopy studies on 12CrMoV steels indicate that nickel content has a significant effect on the

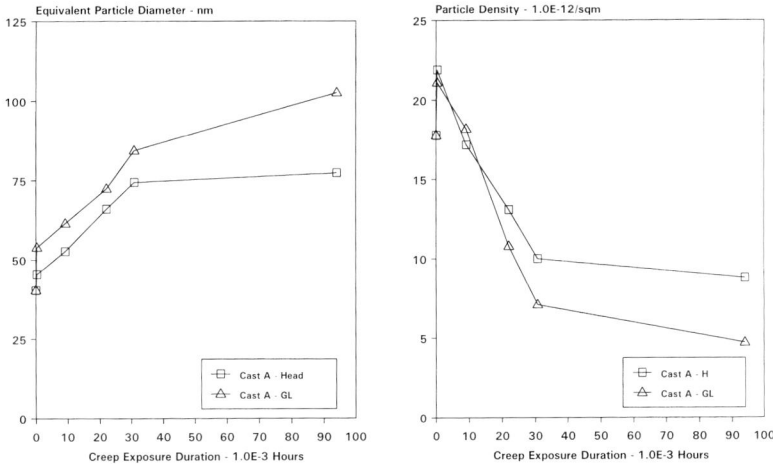

Fig. 3a–b Effect of creep exposure at 600°C on the size and distribution of $M_{23}C_6$ precipitates in 12CrMoVNb steel, Cast A

phases precipitated following long-term creep exposure with M_6X being a likely equilibrium phase forming in steels with nickel contents greater than about 0.8 wt%. At present it is not possible to predict the effects of nickel on the equilibrium phases likely to be formed in creep resistant 9–12%Cr steels using MTDATA or THERMOCALC.

Particle Coarsening Studies
The results of measurements of the equivalent diameter (D_{eqv}) and area density (N_A) of the $M_{23}C_6$ precipitates in the 12CrMoVNb steels after creep exposure at 600°C are shown in Figs 3a–b. In each of the materials examined, particle coarsening was found to be consistently greater in the creep strained gauge lengths of the testpieces compared with their heads, where precipitate growth was primarily due only to the effects of thermal exposure. The results indicate that initially the $M_{23}C_6$ precipitate growth increases at an approximately linear rate with exposure time. However beyond the sigmoidal inflexion point an abrupt reduction in the coarsening rate is observed suggesting a tendency towards a limiting particle size at the longest test durations, Fig. 3a. Furthermore, as the particle size increases, the particle density distribution continuously reduces, at first rapidly and then at a slower rate. These results suggest that there is also likely to be a limiting equilibrium particle size and interparticle spacing for this phase, Fig. 3b. $M_{23}C_6$ precipitate coarsening was also observed in these steels following creep exposure at 550°C, however the coarsening rates were consistently lower at this temperature.

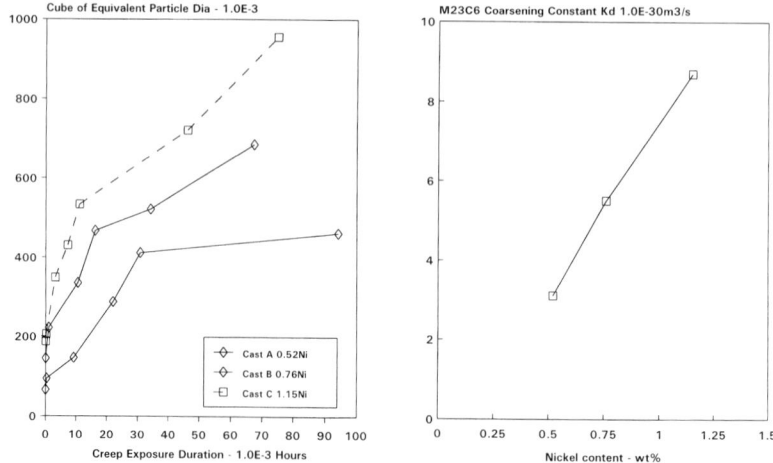

Fig. 4a–b   Effect of nickel content on the coarsening kinetics of $M_{23}C_6$ precipitates in the 12CrMoVNb steels following creep exposure at 600°C

In addition to the observed effects of thermal exposure and creep strain described above, increased $M_{23}C_6$ growth rates were also observed in those casts of steel containing the higher nickel contents, Fig. 4a.

DISCUSSION

The effects of the precipitation sequences on the creep rupture strengths of the 12CrMoVNb steels at 550°C and 600°C can be summarised as follows. In the as-received condition the material is precipitation strengthened by the combined effects of finely dispersed M_2X and/or MX precipitates within the matrix, together with coarser $M_{23}C_6$ particles located at the prior austenite and martensite lath boundaries. With progressive creep exposure at 550°C and 600°C, dissolution of the fine M_2X and secondary (V,Nb)X phases occurs with the simultaneous precipitation of Z–phase and coarsening of the $M_{23}C_6$ precipitate. These processes lead to progressive reductions in both the hardnesses and creep rupture strengths of these steels. The dissolution of the fine matrix M_2X and secondary MX phases, coupled with the precipitation of large Z–phase particles and progressive coarsening and increased interparticle spacing between the $M_{23}C_6$ precipitates, results in the long term creep rupture strength of the material being solely dependent on solid solution hardening. It is this transition from precipitation strengthening to solid solution hardening which accounts for the sigmoidal creep rupture behaviour observed in these steels. Whilst these effects are also observed in tests conducted at 550°C, they occur at much longer test durations at this lower temperature. Finally, in circumstances where Fe_2Mo Laves and M_6X phases are precipitated, an additional reduction in the solid solution hardening elements in the matrix occurs which leads to further long term weakening of the material.

These studies have also shown that in the 12CrMoVNb steels, increased nickel content promotes microstructural degradation by increasing the dissolution rates of M_2X and secondary MX, as well as the coarsening rate of $M_{23}C_6$, thus leading to an increased reduction in creep rupture strength. This is supported by complimentary studies conducted on 12CrMoV steels with nickel contents in the range 0.3 to 1.3 wt%.[17] Furthermore, in the 12CrMoVNb steels, Z–phase precipitation occurs at progressively shorter test durations with increasing nickel content, Tables 7 and 8. In addition, in creep rupture tests conducted at 550°C, M_6X forms in those casts containing the higher nickel contents, apparently in preference to the Fe_2Mo Laves phase. This is also supported by the work referred to above on the 12CrMoV steels.[17] It is therefore clear that in those steels containing raised levels of nickel, viz., >0.5 wt%, microstructural degradation occurs at a faster rate, thus leading to poorer creep rupture strength at shorter test durations. This effectively lowers the potential upper temperature limit at which these materials can be effectively used in service.

It is also clear from the results described above that the driving force for dissolution of the fine matrix M_2X and MX precipitates in the 12CrMoVNb steels exposed at 600°C is the precipitation of the complex nitride, Z–phase. Whilst this also is a contributory factor during creep exposure at 550°C the precipitation of M_6X must also be considered to be important especially on the basis of the observation already made on the high nickel 12CrMoV steels which have been creep exposed at this temperature.[17] The role of nickel in these processes in not immediately clear except perhaps in the case of the M_6X phase which has been shown by EDX microanalysis, unlike $M_{23}C_6$, in 12CrMoV and the 12CrMoVNb steels studied in this work, to be enriched in this element.[17,18]

The results of the $M_{23}C_6$ particle size measurements shown in Figs 3a and 4a clearly demonstrate that the overall relationship between the equivalent particle diameter and the creep exposure duration is non-linear. Furthermore, beyond the sigmoidal inflexion durations the coarsening rate of the $M_{23}C_6$ decreases with increasing exposure time at 600°C, as the microstructure of the material tends towards its equilibrium condition. Most of the microstructural observations reported to date appear to favour Ostwald ripening as the best description of particle coarsening in these creep resistant alloys. This is supported by the fact that in these and other similar steels the volume fraction of $M_{23}C_6$ does not change significantly with exposure duration at 550°C and 600°C.[16,19] If particle coarsening is controlled by this mechanism then the precipitate size would be expected to increase with the nth root of the exposure time, t, i.e. $t^{1/n}$, in accordance with a general equation of the form,

$$d_t^{\,n} - d_o^{\,n} = K_d t \tag{1}$$

where d_t is the precipitate size at time t, d_o is the precipitate size at $t = 0$ and K_d is a temperature dependent rate constant defined by,

$$K_d = K_o \exp(-Q/RT) \tag{2}$$

where Q is the particle coarsening activation energy, R is the gas constant, T is the absolute temperature and K_o is a constant.

In the case where particle coarsening is dominated by volume or bulk diffusion of solutes in the matrix, the particle growth can be specifically described by the Wagner equation,[20] viz.,

$$d_t^3 - d_o^{\,3} = K_d t \tag{3}$$

i.e., the precipitate size increases with the cube root of the exposure time, t, i.e. $t^{1\backslash3}$.

However if other mechanisms, such as interface diffusion, grain boundary diffusion or dislocation pipe diffusion are dominant, particle sizes will increase with the square, fourth and fifth root of exposure time respectively, viz., $t^{1\backslash2}$, $t^{1\backslash4}$ and $t^{1\backslash5}$. The processes whereby precipitate coarsening occurs during creep exposure will be dependent on factors such as the temperature of exposure, the composition of the material, its initial microstructure, the general stability of the microstructure or its resistance to degradation and the amount of creep strain. It is therefore possible that as ageing progresses, interaction of different particle coarsening mechanisms will occur, especially where several phases are likely to be present at any one time during the creep life of the material. Indeed, in complex materials such as the high temperature creep resistant 12Cr steels, which are known to suffer microstructural degradation during exposure at high temperatures, this is most likely to be the case. In these materials there is a strong probability that the particle coarsening mechanism which is dominant during the early stages of the creep life will be replaced or complemented by another as the microstructure of the material changes.

Evaluation of the $M_{23}C_6$ coarsening data in terms of the various mechanisms described above indicated that whilst none describe all of the data satisfactorily, a reasonable comparison can be made by assuming $t^{1\backslash3}$ growth, Fig. 4a. This reveals the presence of inflexions in the data at similar durations to that found for the sigmoidal inflexions in the creep rupture data. Furthermore, the results show that a $t^{1\backslash3}$ growth law satisfactorily describes the data up to test durations corresponding to the observed inflexions in the creep rupture data with the K_d value for Cast C with the high nickel content being approximately three times that observed for Cast A, viz., $K_{d(Cast\ C)}$ = 8.7 × 10^{-30} m³s⁻¹, $K_{d(Cast\ A)}$ = 3.1 × 10^{-30} m³s⁻¹, Fig. 4b. Beyond the sigmoidal inflexions this mechanism breaks down resulting in much reduced coarsening rates being observed for the $M_{23}C_6$ precipitate in all three casts of the 12CrMoVNb steel. Finally this analysis clearly illustrates the detrimental effects of nickel, with the $M_{23}C_6$ coarsening rates consistently increasing with increasing nickel contents in these steels.

CONCLUSIONS

The results of the microstructural investigations conducted on the 12CrMoVNb steels in the as-received condition and after long term creep exposure in the temperature range 475°C to 600°C can be summarised as follows, viz.,

1. In the as-received condition all of the 12CrMoVNb steels exhibited tempered martensitic microstructures containing $M_{23}C_6$ precipitated at the prior austenite and martensitic lath boundaries, with varying amounts of fine matrix M_2X and MX precipitation depending on the initial heat treatment given to the material. The 12CrMoVNb steels also contain primary NbX precipitates randomly dispersed throughout the matrices of each cast in the as-received condition.

2. Long-term creep exposure causes marked softening to occur in each cast of this steel, with this being greater in the strained portions of the testpieces compared with their heads. This is consistent with the observations made by other workers on high temperature low and high alloy creep resistant ferritic steels.

3. Long term exposure of the 12CrMoVNb steels at 600°C results in dissolution of the fine matrix M_2X and MX precipitates accompanied by precipitation of coarse plate-like Z–phase particles, together with Fe_2Mo Laves phase. Coarsening of the $M_{23}C_6$ precipitates also occurs coupled with a significant increase in the interparticle spacing of this phase. Furthermore, creep exposure at 550°C results in an increased amount of Fe_2Mo Laves or M_6X phases present.

4. Increasing nickel contents cause increased $M_{23}C_6$ coarsening rates in the 12CrMoVNb steel as well as accelerating the dissolution of the fine matrix strengthening M_2X and MX precipitates. Higher nickel contents also

favour the precipitation of Z–phase and M_6X to the detriment of the fine M_2X and MX precipitates.

5. In the 12CrMoVNb steels the $M_{23}C_6$ coarsening rate due to creep exposure at 600°C is approximately three times greater in the highest nickel (1.15 wt%) material compared with the cast with the lowest nickel (0.52 wt%) content.

6. The creep rupture strength of this steel is initially controlled by precipitation strengthening due to the presence of fine M_2X and MX particles in the matrix. Progressive dissolution of the fine M_2X and MX particles occurs during creep exposure driven by the precipitation of Z–phase and M_6X. This coupled with coarsening of the $M_{23}C_6$ precipitate results in the long term rupture strength of this steel becoming finally dependent on solid solution hardening. The transition from precipitation to solid solution strengthening is the primary cause of the observed sigmoidal behaviour in this steel. Increased nickel contents cause accelerated microstructural degradation in both types of steel, resulting in the sigmoidal inflexion occurring at progressively shorter creep exposure durations.

7. In order to minimise microstructural degradation, the nickel content of creep resistant 12Cr steels should be restricted to a maximum of 0.5 wt%. In addition reduction of the Nb content would limit the precipitation of Z–phase and thus promote retention of the fine beneficial M_2X and MX precipitates in the matrix.

REFERENCES
1. A. Wickens, A. Strang and G. Oakes: *Proc. Int. Conf. on the Engineering Aspects of Creep*, Sheffield, 1980.
2. J.H. Bennewitz: *Proc. Joint Int. Creep Conference*, Inst. Mech. Engs., New York/London, 1963.
3. A. Hede and B. Aronsson: *JISI*, 1969, **207**, 1241.
4. T. Marrison and A. Hogg: *J Metal Society*, Sept 1972, **No 151**.
5. B.W. Roberts and A. Strang: *Proc. Conference on Refurbishment and Life Extension of Steam Plant*, Inst. Mech .Engs., London, October 1987.
6. E. Smith and J. Nutting: *British Journal of Applied Physics*, 1956, **7**, 214.
7. R.K. Ham and N.G. Sharpe: *Phil. Mag.*, 1961, **6**, 1183.
8. A. Strang and V. Vodarek: 'Precipitation Process in Martensitic 12CrMoVNb steels During High Temperature Creep', *Microstructural Development and Stability in High Chromium Ferritic Power Plant Steels*, The Institute of Materials, London, 1997, 31–52. Robinson College, Cambridge, UK, June 1995
9. J. Hald: *Private Communication*, T.U. Denmark (ELSAM), 1995.

10. R.C. Thomson and H.K.D.H. Bhadeshia: *Met. Trans. A.*, 23A, April 1992, *23A*.
11. T. Goto: *Mitsubishi Technical Bulletin*, 1985, **169**, 1–11.
12. A. Strang and V. Vodarek: Materials Science and Technology, 1996, **12**, 552.
13. H. Gerlach: *Proc. Int. Conference on High Temperature Properties of Steels*, Iron and Steel Inst., Eastbourne, April 1966.
14. P.G. Stone: *Proc. Int. Conference High Temperature Properties of Steels*, Iron and Steel Inst., Eastbourne, April 1966, 505–511.
15. H.K. Chickwanda: *Microstructural Stability of 12CrMoVNb Power Plant Steels*, PhD Thesis, Imperial College , London, January 1994.
16. H.K. Chickwanda, A. Strang and M. McLean: *Int. Conference on Materials for Advanced Power Engineering*, Liege, October 1994.
17. V. Vodarek and A. Strang: Effects of Nickel on the Precipitation Processes in 12CrMoV Steels During Creep at 550°C Submitted to *Scripta Met.*, Feb 1997.
18. A. Strang and V. Vodarek: *Unpublished Results.*
19. Y. Kadoya, B. Dyson and M. McLean: *Private communication*, 1997.
20. C. Wagner: *Z Electrochemie* 1961, **65**, 243.

Quantitative Comparison of the Microstructures of High Chromium Steels for Advanced Power Stations

P.J. ENNIS,[1] A. ZIELIÑSKA-LIPIEC,[2] AND A. CZYRSKA-FILEMONOWICZ[2]

[1] *Research Centre Jülich, Institute for Materials in Energy Systems 1, Jülich, Germany*
[2] *University of Mining and Metallurgy, Faculty of Metallurgy and Materials Science, Kraków, Poland*

ABSTRACT

The high chromium steels developed for advanced power stations, 12Cr1MoV, P91 (9Cr1MoVNbN) and P92 (9Cr0.5Mo1.8WVNbN), have been investigated. Quantitative microstructural investigations (subgrain width, dislocation density, particle size distribution) have been carried out using transmission electron microscopy (TEM). The importance of the fine precipitates in promoting creep strength was demonstrated by comparing the values for the basic 9Cr1Mo steel with the more highly alloyed steels. In short term tests, P91 and P92 exhibited similar stress rupture strengths at 600°C, while 12Cr1MoV has somewhat lower strength. The strengths in short term tests were correlated with the mean size of the carbide and carbonitride precipitates. In long term tests (creep tests at lower stresses), P92 exhibited a significantly higher stress rupture strength than P91 and the difference increased with increasing test duration. The reasons for the higher strength of P92 are discussed.

INTRODUCTION

Increased efficiency and improved environmental protection and lower costs are the innovative driving forces in the development of power generating plants. The economic requirement for improved exploitation of fossil fuels and the political pressures to protect the environment by significantly decreasing the carbon dioxide emissions of power generating plants have led over the last twenty years to a continuous increase in the thermal efficiency of power stations. This has required application of materials with improved high temperature properties, especially good creep properties.

In the recent years, the 9–12%Cr martensitic steels have been widely used in the power generating industry. Besides the 12Cr1MoV steel, which has been widely used in Germany, a 9Cr–1Mo steel containing 0.2V, 0.05Nb and 0.05N, ASTM designation P91, was developed in the USA.[1] In Japan a new 9%Cr steel (manufacturer's designation NF616, ASTM designation P92)

135

which contains 0.5Mo and 1.8W in addition to 0.2V, 0.06Nb and 0.05N has been introduced.[2,3] The microstructure of the 9–12%Cr steels in the austenitised and tempered condition consists essentially of tempered martensite with a high dislocation density and finely distributed carbides and carbonitrides. For the application of these materials in steam power plants which operate at around 600°C and at pressures up to 300 bar, the microstructural stability during service is of great importance. The following structural features are expected to exert an influence on the creep rupture properties of the steel:

– the dislocation density within the martensite laths, which is characterised by narrow-mesh cell or subgrain structures;
– precipitation of $M_{23}C_6$ on prior austenite grain boundaries and on the subgrain boundaries;
– fine, uniformly dispersed carbides and carbonitrides within the structure;
– formation of intermetallic phases, such as Laves phase.

In the present work, quantitative microstructural investigations (subgrain width, dislocation density, particle size distribution) have been performed by TEM in order to characterise the as received microstructures of the 12Cr1MoV, P91 and P92 steels. The objective was to define and determine quantitatively the structural parameters that influence the creep strength, so that the relationship between microstructure and creep rupture behaviour could be established.

EXPERIMENTAL DETAILS
Details of the test materials are given in Table 1. The 12Cr1MoV and P91 steels were supplied by Mannesmann and the P92 steel by Nippon Steel Corporation.

The microstructures of the materials in the as received conditions were investigated by optical metallography and transmission electron microscopy of thin foils and extraction double-replicas.[7] Thin foils were used for statistical determinations of subgrain width (measured perpendicularly to the long axis of the martensite laths) and for investigation of the dislocation density. The dislocation density within the subgrains was measured using the linear intersection method described by Ham[8] and by Klaar.[9] Extraction double-replicas were examined to provide information on the precipitate composition and to allow quantitative analysis of carbide size distributions from large areas of the specimen. Enlarged TEM photomicrographs were examined using a line-intercept technique on a Quantimet 720 (Cambridge Instruments). For size distribution analysis, separate measurements were made for the larger (mainly $M_{23}C_6$) precipitates and for the smaller MX precipitates.

Table 1 Details of the test materials

element	chemical composition, wt%		
	12Cr1MoV	P91	P92
C	0.20	0.10	0.124
Si	0.28	0.38	0.02
Mn	0.56	0.46	0.47
P	0.012	0.020	0.011
S	0.006	0.002	0.006
Cr	11.60	8.10	9.07
Mo	0.89	0.92	0.46
W	–	–	1.78
V	0.27	0.18	0.19
Nb	–	0.073	0.063
B	–	–	0.003
N	–	0.049	0.043
Ni	0.69	0.33	0.06
Al	0.18	0.034	0.002
Form and dimensions, mm	pipe segment, 65 wall thickness	pipe, φ 159, 20 wall thickness	pipe, φ 300, 40 wall thickness
Heat treatment: austenitisation + tempering	10 min/1050°C + 10 min/770°C, air cooled	0.5 h/1050°C + 1 h/750°C, air cooled	2 h/1070°C + 2 h/775°C, air cooled
100 000 h stress rupture strength at 600°C, MPa*	59	94	115

* values for 12Cr1MoV from ISO Technical Report 7468;[4] for P91 from Canonico,[5] for P92 from Wachter *et al.*[6]

RESULTS AND DISCUSSION

Microstructural Analysis

Qualitative TEM investigations showed similar microstructures of tempered martensite for all three steels examined. Figure 1 shows the microstructures of the P91, P92 and 12Cr1MoV specimens after austenitisation and tempering (the heat treatments applied are given in Table 1). Austenitisation produced a martensitic structure with a high dislocation density within the martensite laths. During tempering, two main processes, recovery of the martensite and precipitation of carbides, nitrides and carbonitrides, seemed to be taking place. Recovery processes led to the formation of subgrains and dislocation networks. Abe *et al*[10] reported that the creep strength of 9%Cr steels is correlated inversely with the martensite lath width and therefore with the subgrain size (the original martensite lath boundaries form the subgrain boundaries). Measurements of the average subgrain width $\bar{w}$ and of the dislocation density within the subgrains are presented in Table 2. It can be seen that $\bar{w}$ is fairly

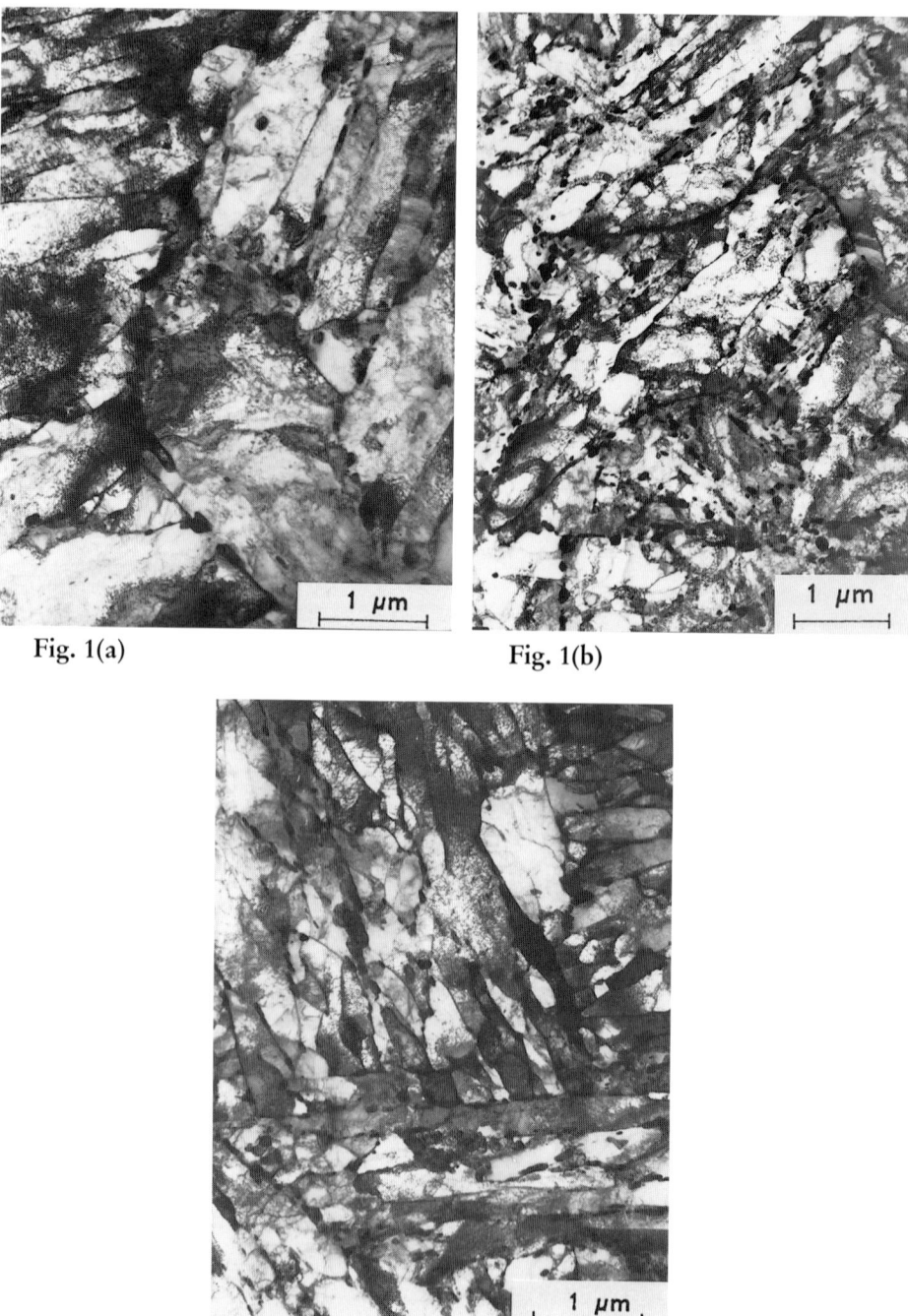

Fig. 1(a)   Fig. 1(b)

Fig. 1(c)

Fig. 1 TEM thin foils of 9–12% chromium steels; (a) P91; (b) P92; (c) 12Cr1MoV.

Table 2 Dislocation density ρ and mean subgrain size w̄ in P91, P92 and 12Cr1MoV steels

steel	dislocation density ρ, m^{-2}	mean subgrain size w̄, μm	standard deviation $S_{\bar{w}}$	number of subgrains
P91	$(7.52 \pm 0.8) \times 10^{14}$	0.40	0.06	64
P92	$(7.46 \pm 0.8) \times 10^{14}$	0.42	0.09	59
12Cr1MoV	$(6.19 \pm 0.8) \times 10^{14}$	0.35	0.04	77

similar in all steels investigated. The small differences in w̄ are connected with different prior austenite grain size. The dislocation densities of P91 and P92 steels were nearly the same, and somewhat higher than that of 12Cr1MoV.

The precipitation of carbides, nitrides and carbonitrides occurred during tempering. In all three steels examined, $M_{23}C_6$ containing Cr, Fe, Mo (W) precipitated preferentially on the prior austenite grain boundaries and on the martensite lath boundaries. These precipitates retard the subgrain growth and therefore increase the strength of the materials. The important precipitates for the mechanical properties of 9–12%Cr steels are the fine MX (where M = Nb or V and X = C and/or N). They pin free dislocations in the matrix leading to increasing creep strength of these steels. In P91 steel mainly spheroidal Nb-rich carbonitrides were observed within the martensite laths. In the P92 steel three types of MX precipitate were present: fine spheroidal Nb-rich carbonitride and plate-like V-rich nitride precipitates were found, as well as some complex carbonitrides consisting of a spheroidal Nb-rich particle to which a V-rich particle was attached, forming 'V-wings'. The 12Cr1MoV steel contained, besides VX precipitates, some particles of M_2X. This phase has also been found in the 9%Cr steels after austenitisation, but during the very early stage of tempering, the M_2X dissolves and is replaced by MX.[6] The particle dimensions were measured from the extraction double replicas, using a Quantimet 720 instrument. The results for the coarse $M_{23}C_6$ and the fine MX are shown in Table 3.

Table 3 Statistical measurements on the precipitates in 12Cr1MoV, P91 and P92 steels

steel	larger particles ($M_{23}C_6$)			smaller particles (MX)			complex M(C,N)		
	d̄	$S_{\bar{d}}$	N	d̄	$S_{\bar{d}}$	N	l̄	$S_{\bar{l}}$	N
P92	90	58	1071	22	12	230	78	15	17
P91	99	72	1659	16	7	686	not present		
12Cr1MoV	170	90	796	63	32	90	not present		

d̄: mean particle diameter; in nm, l̄: length of major axis, in nm, $S_{\bar{d}}$, $S_{\bar{l}}$: standard deviations of d̄ and l̄, N: number of particles measured.

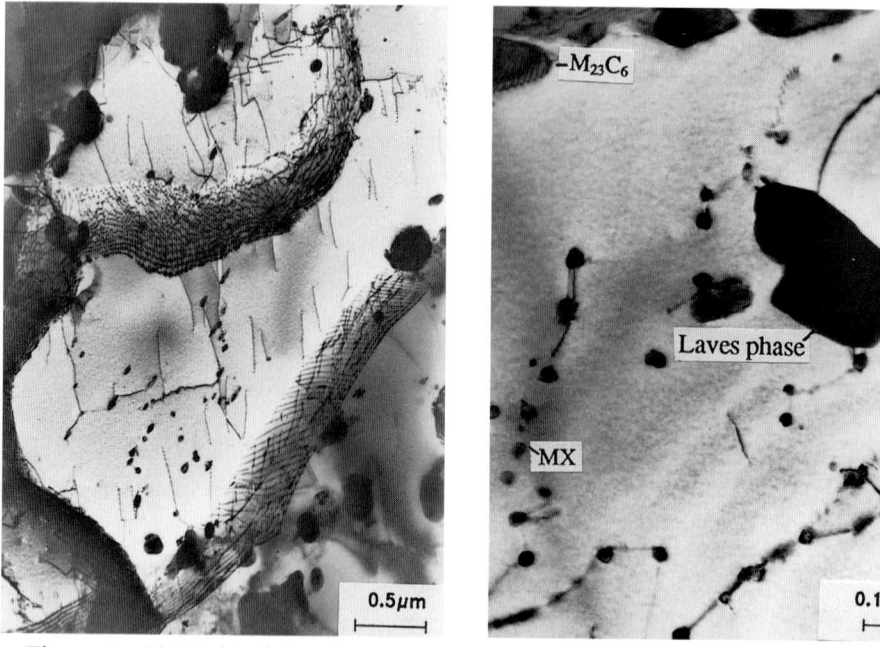

Figure 2 TEM thin foils of P92 creep tested specimens: (a) tested at 600°C, 160 MPa, 9755 h, 22.8%; (b) tested at 650°C, 104 MPa, 6468 h, 21%, showing Laves phase precipitates.

It can be seen that the mean diameter of the fine MX particles is similar in the P91 and P92 steels and much smaller than in 12Cr1MoV steel. The smallest $M_{23}C_6$ were found in P92 steel.

Examination of the microstructure of creep tested specimens showed that with increasing test duration the dislocation density within the subgrains

Table 4 Quantitative metallographic data for P92 specimens creep tested at 600°C

specimen	stress MPa	duration h	creep strain %	dislocation density 10^{14} m^{-2}	subgrain width, μm	mean diameter $M_{23}C_6$, nm	mean diameter MX, nm
AR	–	–	–	7.5 ± 0.8	0.4 ± 0.1	90 ± 12	22 ± 2
creep	100	1014	0.3 T	5.4 ± 0.5	0.6 ± 0.1	nd	nd
creep	160	1502	1.8 T	5.3 ± 0.6	0.7 ± 0.1	119 ± 8	18 ± 1
creep	160	9755	22.8 R	2.5 ± 0.5	1.4 ± 0.1	135 ± 12	21 ± 2
creep	145	17551	19.4 R	2.3 ± 0.5	1.3 ± 0.1	143 ± 12	22 ± 2

AR: as received
T: test terminated before rupture; TEM specimens from specimen middle
R: specimen ruptured; TEM specimens 2–4 mm from fracture face
nd: not determined

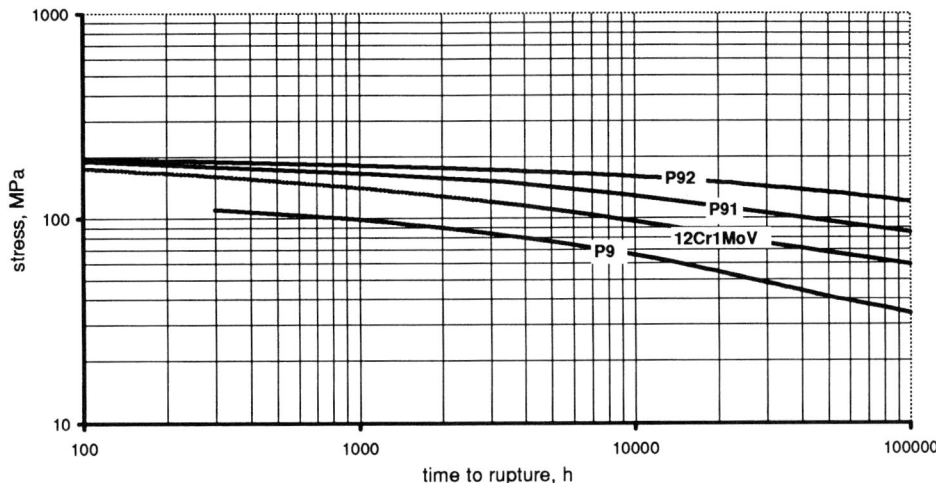

Figure 3 Stress rupture strengths of P9 (9Cr1Mo), P91, P92 and 12Cr1MoV at 600°C.

decreased. The $M_{23}C_6$ precipitates also increased in size but the fine MX precipitates did not appreciably coarsen. The quantitative metallographic data are summarised in Table 4. Figure 2 shows typical microstructures of creep tested specimens. It should be noted that in P92, the Fe–W–Mo Laves phase was found after testing for longer times at 600 and 650°C.

Creep Rupture Behaviour
Figure 3 compares the stress rupture strengths of the three steels. The data for 12Cr1MoV have been taken from the ISO Report 7468,[4] for P91 from[5] and for P92 from.[6] On the basis of the microstructural studies, it could be expected that the 12Cr1MoV steel would exhibit the lowest stress rupture strength due to the larger mean diameter and smaller volume fraction of the MX type precipitates. Figure 3 shows that this is indeed the case.

In short term tests, P91 and P92 exhibited similar stress rupture strengths at 600°C, which correlates with the fact that the fine MX and the coarser $M_{23}C_6$ precipitates in the two 9% chromium steels were very similar. In long term tests (creep tests at lower stresses), P92 exhibited a significantly higher stress rupture strength than P91 and the difference increased with increasing test duration. There are at least two possible explanations for the higher strength of P92. Firstly, the recovery of the martensitic structure with an associated decrease in dislocation density may proceed more slowly in P92 than in P91 due to the addition of W, which increases the A_{c1} transition temperature, producing a more stable martensite. Evidence for this effect is given by the results of the determination of dislocation densities in P91 and P92

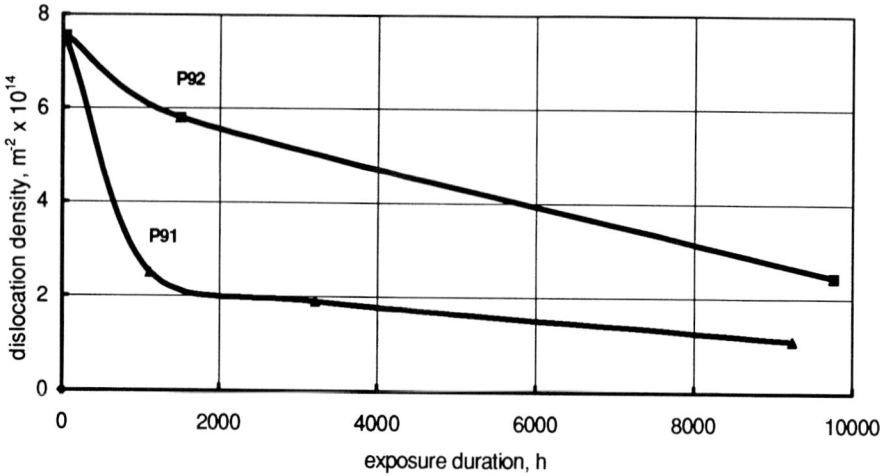

Figure 4 Change in dislocation density with creep test duration for P91 and P92 at 600°C.

after creep testing. Figure 4 shows that the dislocation density in P92 decreases more slowly than that of P91 with increasing test time. After about 10 000 h, the dislocation density in P92 was found to be twice that in P91. Secondly the precipitation of the Laves phase $Fe_2(Mo,W)$ at the prior austenite and subgrain boundaries may be contributing to the strengthening of these boundaries, preventing boundary sliding, which is an important deformation mode at the low stresses typical for actual service.

SUMMARY
Statistical analyses of the microstructures were carried out using both thin foils and extraction replicas. The short term creep rupture properties could be correlated with the size and distribution of fine carbide and carbonitride precipitates in the steels examined. In long term tests, P92 exhibited a significantly higher creep rupture strength than P91 and the difference increased with increasing test duration.

ACKNOWLEDGMENT
The study has been performed under Polish-German bilateral co-operation 'Optimisation and Properties of Materials for Applications in Energy Systems.' The authors appreciate the support of the Polish Committee for Scientific Research (KBN) in Poland and the International Office, BMBF in Germany.

REFERENCES
1. V.K. Sikka, C.T. Ward and K.C. Thomas: *Proc. Conf. Ferritic Steels for*

High Temperature Applications, Warrendale, PA, USA, 6–18 October 1981; ASM 1983, 65–84.

2. K. Fujita: *Conf. Proceedings of the Third International Charles Parsons Turbine Conference on Materials Engineering in Turbines and Compressors*, 25–27 April 1995, Newcastle-upon-Tyne, UK, Vol 2, 1995, 493–516.

3. R. Blum, J. Hald, W. Bendick, A. Rosselet and J.C. Vaillant: 'Newly Developed High Temperature Ferritic-Martensitic Steels from USA, Japan and Europe', *International VGB Conference on Fossil Fuelled Steam Power Plants with Advanced Design Parameters*, 16–18 June, 1993, Kolding, Denmark.

4. 'Summary of average stress rupture properties of wrought steels for boilers and pressure vessels', *ISO Technical Report 7468*, 1981, Ref. No. ISO/TR 7468–1981 (E).

5. D. Canonico: *Second Int. Conf. Interaction of Steels with Hydrogen in Petroleum Industry Pressure Vessel and Pipeline Service*, 19–21 October 1994, Vienna; Proc. vol 2, 607–618.

6. O. Wachter, and P.J. Ennis: *Report of the Research Centre Jülich Jül–3074*, June 1995, ISSN 0944–2952.

7. A. Czyrska-Filemonowicz, K. Spiradek and S. Gorczyca: *Praktische Metallographie*, 1991, **22**, 217–226.

8. K.R. Ham: *Philosophical. Mag.*, 1961, **6**, 1183.

9. H.J. Klaar, P. Schwaab and W. Oesterle: *Praktische Metallographie*, 1992, **29**, 3–25.

Condition Assessment of Long Serviced High Temperature Components

R.D. TOWNSEND, R. TIMMINS, D.M. FINCH AND
J.M. BREAR
ERA Technology Ltd, Cleeve Road, Leatherhead, Surrey KT22 7SA

ABSTRACT

This paper provides a brief review of the methods used to predict the remaining creep life of low alloy steels based on either an assessment of the degree of structural degradation during high temperature exposure or by post service testing. In parent material structures, structural degradation is mostly associated with coarsening of the alloy carbide distributions and causes a reduction in the creep strength. Early models are described whereby the creep properties were related to the interparticle spacing. This correlation however, was experimentally difficult to apply and the specimen hardness, which is a function of the interparticle spacing, was selected as a more practical means of assessing the degradation.

The paper goes on to describe a major nine year Joint Industry Project performed at ERA aimed at improving and evaluating the techniques of condition assessment and life extension for high temperature pressure parts. The object was to develop and validate non-destructive and semi-destructive techniques for post-service condition assessment based on micro-structural changes in new and ex-service 1–1.25Cr0.5Mo and 2.25Cr1Mo materials and relate these to the dominant degradation mechanisms. The ex-service materials had been operated for times out to 246 000 h in the temperature range 500–565°C.

The various assessment methods have been classified into *four categories* according to the accuracy they achieve and the need for direct access to the plant for inspection and sampling:

Category 1 Assessments use standard material data and require details of the component operating history in terms of temperatures and stresses as inputs.

Category 2 Assessments use non-destructive data (i.e. hardness or cavitational damage status) and require access to the plant component for non-destructive inspection.

Category 3 Assessments use heat specific thermal softening data (i.e. hardness reduction with time at several temperatures) to generate the parameters required in a thermal softening model, and thus require small samples removed from the component.

Category 4 Assessments use heat-specific rupture/creep data (i.e. data obtained at several conditions of stress and temperature) to generate the parameters for creep and rupture life equations and thus require the removal of samples from the component large enough for miniature creep samples. This category includes the traditional method of assessing creep life using temperature accelerated iso-stress tests.

Progression to a higher category level assessment places more demanding requirements for access to the plant for inspection and material sampling, with concomitant increase in costs, but provides less conservative, more accurate and generally longer life predictions. Case studies are provided which show that the use of higher category methods can extend the predicted remaining life by up to an order of magnitude.

1. INTRODUCTION

Procedures for predicting creep life were first developed in the CEGB (Central Electricity Generating Board) and applied to critical components such as high temperature steam headers.[1,2] In Generation Operation Memorandum, GOM 101A 'Creep Life Prediction of Headers' a three stage assessment procedure was adopted in order to achieve a conservative but cost effective methodology:

Stage 1 – the creep life consumed was determined through a code-based calculation, review of the plant operating and maintenance history in terms of operating hours, steam temperatures and pressures, minimum stress rupture data and component dimensions as in design.

Stage 2 – as stage 1, but with dimensional checks on the component and measurement of actual metal temperatures.

Stage 3 – based on hands-on inspection of the component involving conventional and advanced NDE, metallographic assessments of life, detailed stress analysis and post service testing of samples taken from the component.

Stages 1 and 2 were regarded as preliminary stages in which the creep lives were calculated using 'Robinson's Life Fraction Rule'.[3] These stages were intended to identify *'problem headers'* predicted to fail within the required life of the boiler which would need the more detailed and less conservative assessments of Stage 3 to improve the accuracy of the life prediction. The creep lives calculated under Stages 1 and 2 were regarded generally as highly pessimistic and therefore conservative. This was due to the use, in the calculations, of minimum stress rupture data and peak temperatures, also because of the inherent conservatism in the stress calculations and that no allowance was made for stress redistribution, and because metal thicknesses frequently exceed design and hence result in lower applied stresses. Stages 1 and 2 were also mainly non intrusive (limited need for direct access to the component) and therefore, of minimum cost. It was recognised therefore that within Stage 3, there is considerable scope for improving not only the accuracy of the assessed creep life, but also of significantly increasing the safe operating period. A large number of different methods have been developed to do this, Townsend.[4] Those considered in this paper are based on life prediction by the quantitative assessment of microstructural degradation in the assessed component. These methods require the development of models relating the creep properties to the micro-structure and the use of post-service testing of samples extracted from the component.

2.   LIFE PREDICTION BASED ON STRUCTURAL EVOLUTION

The most common materials used in high temperature components in power plant and petrochemical plant have been the low alloy steels of the type 1–1.25Cr0.5Mo, 2.25Cr1Mo and 0.5Cr0.5Mo0.25V. These steels have been used extensively world wide because of their good creep resistance and relatively low cost. In all cases these materials enter service in either the normalised and tempered condition or in the annealed condition and their high temperature properties such as creep strength, ductility and crack growth resistance, depend critically on the type and morphological distribution of alloy carbides and to a lesser extent on solid solution strengthening by elements such as molybdenum. The initial micro-structure of these steels is controlled by the composition and by the original heat treatment and subsequently modified during high temperature service. The properties also depend on the purity of the steel which controls the amount of residual elements such as S, P, Sn, As and Sb which have a pronounced effect on the susceptibility to various forms of embrittlement[5] and on the steel making process which controls the presence and distribution of non metallic inclusions such as MnS which determine the propensity to creep cavitation.[6]

The range of structures attained in the low alloy steels after the initial heat treatment can be extremely complex and depend on the composition, the time and temperature achieved during austenitising and on the cooling rate. Thus in the annealed condition the structures will be predominantly ferritic with only small amounts of bainite (10–20%). Increasing the cooling rate or the austenitising temperature, which increases the austenite grain size, increases bainitic hardenability and the amount of bainite can increase up to (40–60%) in a normalised structure or to 100% bainite in the HAZ of a weld.

A common feature with the low alloy ferritic steels is that they enter service in a metallurgical condition in which the major phase changes accompanying the decomposition of austenite are complete and the precipitation of alloy carbides well advanced. However, the steels are in a highly active metastable condition and further carbide precipitation accompanied by changes in composition and coarsening occurs throughout high temperature exposure. These changes in carbide morphology invariably lead to changes in the properties of the steel and in general to a weakening of the structure. In principle they thus provide a 'built-in clock' within the structure from which it should be possible to relate the strength of the material to the morphological state of the carbide distributions.

The role of carbides in low alloy creep-resistant steels has been described comprehensively by Woodhead and Quarell[7] and subsequently by many other workers. Baker and Nutting[8] determined the precipitation sequence in 2.25Cr1Mo during tempering or service at elevated temperatures. Different carbides were found to evolve in the bainitic and ferritic areas of the microstructure.

$$\text{In Ferrite} \qquad \text{In Bainite} \qquad M_7C_3 + M_6C$$
$$\uparrow$$
$$M_2C \rightarrow M_6C \qquad \epsilon \text{ carbide} + M_3C \rightarrow \quad M_3C + M_2C \rightarrow M_{23}C_6$$

In this particular material, M_3C and M_2C are not the equilibrium carbides and marked compositional changes occur until the eventual formation of the equilibrium carbide for this steel M_6C. In 1Cr 0.5Mo steel M_3C is the first formed carbide in the normal ferrite-pearlite structure and M_2C and M_7C after long service exposure. Because of the lower Cr and Mo contents than in 2.25Cr1Mo, the highly alloyed $M_{23}C_6$ carbide does not apparently form in this material, Biss and Wada.[9]

It is generally recognised that all the creep resistant alloys depend primarily on finely dispersed particles for their strength and that the inter-particle spacing is an important factor controlling the movement of dislocations. In ferrite areas the dominant carbide to form in molybdenum steels is Mo_2C, whereas in chromium steels it is M_7C_3 and in vanadium steels V_4C_3. A fine dispersion of these carbides gives rise to secondary hardening peaks during tempering and initially to high creep resistance. During high temperature exposure, solid state diffusion occurs and the particles coarsen to reduce their overall surface energy and for a given volume fraction, this coarsening is accompanied by an increase in the mean inter-particle spacing. It is generally recognised that the reduction in creep strength associated with degradation is responsible for the initiation of tertiary creep in low alloy steels and McLean[10] has calculated the critical spacing above which the creep rate increases rapidly. This concept has been applied to a number of models relating strain rate changes to structural degradation[11-13] in low alloy steels. The model was comprehensively developed and investigated at CERL (the CEGB Central Electricity Research Laboratories) and its validity tested under a project sponsored by EPRI (Electric Power Research Institute).[14,15]

The basis of the model is that the creep rate during tertiary creep is dependant on a changing effective stress σ_{eff} which is itself a consequence of coarsening of the precipitate dispersion and possible accumulated creep
cavitational damage. The model is based on the Dorn expression for creep
rate:

$$\dot{\epsilon} = A\sigma_{eff}^{n} = A(\sigma - \sigma_o)^n \qquad (1)$$

Where A is a temperature dependent term and the effective stress is the difference between the applied stress σ and a threshold term σ_o which has its
physical origin in the interaction between moving dislocations and precipitate
particles. The threshold stress is inversely proportional to the interparticle
spacing, Evans and Knowles:[16]

$$\sigma_o = \frac{\alpha' \mu b}{\lambda} \qquad (2)$$

where α' is a constant, μ the shear modulus and b is the Burger's vector.
The change in (λ) occurs through particle coarsening and follows classical
coarsening kinetics, Greenwood.[17]

$$r^3 = r_o^3 + c't \qquad (3)$$

where r_o is the particle radius at time $t = 0$ and c' a temperature dependent
constant.

Noting for constant volume: $r = G\lambda$

where G is a geometrical constant, eqn 3 becomes

$$\lambda^3 = \lambda_o^3 + ct \qquad (4)$$

Hence, from (1), (2) and (4) the expression for the instantaneous creep rate is
given by:

$$\dot{\epsilon} = A(T) \left\{ \sigma - \frac{\alpha' \mu b}{(\lambda_o^3 + c(T)t)^{1/3}} \right\}^n \qquad (5)$$

This equation thus provides a direct correlation between the creep rate and
the variation of interparticle spacing with time and temperature. To test the
validity of the model, Askins et al tested one cast of 1Cr0.5Mo steel in four
different heat treatment conditions.[14,15] These were selected as: the extremes
of the CEGB specification for this steel (one representing the top and the
other the bottom of the specification), a heat treatment simulating a long ser-

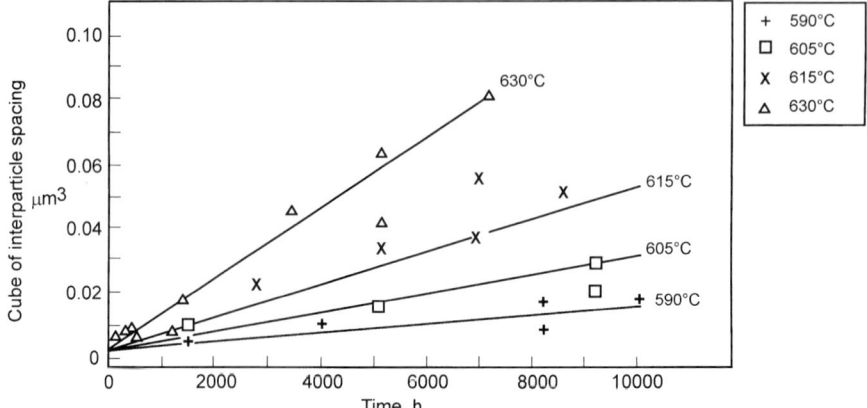

Fig. 1   Particle spacing with time in 1Cr0.5Mo steel. (After Askins[15])

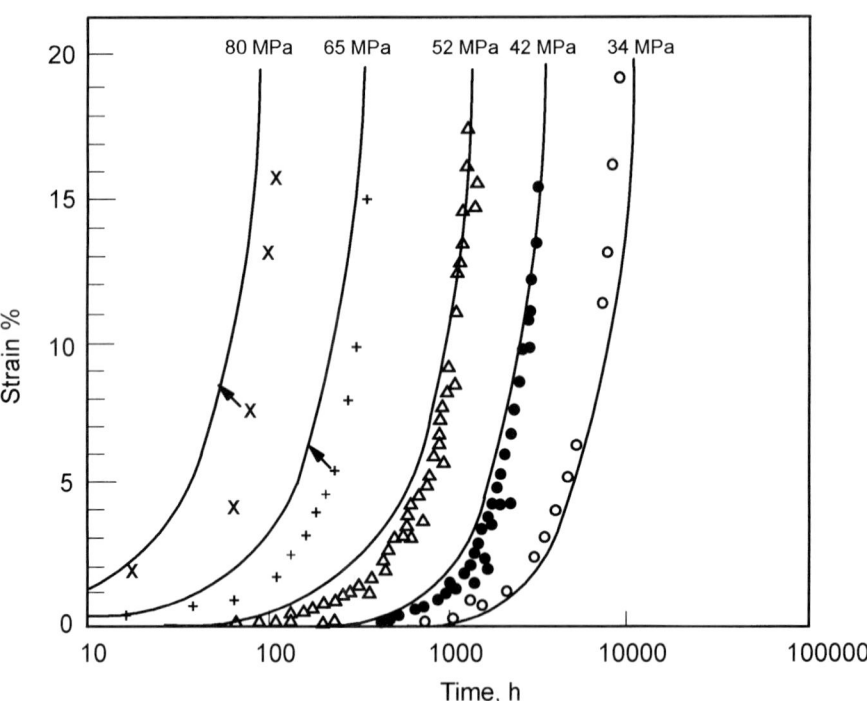

Fig. 2   Creep data and fitted curves for 1Cr0.5Mo in 'upper-bound heat treatment' at 630°C (After Askins[15])

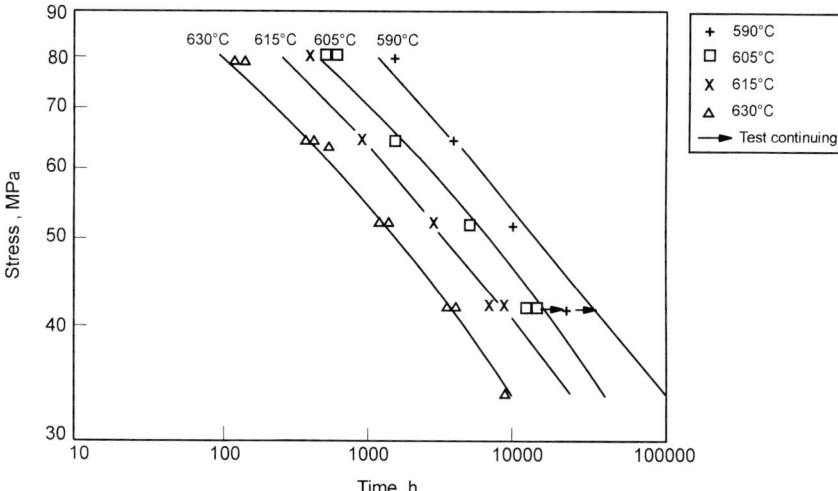

Fig. 3   Stress-rupture data and fitted curves for 1Cr0.5Mo in 'upper-bound heat treatment.' (After Askins[15])

viced (overaged) condition and the fourth a grossly overaged condition to produce a structure with essentially no precipitate strengthening of the matrix. The various constants in eqn (5) were derived from creep tests in the range 590°C–630°C and at stresses between 34 MPa–80 MPa. The metallogaphic input to the model was to determine changes in the interparticle spacing as a function of time and temperature and this was done by analysis of carbon extraction replicas taken from the grip areas of the creep specimens using transmission electron microscopy.

Results for the variation in interparticle spacing displayed the anticipated $t^{1/3}$ dependence Fig. 1. These results were then used to calculate the creep rates and the correlation between the actual creep data and fitted curves at 630°C is shown in Fig. 2. The corresponding stress rupture data were generated using the criterion that rupture occurs when the strain rate in a particular sample reached the strain rate in the totally overaged condition at 10% strain. This was found to provide the fit to the data shown in Fig. 3. In testing the model with data from creep tests on the other heat treatment conditions similar good correlations were obtained between the measured creep rates and stress rupture data and those calculated using the measured interparticle spacings.

The significance of this work is that potentially it provides a reasonably accurate and non-destructive means of determining the remaining life of a component in service simply by taking replicas of the component's structure and measuring the interparticle spacing to characterise the creep and rupture properties. However, in practice, the application of the carbide-

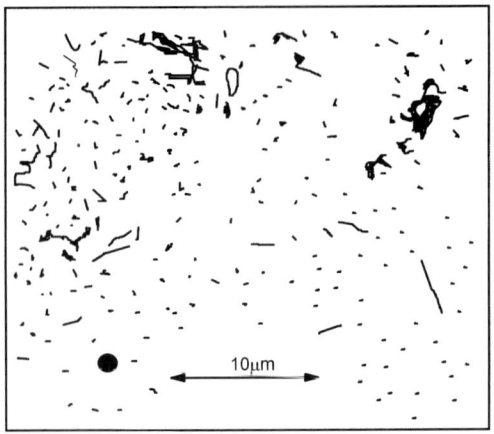

Fig. 4   Heterogeneous distribution of carbides in 1Cr0.5Mo steel.

coarsening model has a number of limitations. Foremost among these is that the carbide distributions in steels are notoriously heterogeneous, Fig. 4 and the starting microstructures for different components are never the same, and the initial carbide spacing λ_o is usually unknown. Therefore it is generally necessary to monitor the carbide spacing as a function of time at at least three temperatures in order to determine the carbide-growth kinetics. Additionally, measurement of carbide spacing from extraction replicas is extremely subjective and requires significant time commitment and sophisticated sampling and analysis techniques to achieve a representative measurement.

To a large extent this latter problem has been overcome by the realisation that the material's hardness could be relatable to the interparticle spacing Askins and Menzies,[18] Cane, Aplin and Brear.[19] Eqn 2 for the threshold stress can thus be related to hardness, H, as follows:

$$\sigma_o = \frac{\alpha' \mu b}{\lambda} = K\alpha(H - H_{ss}) \tag{6}$$

Where α' is a geometric term, H_{ss} is the contribution to the hardness due to solid solution strengthening and K defines the relationship between hardness and tensile strength. As the precipitates coarsen with time, t, at temperature, the hardness and threshold stress decrease:

$$\sigma_o = (H - H_{ss}) \sim \frac{1}{\lambda} \sim t^{1/3} \tag{7}$$

Over the years the basic thermal softening model has been modified by Brear and Cane and a number of co-workers to incorporate the effects of strain softening and primary creep. The basic equation thus becomes

$$\acute{\epsilon} = \acute{\epsilon}_o \left\{ \frac{\sigma - K\alpha'(H - H_{ss})}{KH} \right\}^n \epsilon^m \tag{8}$$

Figure 5 shows predictions obtained using the model, Tack, Brear and Seco.[22] Data obtained from one heat of 2.25 Cr1Mo (hardness = 161 Vickers) tested at 680°C and at stress of 40 MPa are plotted with a series of curves showing the development of the model. Curve (a) represents the basic thermal softening model, curves (b) and (c) show the effects of including allowance for strain softening and primary creep separately and curve (d) shows the combined effect and represents the model in its final form. The agreement with the experimental data is very good. Finally, curve (e) shows the predicted and observed behaviour of the same heat in a heavily overaged condition (hardness = 116 Vickers) which again shows excellent agreement.

Figure 5 thus demonstrates that for a specific cast, the thermal softening model based on measurements of hardness can be applied very effectively to determine the creep properties and thence the remaining life of service components. Further development of the model has been to include an allowance for the effect of creep cavitation by incorporating a generalised damage parameter ω.[21] However, in the low alloy steels under consideration, creep cavitation as a damage mechanism is only important in coarse grained HAZ and inter-critically annealed regions of weldments and does not significantly affect the rupture properties of base materials. The effects of creep cavitation are not therefore further considered in this paper.

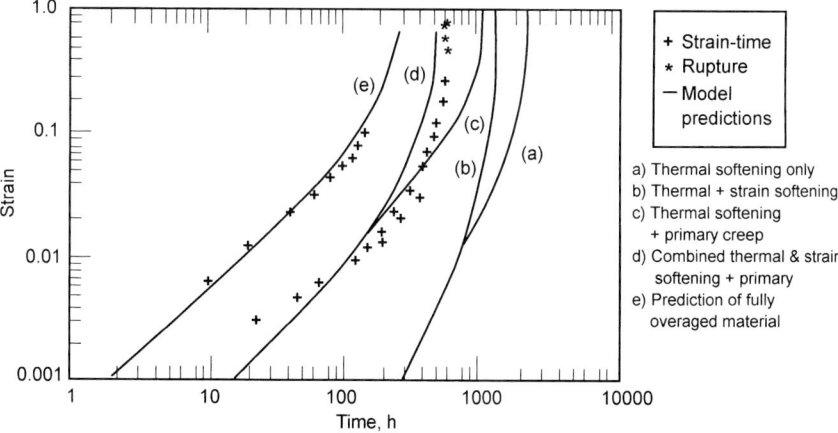

Fig. 5   Creep Model predictions for 2.25Cr1Mo steel at 40 MPa and 680°C. (After Brear, D'Angelo, Seco and Tack[23])

To enable the model to be applied to the practical problem of life prediction of service components it was necessary to explore the cast-to-cast variation and limits of the model parameters which could be achieved/obtained within a given steel specification and the extent to which it is necessary to obtain cast specific values experimentally, to improve the accuracy of the life prediction. The purpose of the recent ERA work under a Joint Industry *Project 2580*[22] described in Sections (3, 4, 5) was to do just this.

3.   THE ERA TEST PROGRAMME

3.1   *Objective*

The overall objective of the project was to develop a non-destructive, post-service condition assessment route for high temperature pressure parts based on the evaluation of the component's metallurgical structure. The principal method of doing this was to provide a predictive assessment methodology based on the application of the creep model outlined above which for the ductile low alloy steel base (parent) materials mainly concerns degradation due to structural softening. An important aspect of the work was to explore the extent to which the accuracy of the life prediction is dependent on achieving heat (cast) specific data as opposed to using data relevant to a steel grade or specification. The significance between these two options concerns the degree of direct interrogation/inspection necessary on a particular component which largely controls the cost of a life assessment campaign.

3.2   *Materials and Test Programme*

Two steel types have been investigated, in the base (parent) material condition:

1–1.25Cr0.5Mo   – 2 new heats and 8 ex-service heats
2.25Cr1Mo   – 1 new heat and 8 ex-service heats

The ex-service heats had been taken from steam pipes, headers and furnace tubes and supplied by the industrial sponsors to the project. These had experienced a range of operating conditions with temperatures in the range 500°C–565°C and for times between 64 000 h to 246 000 h. As expected there was a wide variation in structure between the different casts of each class of steel and as also expected, a wide spectrum of creep properties. It should be noted *that none of the components tested in the project had actually failed in service.*

The test programme included stress-rupture and creep tests to failure, non-destructive and destructive interrupted rupture tests to various life fractions and controlled thermal ageing experiments. These tests were performed over

Table 1   Creep, rupture and ageing test matrix 1–1.25CrMo steel

Temp °C	T_2	T_3	T_4	T_5	T_6	T_7	T_8
Stress (Mpa)	(610)	(625)	(640)	(655)	(670)	(700)	
0	A*	A*	A*	A*		OA*	
σ_1 (31)		CR(1)		CR or ISM CROA*	R		
σ_2 (38.6)		CR or ISM DI*	R NDI*	CR or ISM	CR(2)		
σ_3 (54)	R* NDI*	CR DI*	R	CR			

A	=	Thermal ageing test
OA	=	Thermal overageing tests
R	=	Rupture test
CR	=	Creep test
CROA	=	Creep test of overaged specimens
ISM	=	Interrupted rupture test with strain measurement at each interruption
DI	=	Interrupted test for destructive examination at various life fractions
NDI	=	Interrupted tests for non-destructive examination at various life fractions
*	=	Limited test matrix with only some heats tested

a range of accelerated temperatures (590°C–700°C) and at stress levels typical of those encountered in service (31–54 MPa). The test matrix for the 1–1.25Cr0.5Mo steels was similar to that used for the 2.25Cr1Mo steels and is shown in Table 1. All the stress-rupture and creep tests were conducted in vacuum in order to minimise the effect of oxidation on the creep behaviour.[2,4] Individual test durations ranged from 200–12 000 hours and the data base so obtained is certainly the most extensive achieved worldwide on ex-service material.

The purpose of all these tests was to determine and characterise the creep properties of the individual casts in terms of creep-rates $\dot{\epsilon}$, stress-exponents n, Monkman Grant Constants ϵ_s, times to rupture t_r, activation energies Q and rupture ductilities ϵ_f. These parameters were obtained by analysis of the experimental creep curves and by fitting the data to the Norton law, eqn 9.

$$\dot{\epsilon} = A\sigma^n = A_o \exp\left(\frac{-Q_r}{RT}\right)\sigma^n \tag{9}$$

Where A is a material, cast and temperature dependent constant. Some evidence indicates that it remains relatively constant during the early stages of life and can, therefore, be taken as time independent. The above equation only models the minimum creep rate and as such it is incapable of representing the increasing rate of creep towards failure.

The corresponding rupture life equation can be expressed as :

$$t_r = \frac{\dot{\epsilon}_s}{\dot{\epsilon}_m} = \frac{\dot{\epsilon}}{A}\exp\left(\frac{Q'_r}{RT}\right)\sigma^{-m} \tag{10}$$

where ϵ_s is the Monkman-Grant constant. Here the stress exponent and activation energy for rupture are different from those for creep rate and thus expressed by different symbols m and Q'_r rather than n and Q_r.

As indicated above, an important aspect with these evaluations was to determine the limits of these various parameters within a particular steel (material) specification and the need to achieve heat specific data and whether hardness could be used as a normalising factor to account for heat-to-heat property variations. Long term ageing experiments were also performed and hardness testing used to characterise the thermal softening response of each heat and to determine the ageing parameters required for the model. Finally, each heat was grossly overaged and then creep tested to determine the properties in the absence of precipitate strengthening and with the hardness close to H_{ss} for the solid solution.

4. PROPERTIES OF AS-RECEIVED HEATS

The micro-structures of the as-received heats varied significantly with some showing relatively little degradation such as CAR Fig. 6 and others extensive degradation such as CAU Fig. 7. This clearly was a reflection of differences in the original heat treatments and severity of the service exposures. The range of hardness also reflected the variation in heat/heat degradation and ranged from $Hv_{30} = 136$ to 161 for the 1–1.25Cr 0.5Mo heats and $Hv_{30} = 126$–183 for the 2.25Cr1Mo heats. Concomitant differences were also observed in stress-

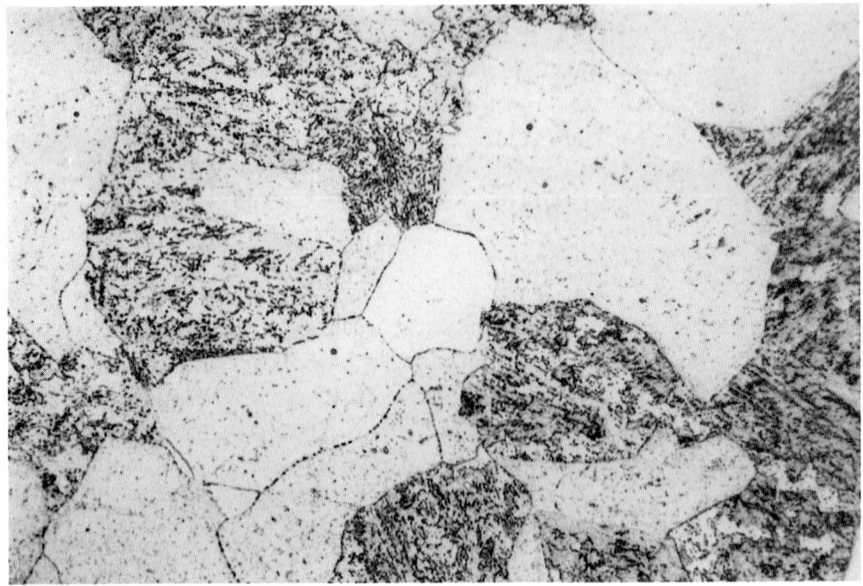

Fig. 6 CAR As received 2.25Cr1Mo cast × 1000, $Hv_{30} = 173$.

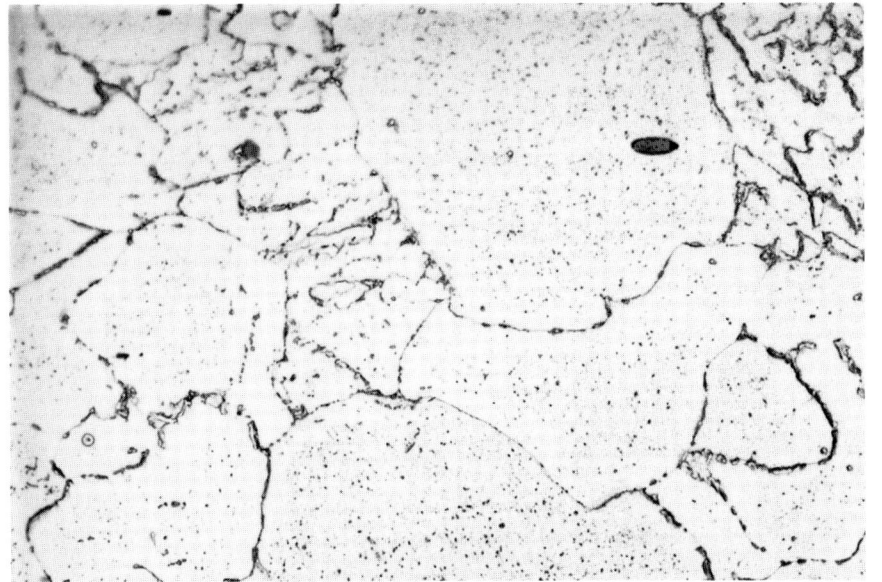

Fig. 7 CAU Ex-service 2.25Cr1Mo pipe (159700 h, 564°C) × 1000, $Hv_{30} = 127$.

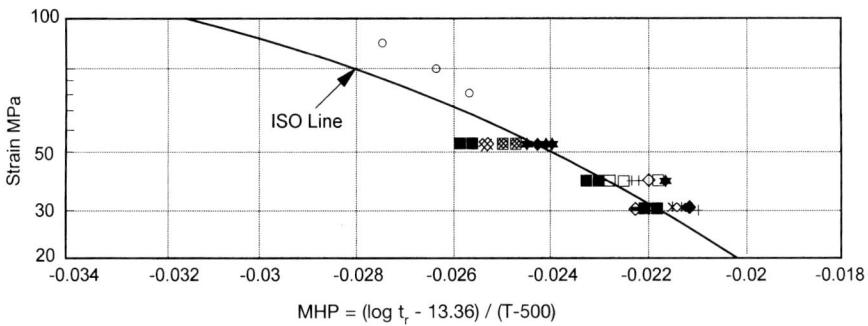

Fig. 8 Comparison of rupture data for new and ex-service materials of 2.25Cr1Mo steel with ISO Parametric equation.

rupture properties with, in some cases, differences of up to an order of magnitude in the times to failure between the weakest and strongest heats, Fig. 8.

5. RUPTURE LIFE ASSESSMENTS FOR THE EX-SERVICE BASE MATERIALS

A number of assessment procedures have been developed and examined. These have been classified into *four* different categories according to the ease of application, the required accuracy and the need in each for direct access to the plant and material sampling.

Category 1 Assessments – use standard materials data and require knowledge of the component's design and operating conditions (i.e. temperatures and stresses) as inputs but do not require direct access to the component. This category is identical to Stage 1 in the original GOM 101. The method uses material data from ISO, BSI, NRIM or ASTM standards as appropriate, service operating parameters ($T(s)$, $\sigma(s)$, $t(s)$) and creep life usage calculated from the 'Life Fraction Rule' (3):

$$\Sigma \frac{t_i}{t_R} = 1 \tag{11}$$

Where t_i is the time of operation at a given stress σ and temperature T and t_R the corresponding minimum rupture time for these conditions from standard data. This is the least accurate assessment category because:

- It is not possible to place the component's material within the scatter-band of the standard data
- Generally there is a +/− 20% scatter in σ in standard data
- It is therefore necessary to assume minimum properties for conservatism

The method therefore grossly underestimates the remaining life. For the materials investigated in the present investigation, four of the ex-service heats (3 × 1–1.25 Cr 0.5Mo and 1 × 2.25 Cr1Mo) were predicted to have life fractions consumed, on the basis of lower bound NRIM standard data, greater than 1, which effectively means they should have failed in service. The life fraction rule was also used to predict the lives of each material from its previous exposure (service) for test conditions actually tested in the programme, ratios of actual to predicted lives ranged from 1.6 to 8.3 again demonstrating the conservative nature of the standard-based life assessment approach.

Category 2 Assessments – use non-destructive inspection data (i.e. hardness and, for brittle materials, an assessment of the degree of cavitation using replicas) which clearly requires access to the plant for non-destructive inspection.

The Norton Law rupture life equation described above represents the rupture life, temperature and stress relationship for a specific heat. This implies that each heat has to be treated separately due to heat-to-heat variation of the activation energy and stress exponent. However, given the success of the model using hardness measurements as a means of predicting the rupture life and creep rates it may be possible to merge the data for all heats of the same material by a simple modification to the model, which reflects the relative flow strength of the various heats.[19] Thus from eqn (9) the minimum creep rate ϵ becomes;

$$\dot{\epsilon} = A'_o \exp\left(\frac{-Q_r}{RT}\right)\left(\frac{\sigma}{\sigma_y}\right)^n \tag{12}$$

where σ_y is the flow stress and can be related to the hardness of the specific heat by

$$\sigma_y = KH \qquad (13)$$

where K is a constant relating hardness to tensile strength. Equation (12) can thus be rewritten as :

$$\dot{\epsilon} = A''_o \exp\left(\frac{-Q_r}{RT}\right)\left(\frac{\sigma}{H}\right)^n \qquad (14)$$

The corresponding relationship for stress rupture would thus be:

$$t_r = B''_o \exp\left(\frac{-Q'_r}{RT}\right)\left(\frac{\sigma}{H}\right)^{-m} \qquad (15)$$

In eqns (14) and (15) H is the current heat-specific hardness and Q_r and m the average values for all heats of the same material. These values were obtained

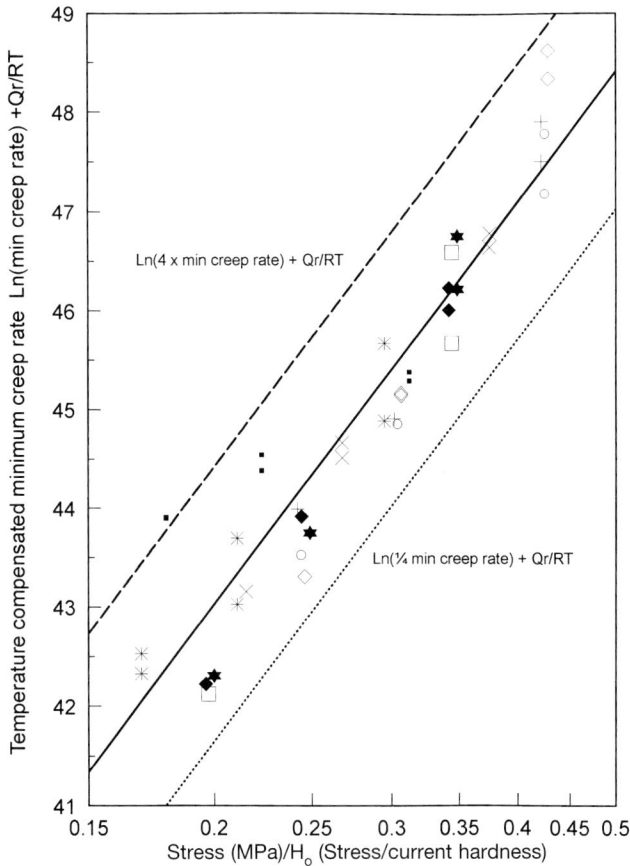

Fig. 9 Minimum creep rates from equation (14) (Hardness-Compensated model) for 2.25Cr1Mo new and ex-service materials.

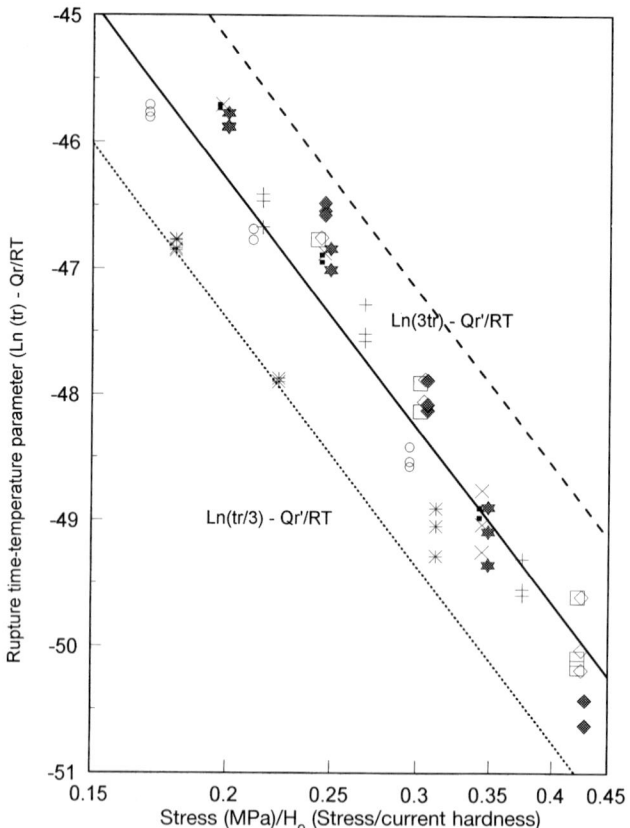

Fig. 10   Rupture lives from equation (15) (Hardness-Compensated model) for 2.25Cr1Mo steel new and ex-service materials.

by analysis of the experimental creep and rupture data. Equations (14) and (15) are plotted for the 2.25Cr1Mo data in Figs 9 and 10 with the stress term normalised by the as-received hardness of each heat. Normalising the data in this way significantly reduced the scatter when compared to the basic rupture and creep equations and demonstrates that all the data for each class of steel can be successfully modelled using hardness compensation to give a scatter band of less than a factor of three for rupture lives and within a factor of four for creep rates.

For the purposes of remaining life prediction, it is recommended therefore that the lower bound should be defined as one third of the rupture lives predicted by eqn (15). Figure 11 shows the remaining life prediction on this basis for 1CrMo at 550°C for several levels of service stress.

Compared to the standard-based approach of Category 1, the hardness compensated approach introduces an additional parameter to the determi-

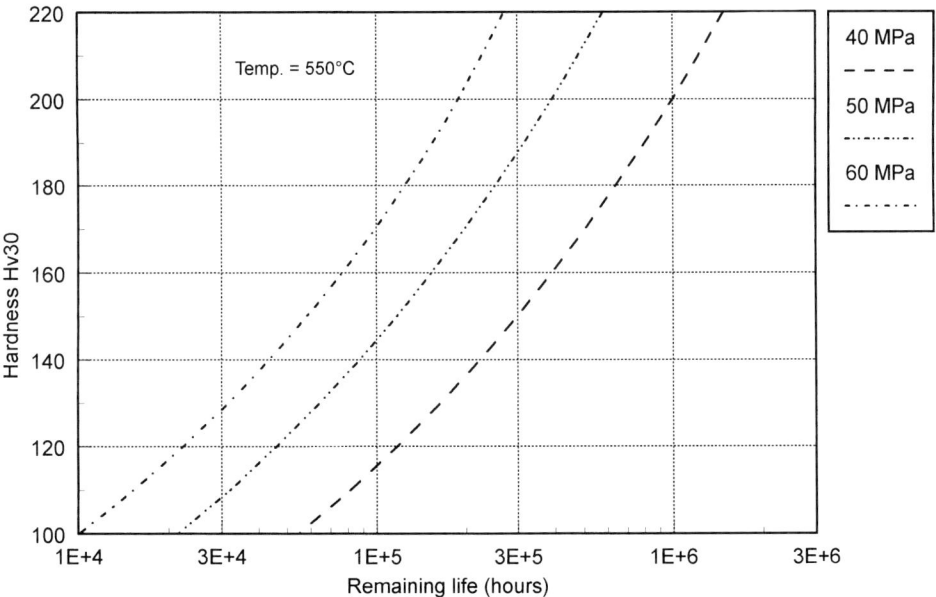

Fig. 11   Remaining life predictions based on the lower bound 'Hardness-Compensated' model (equation (15) for 1Cr0.5Mo.

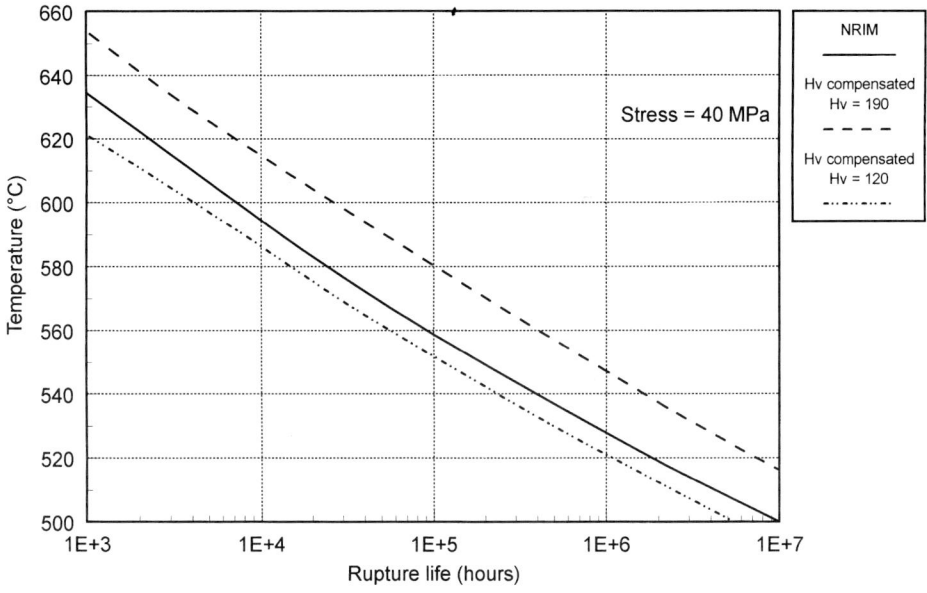

Fig. 12   Comparison of the Lower-bound NRIM material standard life prediction and the Hardness-Compensated model for 1Cr0.5Mo.

nation of remaining life – H, *the current hardness of the component.* Figure 12 shows the lower bound remaining life-temperature plots using the standard based approach and the hardness based approach for two levels of hardness ($Hv_{30} = 190$ and $Hv_{30} = 120$) for an assumed service stress of 40 MPa. Clearly, the hardness compensated approach predicts a range of remaining lives depending on the current hardness level while the standard-based approach cannot account for the effects of hardness or material degradation.

The importance of the hardness compensated approach is that by making non-destructive hardness measurements on a component in service, it is possible to place the component material within the scatterband of material data and predict the remaining life to within a factor of three, which is a marked improvement over Category 1 assessment which can only achieve accuracy within an order of magnitude. The hardness compensated approach however, only provides *material, not heat specific,* information and cannot allow for differences in properties due to composition and the original heat treatment. To improve accuracy of the life prediction by accounting for these it is necessary to increase the interrogation of the component as described in Categories 3 and 4.

Category 3 Assessments – use heat specific thermal softening data (i.e. hardness reduction with time at several temperatures) to generate the parameters required in a thermal softening model and thus require direct access to the component to procure 'heat' specific samples for thermal ageing.

Creep models based on the Norton creep law are incapable of describing the period of accelerating creep towards failure. It is necessary therefore, to invoke a modified effective stress model, Section 2,[22,23] which can describe tertiary creep through changes in the effective stress as a consequence of precipitate coarsening and the possible accumulation of creep cavitational damage. This model can be described as

$$\dot{\epsilon} = A\sigma^n_{eff} = A(\sigma - \sigma_o)^n = A \exp\left(\frac{-Q_r}{RT}\right)(\sigma - \sigma_o)^n \qquad (16)$$

From eqn (16) the threshold stress is inversely proportional to the interparticle spacing and hence hardness through the relationship:

$$\sigma_o \sim \frac{1}{\lambda} \sim (H - H_{ss}) = K'(H - H_{ss}) \qquad (17)$$

Here H represents H_t if thermal softening only pertains and $H_{t,\epsilon}$ if both thermal and strain softening are present. Allowance can also be made in this equation for cavitational damage and constant load effects. The stress term is modified to:

$$\sigma = \frac{\sigma_i(1 + \epsilon)}{1 - \omega} \qquad (18)$$

and combining eqns (16) (17) (18) gives:

$$\epsilon = A \exp\left(-\frac{Q_r}{RT}\right)\left[\frac{\sigma_i(1 + \epsilon)}{1 - \omega} - K'(H - H_{ss})\right]^n \qquad (19)$$

To apply this equation to a specific cast of material, the thermal ageing behaviour of the cast is also required. This was obtained by carrying out thermal ageing tests using small samples taken from *non-critical locations* of the component. The thermal ageing tests were performed under vacuum in order to minimise the possibility of oxidation or decarburisation and performed at at least three different temperatures. Typical data for a 2.25Cr1Mo cast are shown in Fig. 13 from which it was possible to derive the thermal softening parameters C_0, Q_h and H_{ss} from eqn (20)

$$\frac{1}{(H_t - H_{ss})^3} = \frac{1}{(H_o - H_{ss})^3} + C_o \exp\left(\frac{-Q_h}{RT}\right) \bullet t \qquad (20)$$

Comparison with measured and predicted strain data is shown in Fig. 14 and creep rate data in Fig. 15. Figure 16 shows the corresponding comparison of predicted rupture life and measured rupture life obtained assuming a strain

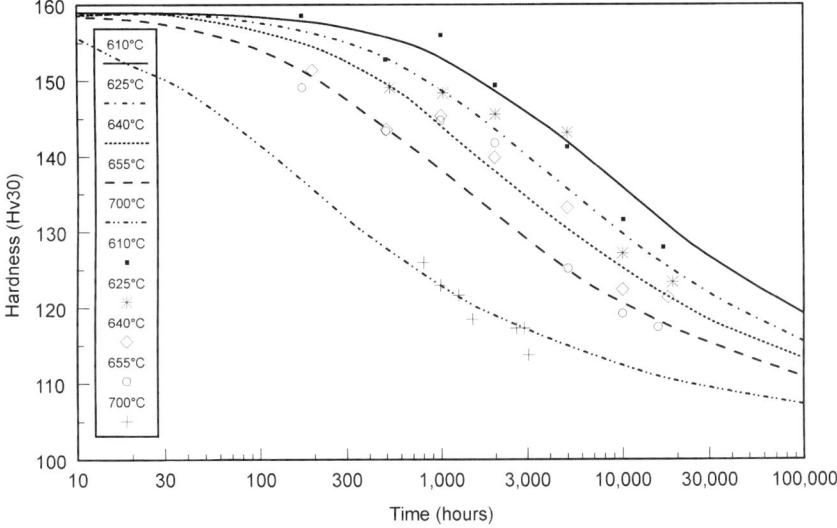

Fig. 13 Comparison of the predicted hardness from equation (20) with aged and over-aged test data on ex-service service 1Cr0.5Mo material (115 000 h, ~535°C).

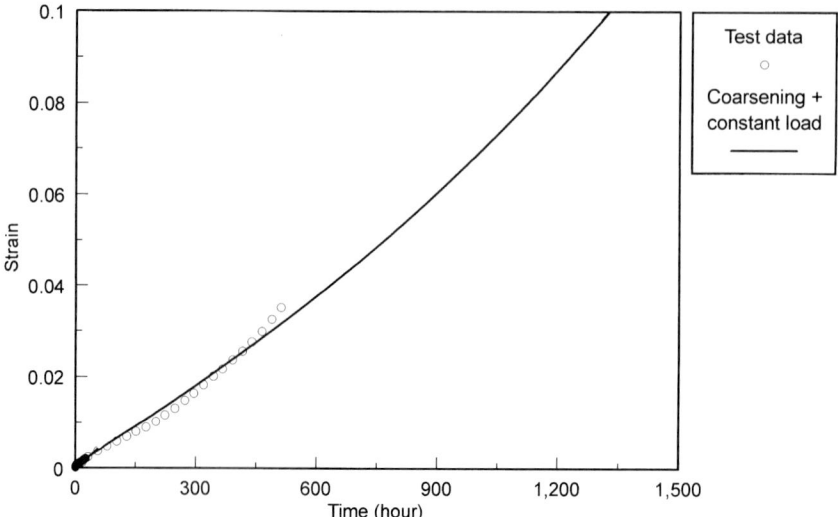

Fig. 14 Comparison of predicted and measured creep strain using the Hardness-based Effective Stress Model, equation (19) for ex-service 2.25Cr1Mo material (85 000 h, ~535°C).

cut-off value of 10% to represent rupture and integrating eqn (19) to that strain.

It is considered that the hardness-based effective stress approach described above provides improved accuracy over the hardness compensated approach described under category 2. However, this method does require the procurement of small samples taken from the component and then the time necessary (up to 10 000 h say) to perform the long term ageing experiments. The samples can however, be taken at any early stage in the life of the component and thus the thermal softening data can be obtained ready for a determination of the life prediction at a later time.

Category 4 Assessments – require the removal of samples large enough to make up into miniature creep and stress-rupture test pieces. These are then used to obtain heat specific creep and stress-rupture data for several conditions of stress and temperature.

This category includes the traditional iso-stress/accelerated temperature test technique which has good accuracy provided the specimens are obtained from the location on the component under investigation.[1,2] Assessments under category 4 can also be made more general (i.e. applicable to a range of temperatures, stresses and other locations), by analysing the creep and rupture data. Three methods were developed during this project: (1) used the hardness-based heat specific effective stress model of category 3, (2) used of the basic Norton Law and (3) used the Θ-projection model.[25] All three methods use heat specific

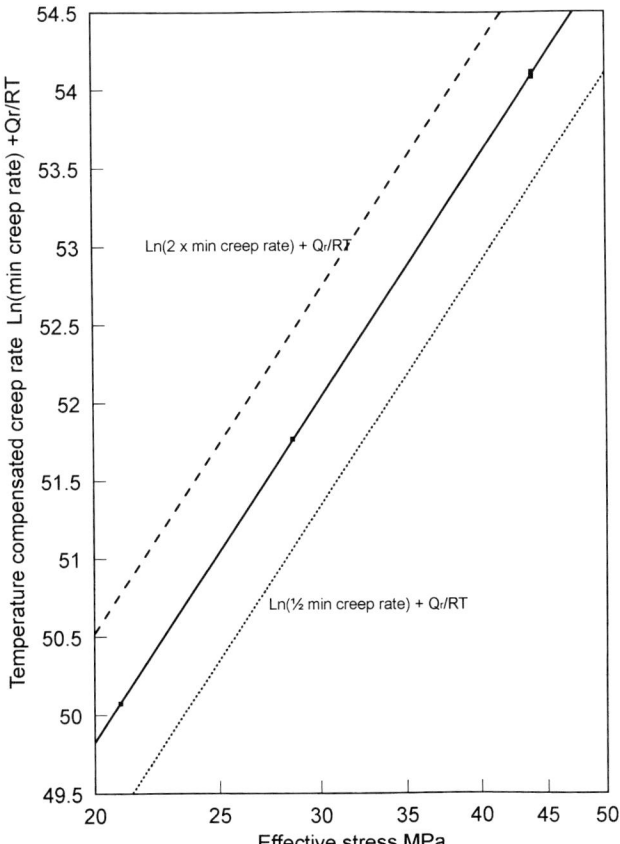

Fig. 15 Temperature compensated minimum creep rates based on the Hardness-based Effective Stress Model, equation (19) for ex-service 2.25Cr1Mo material (85 000 h, ~535°C).

creep and rupture data for at least three conditions of stress and temperature on specimens taken from locations with a material degradation status representative of the location of concern and they achieve similar accuracy. Because these samples are extracted from creep damaged areas there is clearly a limitation in the size of sample removed before the component requires a weld repair. However techniques for sample removal have been reviewed[2] and the use of miniature creep specimens[2,4,14] restricts the area of damage which can often be restored by mechanical blending. The Norton Law was found to be the simplest method to apply and Fig. 17 gives an example of the type of correlation for minimum creep rates in one cast of 2.25Cr1Mo using eqn (9) and Fig. 18 the correlation of rupture life, stress and temperature for 1Cr0.5Mo from eqn 10.

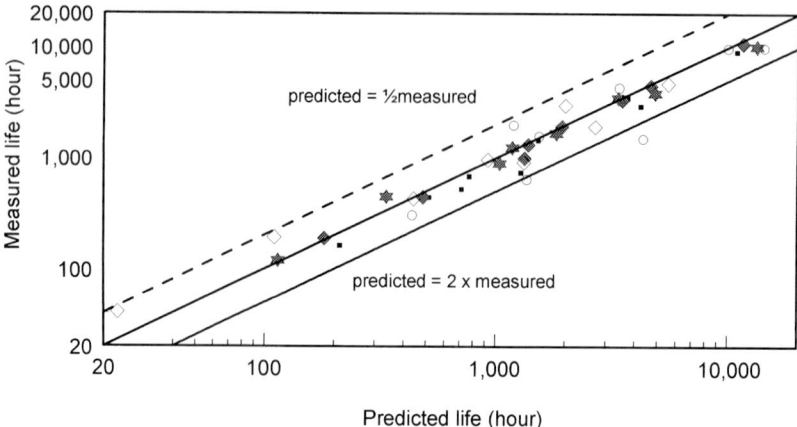

Fig. 16   Comparison of predicted and measured rupture lives for new and ex-service 2.25Cr1Mo materials using the Hardness-based Effective Stress Model, (equation (19).

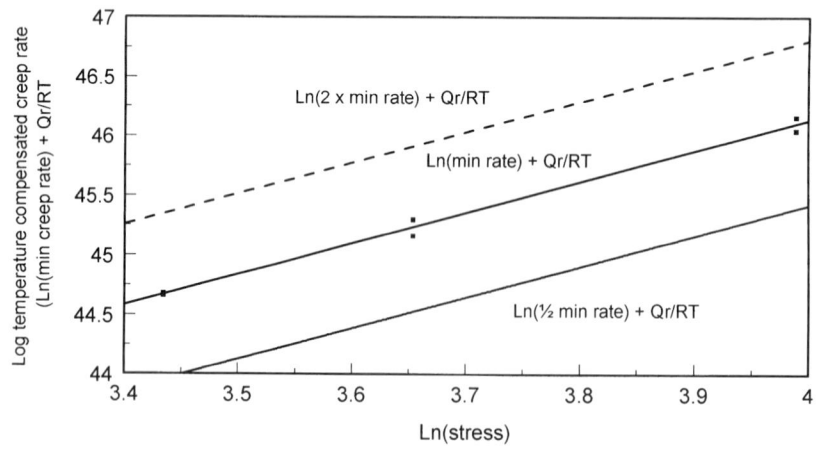

Fig. 17 Minimum creep rate, stress and temperature correlation using Norton's Law equation (9) and experimental data on a new heat of 2.25Cr1Mo steel.

6.   CASE STUDIES

To illustrate the use and life extension potential of the various procedures described above, a number of case studies have been carried out using the ex-service materials tested in this project. Representative casts of each material were chosen for which both thermal ageing data and heat specific creep and rupture data had been generated so that all four categories described above

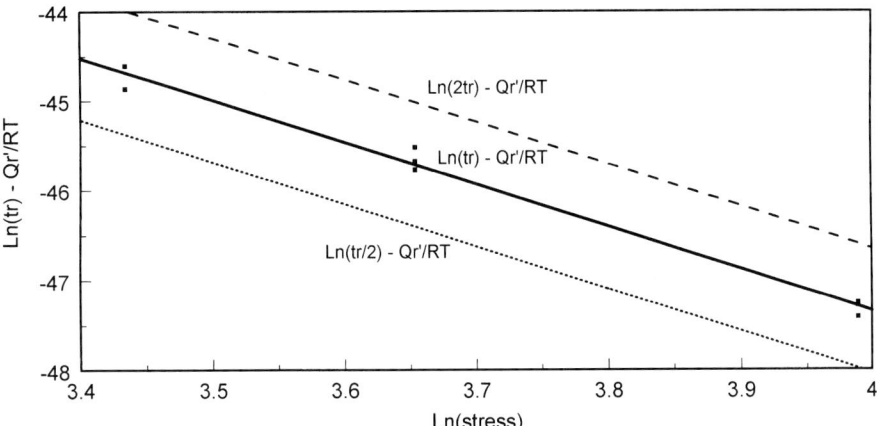

Fig. 18   Rupture Life, stress and temperature correlation using Norton's Law equation (10) and experimental data from ex-service 1.25Cr1/2Mo material (112 000 h, ~541°C).

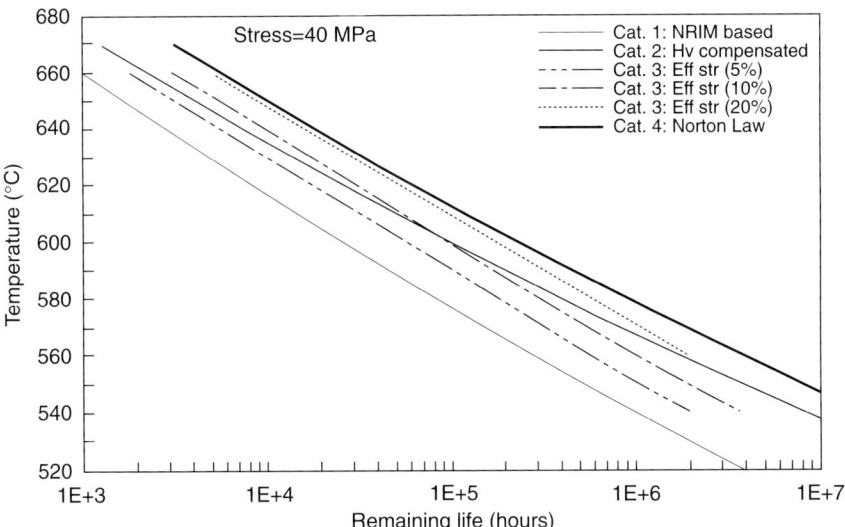

Fig. 19   Comparison of life prediction methods for ex-service 2.25Cr1Mo material (130 000 h, ~535°C).

could be applied. Figure (19) shows a comparison of the four categories for the cast of 2.25Cr1Mo which had experienced 130 000 h in service at ~535°C and Figure (20) the same comparison for a cast of 1Cr0.5Mo which had had 115 000 h in service also at ~535°C.

Figures (19) and (20) clearly show that progression from category 1 to 2 to 4 methods gives significantly longer remaining life predictions. The hardness-

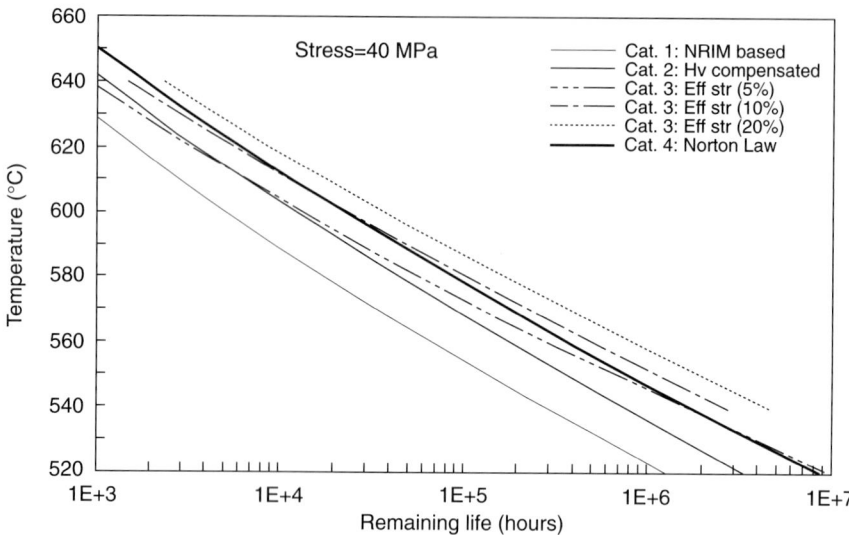

Fig. 20 Comparison of life prediction methods for ex-service 1Cr0.5Mo material (115 000 h, ~535°C).

based effective stress approach of category 3 gives predictions which depend on the level of strain cut-off selected. The case study demonstrates that the use of higher category assessment methods can generally extend the assessed life, *by up to an order of magnitude in time*, by removing the high level of conservatism built into the lower category methods. The reduced conservatism of the higher category methods stems from an increase in the knowledge of the properties of a specific material and its state of structural degradation through metallographic interrogation, sampling and testing.

7.   PROCEDURE FOR IN-SERVICE COMPONENT LIFE ASSESSMENT AND EXTENSION

The options and procedures for life assessment depend on a number of factors including whether access to the plant is possible, the availability of virgin or ex-service material of the same heat and on the previous operating history and maintenance records. With increased availability of the material and access to the plant, the life assessment procedure can become increasely sophisticated leading to more accurate and less conservative estimates of the remaining life.

In the early stages of a component's service when the expected remaining life is still relatively long, a simple conservative life estimate based on the use of standards data is all that is required to guarantee safe operation until the next major outage. When it is anticipated that a large fraction of the component's life has been consumed however, the purpose of the assessment is to

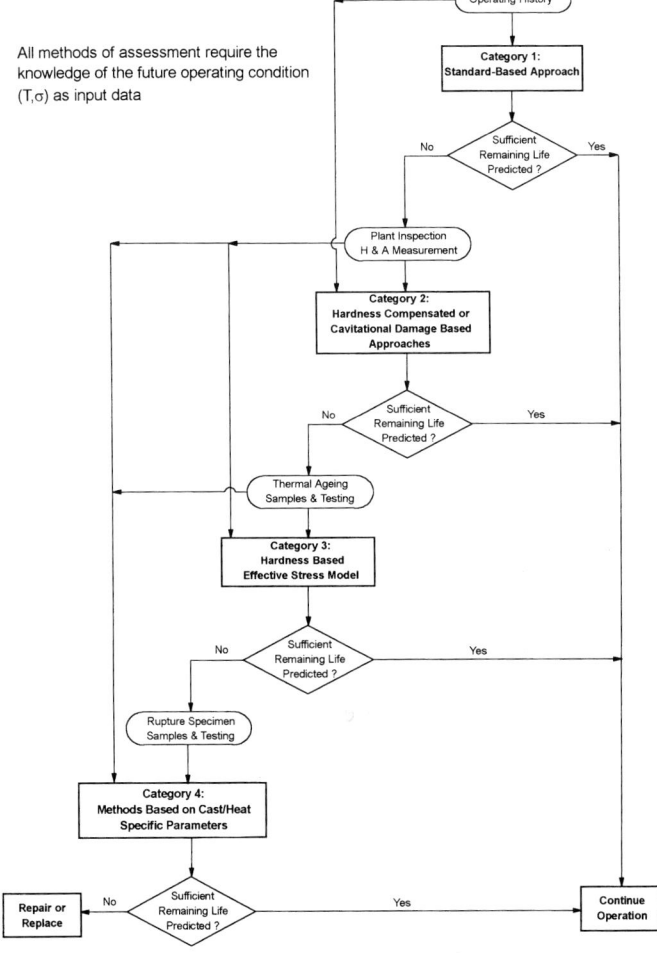

All methods of assessment require the knowledge of the future operating condition (T,σ) as input data

Fig. 21 Procedural Route for applying Category 1–4 Life Assessment Methods.

determine whether any life extension is possible, then the life calculation needs to be refined by using the more sophisticated approaches.

The remaining life assessment methods described in categories 1 to 4 have been incorporated into an overall progressive assessment procedure illustrated in Fig. 21. Progression from the lower to higher levels of assessment places more demands for direct access to the plant component for inspection and sampling thus significantly increasing the cost of the assessment but provides much less conservative, more accurate and generally longer remaining life predictions. The progressive nature of the overall procedure has been devised to justify increased expenditure if the remaining life predicted initially is not adequate to underwrite safe operation to the next scheduled outage with the necessary safety margin.

8.   CONCLUSIONS

1. The creep properties of low alloy steels can be accurately modelled by relating changes in strain rate to structural degradation as determined by measurements of the interparticle spacing, which are time consuming and experimentally difficult, or by hardness measurements which are much easier to perform.
2. A progressive life assessment procedure has been devised comprising *four* categories of assessment dependant on the degree of accuracy required and the need for direct access to the plant for inspection and material sampling.
3. Category 1 assessments are based on the use of prior operational history, lower bound standard material data and require no access to the plant component, but can only provide a lower bound extremely conservative evaluation.
4. Remaining life assessment based on on-site hardness measurements, (the Hardness-Compensated Model, Category 2) provides less conservatism, greater accuracy and generally longer predicted lives.
5. The Hardness-Based Effective Stress Creep Model as used in Category 3, potentially provides improved life assessment capability giving less con-servatism and more accurate prediction than the hardness-compensated model but needs an extensive database on aged material.
6. Remaining life assessment based on post-service heat specific rupture tests provides the most accurate means of life prediction. This method can be applied by analysing the data using the Norton Law or the Θ-projection technique (mathematically complex), or by extrapolation of accelated iso-stress tests. A disadvantage with this technique is that samples of material need to be taken from creep damaged areas which is not always feasible.

ACKNOWLEDGEMENT

This paper is published by permission of ERA Technology Ltd. The authors would like to thank Mr A. Fairman, Mr S. Fujibayashi, Mr A. Chowdhury and Mr D.R. Humphrey for their contributions to Project 2580.

REFERENCES

1. R.D. Townsend: *The CEGB Approach to Remanent Life Assessment*, ECSC Conference Residual Life, Brussels, Nov 1984.
2. B.J. Cane and R.D. Townsend: *Prediction of Remaining Life in Low Alloy Steels*, ASM Materials Science Seminar on Flow and Fracture at Elevated Temperatures, Philadelphia USA, Oct 1983, 279–316.

3. E.L. Robinson: 'Effect of Temperature Variation on the Creep Strength of Steels', *Trans. American Society of Mechanical Engineers*, 1938, **60**, 253.
4. R.D. Townsend: *Materials Integrity; Performance and Remanent Life Assessment*, Proceedings of Joint IERE-EPRI Conference Electricity Beyond 2000, Oct 1991, 249–285.
5. J.M. Brear and B. L. King: 'Residuals, Additives and Materials Properties', *Paper 41*, Royal Society Conference, London, 1979.
6. C.J. Middleton: 'Control of Reheat Cracking and Creep Cavitation in Low Alloy Steels – The Roles of Residual Elements and Sulphur', *ASM R.I. Jaffee Memorial Symposium Clean Steels Technology*, ASM/TMS Materials Week Chicago, USA, 1992, 39–53.
7. J. Woodhead and A. Quarell: *J. Iron and Steel Institute*, 1965, **203**.
8. R.G. Baker and J. Nutting: *J. Iron and Steel Institute*, 1995, **192**.
9. V.A. Biss and T. Wada: *Met. Trans. A.V.*, 1985, **16A**.
10. D. McLean, *Metallurgical Reviews 7*, 1967.
11. K.R. Williams and B. Wiltshire: *Metal Science*, 1973, **7**, 176.
12. P. J. Henderson and M. McLean: *Acta Met.*, 1983, **31**, 1203.
13. B. F. Dyson and M. McLean: *Acta Met.*, 1983, **31**, 17.
14. M.C. Askins, J.M. Brear, B.J. Cane, D.J. Gooch, D.A. Miller, M.S. Shammas, R.D. Townsend: 'Estimating the Remaining Life of Boiler Pressure Parts EPRI Contract 2253–1', *Issued as Volume 3 EPRI Report CS5588*, Nov 1989.
15. M.C. Askins: 'Creep Life Assessment by Measurement of Inter-particle Spacing' *International Conference on Life Assessment and Life Extension*, The Hague, June 1988.
16. H.E. Evans and G. Knowles: 'Threshold Stress for Creep in Dispersion-Strengthened Alloys', *Met. Sci.*, 1980, **262**.
17. G.W. Greenwood: *Particle Coarsening*, Institute of Metals Monograph, No 33, 1969, 103.
18. M.C. Askins and J. Menzies: *Reported in References (4) and (14)*, 1985.
19. B.J. Cane, P.F. Aplin and J.M. Brear: *ASME J. Pressure Vessal Technology*, 1985, **107**, 295.
20. P.F Aplin, J.M. Brear and B.J. Cane: 'Condition Assessment of High Temperature Plant', *Proceedings of the Third International Conference on Creep and Fracture of Engineering Materials*, Swansea, edited by B. Wilshire and R. W. Evans, 1987, 853.
21. B.J. Cane, P.F Aplin and J.M. Brear, 'A Probabilistic Approach to Remanent Creep Life Assessment of Low Alloy Ferritic Components', *Proceedings of the International Conference on CREEP*, Tokyo, Japan, Apr 1986, 447–452.
22. A.J. Tack, J.M. Brear and F.J. Seco: 'Mechanistic Creep Modelling for

Weldment Life Prediction', *Proceedings 5th International Conference on Creep of Materials'*, Orlando, Florida, USA, May 1992.

23. J.M. Brear, D. D'Angelo, F.J. Seco and A.J. Tack: 'Mechanistic Creep Models for 21/4Cr1Mo Welds and Parent Metal', *ERA European Conference Life Assessment of Industrial Components and Structures*, Paper 4.3, Cambridge, 1993.

24. D.M. Finch, R. Timmins and R.D. Townsend 'Condition Assessment and Life Extension of High Temperature Plant', *ERA Report 95–1142*, ERA Project 2580, Part 1: Pressure Vessels and Piping, 1995.

25. R. W. Evans and B. Wilshire: *Creep of Metals and Alloys,* The Institute of Metals, London, 1985, 314.

Service Performance of a 12CrMoV Steam Pipe Steel

J. HALD

Elsam/Elkraft c/o Department of Metallurgy, Technical University of Denmark 204, DK 2800 Lyngby, Denmark.

ABSTRACT

Service experiences with the 12CrMoV steam pipe steel X20CrMoV121 in Danish power plants is reviewed. Surprisingly, no creep damage has been found in welds, but a few incidents of incorrect production heat treatment has led to low creep strength material in components like bends and T-pieces. To avoid such incorrect heat treatment it is recommended to require extra documentation for a correct heat treatment when buying steam lines of martensitic stainless steels like the X20CrMoV121 and the new modified 9Cr steels.

1. INTRODUCTION

An important aspect of all technical development is to work based on previous experience. In the field of creep resistant steels, which behave rather complicatedly during service exposure under creep conditions and where service failure can be catastrophic, the gathering and analysis of service experiences is of special importance both from a service and maintenance point of view and from the point of view of developing better materials.

In the Danish power companies a long tradition of gathering service experiences with steam pipe steels exists. Since 1977 the HP Pipe Committee of the Danish Primary Utilities has systematically gathered service experiences with the low alloyed steels 14MoV63 (0.5Cr0.5Mo0.25V); 13CrMo44 (1Cr0.5Mo) and 10CrMo910 (2.25Cr1Mo), and the results have been published in reports in 1981,[1] 1986[2] and 1991.[3] Recently the Committee has finished work on the 12CrMoV steel X20CrMoV121, which has been used for HP steam pipe systems in Danish power plants since 1965. In this paper the general service experiences with the 12CrMoV steel are summarised and examples of special service incidents are given.

2. THE STEEL X20CRMOV121

Steels with 12%Cr have been used since the 1920s because of their good corrosion properties, and 12%Cr steels alloyed with Mo and V were used in hot parts of the early jet engines because of their high creep strength.[4,5] Use of 12CrMoV steels for steam pipes in power plant was delayed due to weld cracking problems related to the martensitic transformation of the steels. When these problems were overcome in the mid 1950s,[6] a 12CrMoV steel could be standardised for use in steam pipes under the DIN designation X20CrMoV121. In 1963 the first application in steam plant in Germany followed, and the first Danish application of the steel in power plant was in 1965. Since then the steel has become the most widely used for live steam pipes in Danish power plants. The high creep strength of the steel allowed the construction of larger power plants with more advanced steam parameters, and the good thermal fatigue resistance of the steel enabled good cycling ability of the plants. In the mid 1980s the steel was used in the first supercritical steam power plants in Denmark at steam conditions 250 bar/540°C, and in 1992 the limit of the steel was reached with the 400 MW power plant Vestkraft 3 at steam conditions 250 bar/560°C. For plants with more advanced steam conditions, steels with better creep strength must be used. The longest service time for a X20CrMoV121 steam pipe in Danish power plants is currently approximately 160 000 hours at 540°C.

2.1   Metallurgy

X20CrMoV121 is a martensitic stainless steel. The alloy composition with 0.2%C and 12%Cr results in full air hardenability to martensite in section thicknesses up to approximately 100 mm as seen from the CCT-diagram, Fig. 1. The specified chemical composition for the steel is presented in Table 1. For

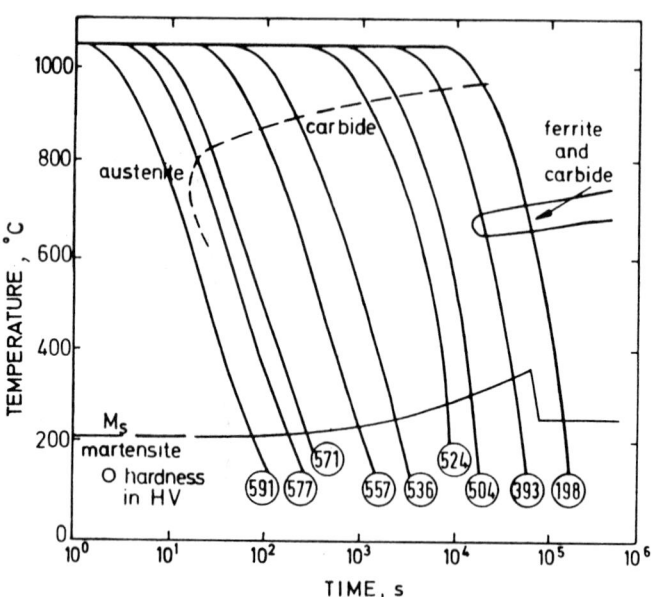

Fig. 1   CCT diagram for the steel X20CrMoV121.[16]

Table 1 Specified chemical composition of the steel X20CrMoV121

mass %, bal. Fe	
C	0.17–0.23
Si	<0.50
Mn	<1.00
P	<0.030
S	<0.020
Cr	10.0–12.5
Ni	0.30–0.80
Mo	0.80–1.20
V	0.25–0.35

steam pipe applications the steel is delivered in a normalised and tempered condition. Normalising at 1040°C–1070°C transforms the steel to austenite, and during subsequent cooling transformation to martensite begins at approximately 300°C and it is complete close to room temperature. During tempering at 730°C–780°C the martensite transforms to a ferritic subgrain microstructure and carbides mainly of the type $M_{23}C_6$ precipitate on subgrain boundaries, Fig. 2. In the normalised and tempered condition the steel has good ductility and tensile strength, and the high stability of the tempered martensite leads to good creep strength. Specifications for heat treatment and

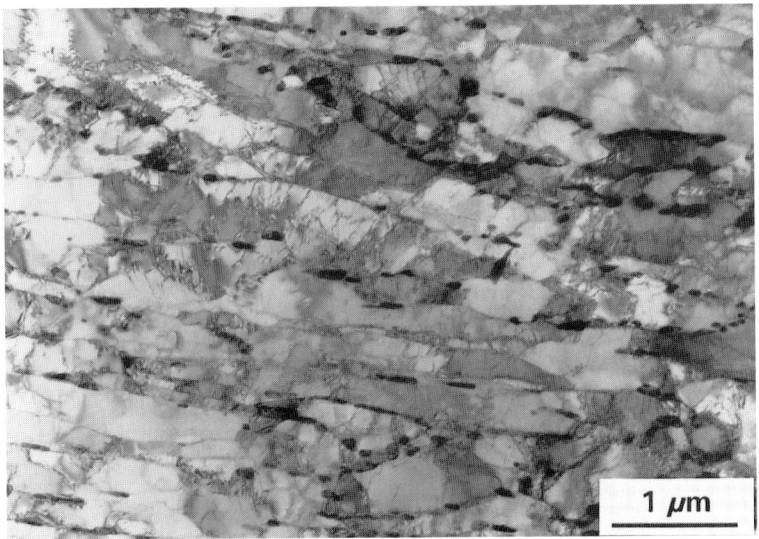

Fig. 2 TEM thin foil micrograph showing the normal ferritic subgrain microstructure with carbides in the steel X20CrMoV121.

Table 2. Specified heat treatment and mechanical properties of the steel X20CrMoV121.

Heat treatment:		
Austenitising: 1020°C – 1070°C/ air cool or oil quench		
Tempering: 730°C – 780°C		
R_{mt}	690 MPa – 840 MPa	
$R_{p0,2}$	> 490 MPa	
$\sigma_{R,100,000\ h}$	500°C	235 MPa
	550°C	128 MPa
	600°C	59 MPa

mechanical properties are shown in Table 2. It should be noted here that all hot forming operations like e.g. pipe bending, should be followed by a full new normalising and tempering heat treatment to restore the mechanical properties. Welding can be done after preheating to 200°C or 400°C. After welding the weldment must be cooled to a temperature in the interval 150°C–100°C to ensure full martensitic transformation of all weld metal before post weld heat treatment (PWHT). It is important that the weld should not be allowed to cool below 100°C before PWHT because of the risk for cold cracking. With such a complicated welding procedure the X20CrMoV121 steel is considered to be difficult to weld.

2.2   Microstructural Stability

The tempered martensite microstructure in X20CrMoV121 consisting of ferritic subgrains with $M_{23}C_6$ carbides is highly stable under normal service exposure in the temperature range 540°C–560°C. Microstructural stability is due to the relatively slow growth of the $M_{23}C_6$ carbides, which support the subgrain microstructure. Stability of the microstructure was studied in a Danish research project by long term isothermal annealing experiments. Results indicated that severe softening of the microstructure would not occur before 500 000–1 000 000 hours of exposure at 540°C–560°C.[7] These results were later supported by microstructural investigations of a sample from a turbine overflow pipe from Germany which had been service exposed for approximately 200 000 hours at 540°C. Only very weak effects of the service exposure could be recognised in the microstructure of this material, Fig. 3. Further investigations of service exposed material from Danish power plants showed that precipitation of vanadium rich MX nitrides can occur during service exposure. This contributes further to the microstructural stability, and it indicates that even though nitrogen is not a specified alloying element in the

Fig. 3 TEM thin foil micrograph taken from X20CrMoV21 material service exposed for 200,000 hours at 540°C.

X20CrMoV121 steel the tramp nitrogen content contributes significantly to creep strength of the steel.[8,9]

3. SERVICE PERFORMANCE

Service experience with low alloyed steam pipe steels such as 14MoV63 (0.5Cr0.5Mo0.25V); 13CrMo44 (1Cr0.5Mo) and 10CrMo910 (2.25Cr1Mo) show that almost all defects are found in circumferential welds (Only seamless pipes are used in high temperature steam pipes in Danish power plants). Damage or defects in the parent metal of steam pipe components are very seldom found, and nearly always related to incorrect production heat treatment or improper chemical composition leading to low creep strength. Inspections of service exposed welds have revealed defects either introduced during the production of the welds or in the form of creep cracks in the intercritically heated zone of the weld HAZ – the so called type IV cracks.[10] The intercritically heated zone of the HAZ has low creep strength due to the heat cycle imposed on the material during welding, which deteriorates the microstructure. Type IV cracks occur because of high stresses across the weld induced by the pipe system, mainly due to improper design or malfunction of the pipe hanger system. Type IV cracks can occur well before 100 000 hours of service.

Microstructural investigations and creep testing of X20CrMoV121 welds have demonstrated that this steel is also susceptible to type IV cracking.[11,12] It has therefore been quite surprising to find that the defect rate of service exposed X20CrMoV121 welds has been extremely low. In fact not a single type IV crack has been found during inspections of X20CrMoV121 steam pipe systems in Danish power plants. Also the number of defects introduced during weld production has been low. A number of reasons can be given for this behaviour. Firstly the fact that X20CrMoV121 is considered to be difficult to weld has meant that more care has been taken during welding and subsequent NDT of the welds, which resulted in a lower frequency of production defects. Secondly the quality of pipe hanger systems have been improved compared to the older power stations with pipe systems made of low alloyed steels. Additionally more attention has been paid to the design of the pipe system in order to reduce system stresses. This has contributed to the suppression of type IV cracks. Finally the fact that a stronger steel can be less sensitive to type IV cracking is of significance in this respect.[13]

The very low defect rate of welds means that the overall service performance of the X20CrMoV121 steel in steam pipes must be characterised as excellent. There have however been a few parent material problems with this steel in combination with errors in production heat treatments, leading to low creep strength material in some components.

3.1   Incorrect Production Heat Treatment

The creep properties of the steel X20CrMoV121 seem to be more sensitive to deviations from the specified heat treatment than found for the low alloyed steels. This has led to adverse service experiences with the steel caused by use of an incorrect production heat treatment. The first reported incident with X20CrMoV121 came from a Belgian power station in the early 1970s, where a turbine overflow pipe burst after 43 000 hours in service at 550°C and a hoop stress of 100 MPa.[14] After several investigations of the material it was finally concluded that the pipe had not been properly heat treated after induction bending, and this had led to low creep strength. In the middle of the 1980s reports came from the UK of X20CrMoV121 superheater tube bursts after a service life of only 16 000 hours. The tube bursts were caused by creep and not by excessive service temperatures. Cause of the low creep strength was a production heat treatment with insufficient austenitisation temperature in the range 900–950°C.[15] The final example from outside Denmark is from a German power plant, where inspections of a live steam pipe system in 1988 and 1989 revealed creep damage in 23 out of 56 inspected pipe bends after a service exposure of approximately 120 000 hours at 535°C/220 bar. In six cases the damage was so severe that the bends had to be immediately replaced. Investigations of the incident showed that the bends had not been cooled to full martensitic transformation during the normalising and tempering heat treatment, which followed the bending operation.[16]

In Denmark the first example of wrong production heat treatment was found in 1987. Replica inspections of a steam pipe after 140 000 hours at 540°C revealed abnormal microstructure in three stub pipes, which came from a special delivery. The material was cut out and creep testing revealed a creep strength, which was 50% of the mean standard value. Microstructural investigations showed large ferritic grains with spheroidised $M_{23}C_6$ carbides. This microstructure is an indication of a heat treatment with an insufficient austenitisation temperature in the range 900 to 950°C. Such a heat treatment is insufficient to transform the steel fully to austenite, and it will leave some of the $M_{23}C_6$ carbides undissolved and spheroidised. The resultant microstructure with large ferritic subgrains and large carbides has low creep strength.

Common to all the materials mentioned in the examples of incorrect production heat treatment is that they can fulfil all requirements on short term mechanical properties set by the delivery standard DIN 17175. This means that even though all of the delivered material fulfils the standard requirements there is still a risk that some components may have a low creep strength. Based on the above examples it was considered that components at risk are those which come from special deliveries like pipe bends and T-pieces. Material out of large production lots like straight pipes were generally not considered to be at risk. In 1987 it was therefore recommended in Denmark to require extra documentation for the production heat treatment of all X20CrMoV121 material delivered to new power plants, and to make replica inspections of all delivered pipe bends to new plants. This recommendation was followed for four new plants built after 1987, and no problems were found. There were however still uncertainties about the material delivered before 1987. It was therefore decided to make replica inspections of all bends and T-pieces in older power plants during normal overhauls made after 100 000 hours of service. Such inspections have been made at four power plants, and no problems were found until 1994.

In 1994 replica inspections were made at the Herningvaerket 90 MW combined heat and power plant after 60 000 hours of service at 525°C. Inspections were made rather early in service life because the plant experienced a large proportion of cyclic load. In seven bends of the same dimension abnormal microstructures were found, Fig. 4. Examination of delivery records revealed that the seven bends were delivered by the same manufacturer, and probably heat treated at the same time after induction bending. This indicated an incorrect production heat treatment. The microstructure of the bends contained 50% of a non tempered martensite microstructure indicating that they had not been cooled to full martensitic transformation after normalising and before tempering. Such a heat treatment mistake is well possible, as air cooling of a bend through the martensitic transformation range from 300°C to 100°C takes several hours.

Fig. 4 Optical micrograph of a replica from a service exposed incorrrectly heat treated bend containing app. 50% of untempered martensite at service start.

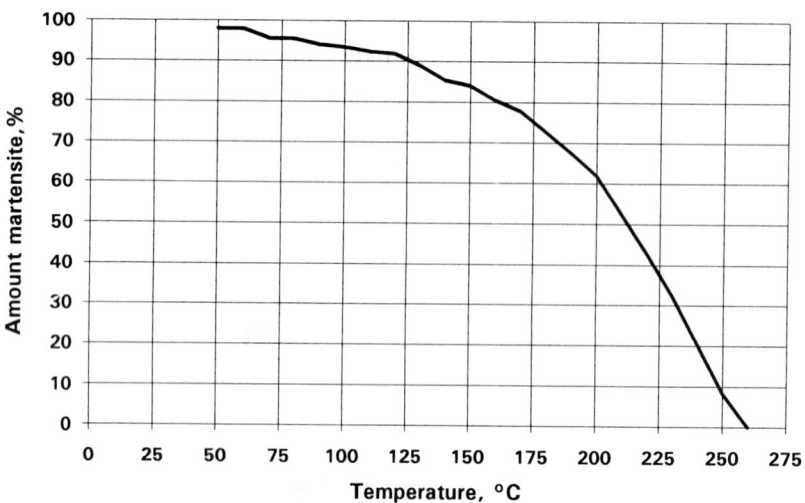

Fig. 5 Martensite transformation in X20CrMoV121.

No creep cavitation was found in any of the inspected bends, and as the microstructure contained a smaller amount of abnormal microstructure than was found in 1988 in the German power plant mentioned above, there was no need for immediate replacement of the bends. A test program was therefore initiated to find the creep strength of the incorrectly heat treated material.

Material from one of the incorrectly heat treated bent pipes was available, as it had been cut out during the production stage. Simulation heat treatments were made to recreate the incorrect microstructure which was present in the power plant at the beginning of service. Dilatometric tests determined the martensite transformation in the steel, Fig. 5, and the heat treatment cycle to recreate the incorrect microstructure could then be chosen:

1050°C/1 h – air cooling to 220°C + 750°C/2 h/AC to room temp.

Further, a fully correct heat treatment was made to test reference material.

Creep tests were made at 100 MPa of the two material conditions and isostress extrapolations to service temperature were made, Fig. 6. The material which had received a fully correct heat treatment showed an extrapolated creep life similar to the mean standard value, whereas the material with incorrect heat treatment showed a strength lower than the standard mean value, but higher than the standard minimum value. The shape of the creep strain curves indicate that the untempered martensite microstructure in the material with incorrect heat treatment is highly unstable, Fig. 7. At the longest testing times carbide precipitation takes place in the untempered martensite, and this leads to low rupture ductility. As the isostress extrapolation method is based on the assumption that the microstructural development is stable over the interesting temperature range, it was uncertain if the extrapolated creep lives of the incorrectly heat treated materials were too optimistic.

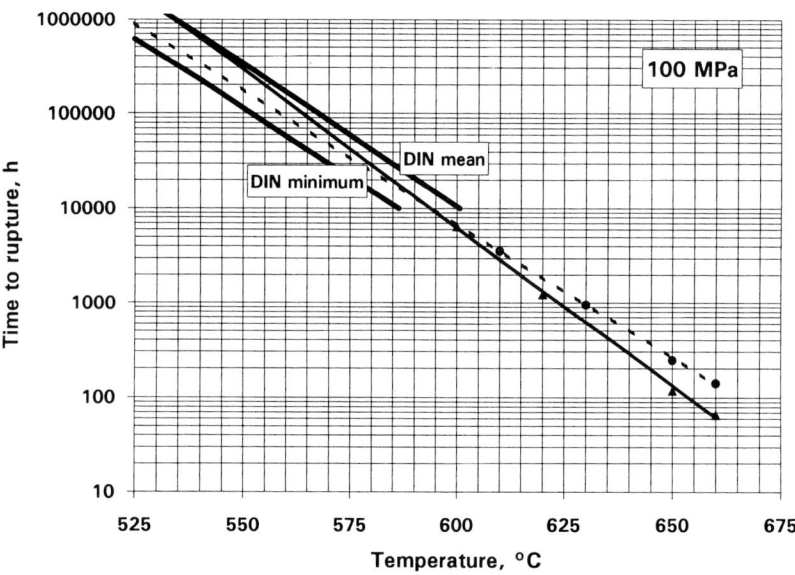

Fig. 6 Creep testing of normal and incorrectly heat treated X20CrMoV121 material.

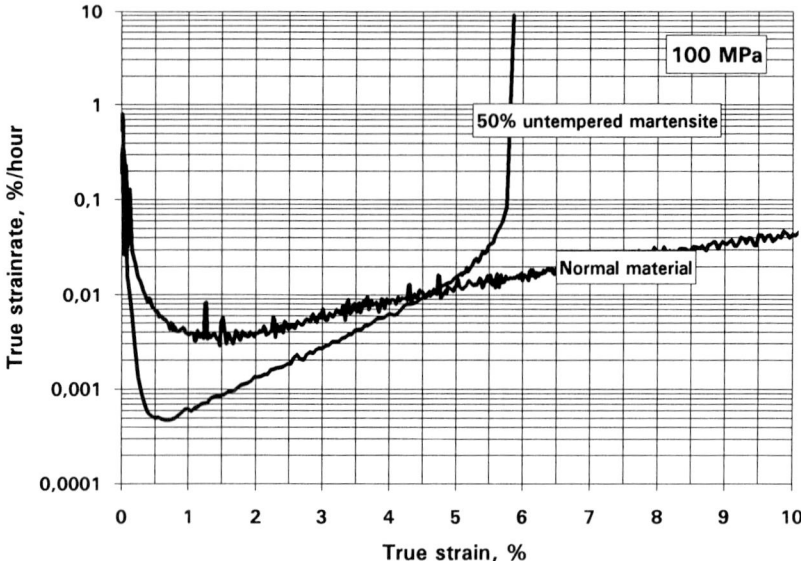

Fig. 7 Creep strain behaviour of normal X20CrMoV121 material at 620°C and incorrectly heat treated X20CrMoV121 material at 610°C.

It was therefore decided to cut out a sample of one of the bent pipes to make isostress creep tests on service exposed material. The microstructure of this material had stabilised during service exposure, and therefore a less uncertain isostress extrapolation could be made. The creep tests on the sample were finished in 1996, and the extrapolated creep life was higher than the standard minimum value. It was therefore not considered necessary to replace the incorrectly heat treated bends.

4. SUMMARY

A review of service experiences with the 12CrMoV steel X20CrMoV121 in steam pipes of Danish power plants has been made. Service experience with the steel has been excellent. Unlike previous experience with steam pipes of low alloy steels the X20CrMoV121 showed no type IV damage in welded joints. This is probably due to improved designs of pipe systems and hangers and to better workmanship during the production stages.

However the creep strength of the X20CrMoV121 steel seems to be more sensitive to deviations from the specified heat treatment than for the low alloy steels. This has led to a few incidents of creep damage or necessary replacements of components with abnormal microstructure. Heat treatments leading to low creep strength are: (1) insufficient austenitisation temperature in the range 900°C–950°C and (2) insufficient cooling after austenitisation leading to only partial martensite transformation.

Examples of both these types of incorrect heat treatment have been found in Danish power plants. Common to both types of incorrectly heat treated material are that they can fulfil all of the standard delivery specification requirements based on short term mechanical properties. It is therefore recommended to require extra documentation for the heat treatment when buying new X20CrMoV121 material and to routinely inspect the microstructure of compoments like bends and T-pieces of older pipe systems.

The excellent service experiences with the X20CrMoV121 steel has facilitated the descision for the Danish utilities to introduce the new modified 9Cr steels like P91 and NF616 at further advanced steam conditions. The new modified 9Cr steels are martensitic stainless steels like the X20CrMoV121, so it is also recommended for these steels to require extra documentation and inspection to assure that the specified heat treatment has been followed.

REFERENCES

1. The HP Pipe Committee: *Damage to pipe systems of 14MoV63*, The Mechanical Engineering Materials Committee of the Danish Primary Utilities, July 1981.
2. The HP Pipe Committee: *Damage to pipe systems of 13CrMo44*, The Mechanical Engineering Materials Committee of the Danish Primary Utilities, June 1986.
3. The HP Pipe Committee: *Damage to pipe systems of 10CrMo910*, The Mechanical Engineering Materials Committee of the Danish Primary Utilities, 1991.
4. J.Z. Briggs and T.D. Parker: *The Super 12% Cr Steels*, Climax Molybdenum Company, New York, 1965.
5. H. Jesper and H.R. Kautz: *Proc. Conf. Werkstoffe und Schweisstechnik im Kraftwerk 1985*, Essen, 1985, Paper 9.
6. E. Kauhausen and P. Kaesmacher: *Schweissen und Schneiden*, 1957, 9(9), 414–419.
7. K. Borggreen and P.B. Mortensen: *Longterm properties of X20CrMoV121*, The Danish Corrosion Center, November 1989.
8. P. Battaini, D. DÁngelo, G. Marino and J. Hald: *Proc. Conf. Creep and fracture of Engineering Materials and Structures*, Swansea, April 1989, The Institute of Materials 1989, 1039–1054.
9. J. Hald: *Steel Research*, 1996, **67**(9), 369–374.
10. R. Blum, J. Hald: *Proc. Conf. Life Extension of Thermal Power Plants*, Moscow, May 1994, VGB Essen, 1994, Paper 16.
11. J. Hald: *Proc. Conf. Rupture Ductility of Creep Resistant Steels*, (Ed. A. Strang), York, December 1990, Institute of Metals, 1991.
12. C. Coussement and M. de Witte: *Proc. Conf. Joining/Welding 2000*, The Hague, July 1991, IIW 1991.

13. C.J. Middleton: *Proc. Conf. Power Plants of the 90's*, I Mech E, London 1990.
14. G. Darmont and C. van Melsen: *Revue de Métallurgie*, Avril 1973, 317–324.
15. D.R. Barraclough and D.J. Gooch: *Materials Science and Technology*, 1985, 1(11) 961–967.
16. O. Wachter, H. Müsch and W. Bendick: *VGB Kraftwerkstechnik*, 1991, **71**(10), 971-979.

Evaluation of the Creep Strength Property from a Viewpoint of Inherent Creep Strength for Ferritic Creep Resistant Steels

K. KIMURA,* H. KUSHIMA,* F. ABE* AND K. YAGI†

* *Environmental Performance Division, National Research Institute for Metals, 1–2–1 Sengen, Tsukuba, Ibaraki 305, Japan.*
† *Planning Office, National Research Institute for Metals, 1–2–1 Sengen, Tsukuba, Ibaraki 305, Japan*

ABSTRACT

Inherent creep strength is defined as a constant creep strength independent of time and microstructure. Sigmoidal inflection of a relation between stress and time to rupture is explainable in terms of a loss of creep strength due to microstructural change followed by the advent of inherent creep strength. Although a large variation of short term creep rupture strength is observed, inherent creep strengths are evaluated to be almost the same for several low alloy Cr–Mo steels. Creep rupture strength of the 12Cr–1Mo–1W–0.3V steels is higher than the other ferritic creep resistant steels investigated in the short term region, however, it decreases abruptly with a decrease in stress and it is almost the same as those of the low alloy Cr–Mo steels at a low stress of 30 MPa. On the other hand, creep rupture strength of the 9Cr–1Mo–V–Nb steel (T91) is higher than the other ferritic creep resistant steels investigated even at a low stress of 30 MPa, since the tempered martensitic microstructure of this steel is stable in comparison with that of the 12Cr–1Mo–1W–0.3V steel.

INTRODUCTION

More than 40 types of creep resistant steels and alloys have been subjected to long term creep test and numerous creep rupture data have been published on NRIM Creep Data Sheets. A concept of inherent creep strength has been previously proposed by the authors[1,2] on the basis of an analysis of these numerous long term creep rupture data.

In this paper, creep rupture strength properties are investigated for ferritic creep resistant steels with a special interest in inherent creep strength. Bolton *et al.*[3] mentioned that changes in microstructure affect a relationship between stress and time to rupture and it may reveal a sigmoidal inflection. Many efforts have been made to predict long term creep strength by extrapolating

185

short term data based on a parametric method,[4-6] however, it is still difficult. Complex relationships between stress and time to rupture may be the reason for this. Such complex relationships which shows sigmoidal inflection are explainable in terms of inherent creep strength. The difference in creep rupture strength properties between 12Cr–1Mo–1W–0.3V steel and 9Cr–1Mo–V–Nb steel (T91) will be described from a viewpoint of inherent creep strength.

EXPERIMENTAL

Creep rupture data of the NRIM Creep Data Sheets[7-18] have been investigated for many types of ferritic creep resistant steels. The Larson-Miller parameter was used for assessment of creep rupture strength and a constant of 20 was applied for all steels investigated. Creep rupture strength properties have been discussed for two high Cr steels of 12Cr–1Mo–1W–0.3V[17] and 9Cr–1Mo–V–Nb (T91)[18] in comparison with those of low alloy Cr–Mo steels. The chemical compositions, processing, thermal history and austenite grain size number of the two high Cr steels are given in Tables 1 and 2. Vickers hardness measurements were conducted for the creep ruptured specimen of the two high Cr steels, at room temperature under a load of 98 N. Microstructures of as tempered and creep ruptured specimen were examined under transmission electron microscope (TEM) for the two high Cr steels. Thin foils for TEM were prepared by spark cutting followed by mechanical grinding and jet-electropolishing at −10°C in a 10% perchloric acid in ethanol solution.

Table 1   Chemical compositions (mass%) of high Cr steels studied

Material	C	Si	Mn	P	S	Cr	Mo	W	V	Nb	N
12Cr–1Mo–1W–0.3V	0.25	0.35	0.70	0.016	0.008	11.9	0.99	0.95	0.28	–	0.021
9Cr–1Mo–V–Nb (T91)	0.09	0.29	0.35	0.009	0.002	8.70	0.90	–	0.22	0.072	0.044

Table 2   Processing, thermal history and austenite grain size number of high Cr steels studied

Material	Processing	Thermal history	Austenite grain size number
12Cr–1Mo–1W–0.3V	Hot rolled	1123K/3h → F.C. 1323K/46min → O.Q. 933K/2h → A.C. 913K/2h → A.C.	4.4
9Cr–1Mo–V–Nb (T91)	Hot extruded and cold drawn	1323K/10min → A.C. 1038K/30min → A.C.	9.2

RESULTS AND DISCUSSION

Sigmoidal Inflection and Inherent Creep Strength
Sigmoidal inflection of a relation between stress and time to rupture is frequently observed for ferritic creep resistant steels. An experimental example is shown in Fig. 1 for the 0.5Mo steels[7] taken from heat exchanger tubes (JIS STBA12) tested over a range of absolute temperatures from 723 and 823 K. Solid curves were obtained by regression analysis using the Orr-Sherby-Dorn parameter.[4] The inflections of the curves were observed at about 200 and 100 MPa. The reason why accurate and reliable predictions of long term creep strength from short term creep testing is difficult, may be due to such sigmoidal inflection. With a decrease in stress, creep rupture life becomes shorter than the dashed lines obtained by linearly extrapolating the short term data. The former inflection may be caused by a decrease in creep strength due to changes in microstructure. In view of the changes in microstructure and hardness of the creep ruptured specimen, it has been speculated that an occurrence of a constant creep strength may lead to the latter inflection at the lower stress. A constant creep strength may appear after long term creep deformation, since strengthening effects which depend on microstructural morphology, such as precipitation strengthening decrease with an increase in exposure time and finally disappear as a result of microstructural change. Such constant

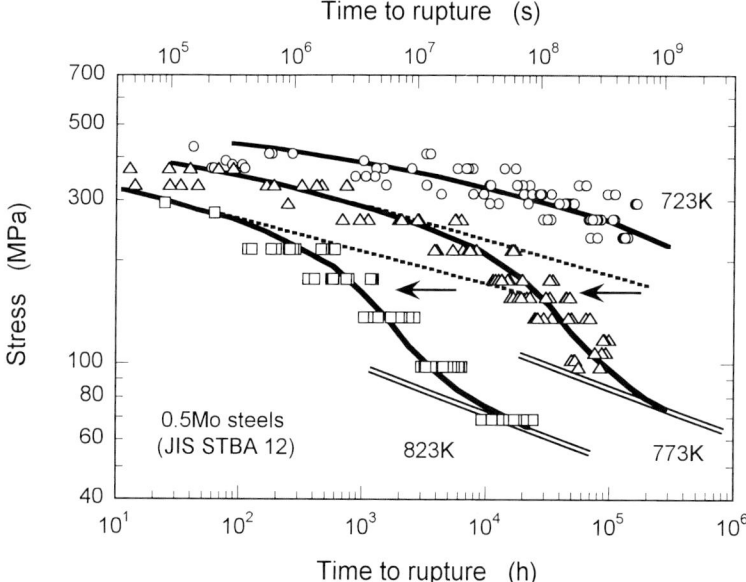

Fig. 1 Stress versus time to rupture for multiple heats of the 0.5Mo steels (JIS STBA12)[7] tested over a range of temperatures.

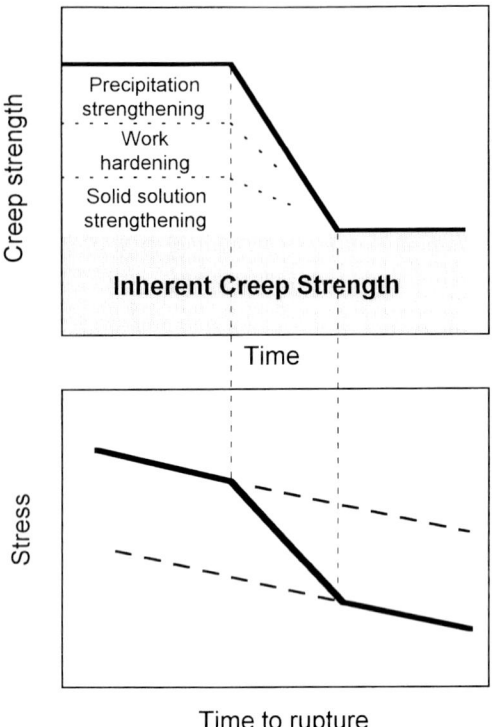

Fig. 2 Schematic illustration on a mechanism of sigmoidal inflection of a relationship between stress and time to rupture.

creep strength has previously been proposed by the authors as an inherent creep strength.[1,2]

Figure 2 shows a schematic illustration of a mechanism of sigmoidal inflection for a relation between stress and time to rupture. Although creep strength can be improved by several strengthening factors, a gradual loss of creep strength is caused by changes in microstructure during thermal exposure and creep deformation. The slope of a stress versus time to rupture curve increases with a decrease in stress as a result of a loss in creep strength. The latter inflection that occurs at long term condition where the creep strength decreases to an inherent creep strength. Sigmoidal inflection may be caused by a loss of creep strength followed by the advent of the proposed inherent creep strength.

Inherent Creep Strength of Low Alloy Cr–Mo Steels
Creep rupture strength data plotted against LMP (C = 20) are shown in Fig. 3 for several low alloy Cr–Mo steels (0.5Cr–0.5Mo,[8] 1Cr–0.5Mo,[9,10] 1.25Cr–0.5Mo–Si[11,12] and 2.25Cr–1Mo[13–15]). The data indicate a large scatter

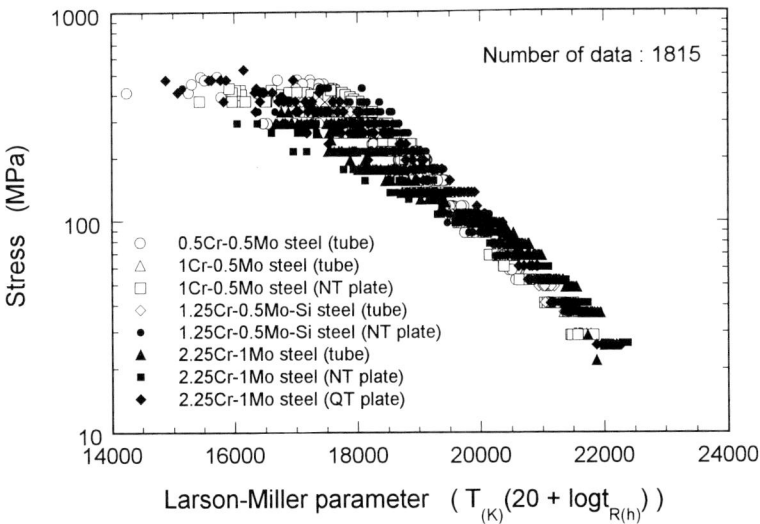

Fig. 3   Creep rupture strength properties for several low alloy Cr–Mo steels. [8–15]

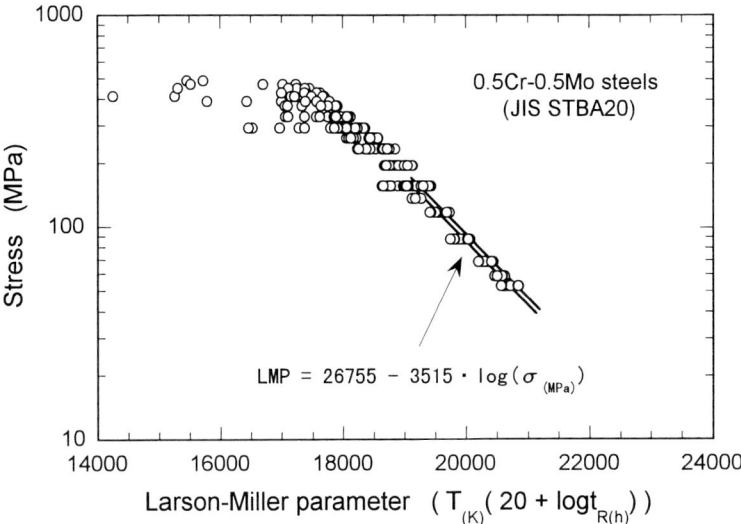

Fig. 4   Creep rupture strength property plotted against Larson-Miller parameter (C = 20) for the multiple heats of the 0.5Cr–0.5Mo steels (JIS STBA20).[8]

of short term creep rupture strength. At 300 MPa, the scatter of the creep rupture strength is more than 3 orders of magnitude in rupture life. With decrease in stress, however, widely scattered short term data have a tendency to converge on a narrow range of creep rupture strength.

Creep rupture data plotted against LMP (C = 20) are shown in Fig. 4 for multiple heats of the 0.5Cr–0.5Mo steels (JIS STBA20).[8] Although the data indicate a large spread of the creep rupture strength at the stresses higher than 200 MPa, convergence of the data to a common linear line is observed at the stresses lower than 100 MPa. The creep rupture data which showing such common linear relationship seems to reflect the inherent creep strength, as mentioned above. Inherent creep strength property of the 0.5Cr–0.5Mo steel, consequently, has been evaluated from a linear relationship as follows,

$$LMP = 26755 - 3515 \cdot \log \sigma \qquad (1)$$

where LMP is a Larson-Miller parameter whose constant is 20 and σ is the stress in MPa. The same analysis on the inherent creep strength properties are conducted for the multiple heats of each low alloy Cr–Mo steels and the results are listed in Table 3. Inherent creep strength properties indicated in Table 3 are illustrated in the stress versus time to rupture diagram at 823 K and shown in Fig. 5. Inherent creep strength of the low alloy Cr–Mo steels have a tendency to increase with an increase in Cr content. However, differences in inherent creep strength of the Cr–Mo steels are less than 5 times in rupture life and those are very small in comparison with the widely scattered short term creep rupture strength (Fig. 3). It may be speculated that large differences in short term creep rupture strength are attributable to strengthening effects which depend on microstructural morphology. The convergence of creep rupture strength to a narrow range may occur in a long term region since the inherent creep strength are almost at the same level for the low alloy Cr–Mo steels.

Creep Strength Properties of High Cr Steels
Creep rupture data for 11 types of ferritic creep resistant steels[7–9, 11, 15–18] are shown in Fig. 6. To reveal the creep rupture strength properties of the 12Cr–1Mo–1W–0.3V steel[17] and the 9Cr–1Mo–V–Nb steel,[18] solid symbols of ● and ▲ are used for the former and latter steels, respectively. The creep rupture strength of the 12Cr–1Mo–1W–0.3V steel is the highest in the short term region, however, it decreases with decrease in stress abruptly and it is almost the same as those of the low alloy Cr–Mo steels at a stress of 30 MPa. Consequently, inherent creep strength of the 12Cr–1Mo–1W–0.3V steel seems to be almost the same as those of the low alloy Cr–Mo steels. On the other hand, short term creep rupture strength of the 9Cr–1Mo–V–Nb steel is lower than that of the 12Cr–1Mo–1W–0.3V steel. However, no abrupt decrease in creep rupture strength is observed for the 9Cr–1Mo–V–Nb steel over a range of stress from 450 to 30 MPa. Although long term creep rupture strength at a range of stresses from 50 to 20 MPa are almost the same for all

Table 3   Inherent creep strength properties evaluated for low alloy Cr–Mo steels.

Material		LMP = A − B·log ($\sigma_{(MPa)}$)	
Nominal composition	Specification	A	B
0.5Cr–0.5Mo steel[8]	JIS STBA20	26755	3515
1Cr–0.5Mo steel[9]	JIS STBA22	27462	3824
1Cr–0.5Mo steel[10]	JIS SCMV2NT	27018	3566
1.25Cr–0.5Mo–Si steel[11]	JIS STBA23	28424	4350
1.25Cr–0.5Mo–Si steel[12]	JIS SCMV3NT	27545	3835
2.25Cr–1Mo steel[13]	JIS STBA24	28593	4232
2.25Cr–1Mo steel[14]	JIS SCMV4NT	28392	4165
2.25Cr–1Mo steel[15]	ASTM A542	26890	3413

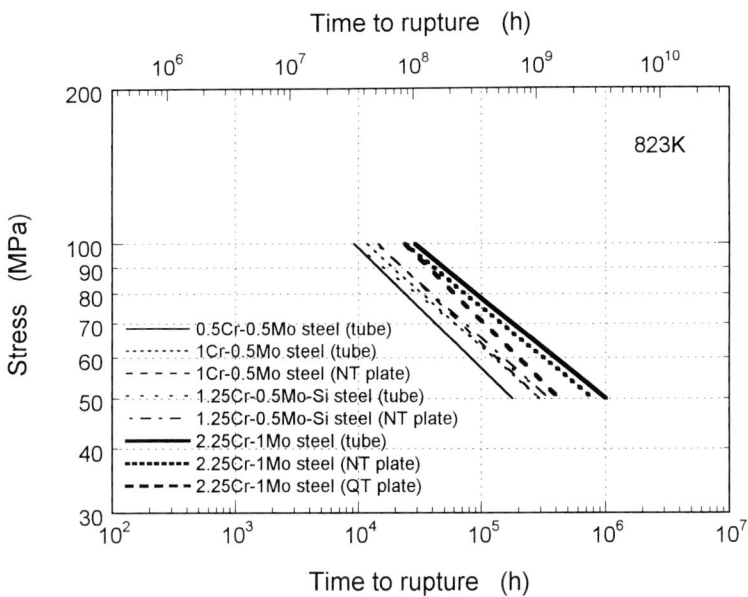

Fig. 5   Inherent creep strength properties plotted on a stress versus time to rupture diagram at 823 K for several low alloy Cr–Mo steels.[8–15]

of the ferritic creep resistant steels, that of the 9Cr–1Mo–V–Nb steel is higher than the others even at such low stress and long term condition.

Vickers hardness of the as tempered and creep ruptured specimens are plotted against LMP and shown in Fig. 7 for the two high Cr steels. The creep ruptured 12Cr–1Mo–1W–0.3V steels indicate very high hardness over H_v = 300 in the short term region and it is remarkably high compared with

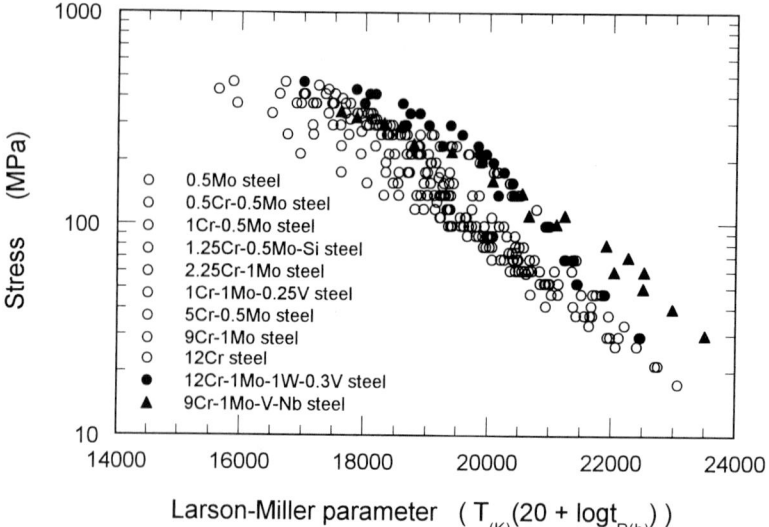

Fig. 6   Creep rupture strength properties for 11 types of ferritic creep resistant steels.[7–9, 11, 15–18]

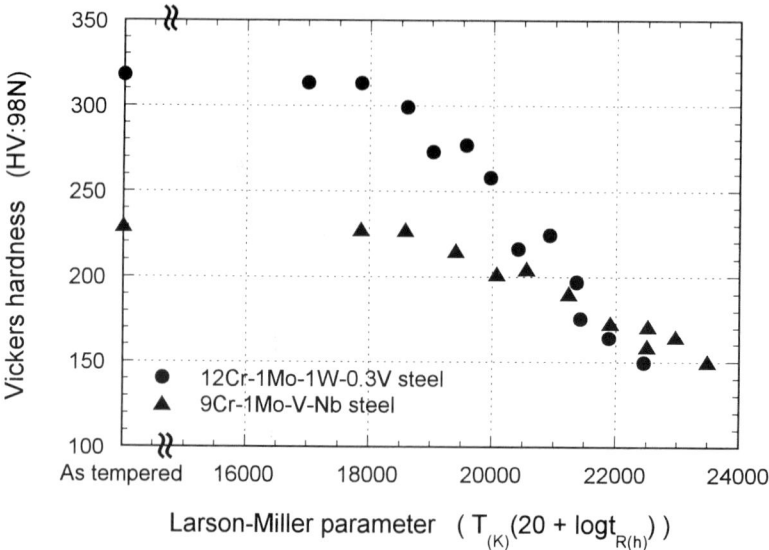

Fig. 7   Comparison of the hardness of the as tempered and creep ruptured 12Cr–1Mo–1W–0.3V steels[17] and 9Cr–1Mo–V–Nb steels.[18]

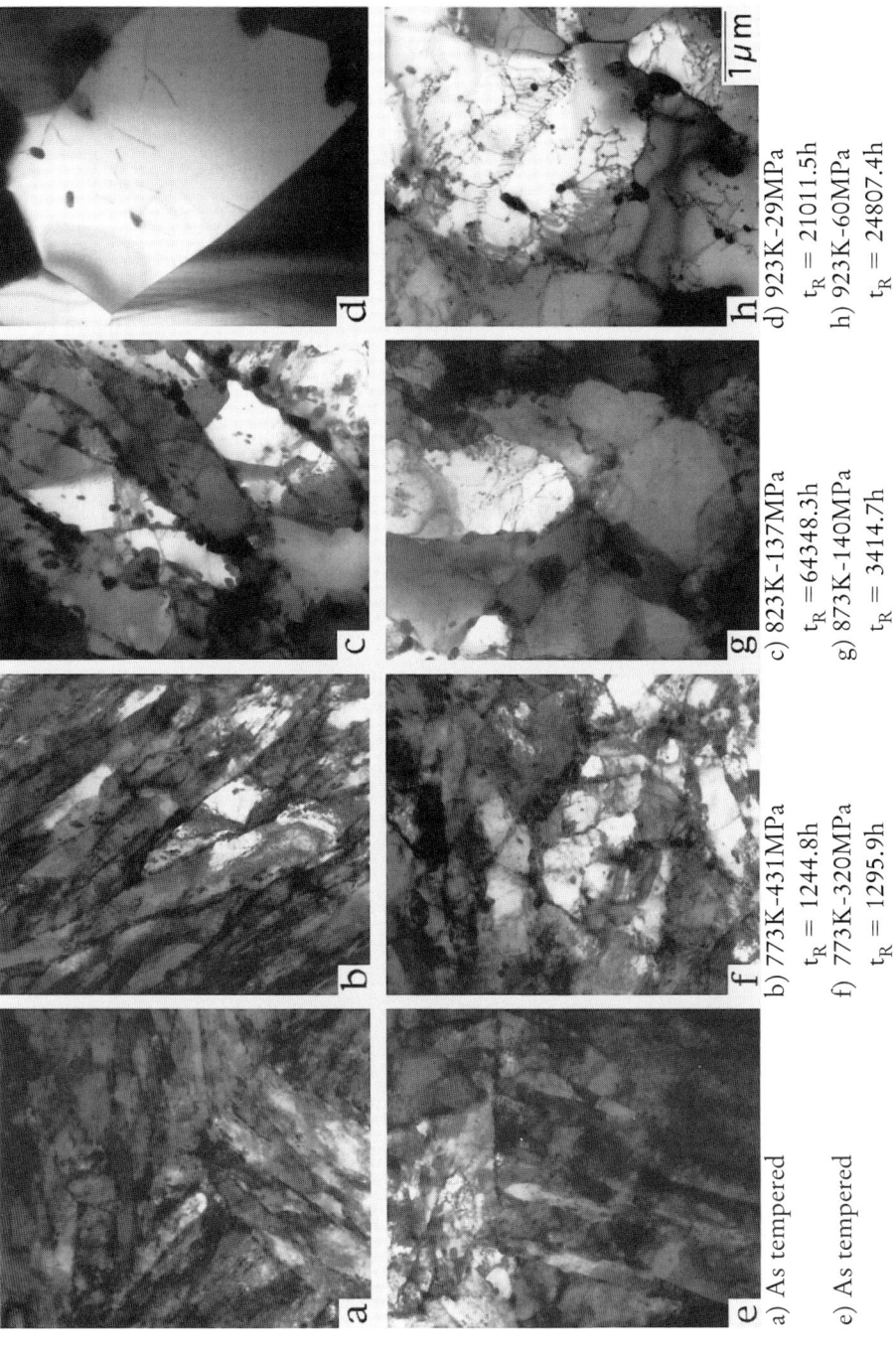

Fig. 8 Transmission electron micrographs of the as tempered and creep ruptured specimen of 12Cr–1Mo–1W–0.3V steels (a)–(d) and 9Cr–1Mo–V–Nb steels (e)–(h).

a) As tempered

b) 773K–431MPa
   t_R = 1244.8h

c) 823K–137MPa
   t_R =64348.3h

d) 923K–29MPa
   t_R = 21011.5h

e) As tempered

f) 773K–320MPa
   t_R = 1295.9h

g) 873K–140MPa
   t_R = 3414.7h

h) 923K–60MPa
   t_R = 24807.4h

that of the 9Cr–1Mo–V–Nb steel which is about $H_v = 230$. The very high hardness of the creep ruptured 12Cr–1Mo–1W–0.3V steels, however, decreases with an increase in LMP abruptly, as well as the creep rupture strength. Hardness of the creep ruptured 9Cr–1Mo–V–Nb steel decreases gradually with an increase in LMP. Contrary to the short term region, hardness of the creep ruptured 9Cr–1Mo–V–Nb steel is higher than that of the 12Cr–1Mo–1W–0.3V steel in the long term region.

Transmission electron micrographs of the as tempered and creep ruptured specimens are shown in Fig. 8 for the 12Cr–1Mo–1W–0.3V steel (a)–(d) and the 9Cr–1Mo–V–Nb steel (e)–(h). LMP of the creep ruptured specimens are (b) 17 853, (c) 20 417, (d) 22 450, (f) 17 866, (g) 20 545 and (h) 22 516, and LMP of the (b), (c) and (d) is almost same as those of (f), (g) and (h), respectively. Tempered martensitic microstructure whose dislocation density is very high is observed for both (a) and (e). The slightly recovered area is observed in only a small portion of the creep ruptured specimens (b) and (f). Directional martensitic lath boundaries are distinctly observed on the specimen (b) rather than (f). Progress in recovery of the microstructure is observed for both (c) and (g). The subgrain size of (c) is almost same as that of (g) and less than 10 μm. Many coarse carbide particles whose size is 100–200 nm are observed on the martensitic lath boundaries for the specimen (c). The microstructure of the specimen (d) is a fully tempered one and there are no fine carbide particles. On the other hand, large amounts of very fine carbide particles less than 100 nm in size and a dislocation network are still observed on the specimen (h). Differences in microstructural features of the two high Cr steels are summarised as follows:

- Recovery of tempered martensitic microstructure of the 12Cr–1Mo–1W–0.3V steel proceeds rapidly compared with that of the 9Cr–1Mo–V–Nb steel.
- Directional martensitic lath boundaries are prominent for the 12Cr–1Mo–1W–0.3V steel.
- Many coarse carbide particles are present at martensitic lath boundaries of the 12Cr–1Mo–1W–0.3V steel.
- A large amount of fine carbide particles exist even at long term condition for the 9Cr–1Mo–V–Nb steel.

Such differences in microstructural stability may be caused by the differences in chemical composition and heat treatment of the two high Cr steels. Large amounts of carbon content, rapid quenching speed due to oil quench and lower tempering temperature of the 12Cr–1Mo–1W–0.3V steel may promote a recovery of the tempered martensitic microstructure. A stable microstructure of the 9Cr–1Mo–V–Nb steel seems to be obtained by the addition of small amounts of Nb which forms a stable carbide, in addition to the low carbon content, slower quenching and a higher tempering temperature. From a viewpoint of inherent creep strength, to improve the stability of the microstructure contributes to prevent a creep strength from dropping into an inherent creep strength and to maintain an excellent creep strength for a longer duration.

CONCLUSIONS

Creep rupture strength properties have been investigated for many ferritic creep resistant steels by considering inherent creep strength and the following conclusions have been drawn.

1. Inherent creep strength is a constant creep strength independent of time and microstructure. Sigmoidal inflection of a relationship between stress and time to rupture is caused by the loss of creep strength due to the microstructural change followed by the advent of inherent creep strength.
2. Short term creep rupture strength of low alloy Cr–Mo steels indicate a large scatter, however, inherent creep strength seems to be almost the same for those steels.
3. Very high creep rupture strength in the short term region of the 12Cr–1Mo–1W–0.3V steel decreases with an increase in the LMP abruptly. Creep rupture strength of the 12Cr–1Mo–1W–0.3V steel is almost the same as those for the low alloy Cr–Mo steels at a range of stress from 20 to 50 MPa, however, that of the 9Cr–1Mo–V–Nb steel is higher than the other steels even at low stresses.
4. Tempered martensitic microstructure of the 9Cr–1Mo–V–Nb steel is stable in comparison with that of the 12Cr–1Mo–1W–0.3V steel. An excellent long term creep strength property of the 9Cr–1Mo–V–Nb steel is obtained by a stable microstructure. Stability and effectiveness of the strengthening effect is clearly evaluated by considering a concept of inherent creep strength.

REFERENCES

1. K. Kimura, H. Kushima, K. Yagi and C. Tanaka: *Tetsu-to-Hagané (J. Iron Steel Inst. Japan)*, 1991, **77**, 667.
2. K. Kimura, H. Kushima, K. Yagi and C. Tanaka: *Proc. Fifth Inter. Conf. on 'Creep and Fracture of Engineering Materials and Structures'*, eds B. Wilshire and R.W. Evans, The Institute of Materials, London, 1993, 555.
3. C.J. Bolton, B.F. Dyson and K.R. Williams: *Mater. Sci. Eng.*, 1980, **46**, 231.
4. R.L. Orr, O.D. Sherby and J.E. Dorn: *Trans. ASM*, 1954, **46**, 113.
5. S. Yokoi and Y. Monma: *Tetsu-to-Hagané (J. Iron Steel Inst. Japan)*, 1979, **65**, 831.
6. R.W. Evans and B. Wilshire: *Creep of Metals and Alloys*, The Institute of Metals, London, 1985.

7. NRIM Creep Data Sheet, National Research Institute for Metals, 1991, **8B**.
8. *ibid.*, 1994, **20B**.
9. *ibid.*, 1996, **1B**.
10. *ibid.*, 1990, **35A**.
11. *ibid.*, 1976, **2A**.
12. *ibid.*, 1994, **21B**.
13. *ibid.*, 1986, **3B**.
14. *ibid.*, 1980, **11A**.
15. *ibid.*, 1991, **36A**.
16. *ibid.*, 1990, **9B**, 1992, **12B**, 1981, **19A**, 1994, **13B**.
17. *ibid.*, 1979, **10A**.
18. *ibid.*, 1996, **43**.

High Resolution Microanalysis of Ferritic Steel HCM12A

M. SCHWIND,* M. HÄTTESTRAND AND H.-O. ANDRÉN

Department of Physics, Chalmers University of Technology, S–412 96 Göteborg, Sweden
**Department of Materials Science and Engineering, Royal Institute of Technology, S–100 44 Stockholm, Sweden*

ABSTRACT

The martensitic chromium steel HCM12A has been investigated using APFIM (atom probe field ion microscopy), a technique capable of chemical analysis down to atomic resolution. The investigated material was isothermally aged at 600°C for times ranging from 0 to 10 000 h. Apart from the matrix, precipitates of type $M_{23}C_6$, MX and Laves phase have been analysed.

HCM12A contains nominally 0.9% copper (by weight) and it was found that the matrix concentration of copper drops during ageing from 0.4% to an equilibrium level at about 0.1%, which is in good agreement with previous thermodynamical calculations. No copper was found in $M_{23}C_6$, MX or Laves phase. The copper instead most probably forms a separate phase. During ageing the amount of tungsten in the matrix drops due to formation of Laves phase. This process is faster in HCM12A compared to steel NF616, indicating a beneficial effect of copper on the nucleation of Laves phase. The composition of Laves phase is close to $(Fe,Cr)_2(Mo,W)$, with additional small amounts of phosphorous, silicon and carbon. The composition of $M_{23}C_6$ is stoichiometric with respect to carbon and does not change much during ageing. Enhanced concentrations of boron were found inside the carbides. Phosphorous was found to segregate to a very narrow region at the carbide/matrix interface. Two different types of MX precipitates were analysed; vanadium nitride and vanadium carbonitride. The vanadium nitrides contained large amounts of chromium. No boron or phosphorous was found in the MX particles. A possible indirect effect of boron on the nucleation of MX on dislocations is discussed.

1. INTRODUCTION

In order to reduce environmental pollution from fossil fired power stations as well as fuel costs it is necessary to increase the thermal efficiency of the plants by raising operating pressures and temperatures. The most important limiting factor to an increase in operating conditions turns out to be the creep and thermal fatigue properties of the materials used in thick section boiler and turbine components. It is desirable to use ferritic alloys because of their higher thermal conductivity and lower thermal expansion coefficient com-

197

pared to austenitic alloys. Advanced 9–12% chromium steels have been recognised as the key materials.[1–4] For the most advanced ultra supercritical plants in use today, operating at 580°C/285 bar, grade P91 can be used. However to improve operating conditions beyond this, there has been a need for further alloy development. The two grades NF616 (P92) and HCM12A (P122) have creep properties superior to P91 and have both been approved by ASME.[5] These two steels contain about 2% tungsten, 0.5% molybdenum and small amounts of boron, which distinguish them from P91 regarding composition. The main difference between HCM12A and NF616 is the addition of copper to HCM12A with the intention to suppress δ-ferrite formation and allow a higher content of chromium.

This work concerns the microstructural evolution of steel HCM12A during isothermal ageing at 600°C up to 10 000 h. Atom probe field ion microscopy (APFIM) has been used to enable chemical analysis of the different phases present with the highest possible resolution. Hald has performed a TEM investigation of the same material as studied in this work.[6] He showed that the steel in the as-received condition (annealed and tempered) had the normal microstructure of tempered martensite consisting of ferritic subgrains with precipitates of the type $M_{23}C_6$ on subgrain boundaries and precipitates of the type MX inside the subgrains. Only very few regions of δ-ferrite were seen as expected from the copper addition. During ageing coarsening of $M_{23}C_6$ and MX took place but the most significant change in microstructure was the precipitation of Laves phase particles of relatively large size. After preparation of thin foils, holes were found in the foils indicating precipitation during tempering and ageing of copper particles, that were lost during the specimen preparation due to preferential etching. Hald also performed equilibrium calculations using the software package *Thermo-Calc* and the result was consistent with the observed microstructure.

To obtain a deeper understanding of the microstructural changes in the material, the unique properties of the atom probe have been used in this work to monitor the redistribution of elements during tempering and isothermal ageing.

2.   ATOM PROBE FIELD ION MICROSCOPY

Atom probe field ion microscopy (APFIM) is a powerful analytical tool in materials science, the most significant feature being the capability of nanometer scale chemical analysis.[7] Other advantages are that all elements can be analysed and that no calibration is involved in the measurement. In the field ion microscope (FIM), field ionisation of inert gas atoms is used to image single atoms on the surface of a needle shaped specimen. At atom probe (AP) consists of a mass spectrometer, with single ion sensitivity, connected to a FIM.

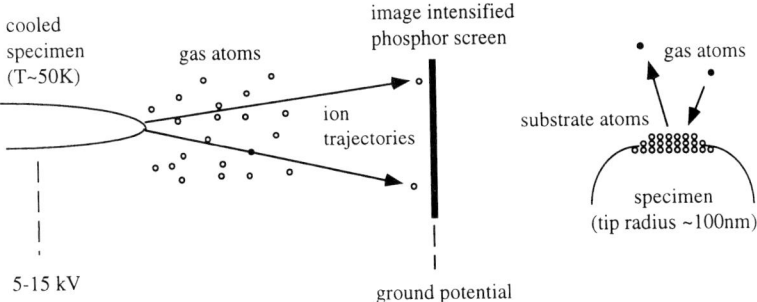

Fig. 1 Basic principles of field ion microscopy (FIM). Surface atoms on the tip of the sharp specimen are imaged on a phosphor screen by field ionised gas atoms (left picture). The gas atoms are ionised in the vicinity of prominent surface atoms, where the electric field is extremely high, and then repelled towards the phosphor screen (right picture).

The basic principles of the field ion microscope are shown in Fig. 1. The needle shaped specimen, with a tip radius of about 100 nm, is placed in a vacuum chamber, which is filled with an inert gas, e.g. neon or helium. A phosphor screen at ground potential is located a few centimetres away from the specimen tip. If the specimen is raised to a high enough positive potential, the field strength above the specimen surface will induce *field ionisation* of the inert gas atoms. This process will take place at sites above prominent surface atoms. After ionisation, the positively charged ions will be repelled from the specimen towards the phosphor screen, where a magnified projection of the surface atoms is obtained. Normally the magnification is about one million times and the spatial resolution of the image can be as good as a few Angstrom. The FIM image contains both crystallographic and chemical information and precipitates can often be seen as either bright or dark areas.

The basic phenomenon involved in operation of the atom probe is *field evaporation*, when atoms from the specimen surface are removed and ionised due to the strong electric field. In practise, the process takes place when high voltage pulses are applied to the specimen already exposed to the high DC voltage. The evaporated ions are then analysed with a time-of-flight mass spectrometer. The analysis procedure is straightforward (see Fig. 2). Ionised atoms from the specimen surface accelerate towards the phosphor screen. There is a small hole in the screen, which only permits ions from a small area on the specimen surface to enter into the mass spectrometer. The analysed area can be chosen, while watching the FIM image, by tilting the specimen until the hole in the screen falls over the image of the area of interest. The mass-to-charge ratio of each ion entering the spectrometer is derived from the

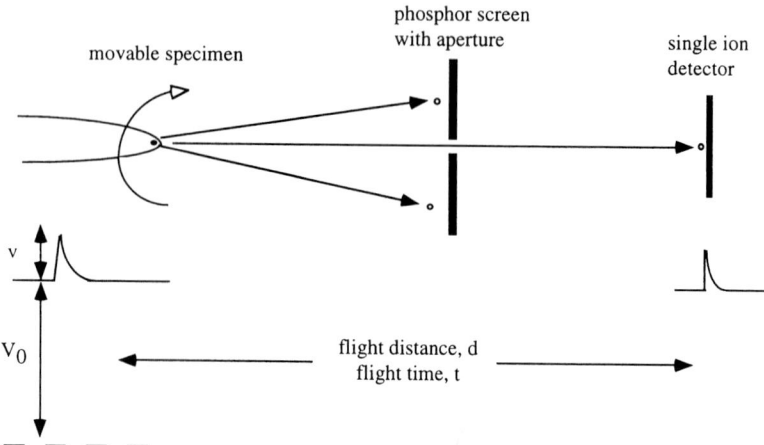

Fig. 2 The atom probe technique. Surface atoms from the specimen are field evaporated when high voltage pulses are applied to the tip. An aperture allows only ions from a selected area on the specimen to enter the spectrometer.

flight time needed to travel from the specimen to the detector. Conservation of energy yields the following relationship:

$$neV = m(d/t)^2/2 \tag{1}$$

where n is the charge state of the ion, V the applied voltage during the pulse, m the ionic mass, d the flight path from the specimen to the detector and t the time of flight.

3. EXPERIMENTAL

The investigated material came from a thick section pipe (350 mm outer dimension and 50 mm wall thickness) originally supplied by Sumitomo Metals Industries. The pipe had been annealed at 1065°C, air cooled and tempered at 770°C for 2 h. The composition of the steel is shown in Table 1. Ageing experiments were performed by Hald on samples used for mechanical tests.[6] The samples were isothermally aged at 600°C for 1000, 3000 or 10 000 h.

3.1 Experimental Conditions and Specimen Preparation

The instrument used for AP analysis in this work is described in references.[8,9] It is equipped with a Poschenrieder device to improve the mass resolution and a chevron mounted multi-channel plate ion detector. Correction for losses due to pile-up in the detector were carried out using a statistical correction

Table 1   Composition of investigated steel HCM12A (balance Fe).[6]

Element	wt %	at %
Cr	11.00	11.84
Mo	0.42	0.25
W	1.94	0.59
V	0.19	0.21
Nb	0.05	0.03
Cu	0.87	0.77
Ni	0.32	0.31
Mn	0.56	0.57
Al	0.012	0.025
C	0.11	0.51
Si	0.02	0.04
P	0.013	0.023
S	0.001	0.002
B	0.0011	0.006
N	0.053	0.21

procedure.[10] The AP analyses were carried out at different specimen temperatures depending on the purpose of the individual analysis. If copper was present the temperature was kept at 60 K or below to avoid preferential field

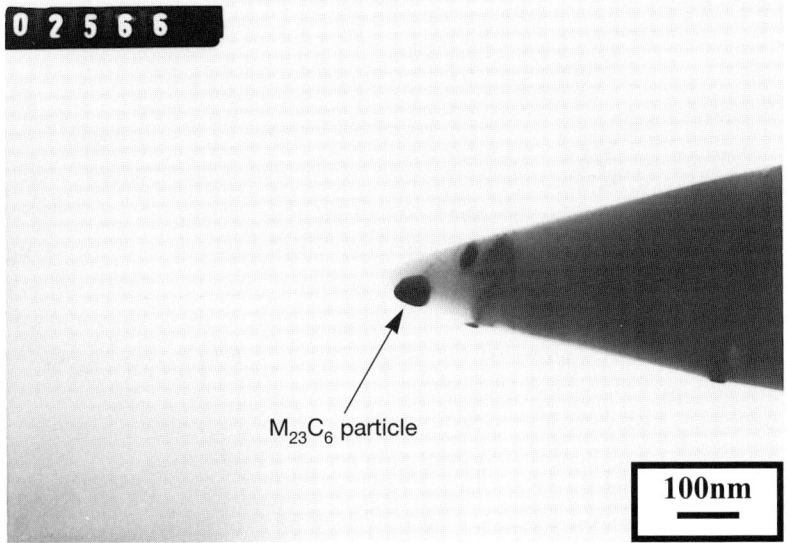

Fig. 3   A specimen for APFIM analysis with a small carbide close to the end of the tip (TEM micrograph).

evaporation of copper. When the absence of copper had been established in a certain phase the temperature was increased to 80 K to increase the lifetime of the specimen. The residual gas pressure in the UHV chamber was kept below 7.10^{-8} Pa and the pulse fraction was 20%.

The specimens were prepared using standard electropolishing methods.[7] The steel was first cut into bars of $0.3 \times 0.3 \times 10$ mm^3 size. Electropolishing was then done in two steps. First a neck was formed in a thin layer of electrolyte (10% perchloric acid in methanol) floating on trichloroethylene. In the second step a weaker electrolyte (2% perchloric acid in butoxyethanol) was used for slow polishing until the lower part of the specimen fell off and a sharp needle was formed. After polishing each specimen was examined in a transmission electron microscope (TEM). At this stage, the outermost part of the tip in general contained only matrix. To place precipitates close to the end of the tip, in a position suitable for analysis, pulsed electropolishing was used to remove material in a controlled manner. Figure 3 shows a specimen with a carbide positioned for analysis.

3.2   Analysed Materials
APFIM analyses of matrix and other phases present were performed for each ageing condition (including unaged material) as summarised below:

Ageing condition	Analysed phases				
Unaged	Matrix	$M_{23}C_6$			MX
Aged 1000 h	Matrix	$M_{23}C_6$	Laves		
Aged 3000 h	Matrix	$M_{23}C_6$	Laves		
Aged 10 000 h	Matrix	$M_{23}C_6$	Laves	MX	

4.   RESULTS

4.1 Investigation of Matrix
The results from atom probe analyses of matrix are given in Tables 2a–d. Several analyses were made of each ageing condition and the mean value of the individual analyses are given. Error estimates are given as one standard deviation. At low concentrations it is difficult to separate silicon and nitrogen due to overlap of isotopes in the time-of-flight spectrum. Therefore the total amount of the two elements are given.

It is clear that the matrix concentration of copper and tungsten drops during ageing. In the case of copper, the concentration decreases from about 0.4%, representing about half the total copper content, to an equilibrium value at about 0.1% reached after 1000 hours of ageing. In the case of tungsten, the concentration equilibrates at a level slightly above 0.8%.

The matrix is almost free of carbon. Concerning nitrogen, according to the

Table 2a–b   Composition of matrix in the unaged steel (left) and in the steel aged for 1000 h (right) (balance Fe)

Element	at%	wt%
Cr	9.6 ± 1.0	9.2 ± 1.0
Mn	0.6 ± 0.1	0.6 ± 0.1
Ni	0.3 ± 0.1	0.3 ± 0.1
Cu	0.32 ± 0.17	0.37 ± 0.19
Mo	0.25 ± 0.04	0.43 ± 0.07
W	0.50 ± 0.18	1.6 ± 0.6
V	0.11 ± 0.03	0.10 ± 0.03
Nb	0.01 ± 0.01	0.02 ± 0.02
Al	0.03 ± 0.02	0.01 ± 0.01
C	0.02 ± 0.02	0.004 ± 0.004
Si, N*	0.25 ± 0.06	0.12 ± 0.03
P	0.025 ± 0.013	0.014 ± 0.007
B	–	–

Element	at%	Wt%
Cr	9.9 ± 1.1	9.2 ± 1.1
Mn	0.6 ± 0.2	0.6 ± 0.2
Ni	0.3 ± 0.1	0.4 ± 0.1
Cu	0.09 ± 0.02	0.11 ± 0.02
Mo	0.18 ± 0.03	0.30 ± 0.05
W	0.29 ± 0.15	0.95 ± 0.5
V	0.17 ± 0.03	0.16 ± 0.02
Nb	0.01 ± 0.01	0.02 ± 0.02
Al	0.02 ± 0.01	0.01 ± 0.01
C	–	–
Si, N	0.2 ± 0.2	0.1 ± 0.1
P	0.05 ± 0.01	0.03 ± 0.01
B	–	–

* wt% calculated assuming Si only

Table 2c–d   Composition of matrix in the steel aged for 3000 h (left) and 10 000 (right) (balance Fe).

Element	at%	wt%
Cr	10.0 ± 1.1	9.3 ± 1.0
Mn	1.0 ± 0.4	1.0 ± 0.4
Ni	0.3 ± 0.1	0.3 ± 0.1
Cu	0.13 ± 0.02	0.15 ± 0.03
Mo	0.10 ± 0.02	0.17 ± 0.03
W	0.28 ± 0.03	0.95 ± 0.07
V	0.21 ± 0.05	0.19 ± 0.05
Nb	–	–
Al	–	–
C	–	–
Si, N	0.2 ± 0.2	0.1 ± 0.1
P	0.025 ± 0.013	0.014 ± 0.007
B	–	–

Element	at %	wt %
Cr	9.9 ± 1.1	9.3 ± 1.0
Mn	0.67 ± 0.07	0.66 ± 0.07
Ni	0.3 ± 0.1	0.3 ± 0.1
Cu	0.08 ± 0.01	0.08 ± 0.01
Mo	0.15 ± 0.05	0.26 ± 0.10
W	0.25 ± 0.03	0.82 ± 0.11
V	0.12 ± 0.02	0.11 ± 0.02
Nb	–	–
Al	–	–
C	0.01 ± 0.01	0.002 ± 0.002
Si, N	0.20 ± 0.04	0.10 ± 0.02
P	0.025 ± 0.013	0.014 ± 0.007
B	–	–

Table 3a–b Composition of $M_{23}C_6$ in the unaged steel (left) and in the steel aged for 1000 h (right).

Element	at%	wt%
Cr	52.6 ± 0.8	55.6 ± 0.8
Fe	20.0 ± 0.4	22.7 ± 0.5
Mo	1.8 ± 0.1	3.5 ± 0.1
W	3.0 ± 0.1	11.3 ± 0.2
V	0.5 ± 0.1	0.5 ± 0.1
Nb	0.04 ± 0.04	0.07 ± 0.07
Ni	0.3 ± 0.1	0.4 ± 0.1
Mn	0.7 ± 0.1	0.8 ± 0.1
Al	0.01 ± 0.01	0.005 ± 0.005
C	20.8 ± 0.3	5.0 ± 0.1
Si	0.07 ± 0.03	0.04 ± 0.01
P	0.04 ± 0.02	0.03 ± 0.01
B	0.10 ± 0.03	0.02 ± 0.01

Element	at%	wt%
Cr	51.2 ± 0.2	52.8 ± 0.2
Fe	19.7 ± 0.7	21.8 ± 0.7
Mo	1.8 ± 0.1	3.5 ± 0.2
W	4.1 ± 0.1	15.1 ± 0.2
V	0.5 ± 0.2	0.5 ± 0.2
Nb	0.02 ± 0.01	0.03 ± 0.01
Ni	0.3 ± 0.1	0.4 ± 0.1
Mn	0.7 ± 0.1	0.8 ± 0.1
Al	0.01 ± 0.01	0.005 ± 0.005
C	21.4 ± 0.7	5.1 ± 0.2
Si	0.02 ± 0.01	0.01 ± 0.01
P	0.01 ± 0.01	0.005 ± 0.005
B	0.09 ± 0.02	0.02 ± 0.004

distribution of isotopes, most ions appearing at the mass-to-charge ratio fourteen should be doubly charged silicon ions. However, due to the low number

Table 3c–d Composition of $M_{23}C_6$ in the steel aged for 3000 h (left) and 10 000 h (right).

Element	at%	wt%
Cr	53.2 ± 1.3	56.3 ± 1.2
Fe	19.0 ± 1.3	21.7 ± 1.6
Mo	1.6 ± 0.3	3.1 ± 0.6
W	3.2 ± 0.4	11.9 ± 1.5
V	0.4 ± 0.1	0.4 ± 0.1
Nb	0.1 ± 0.1	0.2 ± 0.2
Ni	0.3 ± 0.1	0.4 ± 0.1
Mn	0.7 ± 0.1	0.8 ± 0.1
Al	–	–
C	21.2 ± 0.3	5.2 ± 0.1
Si	0.1 ± 0.1	0.05 ± 0.05
P	0.03 ± 0.01	0.02 ± 0.01
B	0.10 ± 0.06	0.02 ± 0.01

Element	at%	wt%
Cr	56.0 ± 1.1	58.9 ± 1.2
Fe	15.9 ± 0.7	17.9 ± 0.8
Mo	1.4 ± 0.2	2.7 ± 0.4
W	3.6 ± 0.3	13.5 ± 1.3
V	0.3 ± 0.1	0.3 ± 0.1
Nb	0.14 ± 0.1	0.26 ± 0.17
Ni	0.3 ± 0.1	0.4 ± 0.1
Mn	1.1 ± 0.2	1.2 ± 0.2
Al	–	–
C	21.3 ± 1.6	5.2 ± 0.4
Si	0.03 ± 0.03	0.02 ± 0.02
P	0.06 ± 0.04	0.03 ± 0.02
B	0.15 ± 0.07	0.03 ± 0.02

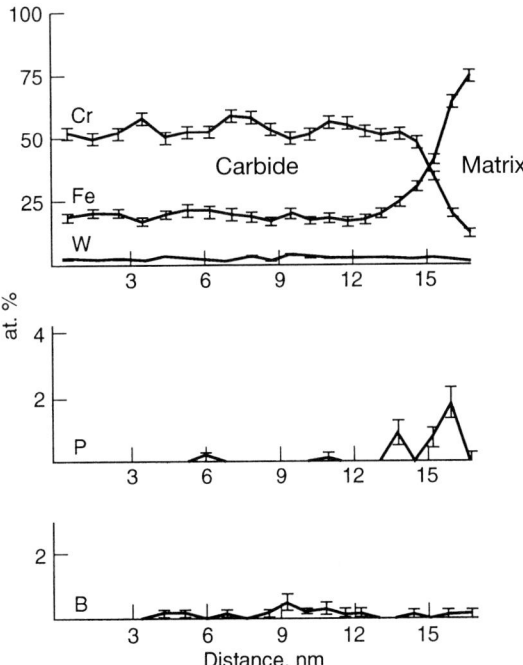

Fig. 4a Concentration profiles through a carbide/matrix interface (unaged material). Phosphorous segregates to the interface. The boron concentration is enhanced within the carbide in a region close to the surface.

of ions, it cannot be excluded that the matrix contains a certain amount of nitrogen.

Although a number of analyses contained some 100 000 ions, no boron was detected in the matrix. If boron would have been evenly distributed in the steel, about five boron atoms should have been present in the analysed volume.

4.2   Investigation of $M_{23}C_6$

The composition of $M_{23}C_6$ precipitates estimated by atom probe analysis are given in Tables 3a–d. The carbon concentration was in all cases found to be close to the stoichiometric composition, and the composition does not change considerably during ageing. Apart from chromium, the carbides contain relatively large amounts of iron, tungsten and molybdenum. No copper was found in this phase.

At most interfaces, segregation of boron and phosphorous was found. This is illustrated in the concentration profiles in Figs 4a–b. The phosphorous concentration is enhanced within a narrow volume in the interface.

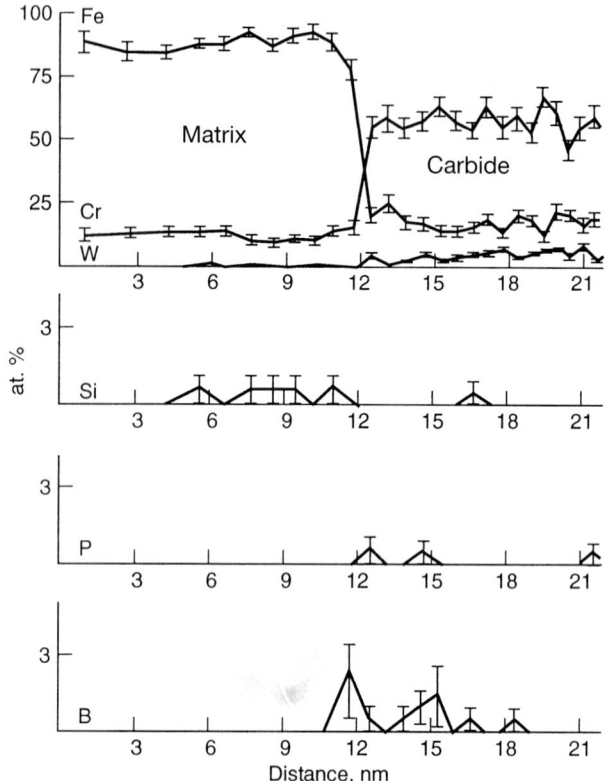

Fig. 4b   Concentration profiles through a matrix/carbide interface (material aged for 10 000h).

This can be interpreted as that the phosphorous is constricted to one single atomic layer between the carbide and the matrix. Boron is enriched within a larger volume in the near surface region of the carbide. Measurements deep inside large carbides suggest that the 'bulk' concentration of boron is low. It should be pointed out that the inhomogeneous distribution of boron inside the carbides makes the estimation of the average boron concentration uncertain. As can be seen in Fig. 4b, the concentration of silicon is lower in the carbides compared to the adjacent matrix.

Figure 5 shows field ion micrographs of pure matrix (upper) and an $M_{23}C_6$ precipitate (lower). The carbide is seen as a bright area in the FIM image.

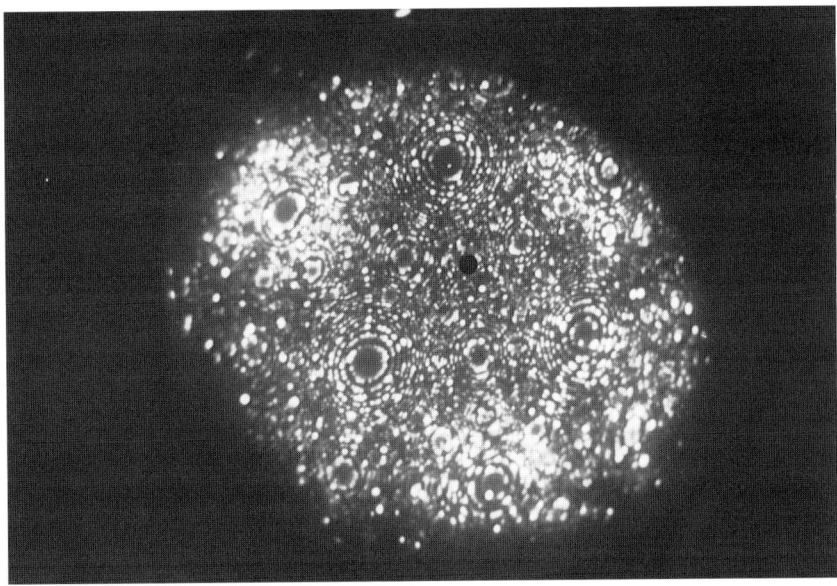

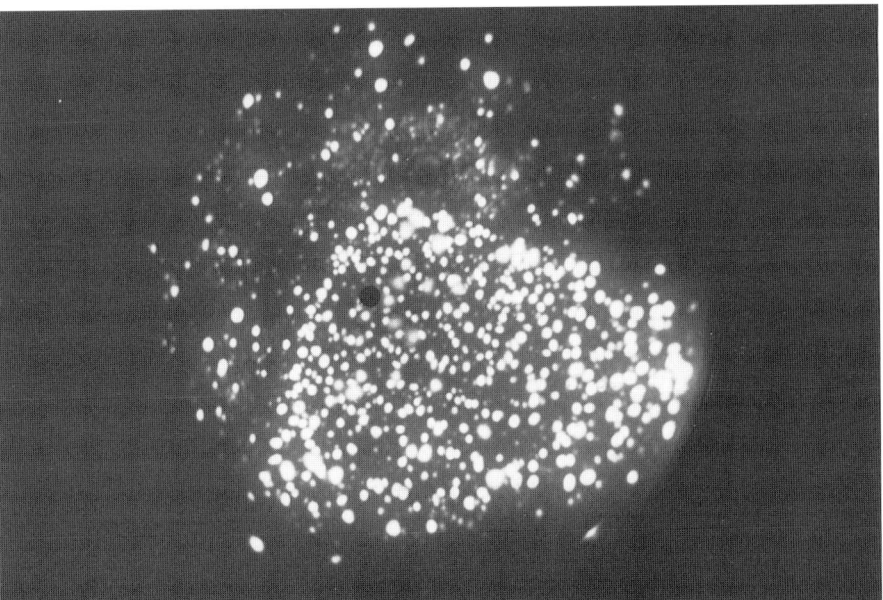

Fig. 5   Field ion micrographs of matrix (upper) and $M_{23}C_6$ (lower). The diameter of the images is approximately 100 nm.

4.3   Investigation of MX

A total of three MX particles were analysed; two vanadium-rich nitrides in the unaged material and one niobium-rich carbonitride in the material aged for 10 000 hours. The composition of the particles are found in Tables 4a–b.

Table 4a–b Composition of MX in the unaged steel (left) and in the steel aged for 10 000 h (right)

Element	at%	wt%
V	42.3 ± 2.3	57.4 ± 3.2
Nb	4.3 ± 0.6	10.6 ± 1.4
Cr	10.9 ± 0.9	15.2 ± 1.3
Fe	0.9 ± 0.3	1.3 ± 0.4
N	41.6 ± 2.2	15.5 ± 0.8
C	–	–

Element	at%	wt%
Nb	53 ± 5	86 ± 8
V	4.5 ± 1.1	4.0 ± 1.0
N	28 ± 4	6.9 ± 1.0
FC	14 ± 3	2.9 ± 0.7

Table 5a–b Composition of Laves phase in the steel aged for 1000 h (left) and 3000 h (right) (balance Fe)

Element	at%	wt%
Cr	14.0 ± 1.9	8.3 ± 0.3
Mo	8.4 ± 0.7	8.3 ± 1.3
W	23.5 ± 1.7	49.2 ± 2.5
V	0.3 ± 0.1	0.2 ± 0.1
Nb	0.5 ± 0.2	0.5 ± 0.2
Mn	1.0 ± 0.1	0.6 ± 0.1
Ni	0.3 ± 0.1	0.2 ± 0.1
Si	3.1 ± 0.5	1.0 ± 0.1
P	0.7 ± 0.1	0.24 ± 0.03
C	0.6 ± 0.1	0.08 ± 0.02

Element	at%	wt%
Cr	16.2 ± 1.0	10.0 ± 0.6
Mo	10.7 ± 0.8	12.1 ± 0.9
W	20.3 ± 1.0	44.0 ± 2.1
V	0.2 ± 0.1	0.12 ± 0.06
Nb	0.4 ± 0.1	0.4 ± 0.1
Mn	1.6 ± 0.5	1.0 ± 0.1
Ni	0.3 ± 0.1	0.2 ± 0.1
Si	2.2 ± 0.5	0.7 ± 0.2
P	1.0 ± 0.2	0.4 ± 0.1
C	0.3 ± 0.1	0.04 ± 0.02

Table 5c Composition of Laves phase in the steel aged for 10 000 h (balance Fe)

Element	at%	wt%
Cr	11.4 ± 0.3	6.4 ± 0.5
Mo	8.3 ± 0.5	8.6 ± 0.7
W	27.2 ± 2.6	54.0 ± 3.5
V	0.2 ± 0.2	0.1 ± 0.1
Nb	0.3 ± 0.3	0.3 ± 0.3
Mn	1.0 ± 0.1	0.6 ± 0.1
Ni	0.3 ± 0.1	0.2 ± 0.1
Si	3.1 ± 1.0	1.0 ± 0.4
P	0.6 ± 0.3	0.2 ± 0.1
C	0.25 ± 0.05	0.04 ± 0.01

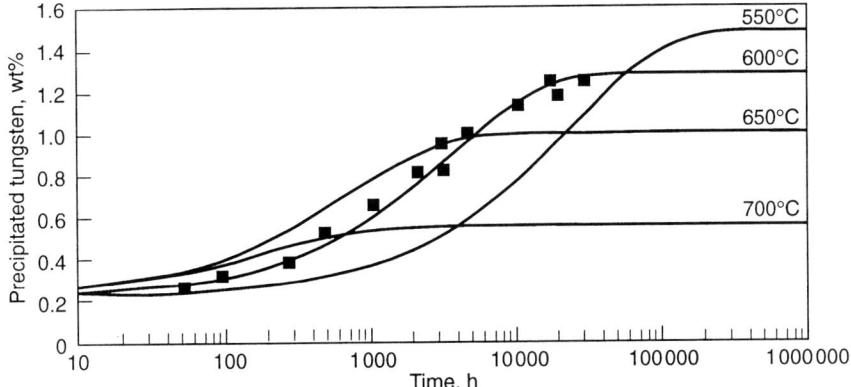

Fig. 6   Calculated amount of tungsten precipitated in steel NF616. Squares are measured values at 600°C used to calibrate the model.[6]

In the case of the single analysis of the carbonitride (Table 4b), the error estimation is derived from the counting statistics.

As can be seen from the low nitrogen and carbon contents, both kinds of particles are considerably unstoichiometric. Note the high chromium content of the vanadium nitrides found in the unaged steel. No boron, phosphorous or copper was found, neither inside the particles nor at interfaces between the particles and matrix.

4.4   Investigation of Laves Phase

The results from the atom probe investigations of Laves phase are found in Tables 5a–c. The composition is close to $(Fe,Cr)_2(Mo,W)$, and does not change during isothermal ageing. No copper, but high concentrations of silicon and phosphorous were found in these precipitates. Also, it can be seen that some carbon is dissolved in the Laves phase. No enhanced concentration of boron was found in the Laves/matrix interface.

5.   DISCUSSION

5.1   Matrix

The low matrix concentration of carbon indicates that almost all carbon is located to carbides. Using a chromium to carbon ratio in the carbides of about 2.5 (by weight) along with the total carbon concentration of 0.5% implies that 1.3% of chromium should be precipitated in carbides. This amount along with the unaged matrix concentration of 9.6% yields a total chromium content of about 10.9%, indicating self consistency of the measurements.

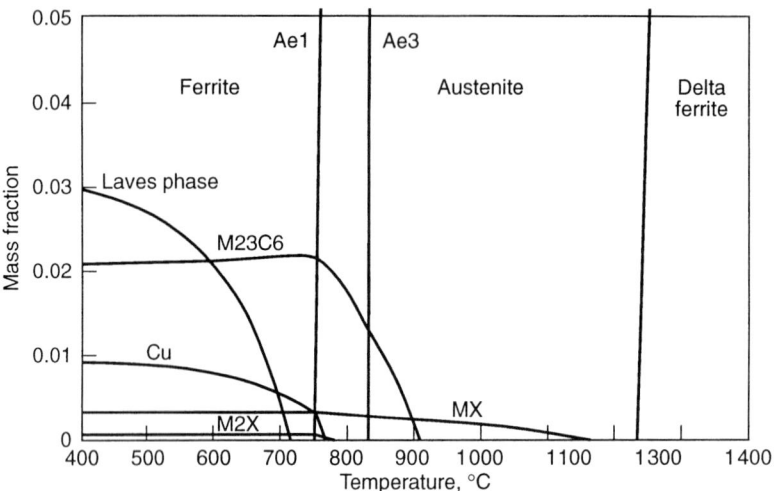

Fig. 7   Equilibrium phases in steel HCM12A calculated by Thermo-Calc.[6]

Because of overlap with silicon in the time-of-flight spectrum, it is difficult to make a direct estimation of the matrix concentration of nitrogen. There is, however, an approximate one to one relation between nitrogen and nitride forming elements (vanadium and niobium) both in the overall composition and in the MX precipitates. Therefore the matrix content of nitrogen would be in the same order of magnitude as the concentration of nitride forming elements in the matrix. About half the amount of vanadium and niobium remains in the matrix after tempering and also after ageing for 10 000 h. This should be true also for nitrogen.

The drop in tungsten concentration in the matrix during the first 1000 hours of ageing is in the order of 0.6% by weight. This value can be compared with tungsten precipitation in the steel NF616. Hald modelled this process and the result is shown in Fig. 6.[6] Squares are measured values used to calibrate the model. The drop in tungsten concentration in NF616 is about 0.3% by weight, i.e. half the corresponding drop in HCM12A. Copper might be responsible for this deviation by increasing the activity of tungsten in the matrix. Another possibility is that the copper addition influence the nucleation of Laves phase. This can be true if the surface energy of the copper/Laves inter-face is lower than the surface energy of the matrix/Laves interface.

5.2   Copper

The atom probe technique has been previously used to study steels alloyed with a certain amount of copper.[11,12] Maraging steels and neutron irradiated pressure vessel steels have been investigated. It was shown that the solubility of copper in ferrite is low and that copper precipitates as a separate phase. In the case of the maraging steels, it was also found that the copper particles play an important role for the nucleation of other phases.[12]

This investigation has showed that no copper is contained in $M_{23}C_6$, MX or Laves phase. Since only a small amount of copper was found in the matrix it can be concluded that copper forms a separate phase, probably consisting of almost pure copper. This statement is also supported by equilibrium calculations performed by Hald with the *Thermo-Calc* software (see Fig. 7).[6] The calculated amount of precipitated copper is in fair agreement with the measured matrix concentration of copper in material aged at 600°C.

5.3 $M_{23}C_6$

Regarding the behaviour of the trace elements phosphorous and boron at the matrix/carbide interface, it is interesting to notice that a similar investigation have been carried out earlier, on a boron containing 9% chromium steel, with a somewhat different result.[13] Lundin and Richarz found pronounced segregation of phosphorous while boron was evenly distributed inside the carbide. The interface segregation of phosphorous can probably be considered as an equilibrium segregation process, i.e. the presence of phosphorous lowers the surface energy. The uneven distribution of boron inside carbides can be understood as a non-equilibrium segregation process that takes place during cooling. This mechanism has been observed and modelled in boron containing austenitic steels.[14] If this explanation is correct it is reasonable to believe that boron at elevated temperatures is evenly distributed in carbides.

Due to the inhomogeneous distribution of boron it is hard to estimate the total amount of boron contained in $M_{23}C_6$. If it is assumed that all carbon is tied up in carbides and the average boron concentration in $M_{23}C_6$ is used, a straightforward calculation yields that about half of the total amount of boron is contained in carbides.

5.4 MX

It is interesting to note the high chromium content and the low nitrogen content in the vanadium nitride particles. This composition of MX precipitates has also been reported by Lundin.[15] A possible explanation for the observed composition is that the chromium diminish the lattice parameter of MX, which implies smaller misfit between the coherent precipitates and matrix.

The dispersion of MX particles inside the subgrains is critical for the creep properties and a key question is the role of boron in this case.[16] If boron effects nucleation and coarsening of MX particles, this could explain the positive effect of boron on long term creep properties. Lundin has suggested a mechanism called 'latent creep resistance', where small MX precipitates are dynamically nucleated on dislocations and dissolved during creep.[15] Boron could possibly take part in this process if it segregates to dislocations. Even if part of the boron is tied up in carbides there might still be some available in the matrix as discussed in section 5.3. No boron was found neither inside the MX precipitates nor at the MX/matrix interface but this fact does not exclude the possibility that boron plays a role in the nucleation of MX phase on dislocations.

5.5   Laves Phase

The composition of Laves phase close to $(Fe,Cr)_2(Mo,W)$, with additional small amounts of silicon, phosphorous and carbon, is in agreement with earlier investigations of this phase.[17] The relatively high concentration of phosphorous in Laves phase is worth noticing. A beneficial effect of Laves phase might be to remove phosphorous from different interfaces during ageing.

6.   CONCLUSIONS

Chromium steel HCM12A, isothermally aged at 600°C, have been studied regarding microstructural changes and redistribution of elements during ageing.

- Copper is present in fairly high concentrations in the unaged matrix, but the concentration drops rapidly during isothermal ageing. The equilibrium concentration at 600°C is in good agreement with the value predicted by thermodynamical calculations. No copper is contained in $M_{23}C_6$, MX or Laves phase. The copper instead most probably forms a separate phase.
- During ageing the amounts of molybdenum and tungsten in the matrix drop due to formation of Laves phase. The tungsten concentration equilibrate faster in HCM12A compared to steel NF616, a fact that might be related to the presence of copper.
- The composition of $M_{23}C_6$ is stoichiometric with respect to carbon and does not change much during ageing. Enhanced concentrations of boron were found inside the carbides close to the carbide/matrix interface. This observation can be explained in terms of non-equilibrium segregation during cooling, in which case boron should be evenly distributed inside the carbides at high temperatures. Phosphorous segregates to a very narrow region at the carbide/matrix interface.
- Two different types of MX particles were analysed; vanadium nitride and vanadium carbonitride. Both kinds of particles are sub-stoichiometric with respect to nitrogen and carbon. The vanadium nitride contains a relatively large amount of chromium. No boron or phosphorous was found neither inside the precipitates nor at the MX/matrix interface.

- The composition of Laves phase is close to $(Fe,Cr)_2(Mo,W)$ with small additions of phosphorous, silicon and carbon. No boron was found in this phase.

REFERENCES

1. C. Berger *et al.*: *Materials for Advanced Power Engineering*, D. Coutsouradias *et al.*, eds, Kluwer Academic Publishers, Dortrecht, 1994 47–72.
2. D.V. Thornton and R.W. Vanstone: *Materials engineering in turbines and compressors*, R.D. Conrey *et al.*, ed, The Institute of Materials, London, 1995, 135–153.
3. T. Fujita: *Materials engineering in turbines and compressors*, R.D. Conrey *et al.*, ed, The Institute of Materials, London, 1995, 493–516.
4. E. Metcalfe: *New Steels for Advanced Plant up to 620°C*, E. Metcalfe, ed., EPRI/National Power, London, 1995, 1–7.
5. F. Masuyama: *New Steels for Advanced Plant up to 620°C*, E. Metcalfe, ed., EPRI/National Power, London, 1995, 98–113.
6. J. Hald: *New Steels for Advanced Plant up to 620°C*, E. Metcalfe, ed., EPRI/National Power, London, 1995, 152–173.
7. M.K. Miller *et al.*: *Atom Probe Field Ion Microscopy*, Clarendon Press, Oxford, 1996
8. H.O. Andrén and H. Nordén: *Scand. J. Metallurgy*, (1979), **8**, 147–152.
9. H.O. Andrén: *J. Phys. (Paris)*, 1986, **47** (C7), 483–488.
10. U. Rolander and H.O. Andrén: *J. Phys. (Paris)*, 1989, **50** (C8), 529–534.
11. M.K. Miller and M.G. Burke: *J. Nucl. Mater.*, 1992, **195**, 68–82.
12. K. Stiller *et al.*: *Appl. Surf. Sci.*, 1996, **94/95**, 326–333.
13. L. Lundin and B. Richarz: *Appl. Surf. Sci.*, 1995, **87/88**, 194–199.
14. L. Karlsson *et al.*: *Acta metall.*, 1988, **36**, 1–48.
15. L. Lundin: 'High Resolution Microanalysis of Creep Resistant 9–12% Chromium Steels, *Ph.D. Thesis*, Chalmers University of Technology, 1995.
16. V. Foldyna *et al.*: 'Creep Resistant Metallic Materials', *Conf. Proc. Hradec nad Moravici, Chech Republic, 1996*, 203–216.
17. L. Lundin: *Scripta Metall. et Material*, 1996, **34**, 741–747.

Dislocation-Particle Interactions During Deformation of the Fe Cr Al Oxide Dispersion Strengthened INCOLOY MA956 at RT–1050°C

B. DUBIEL,* A. CZYRSKA-FILEMONOWICZ* AND P.J. ENNIS†

*University of Mining and Metallurgy, Faculty of Metallurgy and Materials Science, Kraków, Poland

†Research Centre Jülich, Institute for Materials in Energy Systems, Jülich, Germany

ABSTRACT

In this paper the results of microstructural analyses of tensile and compression deformed INCOLOY MA956 are outlined. The investigations revealed that there are two different mechanisms of stengthening depending on the deformation temperature of the alloy. At temperatures in the range 20–400°C, dislocation loops around the dispersoid particles were observed, indicating an Orowan type of strengthening effect. During high temperature tensile and compression deformation the shape of dislocations pinned at the departure side of particles indicated an attractive dislocation-particle interaction.

INTRODUCTION

INCOLOY alloy MA956 is an Fe20Cr5Al0.5Y$_2$O$_3$ alloy produced by the mechanical alloying process which provides a fine dispersion of the oxide phase in the ferritic matrix. The alloy exhibits outstanding strength and corrosion resistance at temperatures even above 1000°C. Calculations based on the difference in the elastic moduli of the oxide particles (dispersoids) and the ferritic matrix[1,2] led to the conclusion that the repulsive dislocation-dispersoid interaction can be the source of the strengthening due to Orowan bowing of dislocations between particles. However, after high temperature deformation an attractive interaction between dislocations and hard incoherent oxide particles has been observed.[3,4]

The aim of the present investigations was to reveal the nature of interaction between dispersoids and dislocations developed in INCOLOY MA956 during plastic deformation at temperatures up to 1050°C.

EXPERIMENTAL DETAILS

MA956 bar (20 mm diameter) was supplied by INCO Alloys International

215

in the hot extruded and recrystallised condition (1330°C/1 h). The nominal composition of the alloy investigated was (in wt %): Fe–20Cr–4.5Al–0.5Ti–0.5Y$_2$O$_3$. The bar exhibited a particularly coarse grain size that allowed single crystal rods of 6 mm diameter to be spark machined. For all the single crystal specimens, the <111> direction was parallel to the extrusion axis. The tensile tests were performed on polycrystalline specimens (25 mm gauge length and 6 mm diameter) in air at constant deformation rate of 10^{-3} s^{-1} in the temperature range 20–950°C. After testing in tension the specimens were furnace cooled to about 100°C, due to experimental limitations. The compression tests were carried out on single-crystal cylindrical specimens (6 mm diameter, 9 mm height) at temperatures of 400, 600 and 800°C at a strain rate of 10^{-3} s^{-1}. After testing, the specimens were quickly cooled down in liquid nitrogen. Structural analysis was performed using optical metallography and transmission electron microscopy (TEM). The TEM investigations were carried out on thin foils and on extraction double replicas using JEM 100C and JEM 2010 ARP microscopes equipped with an energy dispersive X-ray system (EDS) of Link. Thin foils were prepared by double-jet electropolishing in 10% solution of the perchloric acid in glacial acetic acid at about 10°C and 50 V. The replicas were prepared according to Ref. 5.

RESULTS AND DISCUSSION

The microstructure of as received material consisted of oxide dispersoids with some pure alumina and titanium carbonitride particles of up to few μm diameter in a ferritic matrix. The analyses of extracted dispersoids showed that

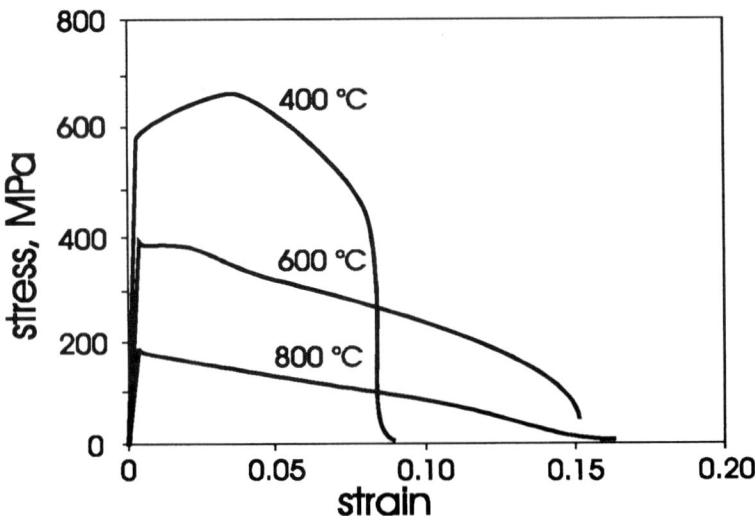

Fig 1. Tensile curves of single crystal specimens.

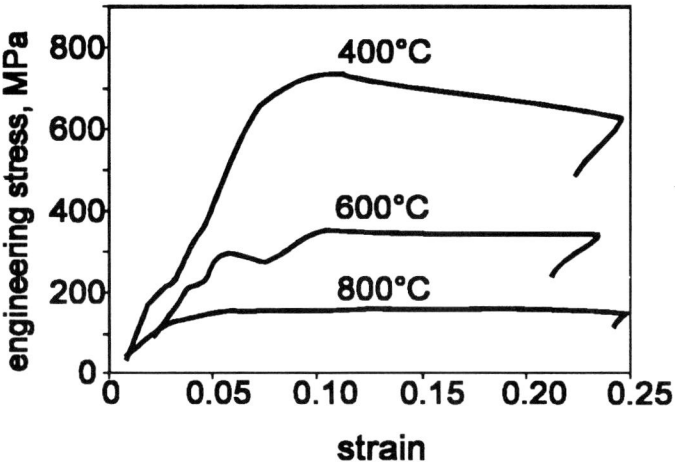

Fig. 2 Engineering compression stress–strain curves.

(a)

(b)

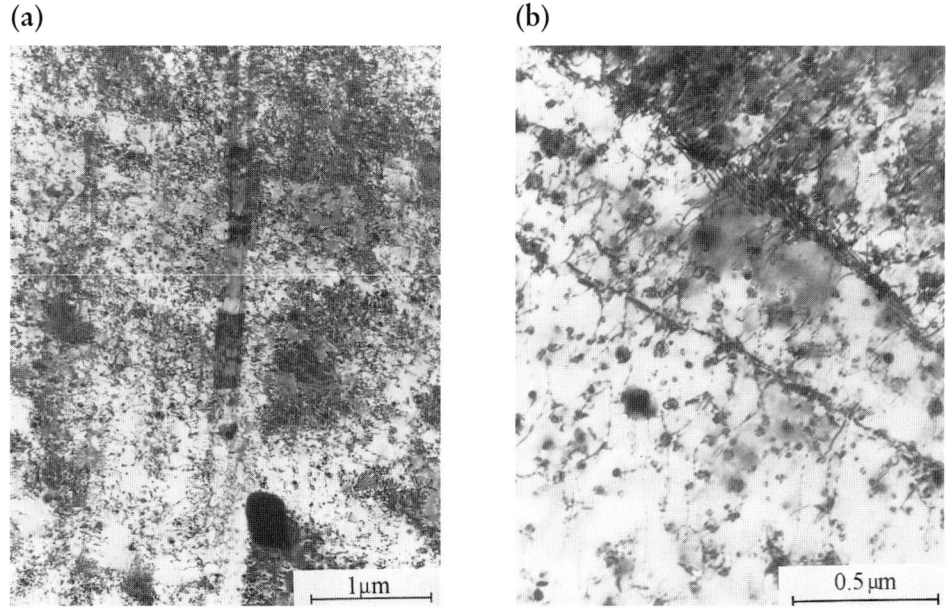

Fig. 3 Microstructure of the specimen tested at 400°C, showing (a) microshear bands in the region of localisation of plastic deformation and (b) subgrains outside the localisation region.

the most of the yttrium-aluminium oxides were tetragonal $Y_3Al_5O_{12}$ (YAT). Statistical measurements of the dispersoid size using the TEM micrographs showed the existence of three size distributions of the Y–Al oxides in the size range 5–300 nm. The mean diameter of the dominant finest dispersoids, which influence the creep resistance, was 9 nm.

Figure 1 shows typical tensile curves of single crystal specimens. In the temperature range 20–400°C the tensile curves revealed strain hardening. At temperatures above 600°C an early onset of necking was observed. The results of tensile tests are discussed detailed in Refs 6 and 7.

Figure 2 shows the engineering stress–strain curves plotted during compression tests. The strength decreased from 750 MPa to 149 MPa as the test temperature increased from 400 to 800°C. The specimens deformed at 400°C were characterised by strong localisation of plastic deformation, with the appearance of thin shear bands. The tested specimens sheared, the two parts of the specimen being displaced by up to 20% of specimen diameter. With increasing test temperature, the tendency for deformation localisation decreased.

Structural analysis performed by TEM on thin foils prepared from the different places in the longitudinal section of tested specimen showed significant differences in the dislocation structure between plastic deformation localisation area and other parts of the tested specimens. The deformed specimens exhibited non-uniform dislocation density. Figure 3 shows the microstruc-

(a)   **(b)**

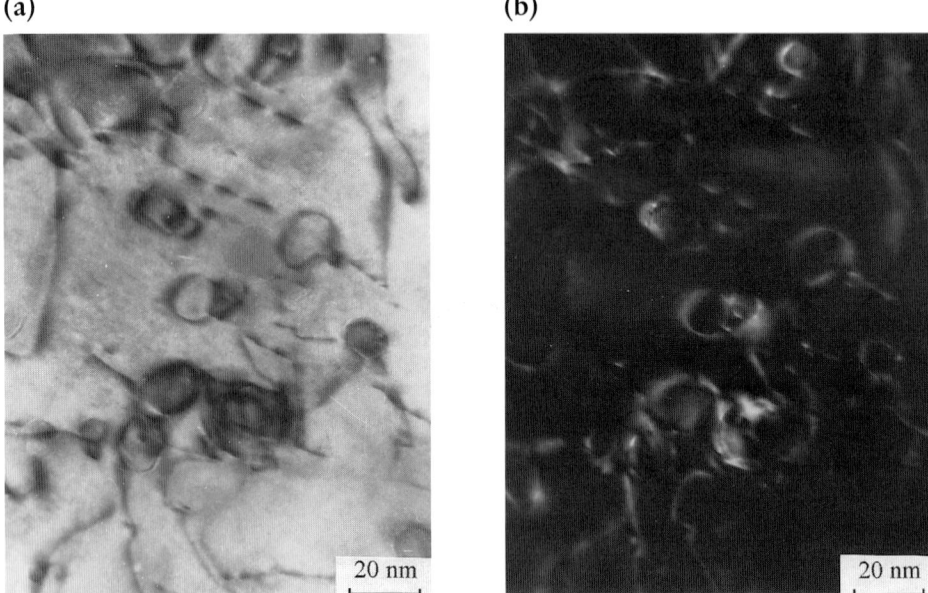

Fig. 4   Microstructure of the specimen tested at 400°C, showing dislocation loops around dispersoids; (a) bright field and (b) dark field images.

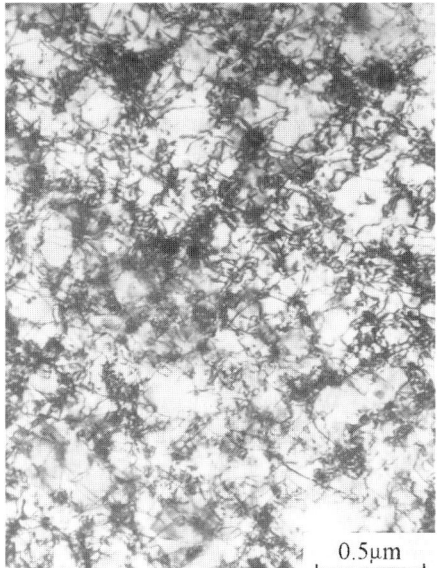

Fig. 5 Microstructure of the specimen tested at 600°C.

ture of a specimen tested at 400°C. In macroscopic deformation localisation area the high dislocation density of $\approx 3.5.10^{15}$ m^{-2} and the microshear bands

(a)   **(b)**

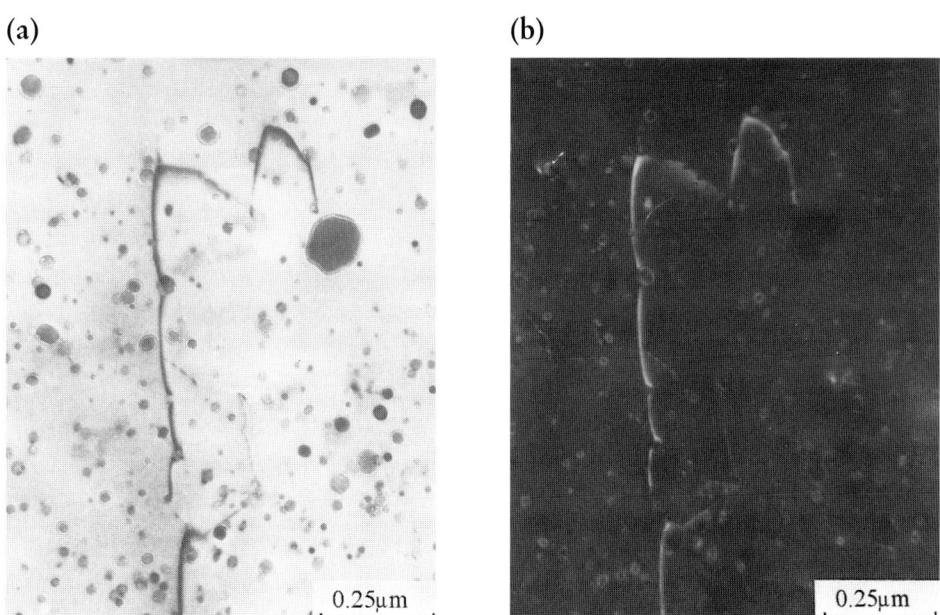

Fig. 6 Microstructure of the specimen tested at 800°C, showing dislocations in the vicinity of dispersoids; (a) bright field and (b) dark field images.

were observed. In the area outside localisation, the dislocation density was lower ($\approx 4.6.10^{13}$ m^{-2}) and moving dislocations often formed subgrains.

In the microstructural analyses attention was focused on the examination of the dislocation substructure, especially dislocation configurations in the vicinity of dispersoids. This allows the nature of dislocation–dispersoid interactions to be investigated, which limit the mobility of dislocations and therefore influence the strength of the material. Figure 4 shows the dislocation-dispersoid configurations in the microstructure of a specimen compression tested at 400°C. TEM bright and dark field images revealed the contrast of dislocations in the vicinity of dispersoids, which suggests Orowan bowing of dislocations between particles.

The structure of a specimen compression tested at 600°C (Fig. 5) exhibited more uniform dislocation density of $\approx 5.5.10^{13}$ m^{-2} and and the tendency to form dislocation cell structures. The specimen deformed at 800°C was characterised by a very low dislocation density ($\approx 1.4.10^{12}$ m^{-2}). Very often the dislocations were pinned at the departure side of particles (Fig. 6). Similar configurations of dislocation and dispersoids were also observed in the structure after creep deformation at 950 and 1050°C.[8] This type of the structure gave the impression of an attractive dislocation-particle interaction as was postulated in the literature for austenitic[9–12] and ferritic[3,8,13] ODS alloys. Such dislocation configurations suggesting an attractive interaction between dislocations and dispersoids appear to be characteristic for the structure of INCOLOY MA956 after high temperature deformation.

SUMMARY

The results of tensile and compression tests showed that there are two different mechanisms of strengthening depending on the deformation temperature. The microstructural investigations revealed the different dislocation substructure, especially dislocation-dispersoid interaction, in the specimens deformed in low and high temperatures. At temperatures in the range 20–400°C, dislocation loops around the dispersoid particles were observed, indicating an Orowan type of strengthening effect. During high temperature tensile, compression and creep deformation, the shape of dislocations pinned at the departure side of particles indicated for an attractive dislocation-particle interaction and Rösler-Arzt model[12] of strengthening developed for austenitic ODS alloys.

ACKNOWLEDGMENT

The authors wish to thank the Foundation for Polish Science for funding the JEOL JEM 2010 ARP microscope.

REFERENCES

1. A. Kelly and R.B. Nicholson: *Progr. Mat. Sci.*, 1963, **10**, 51.
2. L.M. Brown and R.K. Ham: *Strengthening Methods in Crystals*, A. Kelly and R.B. Nicholson, Elsevier Publishing Company, Amsterdam, 1971, 9.
3. J. Preston, B. Wilshire and E.A. Little: *Scripta Met. et Mat.*, 1991, **25**, 183.
4. A. Czyrska-Filemonowicz, M. Wróbel, B. Dubiel and P.J. Ennis: *Scripta Met. et Mat.*, 1994, **32**, 331.
5. A. Czyrska-Filemonowicz, K. Spiradek and S. Gorczyca: *Prakt. Metallographie*, 1991, **22**, 217.
6. B. Dubiel, W. Osuch, M. Wróbel, P.J. Ennis and A. Czyrska-Filemonowicz: *Journal of Materials Processing Technology*, 1995, **53**, 121.
7. M. Wróbel, D. Schwarze, B. Dubiel, P.J. Ennis and A. Czyrska-Filemonowicz: *Archives of Metall.*, 1995, **40**, 447.
8. A. Czyrska-Filemonowicz, M. Wróbel, B. Dubiel and P.J. Ennis: *Scripta Met. et Mat.*, 1994, **32**, 331.
9. D.J. Srolovitz, R.A. Petkovic-Luton and M.J. Luton: *Acta Metallurgica*, 1983, **31**, 2151.
10. E. Arzt and D.S. Wilkinson: *Acta Metallurgica*, 1986, **34**, 1893.
11. E. Arzt and J. Rösler: *Acta Metallurgica*, 1988, **36**, 1053.
12. J. Rösler and E. Arzt: *Acta Metallurgica*, 1990, **38**, 671.
13. H. Schuster, R. Herzog and A. Czyrska-Filemonowicz: *Metallurgy and Foundry Engineering*, 1995, **21**, 273.

The Role of Microstructure in Influencing the Properties of Service Degraded Superalloy Components

M.I. WOOD

ERA Technology, Cleeve Rd, Leatherhead, UK

ABSTRACT

One of the standard tools in assessing the condition of service run superalloy gas turbine blades is an evaluation of the microstructural condition. However, many of the relationships between the degraded microstructures and their associated properties are qualitative, and it is sometimes unclear whether the associations are actually causal or just correlations. Three different areas are considered to highlight the requirement for more detailed studies. These are creep rupture and microstructure, the simulation of service degradation, and the interpretation of near surface features produced by oxidation.

INTRODUCTION

Considerable effort has been expended by metallurgists over the years in establishing both general and specific relationships between properties of engineering interest and microstructural features (which are a function of the material processing route and the alloy's intrinsic chemistry). Considerable advances have been made through exploiting this understanding (as well by means of the well tested trial and error methods) to develop new materials and improving or modifying process control to produce materials within tighter tolerances. The evaluation of these materials then proceeds through the application of 'conventional' testing techniques, as enshrined in design methodologies or codes.

Once decisions have been taken on the material and manufacturing route for a component, various microstructural features have become fixed, i.e. fixed within normal manufacturing tolerances Whilst in service some of these microstructural features alter with time, temperature and stress (deformation). An assessment of the condition of the component then requires that the significance of these changes be considered. However, the effect of these changes are not explicitly visible within the conventional materials database. Because of this, much effort has been expended in developing a suitable database and assessment procedures for service exposed high temperature steels which relates residual properties with the alloy's current microstructure, and the component's thermal, mechanical and environmental history.

In common with the assessment of high temperature steel structures, the assessment of service run gas turbine hot section components requires an evaluation of the condition of the observed microstructure, and the significance of any changes which have occurred. However, somewhat in contrast to the case for steels, far less work has been carried out to investigate these relationships. Whilst this has been the case for a variety of technical and commercial reasons, the situation is now changing. Examples of the somewhat more quantitative relationships that have been demonstrated can be found in the work of Ref. 1, where a power law dependence was found between the γ' density and the hardness and yield strength for IN738 aged out to ~24 000 h. Another example is the dependence of residual rupture life of IN713 on the γ' aspect ratio of service exposed components.[2]

The importance of these types of relationships, if they can be substantiated, lies in the nature of many gas turbine components and the problems that they cause when a material condition assessment is required. Turbine blades tend to be small and of a complex geometry. All recent designs employ hollow thin walled castings (0.5 to 2 mm depending on the application), which not only means that they have a high surface to volume ratio, but there are steep stress and temperature gradients in the regions of interest. The regions of greatest distress are frequently at the leading and trailing edges where it is not feasible to extract mechanical test pieces. The only current way of assessing the condition of this material is by means of its microstructure. This statement is nothing new, and many components have been judged on this basis. However what has been somewhat lacking has been a sound basis which links what can be seen microstructurally and the actual properties of the volume of material of interest. These links need to be due to a causal relationship between the parameters concerned, not merely a correlation.

The work described here is part of a continuing programme addressing this area. Examples for three very different areas are considered here. The first considers the relationship between residual rupture properties, test acceleration and microstructure. The second concerns the simulation of service degradation, whilst the third considers the practical interpretation of near surface features.

EFFECT OF TEST ACCELERATION AND APPROACH

Small test pieces were machined from ex-service blades manufactured in the wrought superalloy Nimonic 115 and the directionally solidified alloy PWA1422. The former were from solid blades, the latter from cooled components (further details can be found in Ref. 3). Stress rupture tests were then carried out at the operational service stress level (calculated for the forged blade, and estimated for the cooled ones) at a range of temperatures (the isostress approach). Additional tests were also carried out at higher stress levels on the Nimonic 115 material.

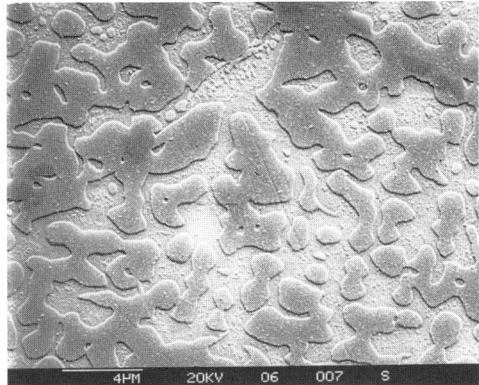

Fig. 1 Degraded γ′ structure for Nimonic 115 after 19 000 h service.

Fig. 2 Virgin Nimonic 115.

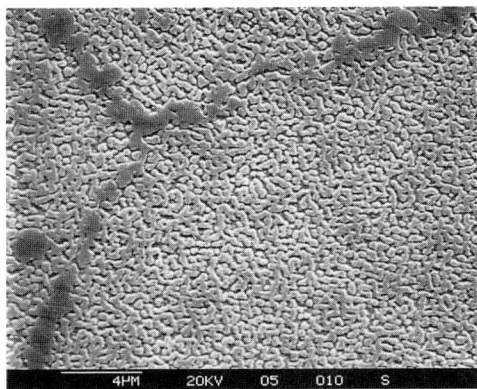

Fig. 3 Overtemperatured Nimonic 115.

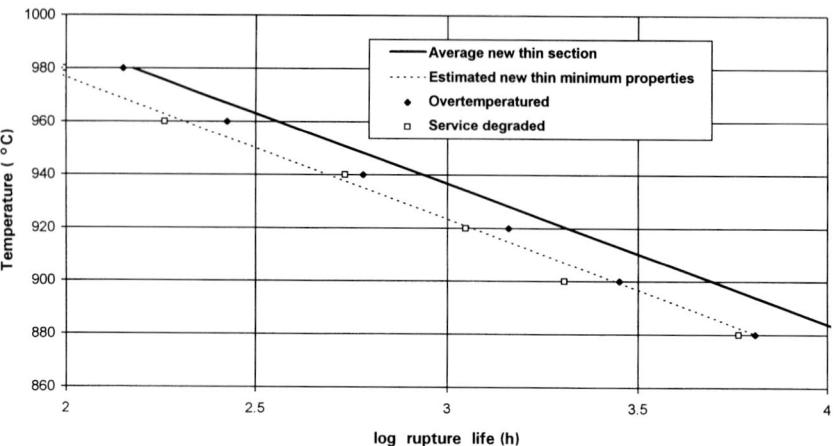

Fig. 4 Isostress plot for new and service run Nimonic 115 at 100 MPa.

Two different sets of Nimonic 115 components were examined. One set had been in service for 19 000 h (notional design life 25 000 h). The structure in the central region of the aerofoil showed the expected degradation of the γ´ as well as $M_{23}C_6$ grain boundary precipitation (Fig. 1: Fig. 2 shows the virgin γ´ structure). Not only had the γ´ coarsened but the precipitate morphology was now more irregular. The second set had been overtemperatured in service after 16 500 h (Fig. 3). The γ´ appeared to have reprecipitated in a convoluted semicontinuous form. The grain boundaries,which formerly had been wavy (as manufactured) were now generally straight.

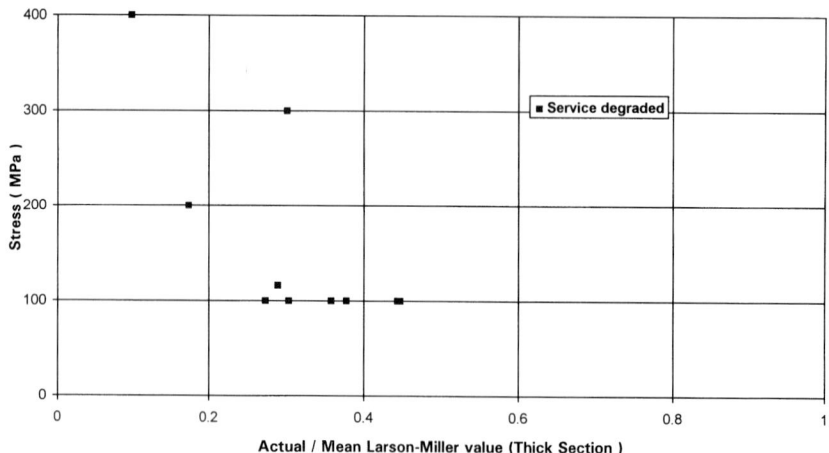

Fig. 5 Effect of stress on the relative residual life of service run Nimonic 115 (19 000 h).

Isostress test on both sets of Nimonic 115 blades showed that their rupture properties (Fig. 4) lay above the minimum properties expected for thin section material (thin in this context means 0.7 mm). This despite the visible degradation of the (conventionally) visible microstructure, the rupture durations were higher than minimum design values. This is not to say that the overtemperatured material was fit for service, since the elongation at failure was low, being about 1–3%.[3]

When the sensitivity of the rupture properties to the evaluation stress was determined, the two microstructures responded very differently. Rupture tests were carried out at the service stress level (100 MPa) and at double, treble and quadruple this level. Different temperatures were used at each stress. Under these conditions average thick section material would be expected to have a life of 700 h. The effect of these different stress levels on the test piece's life is shown in Fig. 5 for the γ' degraded material. The lives at the different stress conditions have been normalised by the average thick section lives under these test conditions so as to facilitate comparisons. The effect that increasing the test stress has, i.e. decreasing the relative properties of the degraded material, is clear – the higher the test stress, the shorter the apparent residual life.

A contrasting situation exists for the overtemperatured material as is shown in Fig. 6. In this case, the relative life of the material is independent of the test stress.

The effect of test stress is in accord with other published work e.g., that on IN738 and U720.[4] Increasing the test stress reduced the apparent residual life of γ' degraded material (relative to what it would be if it had been evaluated

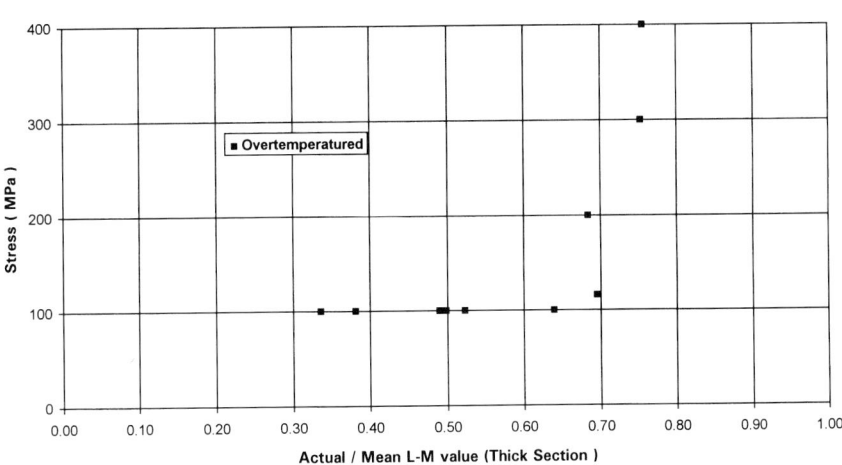

Fig. 6   Effect of stress on the relative residual life of service run and overtemperatured Nimonic 115.

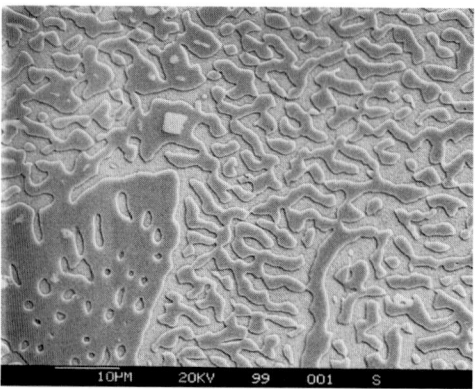

Fig. 7 Rafted γ´ structure in PWA1422 after 6000 h service.

Fig. 8 New γ´ structure in PWA1422 (structure in blade root).

at service stress levels). Various reasons have been proposed for this change. Most authors attribute it to a change in the operative deformation mechanism with stress eg, a change from cutting to climb with reducing stress for dislocation based creep,[5] or to an increasing role for grain boundary sliding.[4] Whichever explanation is most correct, it emphasises that the relationship between residual properties and microstructure is not straightforward. Indeed, at low stress levels representative of service, the functional relationship between properties and the visible microstructure is not a strong one. This type of conclusion is in line with current thinking as to creep deformation processes in superalloys. The strain softening model of Dyson et al[6] lays emphasis, not on the evolution of the conventionally visible microstructure, but the evolution of the dislocation substructure with increasing strain.

Although not so extensively studied, the low sensitivity of rupture behaviour and γ´ structure directionally solidified alloys such as PWA1422.

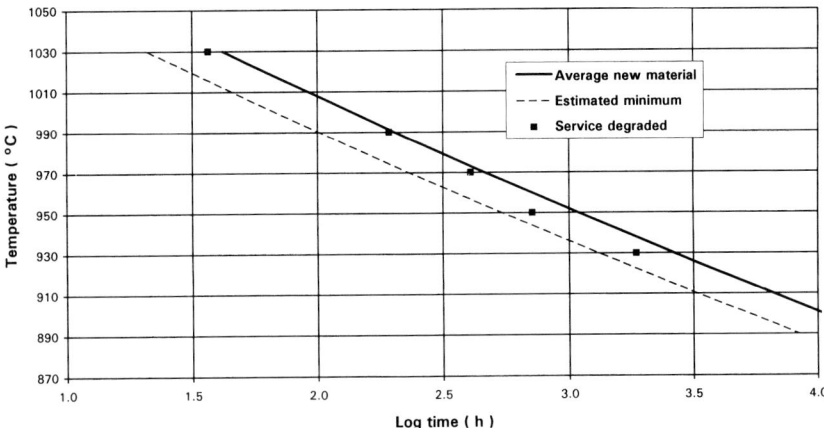

Fig. 9   Isostress plot for service run PWA1422 at 150 MPa.

Components retired from service after 6000 h exhibited a rafted γ' structure in the aerofoil (Figs 7 and 8 show the γ' structure in the root). However, the rupture test results (at 150 MPa) again lay above the expected minimum for new material (Fig. 9). Ductilities were also still high, (~20%).

SIMULATION OF SERVICE DEGRADATION

Much of the work which has been carried out to date to determine the effect of structural degradation on properties has used material which has been just thermally aged (i.e., in the absence of stress). Whilst it is to be expected that this will adequately reproduce effects such as γ' coarsening (but not rafting), and carbide precipitation on grain boundaries, it cannot mimic the build up of creep damage e.g., the dislocation substructure. As noted above, some current models of creep deformation consider this substructure to control the development of creep deformation.

The existence of a difference at this level of the microstructure can be seen in the effect that low levels of creep strain accumulation has on X-ray line broadening, (i.e., at the levels of strain of interest to engineering components such as turbine blades). Nimonic 115 material was crept at low stress with the objective of accumulating ~1% strain in 7500 h at ~850°C. The tests were periodically interrupted for line broadening work. Specimen which were only aged were included for comparison, (more experimental details can be found in Ref. 7).

The results are shown in Fig. 10 in terms of the 'full width at half maximum intensity' (FWHM) of the (311) reflection as a function of accumulated strain. The point to note in the current context is that, even at the low strain of

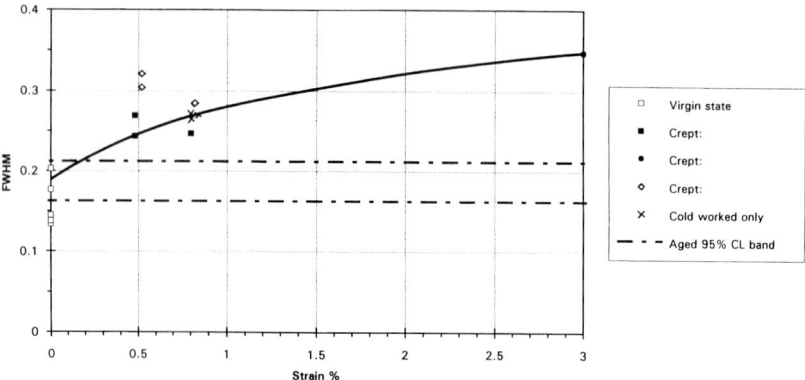

Fig. 10   X-ray line broadening in crept and aged Nimonic 115.

~0.5%, the aged and deformed material had different substructures. Indeed, as might be expected, ageing in isolation had no effect on the peak width.

The significance of this differential between typical simulation material and real components is yet to be determined, although work is in hand to experimentally evaluate this aspect of behaviour.

NEAR SURFACE MICROSTRUCTURAL FEATURES

As noted in the introduction, modern gas turbine blades have a high surface to volume ratio, and work in a chemically aggressive environment. This does not have to be a corrosive in the conventional sense of the word (sodium salts etc.), oxidation can be quite severe even in some 'heavy duty' industrial turbines because of high metal temperatures on the inside of the components. Just as with the case of γ' degradation, the significance of changes in the surface region need to be correctly assessed, and the linkage between several concurrent physical processes correctly identified.

The case in questions concerns a heavy duty row 1 blade manufactured from a cast polycrystalline superalloy with internal cooling by means of holes drilled through the longitudinal axis of the blade (see Ref. 8 for more details). Because of in-service blade failures, (after ~15 000 h–20 000 h) attention had concentrated on the microstructure around the trailing edge cooling holes. Both deep internal oxidation and nitridation were observed (Fig. 11), extending to 200–300 mm in places, along with cracking and voiding at the surface and subsurface (Fig. 12). The manufacturer attributed the failure mechanism to an enhancement of local stresses due to excessive levels of oxidation, and proposed to coat the holes to eliminate the oxidation problem.

However, it has been shown on several occasions that coatings do not significantly affect the creep life (or strain rates) of superalloys (in air), providing that any loss in load bearing cross-section due to the formation of the

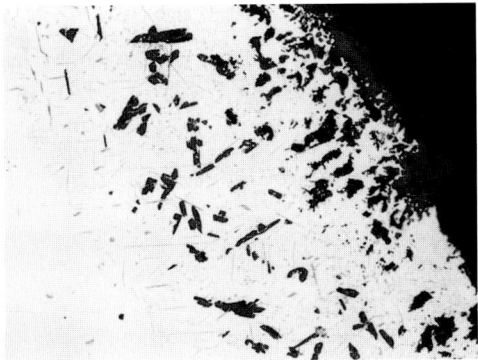

Fig. 11   Internal oxidation in uncoated cooling passages of industrial blade after 20 000 h.

Fig. 12   Surface crack formation in uncoated cooling passages of industrial blade after 20 000 h.

coating is taken into account when determining the effective stress on the test piece.[9] Coatings will often delay the onset of surface cracking, but since the cracks are indicative features of strain accumulation, not the factor governing directly the strain rate and accumulation of strain, they do not significantly extend the time to failure. A complicating factor in this case was the possible role of local embrittlement due to the inward diffusion of oxygen and nitrogen from the uncoated inner surface. There is laboratory evidence for the existence of such an effect, and even experimental demonstrations possibly showing its localised effects in components.[11] The consequence of the manufacturer's interpretation of the microstructural information was that the 'solution' which was implemented was ineffective and there were further engine failures. The manufacturer then redesigned the component to reduce the local metal temperatures.

OVERALL DISCUSSION

The three particular areas highlighted here are ones from a longer list of microstructural instability effects which take place in superalloy blading during their time in service. These need to be taken into account when assessing the engineering integrity of such components. As well as obvious changes such as γ' degradation, carbide coarsening and grain boundary precipitation, with their consequential effect on properties such as ductility and impact toughness, there are other, somewhat more controversial, issues. An example of the latter is the question of oxygen and nitrogen embrittlement cited in the previous section. In addition to the question of whether it is technically significant, it has the characteristic that it is not readily observable because its microstructural effects are at a finer level than it is normal to be looked at. Current assessment methodologies do not adequately take account of such 'invisible' microstructural changes. There is also the issue as to which of the relationships that are observed are merely correlations, and which are actually linked together mechanistically. Whilst some success has been shown for the use of γ' microstructural features as an indicator of life consumption in military aeroengines[2] their limitations are also known. This may be because the linkage is actually weak, or it could be that the parameters looked at are not linked directly at all, but are merely correlated because of their relationships to a third factor. However, these comments reflect a general feature of the technical area, the relative lack of appropriate databases and assessment techniques which have been developed for, and intended specifically to be applied to, degraded superalloy components.

ACKNOWLEDGEMENTS

Some of this work was funded by the UK Department of Trade and Industry under its MTS programme. This is gratefully acknowledged.

REFERENCES
1. U. Yoshioka, D. Saito, K. Fujiyama and N. Okabe: 'Effect of material degradation on mechanical properties of cast nickel based superalloy IN738LC', *Advanced Materials and Coatings for Combustion Turbines*, ASM International, 199, 53–58.
2. G. Blickstad, C. Persson, P-O. Persson and C-G. Samuelsson: 'Increased

service life by use of life time prediction methods and reuvenation treatments', *High Temperature Materials for Power Engineering* Kluwer Academic Press, Dordrecht, 1990, 1131–1140.

3. M.I. Wood: 'The assessment of service induced degradation of nickel based superalloy gas turbine blading', *Material and Manu. Processes,* 1995, **10**, 903–923.

4. A.K. Koul and R. Castillo: 'Creep behaviour of industrial turbine blade materials', *Advanced Materials and Coatings for Combustion Turbines,* V.P. Swaminathan, ed., ASM, Ohio, 1994, 75–88.

5. W. Hoffelner: 'Creep dominated damage processes high temperature', *High Temperature Alloys for Gas Turbines and Other Applications,* D. Reidel Pub. Co, Dordrecht, Netherlands, 1986, 413–439.

6. B.F. Dyson and M. McLean: 'Creep deformation of engineering alloys: developments from physical modelling', *ISIJ International,* 1990, **30**, 802.

7. M.I. Wood and D. Raynor: 'Condition assessment techniques for degraded gas turbine superalloy material', *Int. J., Pres. Ves., and Piping,* 1996, **66**, 341–350.

8. M.I. Wood: 'Internal damage accumulation and imminent failure of an industrial gas turbine blade: Interpretation and implications', *ASME Conference preprint 96-GT-510,* American Society of Mechanical Engineer, NY.

9. A. Strang and E. Lang: 'The effect of coatings on the mechanical properties of superalloys', *High temperature alloys for gas turbines,* D. Reidel Pub. Co., Dordrecht, 469–506.

10. D.A. Woodford: 'Environmental damage of a cast nickel base superalloy', *Met. Trans.,* 1981, **12A**, 299–308.

11. D.A. Woodford: 'The Design for Performance concept applied to life management of gas turbine blades', *Baltica III: Plant Condition and Life Management,* 1995, 319–332.

Microstructure and Material Properties Control by Advanced Melting and Alloy Processing Procedures

G. OAKES, L.H. SHAW AND W. COULSON
Special Melted Products, Atlas House, Attercliffe Rd, Sheffield, S4 7UY, UK

ABSTRACT

Since the birth of the gas turbine, 1938, continuous improvements have been made in the properties of metallic materials used. These have arisen through advances in alloy chemistry, cleanness and homogeneity which are a direct result of improved processing methods such as melting, remelting, homogenisation and thermo-mechanical processing.

The evolution of alloys and, more appropriately, processing methods from those used in the early aerospace gas turbine engines to their modern derivatives will be reviewed. The materials covered range from low alloy, ferritic and austenitic steels, nickel base and to titanium alloys.

In particular the changing role of the melter and primary converter is highlighted with reference to the structure and property improvements achieved after melting, remelting and subsequent thermo-mechanical processing.

The need for structure and property control has in all cases been driven by the engineers desire to obtain enhanced high temperature and associated properties and increase inspection requirement for safety purposes. In addition the current engineering trend requires the manufacturer of materials to demonstrate enhanced process capability and reproducibility.

The information gained via the advanced aero engine gas turbine developments is currently being transferred to industrial power plant including both industrial gas and steam turbine components.

This paper details some, but not all, advances made in the production of advanced superalloys and offers examples. Finally the paper looks at developments in structural control by use of non ingot metallurgy starting stock, i.e. powder metallurgy and sprayforming.

INTRODUCTION

This paper describes the methods by which modern day material manufacturers have achieved microstructural, and therefore property control of high temperature creep resistant superalloys for gas turbine applications. These processes have evolved over the last sixty years and this will be reviewed in order to put the modern practices into context.

Because of the interrelationships which exist between design requirements,[1] alloy chemistry[2-7] and process route, it is not possible to discuss alloy production without mentioning concurrent alloy development.[8-11] Consequently, alloy developments will also be discussed in this paper.

Of the two main areas of gas turbine manufacture,[12] land and aero-engine, the stringent demands of safety and performance of the latter have driven the development of materials and processes.[13] Therefore, to illustrate the advances made, the early engines of Sir Frank Whittle will be compared to modern high thrust engines. Table 1 itemises the materials used and engine specification of an early Whittle engine and a modern civil high thrust turbo-fan engine.

In order to achieve the above improvements in thrust, fuel consumption and turbine entry temperature the design engineer has been driven to find materials of ever improving:

Table 1

Whittle W1	Civil high thrust turbo-fan
Compressor: Aluminium	Titanium 6Al–4V, Ti IMI 834, Alloy 718
Turbine: Austenitic Stainless	Ni base Waspaloy, Alloy 718, CMSX–4
Shafts: 1%CrMoV	Triple melt CrMoV, Alloy 718, Waspaloy
Bearings: White Metal	M50, M50 nil, T1
Thrust: 400 Kg (850 lbs)	40,000 Kg (W1 × 100)
Specific fuel consumption:	1/2 × W1 engine
Turbine entry temp:	2 × W1 engine

Note: Typical chemical analyses of all alloys in the above table, and those mentioned later, can be found in Appendix 1.

1. Strength and durability, at temperature.
2. Low cycle fatigue resistance.
3. Creep resistance.
4. Corrosion and oxidation resistance.
5. Specific density.
6. Cost effectiveness.
7. Capability to be manufactured and examined reliably.

Materials and Manufacturing Route for the W1 Engine
The W1 engine used materials largely derived from the steam turbine technology of the day, see table 1. All the above materials were processed via an air melt–forging–machining route. At that time the influence of grain size (structural control), residual elements and non-metallic inclusions were not clearly understood. The effects of creep were appreciated but, due to war, a

short component life philosophy reigned, i.e. a quick replacement/rebuild was considered more acceptable.

After the war a more scientific approach was adopted by material researchers and a greater understanding of creep evolved. This lead to the separate development of a number of alloy families in the 1940s and 1950s. The main groups were as follows:

12% Cr steels (S61 & S62) → Jethete → FV448 → FV535 (All UK).

Ni-Cr alloys → Nimonic 75 → add Al and Ti for precipitation strengthening to give Nimonic 80 (1941) and Nimonic 80A (1944) → add Co for increased stress rupture capabilities above 800°C to give Nimonic 90 (1945) (All UK).

410 series (USA) → Greek Ascoloy (USA) → Tinidur (Germany) → V57 (USA) → A286 (USA) → then utilised Nimonic alloys.

In the early days all alloys were air melted, wrought and machined. Many of the above alloys had additions of reactive elements such as titanium, aluminium, zirconium and niobium to further improve high temperature strength and creep resistance. This, however, increased the propensity for deleterious alloy segregation, and, because they were airmelted, formed harmful reaction products, oxides/nitrides, that tended to segregate during processing which adversely affected fatigue properties. This problem was overcome by the adaptation of an existing technology then used to commercially melt titanium alloys, namely vacuum arc refining.

DEVELOPMENT OF REMELTING TECHNOLOGY

The remelting processes, vacuum arc refining (VAR) and electroslag refining (ESR), are essentially zone refining techniques, which incrementally melt and solidify a controlled amount of molten metal to give a more structurally refined and consolidated ingot. The specific advantages of remelting by either process are:

1. Low oxygen/better cleanness
2. Increased uniformity, top to bottom and centre to edge of ingot
3. Reduced anisotropy
4. Reduced micro/macro segregation
5. Greater cast to cast reproducibility
6. Improved forgeability
7. Less forging required to achieve consolidation

These technologies were introduced some 30 years ago but have continued to be developed and are still used extensively for many highly alloyed and critical components.

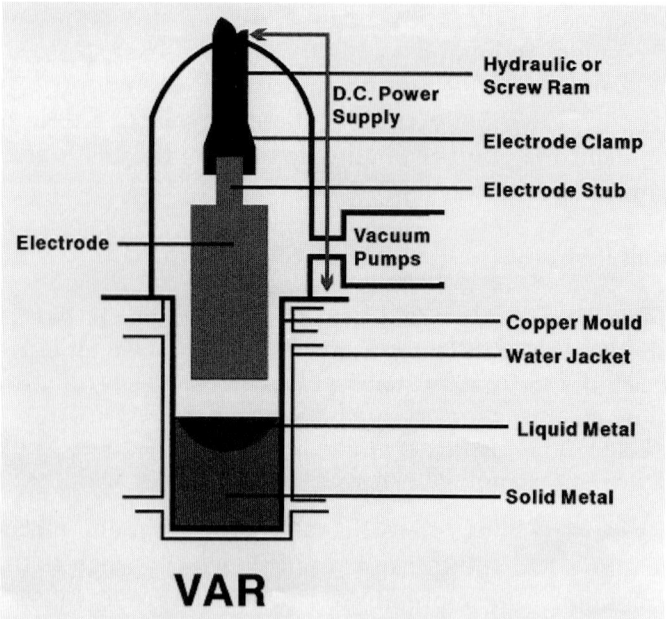

Fig. 1   Schematic of the Vacuum Arc Refining Process.

Vacuum Arc Refining (VAR)

Vacuum arc refining was first utilised in 1905 to melt tantalum.[14] The process was adapted in the late 1950s/early 1960s to melt titanium sponge. The benefits of remelting air melted nickel and iron based feedstock via vacuum arc refining were quickly established. The VAR process, see Fig. 1, is completely enclosed in a vacuum chamber which continues to pump throughout the process thereby drawing off volatile species, such as Pb, Sb, As and Bi. The feedstock is essentially an electrode (a bar of desired alloy composition) down which high currents are conducted. Arcing occurs between the electrode and the molten pool at the top of the ingot. This continually melts the electrode forming a relatively small liquid pool. The pool solidifies, within a water cooled crucible, to form a fully consolidated, structurally refined ingot. The specific benefits of VAR, in addition to those mentioned in the previous section, are as follows:

1. Reduced level of volatile elements.
2. Low hydrogen levels.
3. Low background oxides.

Electro-Slag Refining (ESR)

The electro-slag refining process, see Fig. 2, was developed to avoid the expense of maintaining the vacuum in the VAR process. The feedstock is

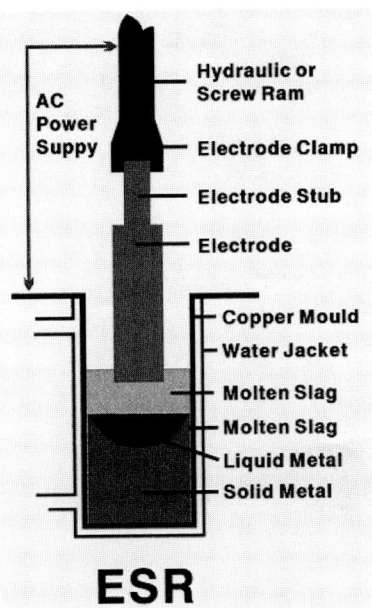

ESR

Fig. 2   Schematic of the Electroslag Refining Process.

again consumed as the bottom of the electrode is just submerged under the slag surface, thereby isolating the molten metal from the atmosphere. The molten slag, heated by AC current resistance heating of the slag itself, melts the electrode forming a liquid metal pool, beneath the molten slag pool. The metal continually solidifies, in the water cooled crucible, to form the ingot whilst being replenished with liquid from the electrode. The nature and chemistry of the slag removes sulphur and breaks up inclusions trapped from primary melting. The specific benefits of ESR, in addition to those general to remelting mentioned earlier, are as follows:

1. Breakdown of large inclusions
2. Retention of high vapour pressure elements, e.g. nitrogen, manganese
3. Improved ingot surfaces
4. Desulphurisation

The Property Benefits of Remelting
Both remelting processes reduce levels of alloy segregation due to the relatively small volume of liquid metal and its more rapid solidification. Certain tramp elements and non-metallic inclusions, mentioned earlier, are also removed, depending upon the process used. The improvement in cleanness of a 4% NiCrMo steel (S82) is shown in Fig. 3. This enhancement in cleanness and homogeneity results in superior mechanical properties such as fatigue life

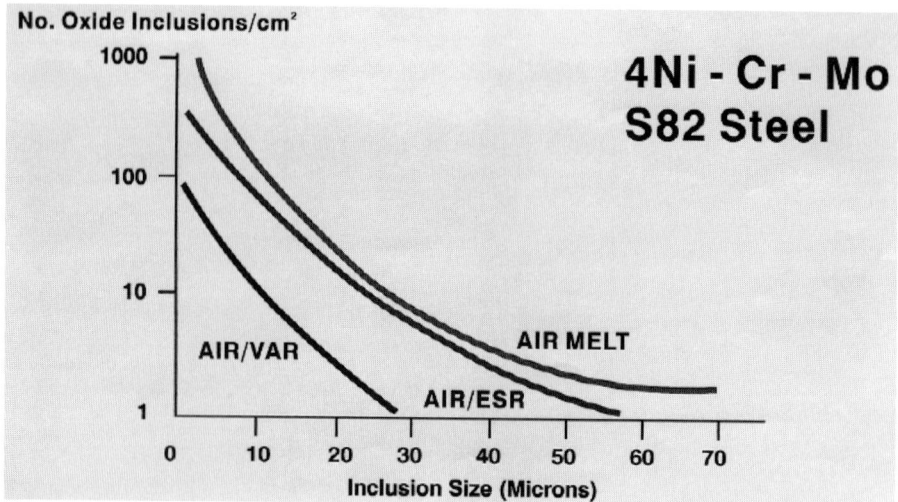

Fig. 3   Cleanness of air melted, air melt/ESR and airmelt/VAR 4%NiCrMo steel.

in a 3% CrMoV steel (Fig. 4) and notch tensile strength in FV535 disc steel (Fig. 5).

Many highly alloyed materials, such as Alloy 718 and Waspaloy, may still possess a highly segregated structure, even after ESR or VAR remelting. In order to dissipate such segregation the material must undergo a heat treatment, termed homogenisation. This heat-treatment generally occurs at a temperature below the liquation point of the offending segregate but high enough

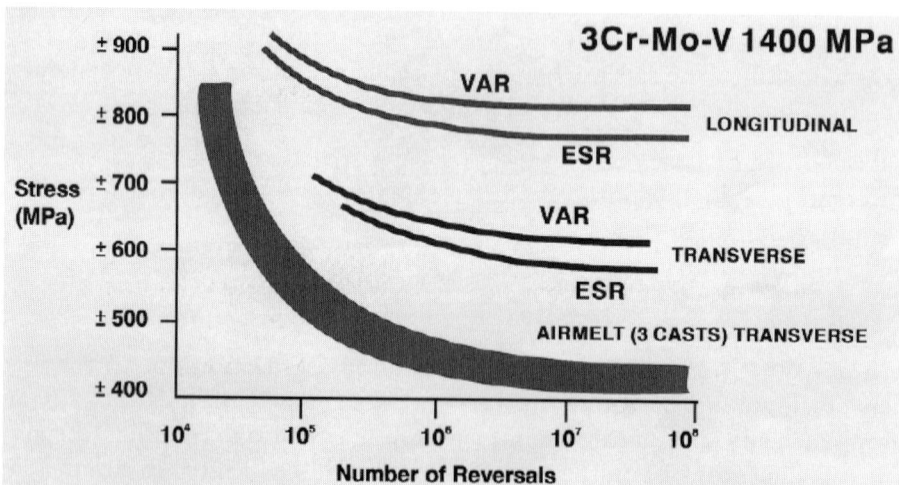

Fig. 4   Fatigue life of air melted, airmelt/ESR and airmelt VAR 3%CrMoV steel.

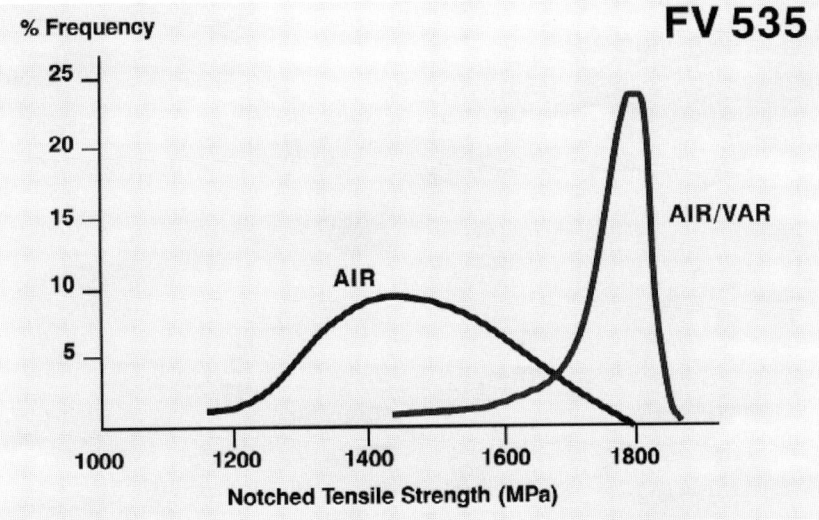

Fig. 5   Notch tensile strength of air melted and airmelt/VAR FV535 disc steel.

to allow its dissolution into the matrix of the alloy in a short a time as possible, as shown in Fig. 6.

Aero-engine designers were, and indeed still are, projecting ever increasingly stressful operating conditions which demanded further material property improvements, above those obtained from remelting an air melted electrode. This came in the form of a process development, namely vacuum induction melting (VIM).

DEVELOPMENT OF VACUUM INDUCTION MELTING TECHNOLOGY

So far, to the early 1960s, the process developments described have involved alloy evolution, thermal treatments and remelting, all involving air melted material, as described earlier. In order to gain further property enhance-

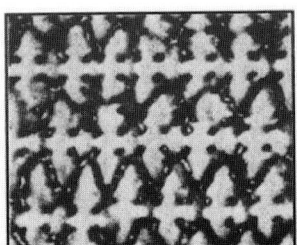 

Fig. 6   Left – As cast microstructure. Right – Post homogenisation microstructures of the same Ni9Mo2Ta13Al alloy.

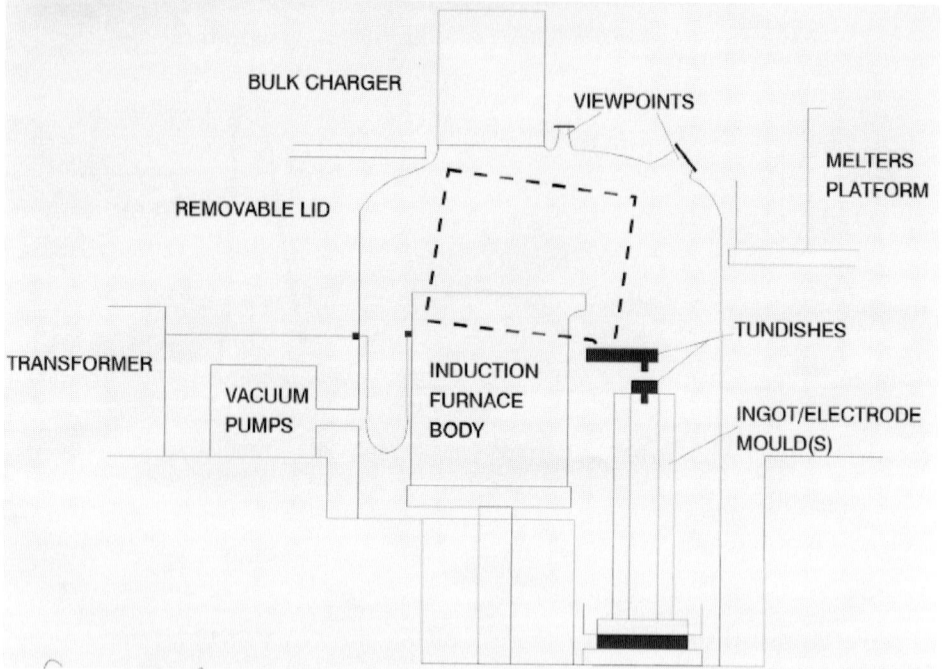

Fig. 7 Schematic of the Vacuum Induction Melting Process.

ments the material was now to be melted in a vacuum induction furnace, see Fig. 7.

Advantages of melting material in a vacuum are:

1. Prevents reaction with oxygen and nitrogen, and can reduce levels present in the charge.
2. Volatilises and removes high vapour pressure tramp elements, such as Pb, Bi, As and Sb.
3. More accurate chemistry control and lower residuals.
4. Reactive trace element additions could now be made, without fear of oxidation control recoveries, to further enhance material characteristics. Examples are illustrated below:

   Zr and B are added as grain boundary modifiers which tend to pin the grain boundaries refining the grain structure therefore improving forgeability and stress rupture properties.

   Ca, Mg and Ce are added to react preferentially with the sulphur to form a more random dispersion of sulphides as opposed to the deleterious low temperature grain boundary precipitating nickel sulphide.

Some Material Property Benefits of VIM
In addition to VIM melting closer scrutiny of raw materials enabled even

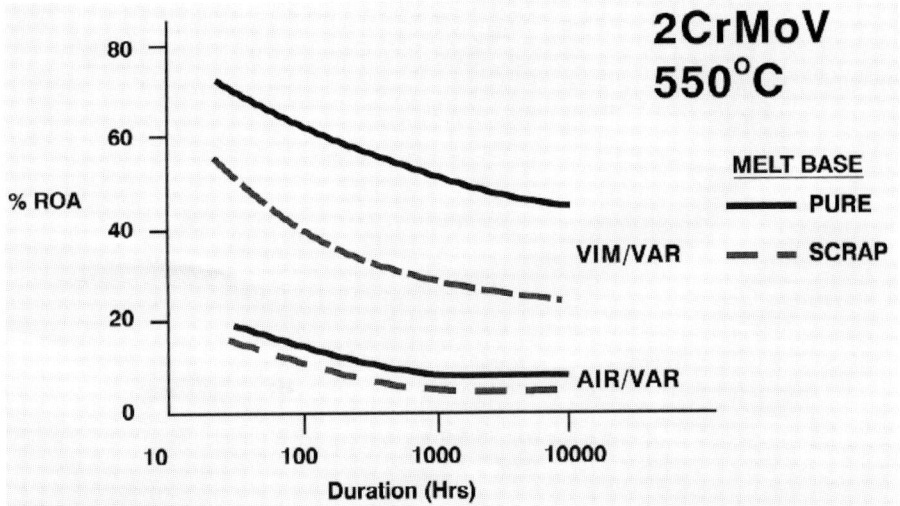

Fig. 8   Creep ductility of air/VAR and VIM/VAR 2%CrMoV steel.

lower levels of undesirable elements to be achieved. The effects of these changes on material properties are illustrated by the following examples.

The creep ductility of a 2%CrMoV alloy at 550°C was substantially enhanced when using VIM feedstock material prior to VAR remelting, see Fig. 8. Even better properties were achieved when pure materials were used instead of scrap metal.

Reducing the level of impurities, such as lead and bismuth, in the same alloy, significantly improved ductility, as can be seen in Fig. 9. Ultimate tensile strength was increased by 100 Mpa in RBD material when a VIM/VAR route was used instead of standard Air/VAR, refer to Fig. 10. This was due to a reduction of inclusions which restricted the number of preferential fracture sites.

The processing of materials, in terms of raw materials control, vacuum induction melting and remelting, had developed as far as it could, at that time. To keep pace with the continued demand for materials which could operate for longer periods at higher temperatures, the materials manufacturer had to further improve cleanness and, in addition, increase the volume fraction of strengthening elements in certain alloys. The latter had the adverse affect of making these highly alloyed materials unformable by conventional casting/forming methods. The next section highlights some of the methods used to further purify and commercially produce these highly alloyed materials.

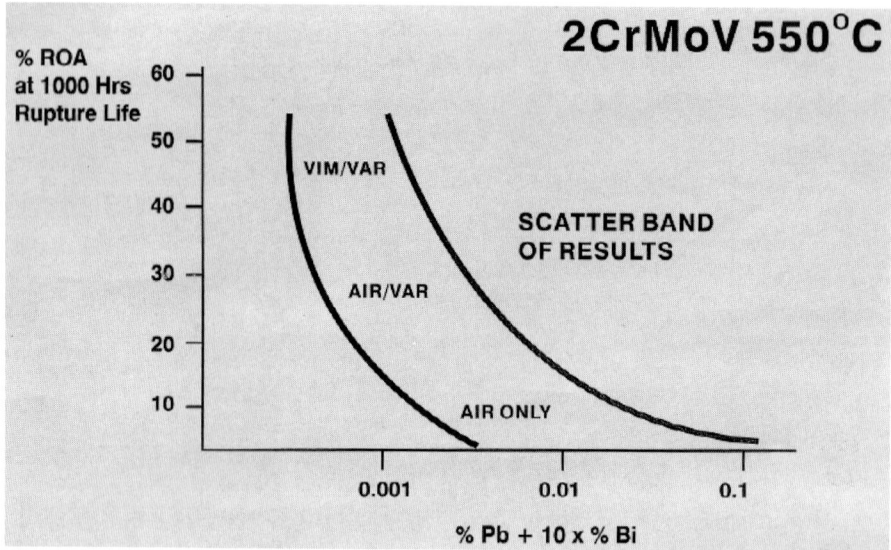

Fig. 9 Effect of impurities on the creep ductility of air melted, air/VAR and VIM/VAR 2%CrMoV steel.

ADVANCED PROCESSING TECHNIQUES

Triple Melted Materials
In order to achieve even further improved performance in parts of the gas turbine under the highest operational stresses the materials need to be defect free

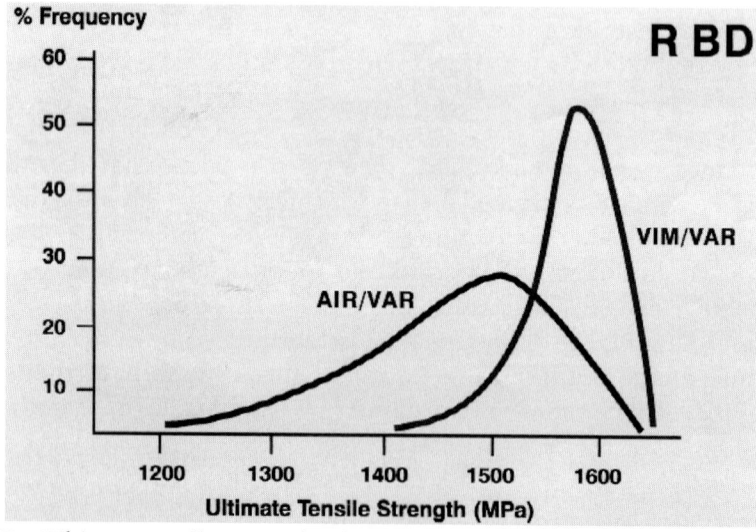

Fig. 10 Ultimate tensile strength of air/VAR and VIM/VAR RBD steel.

and ultra-clean. The benefits of vacuum induction melting and remelting materials have been described. However, even cleaner and more homogenous materials could be achieved when an extra remelting stage was added. These materials were now 'triple melted', i.e. VIM → ESR → VAR, in that order.

Triple melting was developed for highly alloyed materials, such as Alloy 718, Waspaloy and Alloy 720, for critical applications, i.e. engine rotating parts – discs and shafts. The benefits include a reduced risk of macrosegregation, i.e. spot and freckle, ultra-cleanness, low residual practices and a degree of structural control. Alloys less susceptible to segregation, but where cleanness and low tramp elements are paramount, are triple melted. For example, HCM3–3%CrMoV aero-engine shafts by S.M.P. Ltd. Figure 11 shows the increases in strength and toughness of triple melted steel compared to conventional air/ESR and VIM/ESR materials.

Investment casting
This process has been well established as a method to produce turbine blades. The demand for more creep resistant materials spiralled, due to increasing turbine entry temperatures and rotating speeds. During the 1960s the effect of grain boundaries on creep properties had been researched and understood.[15,16] One method of limiting creep was to reduce grain boundary area thereby limiting grain boundary sliding and diffusion. This was achieved by initially moving from forged to cast blades thereby increasing the grain size.

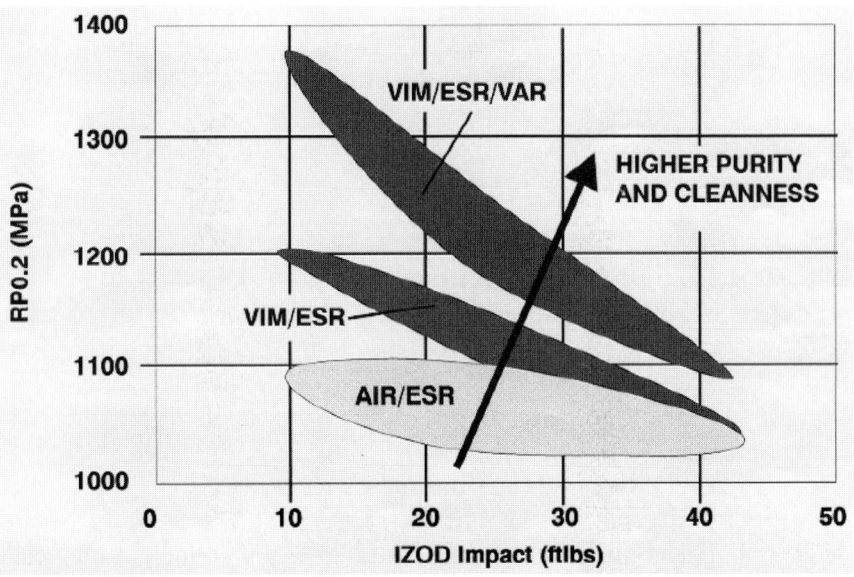

Fig. 11 Strength and toughness of air/ESR, VIM/ESR and VIM/ESR/VAR HCM3.

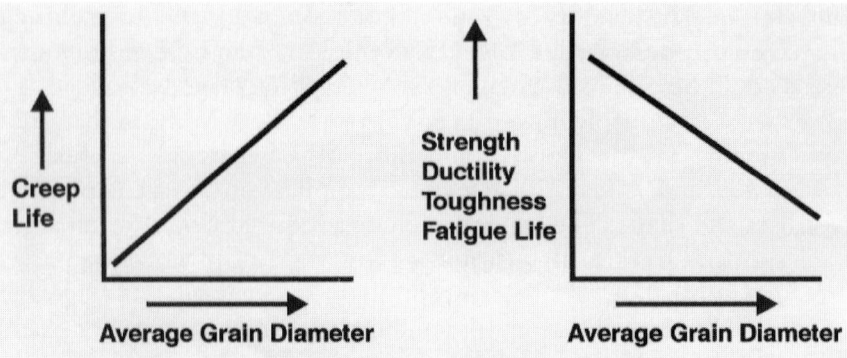

Fig. 12 Equiaxed, directionally solidified and a single grain blade castings.

Subsequently, creep properties were further improved by directional solidification during investment casting the blades. The grains were forced to grow along the direction of loading thereby eliminating grain boundaries perpendicular to the stress axis, see Fig. 12. This also had the beneficial effect of reducing the tendency to form thermal fatigue cracks at the trailing edge of the blade in service.

Continued developments along this theme eventually lead to the removal of all grain boundaries within the turbine blade, which essentially became a single crystal. The elimination of grain boundaries also removed the need for grain boundary modifiers, such as B, Hf, Zr. This increased the incipient melting point of the alloy which permitted a higher temperature to be used for gamma prime solutioning enabling improved precipitate strengthening. For the highest rated applications the addition of intricate air-cooling ducts within the blade have enabled gas operating temperatures to even exceed the melting point of the blade alloy.

Powder Metallurgy

Many highly alloyed materials, such as Alloy 100 and Astrolloy, were developed that could not be successfully manufactured by the conventional cast and wrought route, refer to the 'unforgeable' portion in Fig. 13. These alloys were difficult to cast and work due to the high levels of alloying additions that promoted deleterious excessive segregation and on cooling the castings would fracture. Even when homogenisation was practical forging was difficult due to the increase in gamma prime content which decreased hot workability.

A process was developed which could, by the early 1970s, successfully produce such materials. This technology was known as powder metallurgy (PM). This involved the atomisation of pre-alloyed melts of known composition to produce a powder. These were then uni-axially compressed and sintered to form fully consolidated near-net shape compacts. The properties of PM produced materials could be equivalent or better than those conventionally formed.

To further improve PM material properties, for use as engine discs, the powder was first hot isostatically pressed, or 'HIP'ed'. In this process the powder was placed in a deformable inert container, the shape of which was similar to that required. Heat was applied then intense pressure which deformed the container thereby uniformly compacting the powders. Typical fatigue properties of HIP'ed material, compared to conventionally processed can be seen in Fig. 14.

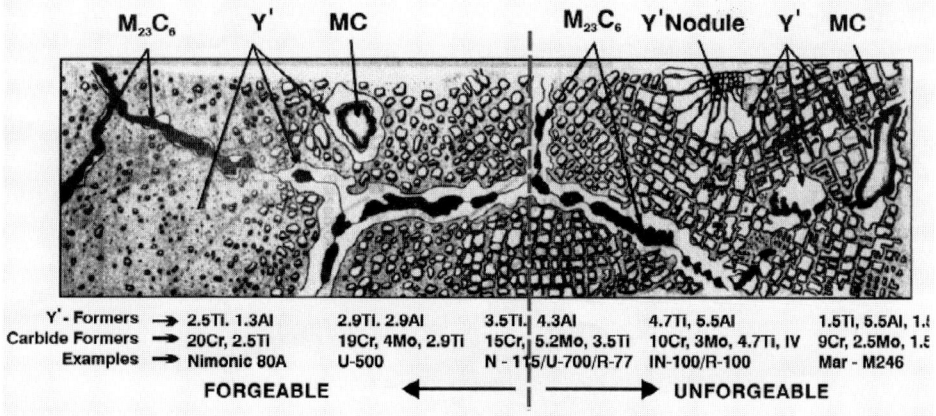

Y'-Formers	➔	2.5Ti. 1.3Al	2.9Ti. 2.9Al	3.5Ti. 4.3Al	4.7Ti, 5.5Al	1.5Ti, 5.5Al, 1.8
Carbide Formers	➔	20Cr, 2.5Ti	19Cr, 4Mo, 2.9Ti	15Cr, 5.2Mo, 3.5Ti	10Cr, 3Mo, 4.7Ti, IV	9Cr, 2.5Mo, 1.5
Examples	➔	Nimonic 80A	U-500	N - 115/U-700/R-77	IN-100/R-100	Mar - M246

FORGEABLE ◄———————————|———————————► UNFORGEABLE

Fig. 13   Examples of alloys and suitability of process route due to gamma prime and carbide content.

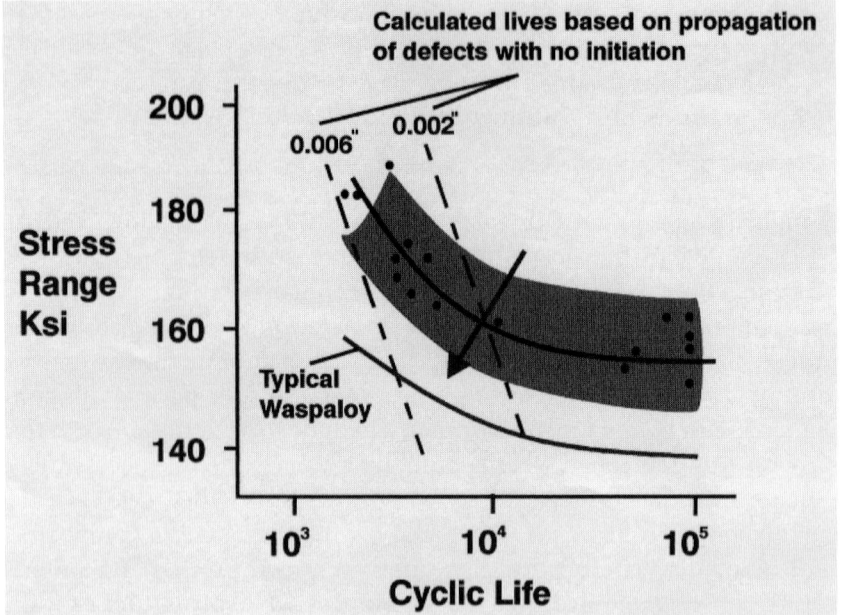

Fig. 14   Fatigue life of conventionally processed and HIP'ed Waspaloy.

These HIP'ed powder compacts yielded excellent static properties, however, low life fatigue failures occurred in service. These failures were unpredictable because they were initiated at defects on prior particle boundaries due to powder contamination. To overcome this the handling of the powder was improved to avoid contamination and was then extruded prior to forging. These additions to the process route reduced the level of powder contamination and broke up any remaining defects, reducing their influence on fatigue life. This process was developed in the early 1980s and was coined 'Gatorizing™'. The disadvantage of these changes was an increase in the cost of PM components for critical applications.

Gatorizing
Gatorizing is a thermo-mechanical processing operation which involves the heat treatment of a HIP'ed, or forged triple melted alloy, to precipitate over-aged gamma prime. This precipitate pins the grain boundaries therefore restricting grain coarsening and recrystallisation. It is then isothermally forged sub-gamma prime solvus which enables 'superplastic' forming.

SPRAYFORMING
The expense of powder metallurgy processing and conventional triple melting and homogenising material provided the impetus to pursue less expensive

production routes. An existing technology developed in the mid 1970s, known as sprayforming, was considered as a possible alternative to both routes. Research starting in the early 1990s demonstrated the potential of sprayforming for use in commercial and aero-space applications. Benefits reported so far include:

1. Inherently fine uniform microstructures
2. Freedom from macrosegregation
3. Low levels of microsegregation
4. Enhanced formability leading to the possibility of more complex shaped product
5. Less complex manufacturing route
6. Benefits for more conventional alloys
7. Viable production route for complex/segregation prone alloys
8. Properties equivalent or better than conventionally produced material
9. Increased property minimum due to reduced scatter
10. Reduced sampling to determine properties
11. Enhanced inspectability
12. Improved heat treatment response

Sprayforming involves melting the feedstock and pouring into a tundish to achieve a steady ferro-static head. One, or two, nozzles at the bottom of the tundish allow the molten metal to form a steady stream which is atomised by nitrogen or inert gas. The semi-molten metal is propelled to the rotating and retracting starter plate and forms a solid preform, see Fig. 15 for a view of horizontal sprayforming. The preform does not require HIP'ing to achieve

Fig. 15   Photo of twin atomised horizontal spray forming.

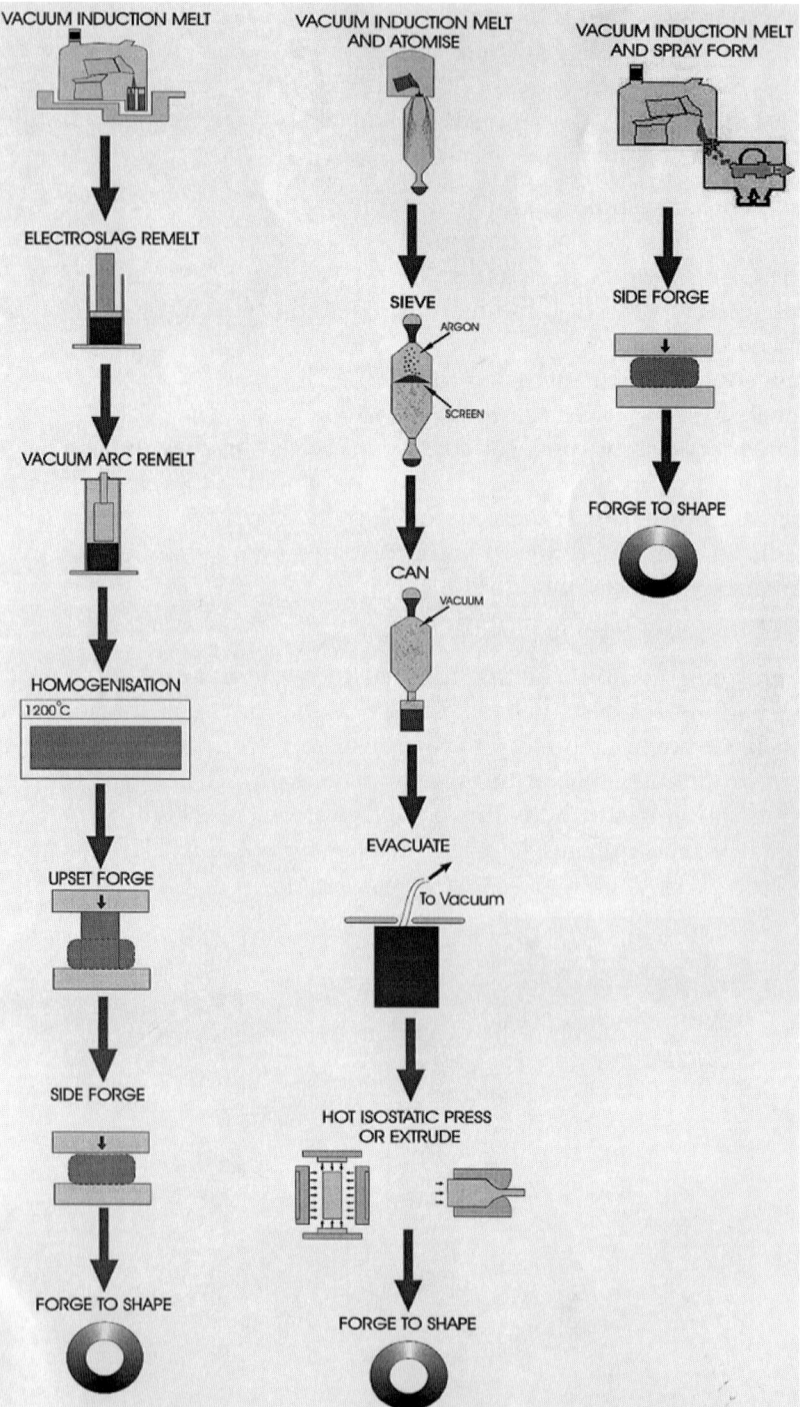

Fig. 16   Process route of a triple melted, powder met and sprayformed component.

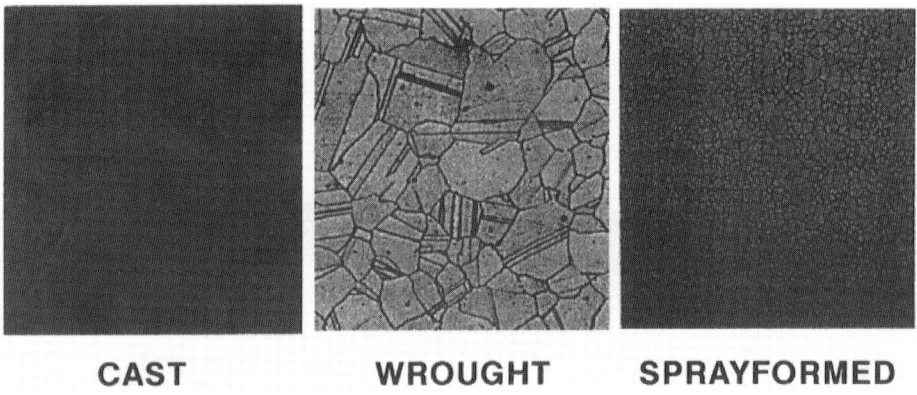

CAST   WROUGHT   SPRAYFORMED

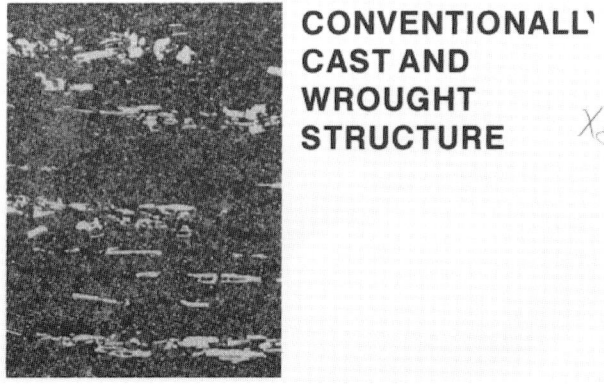

CONVENTIONALL'
CAST AND
WROUGHT
STRUCTURE

SPRAYFORMED
AND
WROUGHT

Fig. 17 Top – Microstructures of as-cast, as-cast/wrought and sprayformed superalloy. Middle – As-cast and wrought tool steel. Bottom – Sprayformed and wrought tool steel.

full consolidation prior to conversion by standard forging practices to the final component.

The simplicity of sprayforming an aero-space component over conventional PM and ingot routes is demonstrated in Fig. 16. This simplicity does not infer a lack of quality. The microstructure of an 'as-sprayformed' nickel based alloy is much finer than conventionally cast and wrought product, see Fig. 17. Sprayformed and wrought tool steel material (D2) contains much finer and more uniformly distributed carbide networks than cast and wrought product. The potential for sprayforming appears considerable, however, only commercialisation and further evaluation of the product in service applications will determine its success.

All the manufacturing methods described are currently used for production of aerospace materials with the exception of sprayforming. Each method has its niche depending on cost and fitness for purpose.

FUTURE REQUIREMENTS

The future of materials for gas turbine applications will continue to need to address a number of key factors as follows:

1. Engine performance – whilst this is a complex topic performance is partially related to the usable temperature/strength capability of the materials under the required service environment. Therefore, to increase performance it follows that temperature and strength capabilities must also increase.

   This will require the development of new alloys or the adaptation and optimisation of existing materials to increase their design limits for use in the hotter parts of the engine. As a consequence, novel manufacturing methods must be found to produce these new alloys or adapt existing alloys and form the components from the relatively unformable higher temperature materials.

2. Increased reliability and better inspectability leading to the possibility of using existing materials under more severe service conditions.

3. And last, but very importantly, reduced manufacturing and through cost (including costs of ownership).

REFERENCES

1. *Materials Engineering in Turbines and Compressors,* Third International Charles Parsons Turbine Conference 1997, R.D. Conroy, M.J. Goulette and A. Strang eds, 1997. The Institute of Materials, 1997.
2. *Materials Development in Turbo-Machinery Design,* Second Parsons International Turbine Conference 1989, D.M.R. Taplin, J.F. Knott and M.H. Lewis eds, The Institute of Metals, London.

3. *High Temperature Materials in Gas Turbines*, Symposium Proceedings held at Brown Boveri, Baden, Switzerland, 1973, P.R. Sahm and M.O. Speidel eds, Elsevier Scientific Publishing Company, 1973.

4. *High Temperature Materials for Power Engineering Parts I and II*, Conference Proceedings organised by CRM (Liege, Belgium) 1990, Kluwer Academic Publishers, 1990.

5. *High Temperature Materials II*, G.M. Ault, W.F. Barclay and H.P. Munger eds, Metallurgical Society Conference, Volume 18, 1961, Interscience Publishers, 1961.

6. *High Temperature Steels and Alloys for Gas Turbines*, Symposium Proceedings held by the Iron and Steel Institute, ISI Publication, London, 1951, Special Report No. 43.

7. *High Temperature Alloys for Gas Turbines*, Conference Proceedings organised by CRM, Liege, Belgium 1978, D. Coutsouradis *et al* eds, Applied Science Publishers Ltd, 1978.

8. *Superalloys*, Proceedings of the 8th International Symposium on Superalloys, Pennsylvania, USA, The Minerals, Metals and Materials Society, 1976.

9. *Superalloy 718 Metallurgy and Applications*, Proceedings of the International Symposium on the Metallurgy and Applications of Superalloy 718, E.A. Loria ed., The Minerals, Metals and Materials Society, 1989.

10. *Superalloys 718, 625, 706 and Various Derivatives*, Proceedings of the International Symposium on Superalloys, The Minerals, Metals and Materials Society, 1994.

11. *Superalloys 718, 625, 706 and Various Derivatives*, Proceedings of the International Symposium on Superalloys, The Minerals, Metals and Materials Society, 1997.

12. 'The Superalloys', *Vital High Temperature Gas Turbine Materials for Aerospace and Industrial Power*, C.T. Sims and W.C. Hagel eds, John Wiley and Sons, 1972.

13. *Rupture Ductility of Creep Resistant Steels*, A. Strang ed., Conference Proceedings organised by the Alloy Design and Behaviour Group of the Institute of Metals, York, England, The Institute of Metals, 1990.

14. Dr W Von Bolton: *Z. Electrochem*, 1905, **45**, 11.

15. *Creep Strength in Steel and High Temperature Alloys*, Conference Proceedings organised by the Iron and Steel Institute, University of Sheffield, England, The Metals Society, 1972.

16. *High Temperature Properties of Steels*, Conference Proceedings organised by the Iron and Steel Research Association, Eastbourne, England, The Iron and Steel Institute, 1966.

Appendix 1 Typical Compositions of all alloys covered

Alloy	Composition																			
	C	Si	Mn	Cr	Mo	Ni	Ti	Al	Nb	Ta	N2	Co	Fe	V	Re	W	Sn	Zr	Hf	B
Ti6Al4V							bal	6.0						4.0						
TiMI834	0.05				0.5		bal	5.6	0.7								4.0	3.6		
Nimonic 75	0.10			20.0		bal	0.4													
Nimonic 80	0.07			20.0		bal	2.4	1.0												
Nimonic 80A	0.07			19.5		bal	2.4	1.4										0.07		0.003
Nimonic 90	0.08			19.5		bal	2.4	1.4				16.5						0.07		0.003
Alloy 720	0.035			18.0	3.0	bal	5.0	2.5				15.0				1.25		0.03		0.033
Alloy 100	0.18			10.0	3.0	bal	4.7	5.5				15.0						0.06		0.015
Astrolloy	0.06			15.0	5.3	bal	3.5	4.0				17.0								0.030
Alloy 718	0.04			17.6	3.0	bal	0.9	0.55	5.2				20.0							
Waspaloy	0.08			19.5	4.3	bal	3.0	1.3				13.5						0.06		0.006
CMSX-4				6.5	0.6	bal	1.0	5.6		6.5		9.7			3.0	6.4			0.10	
M50	0.80			4.0	4.3								bal	1.0						
M50nil	0.14	0.2	0.4	4.1	4.2	3.4							bal	1.15						

Appendix 1 (cont.) Typical Compositions of all alloys covered

Alloy	Composition																			
	C	Si	Mn	Cr	Mo	Ni	Ti	Al	Nb	Ta	N2	Co	Fe	V	Re	W	Sn	Zr	Hf	B
AISI T1(18/4/1)	0.75			4.5	0.4								bal	1.25		18.0				
S61	0.10	0.3	0.7	12.0									bal							
S62	0.22	0.3	0.7	12.0									bal							
M152 (Jethete)	0.12	0.4	0.8	11.5	1.8	2.5							bal	0.3						
FV448	0.15	0.35	0.8	10.5	0.7	0.75			0.3		0.06		bal	0.25						
FV535	0.700	0.5	10.6	10.7	0.7	0.3			0.3		0.02	6.0	bal	0.28					0.005	
410 Series	0.10	0.4	0.7	12.5									bal							
Greek Ascoloy	0.12	0.3	0.4	13.0		2.0							bal			3.0				
Tinidur	0.04	0.73	1.0	15.0		26	2.2	0.15					bal							
V57	0.08	0.75	0.4	14.8	1.3	27	3.0	0.25					bal	0.5						0.010
A286	0.05	0.4	1.4	15.0	1.3	26	2.2	0.2					bal	0.03						0.003
S82	0.15					4.0							bal							
3%CrMoV	0.40	0.25	0.7	3.3	0.9								bal	0.2						
RBD	0.20			3.0		0.75							bal	0.4	10.0					
D2	1.6	0.4	0.4	12.5	0.8								bal	0.8						

Tradename acknowledgements are given to IMI, Cannon-Muskegon, INCO alloys and Special Metals.

Analysis of Strengthening Mechanisms in 9 to 12% Chromium Steels

Z. KUBOŇ, V. FOLDYNA, V. VODÁREK
VÍTKOVICE, a.s., Research & Development, Ostrava, Czech Republic

INTRODUCTION

From the very beginning of the development of modified chromium steels much effort was devoted to finding a way how to improve their creep resistance. The most promising method was found about 20 years ago. The solution was based on increasing the nitrogen content of the steel and adding small amounts of vanadium and niobium. The creep rupture strength (CRS) of the modified 9%Cr–1%Mo steel, called P91, was about 40 to 50% higher than that of the previous generation of chromium steels. Further research has been concentrated on improving the creep properties of the P91 steel. It was believed, that the creep resistance could be improved by increasing the Mo content of the steel or by adding W. Steels with a molybdenum equivalent

Table 1   Chemical composition of selected chromium steels

Elements	P91	E911	Chemical composition in mass %			B2	D3
			P92 (Nf616)				
			heat 1	heat 2	heat C		
C	0.10	0.105	0.124	0.106	0.085	0.17	0.16
Si	0.38	0.20	0.02	0.04	0.06	0.08	0.12
Mn	0.46	0.35	0.47	0.46	0.45	0.06	0.49
P	0.020	0.007	0.011	0.008	0.004	0.07	0.010
S	0.002	0.003	0.006	0.001	0.002	0.001	0.003
Cr	8.10	9.16	9.07	8.96	9.00	9.36	11.30
Mo	0.92	1.01	0.46	0.47	0.50	1.55	0.32
W	–	1.00	1.78	1.84	1.80	–	1.80
V	0.18	0.23	0.19	0.20	0.20	0.27	0.22
Nb	0.073	0.068	0.063	0.069	0.054	0.06	0.06
B	–	–	0.003	0.001	0.004	0.010	–
N	0.049	0.072	0.043	0.051	0.048	0.015	0.055
Ni	0.33	0.23	0.06	0.06	<0.05	0.12	0.79
Al	0.034	–	0.002	0.007	<0.001	0.014	0.010

(%Mo + 1/2%W) about 1.5% were assumed to be the most creep resistant. Moreover, a beneficial effect from the addition of boron was also expected.[1-5]

The aim of the present paper is to analyse the different strengthening mechanisms occurring in the chromium creep resistant steels. This analysis will be used for the most probable explanation of the observed differences in the creep properties of some advanced 9–12%Cr steels.

MATERIALS AND EXPERIMENTAL RESULTS

The chemical compositions of some selected modified chromium steels are shown in Table I. Steels P91, E911 and P92 (heats 1 and 2) have been examined by Ennis and Wachter.[6]

Steels B2 and D3 have been investigated in the European COST 501 Program on Critical Components for Advanced Steam Cycles.[3,5] Steel P92 (Nf 616–heat C) was tested by the Nippon Steel Corporation.[7] The creep rupture properties determined experimentally, as well as being calculated by the model presented by creep testing are summarised in Table 2.

Table 2   Creep rupture strength at 600°C in 100 000 hours of investigated chromium steels and estimated precipitation strengthening due to VN particles

CRS at 600°C in 10^5 h	P9	P91	E911	P92			B2	D3
				heat 1	heat 2	heat C		
experimental, [MPa]	34	87	105	115	115	112	120	74
calculated, [MPa]	$39^{(1)}$	71–73	100–106	79–84	85	87–93	57	81–96
$c \cdot N_{ss}$ [MPa]	–	27–29	57–63	36–41	41	41–47	8	33–47

$^{(1)}$ On the assumption – $Rp_{0.2} = 390$ MPa

CREEP RUPTURE STRENGTH ESTIMATION

The reliability of published creep rupture strength (CRS) properties depends on the extent and durations of the creep rupture tests as well as on the method of CRS estimation used. Sometimes the quoted differences in CRS may be differences of extrapolation rather than real differences in properties.[8] Therefore, these data should be judged very carefully.

Lately, the creep rupture properties of some 9% chromium steels were compared. The greatest difference in creep rupture strength was found between P9 (9Cr1Mo) and P91. The additional increase in creep rupture strength of E911 and P92 over P91 was attributed to the addition of W. The steel with the highest W content (~1.8%) exhibited the highest creep rupture strength.[6] The creep rupture tests performed on the P92 steels investigated by Wachter and Ennis attained about 10 000 hours. For creep tests of longer

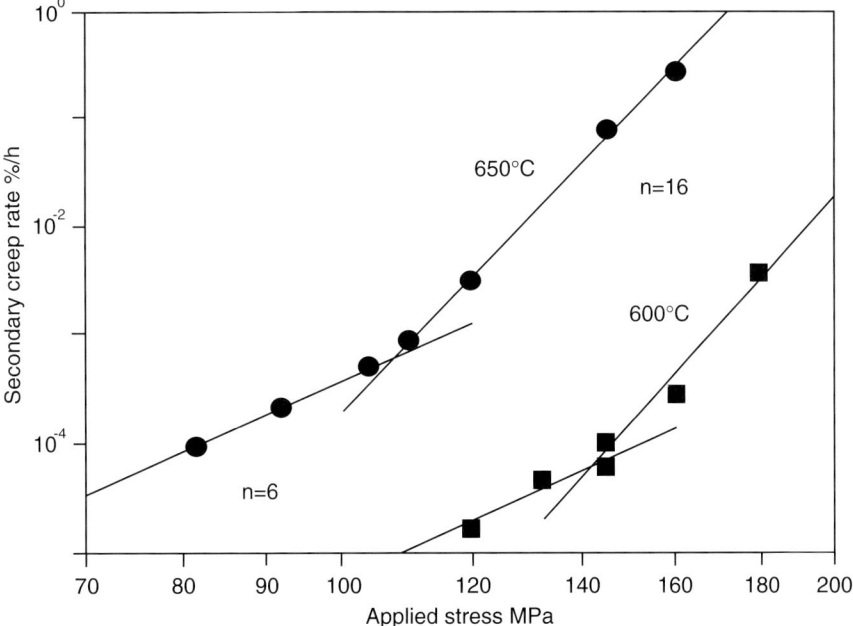

Fig. 1   Secondary creep rate as a function of applied stress for Nf616 steel.[9]

duration, which were still in progress, the creep rupture life was predicted from the secondary creep rate $\dot{\epsilon}_s$ using the Monkman–Grant equation (steel E911 and P92). The predicted times to rupture were near to 100 000 hours. Moreover, the secondary creep rates were plotted as a function of the applied stress (σ). An example for steel 92 is shown in Fig. 1.[9] For tests at high stresses (more than 145 MPa at 600° C and above 110 MPa at 650°C), the stress exponent n_c, defined as,

$$n_c = \left(\frac{\delta \log \dot{\epsilon}_s}{\delta \log \sigma} \right) \qquad (1)$$

was found to be 16. On the other hand, for creep tests carried out at lower stresses the stress exponent was reduced to 6. The observed change in the stress exponent indicate the change in deformation mechanism, which must be considered when the rupture data are extrapolated to long times. Extrapolations based on creep data attained in the high stress region will lead to the considerable overestimation of the long term creep resistance.[10] Two regions with different stress exponents were found for steels P91 and E911, as well.[6]

Creep rupture tests performed on the steel P92 (heat C) attained times to rupture of over 40 000 hours.[7] Nevertheless, the CRS estimation which took into consideration all creep rupture data (most of them performed in the high stress region), led to a significant overestimation. The CRS at 600°C in 100 000 h was estimated to be as high as 132 MPa.[11] Taking into account only creep rupture tests carried out at lower stresses, the estimated CRS is lower, only 112 MPa.[12] This latter and much more reliable value is presented in Table 2.

Creep rupture tests for the duration of nearly 30,000 hours have been available for the 10^5 hrs CRS estimation of steels B2 and D3. The method used for CRS estimation was the same as in the case of steel P92–heat C. Details of this approach have been described recently.[10]

MODELLING OF CREEP RUPTURE STRENGTH

It was shown that the proof stress at room temperature ($Rp_{0.2}$) and the nitrogen content in the solid solution (N_{ss}) are significant factors controlling the creep resistance of 9 to 12% chromium steels. The creep rupture strength in 100 000 hours at temperature T ($R_{mT/100,000/T}$) can be calculated by

$$R_{mT/100,000/T} = a + b \cdot Rp_{0.2} + c \, N_{ss} \qquad (2)$$

The temperature dependent regression coefficients a, b and c were established by analysing experimental results attained on 33 heats of various types of chromium steels. The values of these coefficients at 600°C were: $a = 234$, $b = 0.04$ and $c = 1017$. The constant a represents the *inherent* creep strength of chromium steels. The product $b \cdot Rp_{0.2}$ stands for the contribution of proof stress at room temperature, which reflects mainly the precipitation strengthening by $M_{23}C_6$ and niobium rich *primary* NbX particles. Lastly, the product $c \cdot N_{ss}$ determines the contribution of the *secondary* MX precipitate (i.e. mainly vanadium rich nitride) to the creep rupture strength. N_{ss} represents the nitrogen content in the solid solution which is not bound as AlN and/or TiN and/or Nb(CN). It is clear that N_{ss} is practically equal to the so called available nitrogen (N_{AV}) for formation of VN. Different methods of N_{ss} and/or N_{AV} estimation were described earlier.[13–19]

Very good correlation between the calculated and measured creep rupture strength ($R^2 = 0.791$) can be comprehended as a confirmation that the semi-empirical eqn (2) is based on the correct assumptions and that it represents the main contributions to the creep strength.

Using eqn (2), the creep rupture strength and the contribution of precipitation strengthening due to precipitation of VN were estimated (Table 2). With the exception of heat B2, the product $c \cdot N_{ss}$ represents the most substantial part of the estimated creep rupture strength.

DISCUSSION

When assessing how the creep rupture strength of the investigated chromium

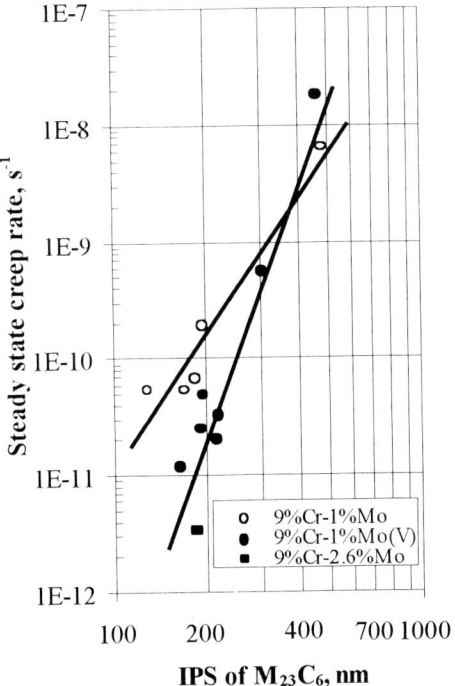

Fig. 2   The dependence of steady state creep rate on the interparticle spacings
(IPS) of $M_{23}C_6$.[20]

steels depends on the different strengthening mechanisms, we must first of all
perform a critical examination of the reliability of the presented creep rupture
strength data (Table 2).

For creep rupture strength estimation of steels E911 and P92 (heats 1 and
2) the Monkman–Grant equation was used.[6] Many years ago it was found
that given identical interparticle spacing of $M_{23}C_6$ in the as received condition, steady state creep rate at 550°C and 100 MPa in the steel
9%Cr–2.6%Mo is at least one order of the magnitude lower than that of the
9%Cr steel with molybdenum content of about 1% evaluated under the same
conditions (Fig. 2).

The respective creep rates were measured in time period between 15 000
and 20 000 hours. Nevertheless, long time creep rupture tests (up to nearly
100 000 h) did not confirm the beneficial effect of molybdenum additions on
the CRS of 9%Cr steels with Mo contents higher than about 1% (Fig. 3).[20,21]
It is clear that the difference in the observed steady state creep rate can be
attributed to the precipitation of Fe_2Mo during the initial period of creep
exposure. Although such a precipitation decreases the creep rate, neither precipitation strengthening due to Fe_2Mo Laves phase, nor solid solution

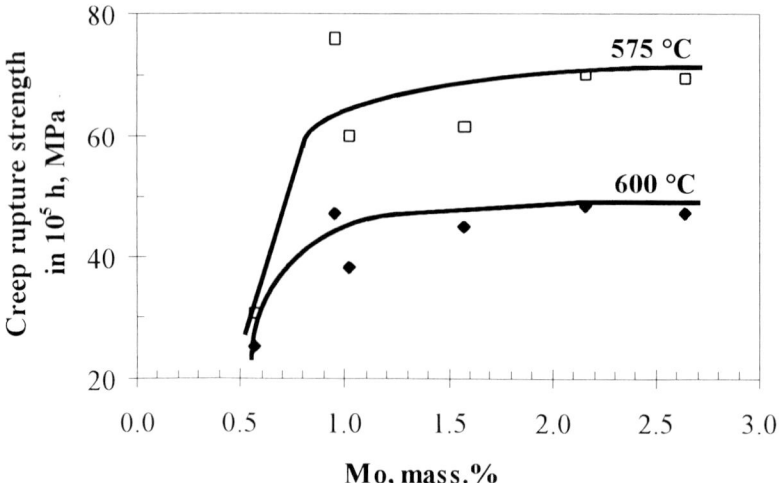

Fig. 3   Dependence of creep rupture strength on molybdenum content in 9% chromium steels.[21]

strengthening due to higher Mo content in the solid solution improves the creep resistance of examined steel 9%Cr–2.6%Mo after long term creep exposure. In this case, the predicted time to rupture can be overestimated when the Monkman–Grant equation is used. In steels with lower Mo or Mo equivalent the overestimation may be less significant. On the other hand, the Monkman–Grant relation enables us to estimate the breakpoints indicating the change in the creep or fracture mechanism and to define the high and low-stress region.

A reliable assessment of the long term CRS seems to be possible provided that the creep data attained in the low stress region are taken into account. The creep tests, and/or creep rupture tests performed at high stresses are useful above all for the indication of relevant breakpoints. From this point of view it is quite necessary to estimate the creep rupture strength of steels E911 and P92 (heats **1** and **2**) based only on tests performed in the low stress region.

The creep rupture strength estimation of the heat **C** (P92) was based on much longer test data than were available for CRS estimation of heats **1** and **2**. Nevertheless the results attained are in good agreement. When assessing the CRS of the steel P91, the Monkman–Grant equation was not used for the prediction of time to rupture from the secondary creep rate, but published data were included.[6] There is no doubt that the 87 MPa presented in Table 2 is quite an acceptable creep rupture strength for the steel P91 at 600°C in 100 000 h, but the estimation by means of eqn (2) indicates an even lower CRS, only about 73 MPa (Table 2).

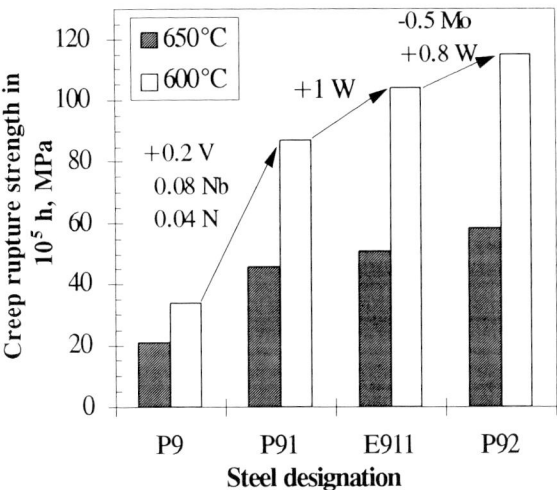

Fig. 4 Comparison of the 10^5 h CRS of various chromium steels at 600 and 965°C.[6]

Precipitation of vanadium nitride plays the decisive role in the creep resistance of advanced chromium steels. The volume fraction of VN is controlled by nitrogen content which is available for the formation of VN. Increasing the Al content in the steel from 0.005 to 0.035% decreases the amount of available nitrogen and in this way the CRS at 600°C in 100 000 h is lowered by about 15 MPa.[16–18] Vanadium carbide does not precipitate in the 9–12% Cr steels due to high solubility of VC in these steels at elevated temperatures. Therefore, the shortage of available nitrogen in the steel cannot be compensated for by carbon.

Most probably the creep tests that are still in progress will not confirm the expected CRS of the steel P91 with a high aluminium content (0.034%). On the contrary, very good agreement was found in the CRS of the steel E911 estimated by creep tests performed at low stresses and calculated by the eqn (2), respectively (Table 2). It should be emphasised that eqn (2) was developed for different 9 to 12% chromium steels containing either 1% Mo or 1.5% molybdenum equivalent (1% Mo + 1% W). In the set of 33 heats also 7 tungsten bearing heats of the steel 12Cr–1Mo–1W–0.3V were included. Very good correlation between the calculated and measured creep strength indicated that the addition of 1% W to the well-known steel 12Cr–1Mo–0.3V did not lead to any improvement of CRS of this steel.[14,15,20,22]

Similarly, the creep rupture properties of tungsten bearing 10% chromium cast steel containing 1%Mo–1%W and tungsten free 10% chromium cast steel containing only 1% Mo were found to be consistent with the steel P91. Aluminium and nitrogen contents were in both steels nearly identical

(0.006%Al, 0.049%N in the W-bearing steel and 0.007%Al, 0.050%N in the W-free steel, respectively).[4]

All these results did not confirm the beneficial effect of tungsten addition on the CRS at 600°C in 100 000 h. Nevertheless, the CRS of the investigated steel E911 was found to be higher than that of steel P91 (Table 2). The observed and in no case negligible improvement in CRS was attributed to the W addition (Fig. 4).[6] With respect to the above discussion, a more reasonable explanation of the observed improvement in CRS of E911 over P91 consists in higher precipitation strengthening of the steel E911 due to higher available nitrogen content in the steel containing tungsten. Steel E911 can be characterised by a $c \cdot N_{ss}$ product which is about twice as high as that of the investigated steel P91 (Table 2).

The additional increase in the CRS of P92 over E911 (Fig. 4) was also attributed to the addition of W.[6] The development of steel P92 was based on the assumption that the replacement of part of the Mo in 9 to 12% Cr steel by W is very effective in increasing the CRS. The best combination was found to be 0.5%Mo–1.8%W. It was assumed that W increases the solid solution strengthening.[2,11] Later it was shown that the precipitation of Laves phase $Fe_2(Mo,W)$ strongly depletes the tungsten and molybdenum contents in the solid solution in the steel P92. After creep exposure at 600°C for 20 000 hours, only one third of the total W content in the steel remained in the solid solution and could contribute to the solid solution strengthening.[11,14,23,24] Comparative tests performed recently on the steel P92 have indicated that the solid solution strengthening effect of W is small but the contribution of W to creep strength was supposed to be via precipitation strengthening due to Fe_2W.[23,24] Tungsten bearing Laves phase is much more thermally stable than Fe_2Mo.[19,25] Nevertheless, with respect to high coarsening rates of Laves phase, precipitation of this phase does not produce a significant strengthening for long time creep exposure. As a confirmation of this assumption, no beneficial effect of higher tungsten content was found in the steel D3 containing 1.8% W and 0.32% Mo. The CRS at 600°C in 100 000 hours of steel D3 was found to be as low as 74 MPa (Table 2),[10] while the CRS of the steel TMK2 with about the same chemical composition was reported to be as high as 130 Mpa.[3] Analysis of different methods of CRS estimation published earlier[10] revealed that the differences in the CRS of both steels could be fully explained by differences in the extrapolation methods use to predict their long term CRS's. The high CRS of steel TMK2 was undoubtedly overestimated very much.

The CRS of steel D3 at 600°C for 100 000 hours calculated by the eqn (2) was found to be somewhat higher than that estimated by creep tests performed at low stresses (Table 2). This observed decline in CRS may be connected with higher Ni content in the investigated steel D3 (Table 1). As early

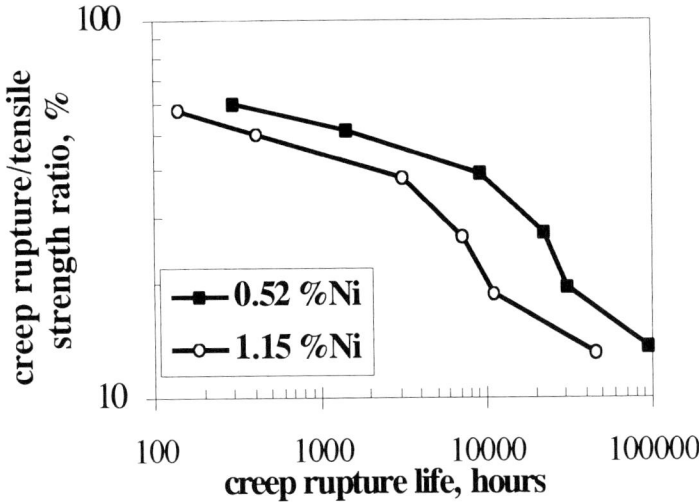

Fig. 5   Indication of sigmoidal inflection as a dependence of CRS/R$_{mt}$ ratio on time to rupture.[28]

as 1972 Marrison and Hogg observed the detrimental effect of nickel on the long-term creep strength of 12%CrMoVNb steel.[26] Lately it was reported that the sigmoidal inflection points appeared earlier in steels with higher Ni content (Fig. 5).

Nickel increases the activity of carbon in the steel and thus accelerates all carbide reactions in the steel. Nickel is also known to promote the precipitation of M$_6$X phase, especially at lower temperatures. The coarsening rate of this phase is very high and brings about dissolution of M$_2$X and MX with corresponding degradation of the precipitation strengthening.[28–30]

As nickel was not incorporated in the above described model of creep rupture strength estimation, it is quite natural that the calculated CRS of the steel D3 with higher Ni content was somewhat higher than that established experimentally.

With respect to the above discussion it can be concluded that neither solid solution strengthening due to a higher W content in the solid solution, nor precipitation strengthening due to Fe$_2$W precipitation during creep exposure, improves the creep resistance of the investigated steel P92 containing about 1.8% W.

The nitrogen contents in the heats of P92 examined were relatively low and therefore the calculated CRS indicated lower creep resistance than was established experimentally. The product c.·N$_{ss}$ for the steel E911 was higher than the relevant c.·N$_{ss}$ products estimated in all heats P92, although the experimentally established CRS of the steel E911 was lower (Table 2).

The very high creep resistance of the boron–bearing steel B2 with low

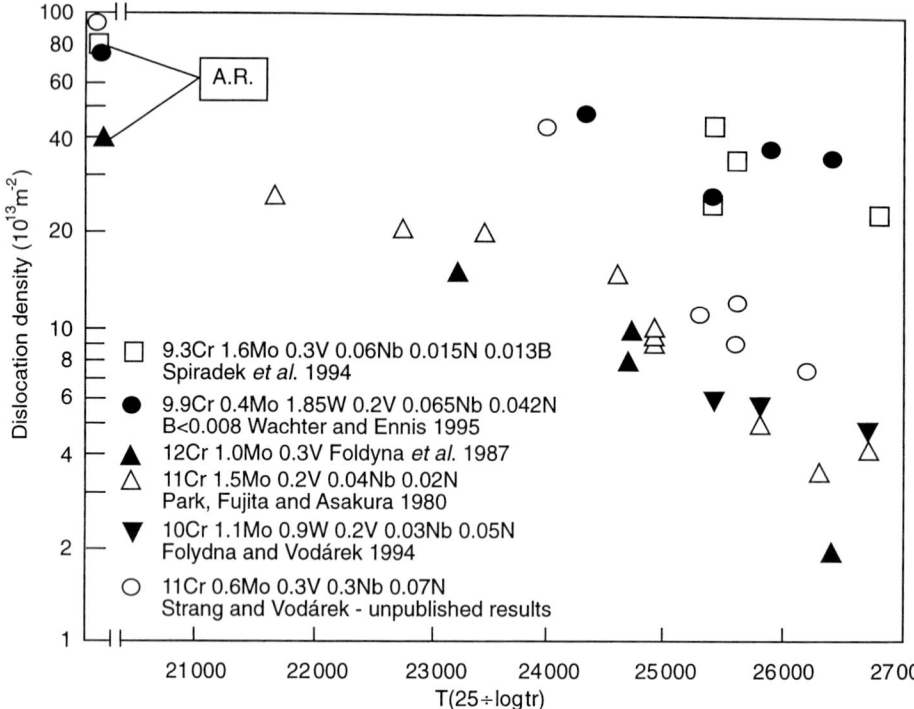

Fig. 6 Dislocation density as a function of Larson–Miller parameter (data after[5, 34–38]).

nitrogen content is remarkable and surprising (Tables 1 and 2). The calculated CRS indicated much lower creep resistance than was established experimentally. It is clear that another strengthening mechanism must operate in boron–bearing steels besides those found in advanced chromium steels without boron. The possible explanation was suggested by Lundin et al. who proposed the model of the 'latent creep resistance'.[31] According to this model precipitation occurs successively during creep as a dynamic process. In this case, a delicate balance between nucleation and dissolution of small precipitates is necessary. The precipitates nucleate on dislocations taking advantage of the strain field round the dislocation core, thus effectively pinning the dislocations. When the dislocations manage to break away, precipitates become unstable and dissolve. Repeatedly precipitating and dissolving particles pin dislocations and at the same time grow very slowly. Thus, all degradation processes during creep are slowed down and decreasing steady state creep rate can be expected.

Boron is known to segregate on the grain and subgrain boundaries and lower their interface energy. Although it was proved that up to about one half of the total boron content in chromium steels is bound in $M_{23}C_6$ carbide,[31,33] it seems to be probable that the rest is still a sufficient amount to modify the interface energy. Boron then could segregate also on dislocations where it probably increases the critical radius of MX nuclei. The higher critical radius of MX nuclei then retards the coarsening process by facilitating particles to dissolve and re-precipitate on dislocations again. This is, of course, only a hypothesis still open to challenge as so far it has not been proved experimentally. The only result obtained by APFIM analysis on the VN–matrix interface did not revealed any enrichment of boron.[32]

At the same time, the observed high dislocation stability in B–bearing steels B2 and P92 is also remarkable,[5,34–38] (Fig. 6). A possible explanation for the observed high dislocation stability is the latent creep resistance. Fine VN particles, repeatedly nucleating and dissolving can act as obstacles for dislocation movement and may reduce the rate of recovery, as well. High dislocation density is of course desirable. This is probably the only explanation for the observed high creep resistance and high dislocation density after long-term creep exposure, although the nitrogen content of steel P92 was relatively low and very low in the case of steel B2. The optimum contents of B and N in the chromium steels is not known and it is not yet possible to incorporate the specific behaviour of B–bearing steels in the model of creep rupture strength estimation.

CONCLUSION

Microstructural instability of chromium modified steels makes the prediction of long-term creep properties extremely difficult. Therefore, it is absolutely necessary to extrapolate the creep data only in the low stress region. The superiority of tungsten-bearing steels originates from the fact that in many cases it is based on short-term creep tests with only a very restricted number of relatively long-term creep tests. Then tungsten still contributes to the creep strength in most tests by solid solution or precipitation strengthening enhanced in the first stage of creep exposure by Laves phase precipitation. The long-term results are overweighted by the great amount of test data with only short duration and unrealistic values of creep rupture strength are obtained. Chromium steels containing boron seem to be more promising as far as long term creep strength is concerned. Smaller VN particles, high dislocation density and unrecovered structure of these steels even after a long time give them high creep strength even when the nitrogen content of the steel is low. On the other hand, only a little is known about the optimum boron and nitrogen concentrations as well as about other properties of boron-bearing chromium modified steels. But this seems to be the way forward of how to improve and optimise the creep properties of the new generation of chromium modified steels.

REFERENCES
1. V.K. Sikka: *Proc. of the Conference Ferritic Alloys for Use in Nuclear Energy Technologies*, Snowbird, Utah, June 1983, 317.
2. H. Masumoto, H. Naoi, T. Takahashi, S. Araki, T. Ogawa and T. Fujita: *2nd Int. Conf. Improved Coal-Fired Power Plants*, EPRI, Palo Alto, 1988, 40.3.
3. C. Berger, B. Scarlin, K.H. Mayer, D.V. Thornton and S.M. Beach: *Proc. of the 5th Conf. Materials for Advanced Power Engineering 1994*, C.R.M., Liége, Belgium, October 1994, 47.
4. B. Scarlin, C. Berger, K.H. Mayer, D.V. Thornton and S.M. Beach: *ibid.*, 73.
5. K. Spiradek, R. Bauer and G. Zeiler: *ibid.*, 241.
6. P.J. Ennis and O. Wachter: *VGB Conference Materials and Welding Technology in Power Plants 1996*, Cottbus, Germany, October 1996, paper 4.
7. *DATA PACKAGE FOR Nf 616 FERRITIC STEEL* (9Cr–0.5Mo–1.8W–Nb–N), Second Edition, Nippon Steel Corporation, March 1994.
8. P. Greenfield: *Proc. High Temperature Materials for Power Engineering 1990*, Kluwer Academic Publishers, 1990, 423.
9. A. Zielinska-Lipiec, A. Czyrska-Filemonowicz, P. J. Ennis and O. Wachter: *Proc. of the IXth Int. Symposium Creep of Metallic Materials*, Hradec nad Moravicí, September 1996, 254.
10. V. Foldyna, Z. Kubon, M. Filip, K.-H. Mayer and C. Berger: *Steel Research*, 1996, **67**, 375.
11. H. Mimura, M. Ohgami, H. Naoi and T. Fujita: *Proc. of the 5th Conf. Materials for Advanced Power Engineering 1994*, C.R.M., Liége, Belgium, October 1994, 361.
12. V. Foldyna, Z. Kuboň and A. Jakobová: *Creep rupture strength assessment of Nf 616*, unpublished report, Research Institute of VÍTKOVICE, a.s., Ostrava, Czech Republic, 1994.
13. V. Foldyna, A. Jakobová, R. Říman and A. Gemperle: *Steel Research*, 1991, **62**, 453.
14. A. Jakobová and V. Foldyna: *Proc. of the 7th JIM Inter. Symposium Aspects of High Temperature Deformation and Fracture in Crystaline Materials*, Nagoya, Japan, July 1993, 317.
15. V. Foldyna, A. Jakobová, V. Vodárek and Z. Kuboň: see [3], 453.
16. V. Foldyna and Z. Kuboň: *Proc. of the Conference Performance of Bolting Materials in High Temperature Plant Applications*, York, UK, April 1994, A. Strang, Ed.The Institiute of Materials, 1995, 175.
17. Z. Kuboň and V. Foldyna: *Steel Research* , 1995, **66**, 389.
18. V. Foldyna and Z. Kuboň: *Proc. 3rd Int. Charles Parsons Conf. Materials Engineering in Turbines and Compressors*, Newcastle upon Tyne, UK,

April 1995, R.D. Conroy, M.J. Goalette and A. Strang eds, The Institute of Materials, 1996, 353.

19. V. Foldyna, Z. Kuboň, A. Jakobová and V. Vodárek: *Microstructural Development and Stability in High Chromium Ferritic Power Plant Steels*, Cambridge, UK, June 1995, A. Strang and D.J. Gooch eds, The Institute of Materials, 1997, 73.

20. V. Foldyna, A. Jakobová, R. Říman and A. Gemperle: *Proc. 2nd Int. Conf. Creep and Fracture of Engineering Materials and Structures*, B. Wilshire and D.R.J. Owen eds, Pineridge Press, Swansea, UK, 1984, 685.

21. V. Foldyna, Z. Kubon, A. Jakobová and V. Vodárek: see [9], 203.

22. 'Data Sheets on the Elevated Temperature Properties of 12Cr–1Mo–1W–0.3V Stainless Steel Bars for Turbine Blades (ASTM A565–616)', *NRIM Data Sheet No. 10A*, Tokyo, 1979.

23. J. Hald: *The EPRI/National Power conf. New Steels for Advanced Plant up to 620°C*, May 1995, The Society of Chemical Industry, London, UK, E. Metcalfe Ed., 152.

24. J. Hald: *Steel Research*, 1996, **67**, 369.

25. J. Hald and Z. Kuboň: see [9], 52.

26. T. Marrison and A. Hogg: *Proc. of the Conference Creep Strength in Steel and High-temperature Alloys*, Sheffield, UK, September 1972, Metals Society, London, 1974.

27. A. Strang and V. Vodárek: see [9], 217.

28. A. Strang and V. Vodárek: see [19], 31.

29. A. Strang and V. Vodárek: *Material Science and Technology*, 1996, **12**(7), 552.

30. V. Vodárek and A. Strang: *Scripta Metallurgica et Materialia*, in press.

31. L. Lundin: 'High Resolution Microanalysis of Creep Resistant 9-12% Chromium Steels', *PhD. Thesis*, Chalmers University of Technology, Göteborg, 1995, paper 4 – accepted for publication in Materials Science and Technology.

32. M. Schwind, M. Hattestrand and H.-O. Andrén: *this proceedings*.

33. L. Lundin and B. Richarz: *Applied Surface Science*, 1995, **87/88**, 194.

34. O. Wachter and P. J. Ennis: 18. Vortragsveranstaltung 'Langzeitverhalten warmfeste Stähle und Hochtemperaturwerkstoffe', Düsseldorf, Germany, December 1995, 120.

35. V. Foldyna, A. Jakobová, Ř. Ríman, A. Gemperle, G. Belka, E. Preissler and G. Münch: *Proc. VI Symposium Warmfeste metallische Werkstoffe*, Zittau, East Germany, November 1987, 62.

36. J.M. Park, T. Fujita and K. Asakura: *Trans. ISIJ*, 1980, **20**, 99.

37. V. Foldyna and V. Vodárek: Minutes of the sixth meeting *Metallography and Alloy Design Group*, COST 501/III, WP 11, Attachment 9, Liege, October 1994.

38. A. Strang and V. Vodárek: *unpublished results*.

Parameters Affecting the Long Term Creep Rupture Properties of Durehete 1055

J. ORR,[1] L. WOOLLARD[1] AND H. EVERSON[2]

[1] British Steel Swinden Technology Centre
[2] British Steel Engineering Steels

ABSTRACT

Durehete 1055, a 1%CrMoVTiB bolting steel has been established for many years. Whereas the strength of the steel has always been sufficient for its intended purpose, user experience has demonstrated that initially the ductility properties were not always adequate for particularly demanding situations.

Several metallurgical developments have been introduced and tested over the years, aimed at improving the creep ductility of this steel. The most successful of these improvements has been the control of the residual element content. Other parameters also investigated include control of composition, within the specified ranges, and prior austenite grain size control. All these parameters are addressed and their significance described in this paper.

The paper also includes results of data assessment carried out in collaboration with the European Creep Collaboration Committee.

1. INTRODUCTION

Several high temperature bolting steels have been developed by British Steel Engineering Steel (BSES) and its predecessors, to meet the demands of the power generation industry over the last 50 years.[1]

The main steels in this category are the Durehete family which are based on the 1%CrMo–1%CrMoV steel types.[1] Other steel types, for example warm worked Esshete 1250, have been pioneered by the same company.[2]

This paper deals with the developments which have occurred and been influential in determining the high temperature properties of the Durehete 1055, which is used for the most arduous bolting service regimes in steam turbines.

Factors discussed include composition development, grain size control, steelmaking practice and residual element control, with particular reference to creep rupture ductility and notched strength of Durehete 1055.

2. DEVELOPMENT HISTORY

Before the advent of the gas fired Combined Cycle Gas Turbine (CCGT) units, the most efficient power stations operated by CEGB (predecessor to National Power and PowerGen) were the 500 MW(e), coal fired units operating with steam temperatures up to 565°C. There is a direct (thermodynamic) relationship between steam temperature and efficiency. Therefore to improve the latter, the demands on the steel components became more arduous. A peak in this activity was reached in the 1950s at which time a steel grade based on 1%CrMoV, Durehete 1050, was developed and characterised.[3]

The strength of Durehete 1050 was found to be adequate for the service regime, but low rupture ductility values developed fairly quickly, thus making it sensitive to stress concentrations such as notches, e.g. thread roots in bolts, Fig. 1.

A detailed metallographic explanation of the reasons responsible for low creep rupture ductility values in Durehete 1050 is available in Ref. 4. In summary, the ductility is determined by the near grain boundary conditions, which influence the propensity to intergranular cracking.

Experimental work showed that by adding titanium and boron and decreasing the austenitising temperature (to restrict solubility of V_4C_3), significant improvements in rupture ductility were obtained.[4] Thus, Durehete 1055 was developed, Table 1. Other workers have shown also that similar low alloy steels used in humid environments can have improved resistance to 'delayed fracture' (arising from corrosion) by additions of titanium and boron.[5] In Durehete 1055 the addition of titanium and boron resulted in fine precipitates in the zones adjacent to the prior austenite grain boundaries, which in Durehete 1050 had become denuded areas during testing. This fine precipitate was not uniquely identified but considered to be TiC.[4] Japanese research also postulated a similar effect.[5]

More recent evidence, using Energy Dispersive Analysis (EDA) during examination of Durehete 1055 samples, confirms that fine titanium-rich particles are formed during the tempering state but transform during testing to V_4C_3.[6]

3. STRUCTURE AND PROPERTIES OF DUREHETE 1050/1055

The optimised steel composition for Durehete 1050 was found to be approximately 0.2%C, 1%Cr, 1%Mo, 0.75%V, Table 1. The high creep strength developed arose from solid solution strengthening by chromium and molybdenum and precipitation of V_4C_3, Fig. 2. The latter was maximised by a vanadium to carbon ratio close to stoichiometry, in the presence of 1%Mo, Figs 2 and 3. Significantly higher molybdenum contents led to a decrease in strength because of an increase in solubility of V_4C_3 (which led to coarser grain sizes) and competition for carbon leading to the formation of molybdenum-rich

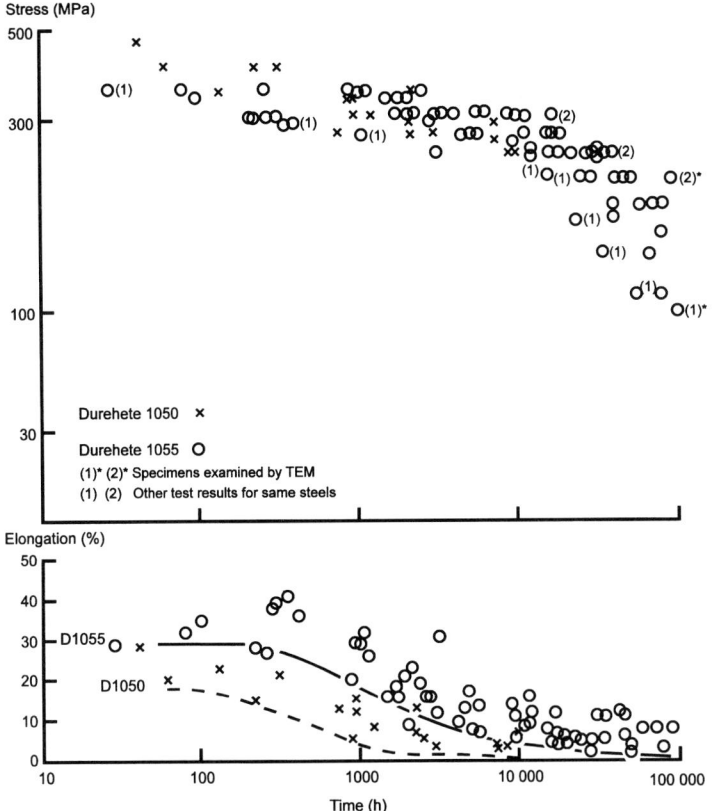

Fig. 1   Rupture Properties of Durehete 1050 and Durehete 1055 at 550°C.

carbides, such as M_6C, which are not as effective as for V_4C_3 in terms of conferring creep strength, Fig. 3. Thus it was confirmed that ~0.7%V was the optimum addition to produce the highest creep strength by forming almost predominantly V_4C_3 without the formation of Fe_3C.[4] It is possible that 1%Mo in Durehete 1055 further increases the stability of V_4C_3.[7]

Studies of the constitution of carbides in CrMoV steels, show that for steels with 1%Cr, 1%Mo and 0.5–1.0%V, in the equilibrium condition, the predominant carbide type is V_4C_3 with possible small amounts of Mo_2C (or M_6C).[8]

However, the high strength creep obtained by optimising V_4C_3 precipitation led to low creep ductility and hence it was found that Durehete 1050 could behave in a creep brittle manner, Fig. 1. As described above, it was concluded that this was due to the formation of precipitate-free zones adjacent to grain boundaries where the creep strain could concentrate and hence cause failure with low total extension values.[4]

Durehete 1055 has the same basic alloy content and balance as Durehete

1050, Table 1 and hence it is not surprising that the same strengthening mechanisms operate. Transmission electron metallography (TEM) using diffraction patterns confirms the presence of fine particles of a face centred cubic

Table 1   Typical Compositions of Steels (wt.%)

Grade Name	C	Si	Mn	Cr	Mo	V	Ti	B
Durehete 1050	0.21	0.20	0.48	1.00	1.00	0.70	–	–
Durehete 1055	0.20	0.30	0.50	1.00	1.00	0.70	0.09	0.003

structured carbide, Fig. 4(a) with a lattice parameter of ~4.20 Å which is within the range of 4.15–4.30 Å which applies for V_4C_3.

Further work including use of Energy Dispersive Analysis (EDA) indicates that two main precipitate types may be present as fine particles which can influence creep strength.[6] These are a VC type, containing ~80%V and ~20%Mo, and a TiC type. This work has also shown that there is transformation of the TiC type towards V_4C_3 in long test durations at 550–600°C.

Electron metallographic examination of two test pieces, representing the longest tested material (~100 000 h at 550°C) to date for Durehete 1055 (from ERA Project 2021* and the British Steel Engineering Steels programme at Swinden Technology Centre) has been carried out. The two specimens derive from different casts of Durehete 1055 but of similar bar size and heat treatments, Table 2. The cast compositions, bar size and heat treatment data, for the two materials are given in Table 2.

The relative weakness of specimen 1, shown with its counterpart results in Fig. 1, indicates that its strength has decreased more quickly at 550°C, particularly for durations greater than 10 000 h, than has that of specimen 2, Fig. 1. Furthermore, there is evidence of sigmoidal behaviour in the strength-time curve for steel 1, Fig. 1. Similar features have been observed in steels of this type tested at 500 and 575°C. A schematic description of the changes in strengthening mechanisms occurring during creep testing which result in a sigmoidal strength-time curve is given in the Appendix.

Electron metallographic data in Fig. 4 and Table 3, shows that both specimens contain significant amounts of fine V_4C_3 precipitates. However, the general structure of specimen 1 indicates that these particles are coarser, and specimen 2 has also retained more evidence of the original lath boundaries, Figs 4(a) and (b). These features accord with the difference in strength of the two specimens, Fig. 1.

---

* ERA's Project 2021: 'Creep of Steel' was the large UK national collaborative research programme performed on behalf of alloy producers, plant manufacturers and power generators, into the high temperature mechanical behaviours of steels and alloys for power and other high temperature plant'.

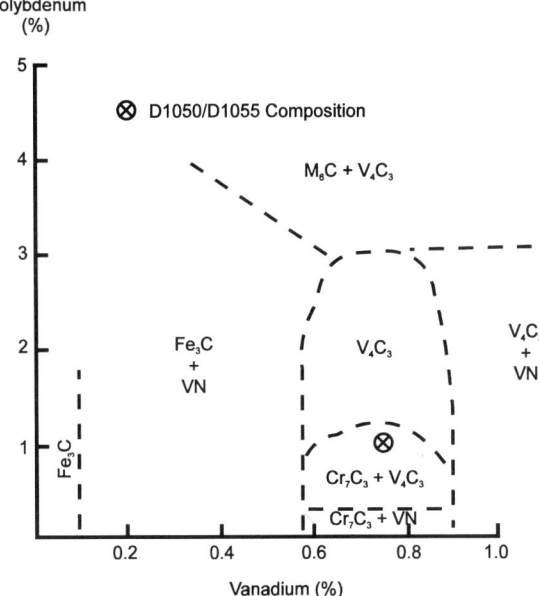

Fig. 2   Precipitation in Vanadium/Molybdenum Steels (ex Ref. 3).

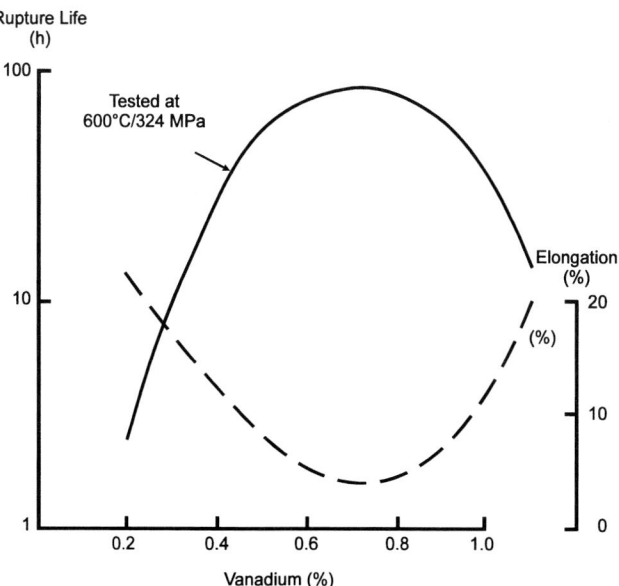

Fig. 3   Effect of Vanadium on Stress Rupture Properties of Durehete 1055 (ex Ref. 4).

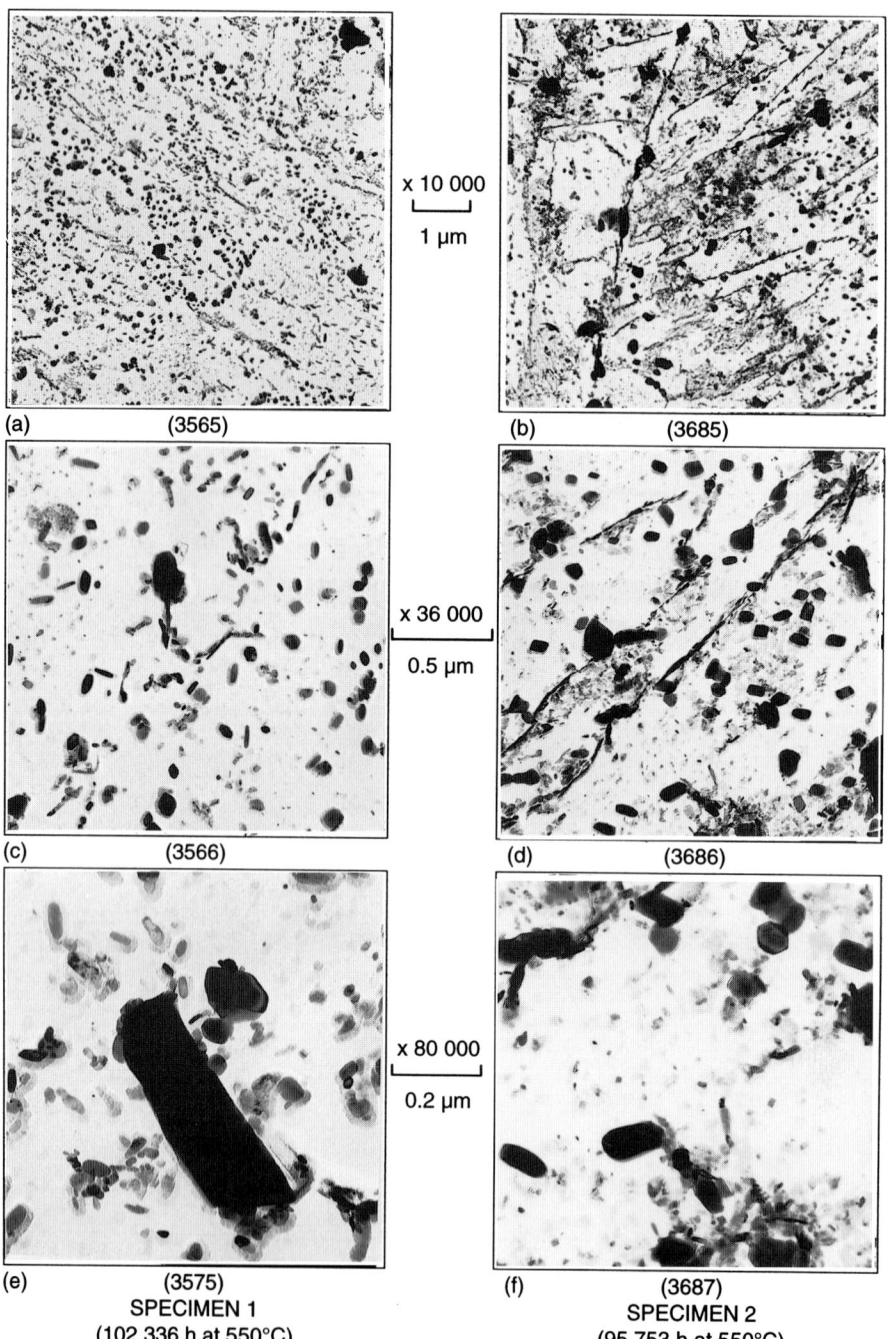

x 10 000
1 μm

(a)   (3565)

(b)   (3685)

x 36 000
0.5 μm

(c)   (3566)

(d)   (3686)

x 80 000
0.2 μm

(e)   (3575)
SPECIMEN 1
(102 336 h at 550°C)

(f)   (3687)
SPECIMEN 2
(95 753 h at 550°C)

Fig. 4   Precipitation in Durehete 1055 after ~100 000 hours at 550°C.

Table 2   Cast Compositions and Archive Details of Specimens 1 and 2 (wt.%)

Specimen	C	Si	Mn	P	S	Cr	Mo	Ni	Al	As	B	Cu	Sn	Ti	V	Sb
1	0.20	0.21	0.48	0.030	0.012	1.04	1.03	0.33	0.070	0.041	0.005	0.20	0.010	0.11	0.68	0.0020
2	0.21	0.25	0.48	0.010	0.012	1.00	0.93	0.05	0.054	0.012	0.004	0.06	0.011	0.10	0.69	0.0016

Specimen	Bar Size (mm dia.)	Original Heat Treatment
1	57	980–1000°C h WQ+680–700°C 6–8 h AC
2	76	980°C 2 h WQ+690°C 6 h AC

Other metallographic features of the two specimens found are:

(i)   The V_4C_3 type precipitates are present in two size ranges as small particles, see Table 3, and also present as larger particles on grain boundaries, Figs 4(c) and (d).

(ii) Large rhombic shaped, molybdenum-rich particles, are present, which are more evident in specimen 1 than in specimen 2, Figs 4(e) and (f).

The V_4C_3 based particles in both specimens contain significant amounts of molybdenum, at 30–35% and small amounts of chromium, Table 3. However, the chromium content, at ~12%, is higher, and consequently the vanadium content lower, in the V_4C_3 particles of specimen 2 than those of specimen 1, Table 3. This is at variance with the basic compositions of the two steels, Table 2. Other compositional differences are the higher molybdenum, nickel and residual element contents of specimen 1. In classical terms, it might be expected that a higher alloy content would give higher strength in an as tempered martensite/bainite structure. This is obviously not the case for these two specimens so other factors outside of compositional differences, bar size and heat treatment must have had a significant effect on the difference in creep strength. It is known, for example, that specimen 1 would have been subjected to several interruptions during the test campaign, for the purposes of strain measurement. Such interruptions may have reduced the apparent creep life of some steels and alloys.

The metallographic data indicate that the V_4C_3 particles formed during the original tempering process have not grown very quickly and, furthermore, the very fine particles indicate that additional precipitation had probably occurred during testing. The overall effect, despite the difference in strength indicated in Fig. 1, is that Durehete 1055 depends for its high strength, rela-

Table 3 Composition of Precipitates in Durehete 1055

Precipitate Form	Typical Size	Composition Range %				
		Cr	Mo	V	Ti	Fe
Specimen 1: (102 336 h at 550°C)						
Fine Needle	0.1 × 0.04 μm	5–6	32–37	58–65	2	2
Square/cuboid/round	0.6 μm	6	19–28	66–73	2	1
Large Rhombus	0.6 × 0.2 μm	5	71–72	24–27	–	1
Specimen 2: (95 753 h at 550°C)						
Needle (in grain)	0.08 × 0.05 μm	11–13	31–38	45–55	1	2–3
Rod/cuboid	0.5× 0.2 μm	12–13	66–73	14–19	1	1–2
Cuboid	0.15× 0.07 μm	13	42	44	–	1
Ovoid/blocky on g.b.	0.1 × 0.1 μm	11–12	23–26	59–60	1–7	1

tive to that of other steels used for bolting, on the stability of the V_4C_3 precipitates. Hardness data in Fig. 5 demonstrate further how Durehete 1055 retains its strength over long service/test durations. The data for Materials A and B in Fig. 5 are identified to indicate the trend behaviour.

The relatively large, molybdenum-rich particles, Table 3, may be evidence of transformation of some V_4C_3 based particles to less effective strengthening particles, as suggested by Ref. 6. Since these were most apparent in specimen 1, this could explain its relative weakening beyond ~10 000 hours of testing.

More detailed analysis work is still required however to determine fully the metallographic factors controlling this behaviour in Durehete 1055 and the role of small changes in composition, such as residual element content.

The other change which occurred when Durehete 1055 was developed, was a reduction in the austenitising temperature from 1050°C initially to 1000°C and more recently to approximately 980°C. This was effective for two reasons: restriction of grain growth giving a greater grain boundary area per unit volume and a slightly lower dissolution of vanadium and carbon. The former effect leads to the capability of higher strain accommodation before ultimate failure and the latter a reduction of strength by lower martensitic hardness and a smaller volume fraction of V_4C_3 on tempering, which although causing some reduction in strength gives further benefit to ductility.

However, the data in Fig. 1 also indicate that the improved ductility values in Durehete 1055 arising from the addition of titanium and boron (and reduction in austenitising temperature) were not maintained for very long durations. This was one of the contributory factors in a number of bolt failures which occurred in CEGB operated turbine units in the 1970s, although operational and design factors were major contributors.

The metallurgical factors considered to be responsible, since the low ductility failures were obviously intergranular, were prior austenite grain size and grain boundary precipitation/segregation conditions.

Thus, it is indicated that whilst the changes instituted, which led to the initial development of Durehete 1055, were on the 'macro' scale, further finer scale changes were needed. Some of the changes which were introduced from about 1980 onwards required significant further research, largely because of two factors:

(i)   improvements in ductility were required in long test durations

(ii) the metallographic area of importance was on a micro, if not sub-microscopic scale, i.e. at/near/within grain boundaries.

Research work at various establishments, mainly within British Steel and CEGB, and at various turbine makers, focused on two factors, namely prior austenite grain size and residual element content. The latter became the pre-

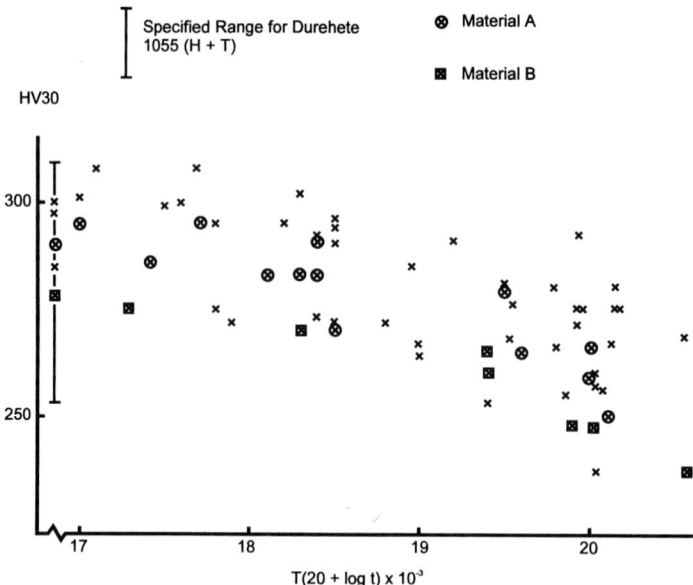

Fig. 5 Hardness of Durehete 1055 after Testing.

dominant issue because of the known segregation behaviour of residual elements, such as phosphorus, arsenic, tin etc. to grain boundaries and the consequent effect on embrittlement/ductility parameters of CrMoV steels.[9]

4. PRIOR AUSTENITE GRAIN SIZE

All other factors being equal, it is generally expected that coarser grain sizes lead to increased creep strength, at least in the short term.

Data for Durehete 1055, for example from stress relaxation tests at 565°C for 150 h, demonstrate that grain size does influence the short term properties Fig. 6(a). However, the range of creep strength values, in this case residual stress, due to the range of grain sizes, Fig. 6(a), diminishes with increasing test duration so that after 10 000–20 000 hours, the remaining effect of grain size on 'creep strength' is quite small. This is because the dependence of long term creep strength is on much finer microstructural features, as discussed above.

However, a role for prior austenite grain size was implicated in the bolt failures which occurred in the 1970s, it being considered that coarse grain sized material could be less ductile than finer grained material. Despite the fact that by austenitising at ~980°C this would be below the 'normal' grain coarsening temperature for Durehete 1055, cases were found where coarse grained or more particularly mixed coarse and fine grained structures existed in failed

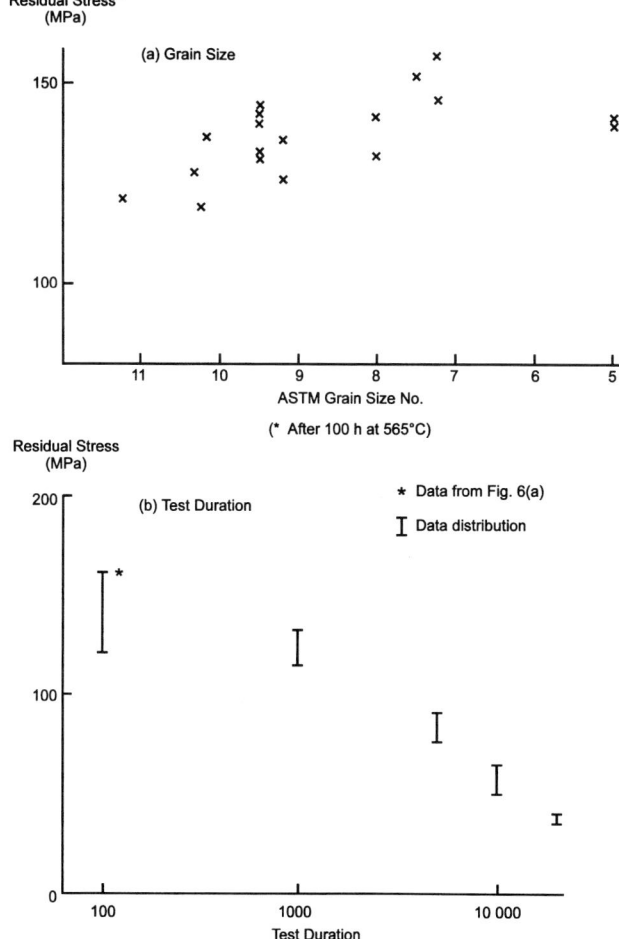

Fig. 6   Residual Stress, Grain Size and Test Duration.

bolts. It has been demonstrated that NiCrMoV and CrMoV steels can exhibit a 'memory' effect, whereby on austenitising at just above the (relevant) Ac_3 temperature, where normally a fine austenite grain size would form, a coarse grain size can be formed, this being similar to that found before heat treatment, for example as the as-rolled bar.[10]

The austenite 'memory' effect, as it is termed, arises from growth of 'acicular' austenite from preferred nuclei, believed to be retained austenite or fine carbides at martensite/bainite lath boundaries.[10] A treatment at ~700°C prior to austenitising at ~1000°C has been found to remove or significantly reduce the memory effect in Durehete 1055, by transforming the retained austenite.[10]

Such a treatment has been included in the heat treatment route for Durehete 1055 since around 1985.

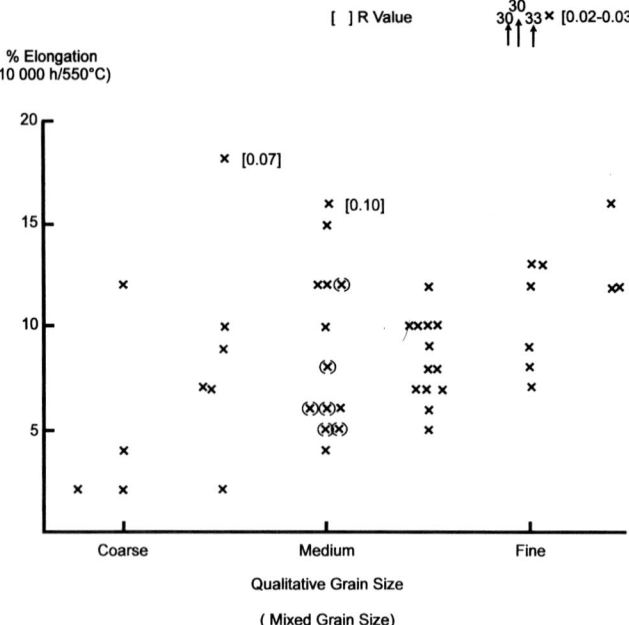

Fig. 7   Relationship of Prior Austenite Grain Size and Rupture Ductility.

Thus a reason for the presence of coarse grained structures in some Durehete 1055 bars can be given. However, it is believed that this particular feature was peculiar to a limited range of bar diameters, arising from a process route which allowed coarse grained structures to arise after hot rolling. This process route is no longer in use.

A qualitative survey of prior austenite grain sizes for past and present commercially produced and heat treated materials in the creep test programmes at Swinden Technology Centre, shows that most were in the medium-fine category but some had coarse or mixed grain sizes. A few spot assessments show the finer grained material to have grain sizes in the range ASTM 5 or finer, see also Fig. (6a).

The relationship between the qualitative grain size data and creep rupture ductility values (in the case of Fig. 7, % elongation after 10 000 h at 550°C),. show that very low ductility values only occur when coarse grain sizes exist and none for fined grained material. However, there is not a significant correlation between these two parameters especially in the medium-fine grain size area. Data are also included in Fig. 7 for fine grain sized materials, where ductility values of ≥30% were obtained, which relate more to chemical composition than to the microstructural parameters – in this case the residual element content. Other data points are also shown where the 'R' value (see Section 5) is known and these indicate the greater importance to creep ductil-

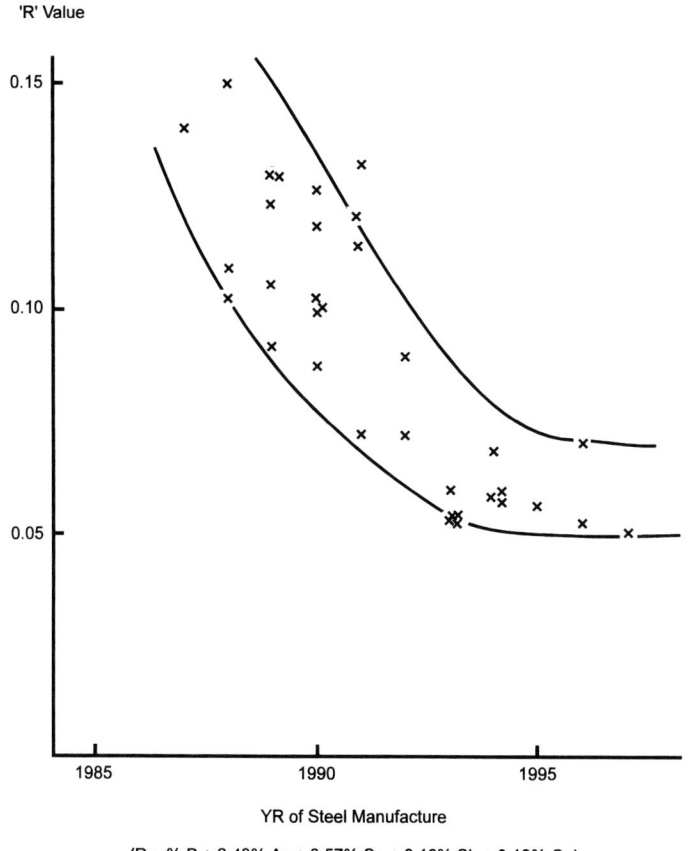

Fig. 8   'R' Value for Durehete 1055 and Year of Steel Manufacture at BSES.

ity of the residual element content than prior austenite grain size – see for example where 'R' values of 0.07–0.10 in coarse–medium grain sized material give elongation values of $\geq 15\%$, i.e. equivalent to that achieved in higher residual material with fine grain sizes, Fig. 7.

Thus it can be concluded that the majority of Durehete 1055 has medium to fine prior austenite grain sizes. The variation in ductility shows only a weak correlation with grain size, there being other compositional factors, in particular residual element content, which play a more significant role.

5.   RESIDUAL ELEMENT CONTENT
The effects of residual element content on the creep rupture ductility of low alloy steels have been recognised since the 1970s. For Durehete 1055 this effect became focused around 1980 and a number of papers and reports are

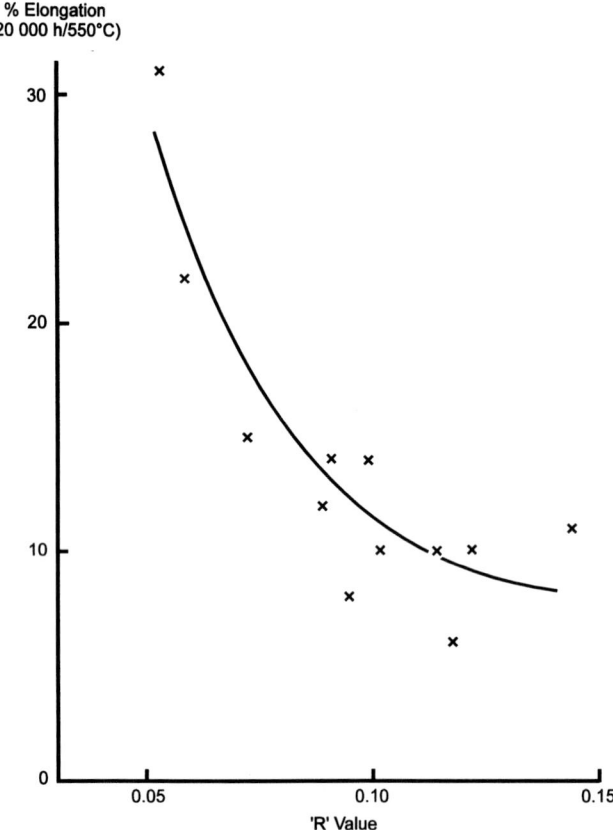

Fig. 9   Effect of Residual Element Content on Rupture Ductility of Durehete 1055.

available to give details of the developments.[11–13] It is usual to reflect the residual element content by a cumulative 'R' factor which sums the respective effects of phosphorus, tin, arsenic, antimony and copper.[9]

where R = %P + 2.43 (%As) + 3.57 (%Sn) + 8.16 (%Sb) + 0.13 (%Cu)

The equation used gives a predominant effect to that of phosphorus with arsenic, tin and antimony shown to be 2.5, 3.5 and 8 times less effective respectively on a weight percentage basis. Experimental work has indicated that whilst concentrations of phosphorus and tin have been found at grain boundaries in Durehete 1055, this is not the case for arsenic.[14] However, the role of antimony is not clear, since there is evidence in 1CrMoV steels that it occurs only on already established 'free' surfaces, e.g. creep cavities, whilst in CrMo steel it was considered to act like phosphorus, tin and copper in reduc-

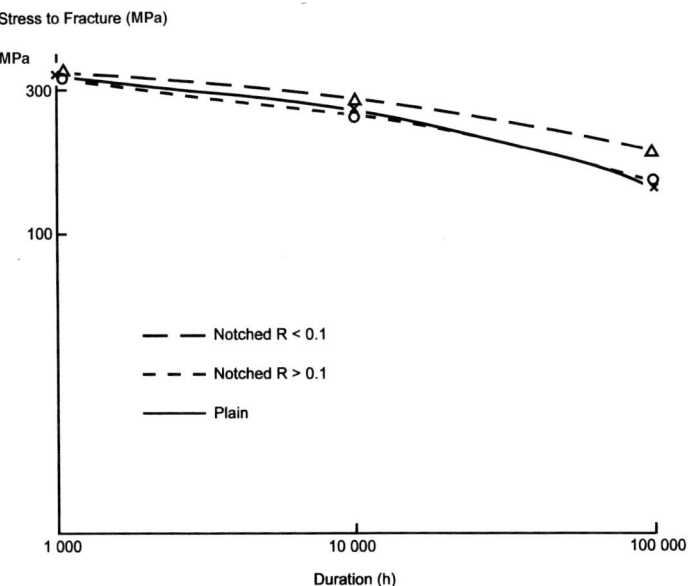

Stress to Fracture (MPa)

— — Notched R < 0.1

— — — Notched R > 0.1

——— Plain

Duration (h)

Fig. 10   Durehete 1055–550°C–Data Assessment Results.

ing the grain boundary energy and thus increasing the cavity nucleation rate.[15]

The proposition, therefore, to the steelmaker was to reduce the residual element content of Durehete 1055. Double vacuum melting, involving VIM and VAR, using virgin materials has been examined and assessed by BSES and the capability to produce steel with 'R' values of well below 0.05 has been demonstrated. However, such a steelmaking route is not economic for the intended product application. Furthermore, it was not a practical proposition to try to achieve reductions of specific residual elements by certain amounts as indicated by the science.[14,15]

Therefore, a low residual, clean steel, air melt practice has been adopted by BSES and over the years, since about 1985, significant creep reductions in the 'R' value for Durehete 1055 have been achieved, Fig. 8, by scrap and ferro-alloy selection, extending the dephosporisation period during steelmaking and using selected steelmaking fluxes.[13] This practice is now well established at BSES so that since 1993 all casts made have 'R' values ≤0.1, which is recognised to be a practical limit below which high rupture ductility, can be achieved, see Fig. 9. This leads to a notched strength equal to or above plain strength (see Fig. 10).

It is planned to introduce an 'R' value of 0.1 maximum into the specification when in-line apparatus becomes available for rapid determination of the full range of relevant residual elements.

The significance of 'R' values above and below 0.1 has been confirmed by a recent assessment exercise of data for Durehete 1055, carried out by a working group of the European Creep Collaborative Committee (ECCC).[16] The results of this work, based on test results extending to ~100 000 h at 550°C, showed that the difference in residual element content, above and below R = 0.1 had no effect on the strength of plain specimens. However, the strength of notched specimens was found to depend on the 'R' value being above or below 0.1 as summarised in Fig. 10. Data on notched specimens for the lower residual element content material extend currently to 70 000 h at 550°C. Rupture ductility values for material for R ≤0.1 range from 8 to over 30%, Fig. 9 and thus confirm the long term benefit to creep ductility/notched strength gained by the practically obtained reduction in residual element content to current typical 'R' values of 0.05–0.08 in Durehete 1055, Fig. 8.

6. CONCLUSIONS

The development of Durehete 1050 in the 1950s and its successor, Durehete 1055, has provided a bolting steel with high and stable creep strength, due to the precipitation of fine V_4C_3 particles, which is enhanced by the presence of 1% molybdenum.

Long term testing, e.g. to 100 000 hours at 550°C, demonstrates the stability and probably further precipitation of V_4C_3 but with some coarsening and transformation towards molybdenum rich particles. However, this trend occurs only slowly in Durehete 1055 and hence a high creep strength is maintained for long periods.

Low creep ductility values were the reason why Durehete 1055 was developed, through the addition of titanium and boron, which removed the tendency to form denuded zones adjacent to grain boundaries. This retarded the onset of low ductility values to beyond 10 000 hours at 550°C. Low ductility values, possibly enhanced by coarse grain sizes, were implicated in several bolt failures after longer durations in service but other operational factors were considered to be main contributors.

Further work by steelmakers has demonstrated that by reducing the content of residual elements, such as phosphorus, tin and copper, the cavity nucleation rate at grain boundaries in Durehete 1055 is reduced significantly. Arsenic is not believed to play a role in this mechanism and that of antimony is not fully understood.

A reduction in the total residual element content to 'R' values <0.1, has meant that creep ductility values remain sensibly high in long test durations and as a direct consequence the notched strength of Durehete 1055 is higher than that of plain specimens with data available now to 70 000 hours at 550°C.

ACKNOWLEDGEMENTS
The authors acknowledge the help and assistance of other staff at Swinden Technology Centre. They also thank Dr. K.N. Melton, Research Director, Swinden Technology Centre and Dr. I.G. Davies, Technical Director, British Steel Engineering Steels, for permission to publish this paper.

REFERENCES

1. H. Everson, J. Orr and D. Dulieu: 'Low Alloy Ferritic Bolting Steels for Steam Turbine Applications. The Evolution of the Durehete Steels', Paper in Conference *Advances in Material Technology for Fossil Power Plants*, 1–3 September 1987, Chicago, Illinois, USA, ASM International + EPRI.

2. J. Orr, H. Everson and G. Parkin: 'Warm Worked Esshete 1250: a High Strength Bolting Steel', *Ironmaking and Steelmaking*, 1994, **21** (5), 345–352.

3. M.G. Gemmill and J.D. Murray: 'Bolting Materials for High Temperature Steam Plant Service', *Engineering*, December 1955, 824–831.

4. P.G. Stone and J.D. Murray: 'Creep Ductility of Cr–Mo–V Steels', *Journal of the Iron and Steel Institute*, November 1965, 1094–1107.

5. I. Kimura, T. Watanabe, M. Honda, R. Hiroi and M. Usa: 'Development of Steels with Excellent Resistance to Delayed Fracture for High Tension Bolts', Nippon Steel, *Technical Report Overseas No. 3*, June 1973.

6. M.J. Collins, 'Carbide Stabilities in a Cr–Mo–V Bolting Steel', *Material Science and Technology*, 1989, **5** (4) 323–327.

7. W. Crafts and J.L. Lamont: 'Carbides in Long Tempered Vanadium Steels', *Mining and Metallurgical Engineer*, 1950, **188**, 561.

8. K.W. Andrews, H. Hughes and D.J. Dyson: 'Constitution Diagrams for CrMoV Steels', *Journal of the Iron and Steel Institute*, May 1972, 337.

9. B.L. King: 'Intergranular Embrittlement in CrMoV Steels: An Assessment of the Effects of Residual Impurity Elements on High Temperature Ductility and Crack Growth', *Phil. Trans. R. Soc. A*, London, 1980, **295**, 235.

10. S.T. Kimmins and D. Gooch: 'Austenite Memory Effect in 1Cr–1Mo–0.75V (Ti,B) Steel', *Metal Science*, 1983, **17** (11), 519.

11. H. Everson, J. Orr, D. Dulieu and D. Burton: 'Improvements in Rupture Ductility of Durehete 1055', Paper 21 in Conference *Rupture Ductility of Creep Resistant Steels*, York, December 1990, The Institute of Metals.

12. J. Orr, H. Everson, D. Burton and J. Beardwood: 'The Continuing Development of Durehete 1055 and Other High Integrity Bolting Steels', *Performance of Bolting Materials in High Temperature Plant Applications*, The Institute of Materials, London, 1995, 138.

13. H. Everson and J. Orr: 'Low Residuals in Bolting Steels', *Clean Steel: Superclean Steel*. The Institute of Materials (on behalf of EPRI), London, 1996, 213.
14. N.G. Needham, K. Green and T. Gladman: 'Elevated Temperature Cracking of High Temperature Steels', *ECSC Report EUR 10749EN*, 1987.
15. M.P. Seah: 'Impurities, Segregation and Creep Embrittlement', *Phil. Trans. R. Soc. A*, London, 1980, **295**, 265.
16. J. Orr: *Report to Working Group 3.4 of European Creep Collaborative Committee.*

APPENDIX

SIGMOIDAL STRENGTH REACTIONS

Sigmoidal behaviour in strength/time curves, shown schematically in Fig. A1, arises from the situation that the initial strength, e.g. following the quality heat treatment, is made up of at least two components. These components form the base strength of the steel derived from solid solution strengthening and precipitation strengthening. Whilst the base strength decreases only slowly over time at temperature, the precipitation strength will, after some time, eventually decrease at a relatively fast rate, e.g. by Ostwald ripening and precipitate transformations, unless additional strengthening processes occur. It is the decay of the latter which leads to the sigmoidal shape appearing in the strength/time relationship, Fig. A1.

In the case of Durehete 1055, the base strength derives from the strength of the tempered martensite with solid solution strengthening principally by chromium and molybdenum and precipitation strength by the formation of V_4C_3.

There is some evidence to suggest that further precipitation of V_4C_3 occurs during testing which counterbalances some of the strength loss from the coarsening of the initially formed V_4C_3 and also its transformation into less effective, molybdenum rich, particles.

The outcome, as indicated by the electron metallographic and creep data, is a slow decline in the overall creep strength of Durehete 1055, maintaining an acceptably high level.

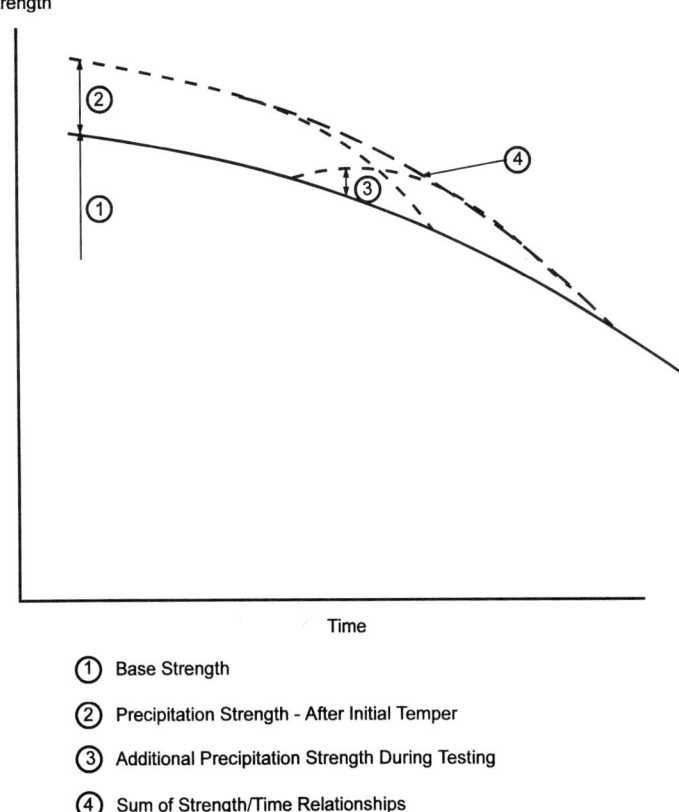

Strength

Time

① Base Strength

② Precipitation Strength - After Initial Temper

③ Additional Precipitation Strength During Testing

④ Sum of Strength/Time Relationships

Fig. A1   Schematic Sigmoidal Strength/Time Relationship.

Structure and Properties of Large 10% CrMoVNbN Forgings Developed for Long Term High Temperature Steam Turbine Applications

M.A. WALSH AND S. PRICE

Forgemasters Steel and Engineering Limited, P.O. Box 286, Brightside Lane, Sheffield, S9 2RW

ABSTRACT

Over the last decade, the Power Generation industry has seen a substantial investment in improved power plant designs geared towards reducing costs by increased thermal efficiencies. An improvement in power plant efficiency can be achieved by increasing operating temperature and pressure, and supercritical power stations have achieved a competitive efficiency edge in this way.

Forgemasters Steel and Engineering Limited has been involved in a collaborative European project, organised within COST, to develop advanced 9–12%Cr turbine rotor forgings, for use in high temperature, advanced steam cycles.

One of the aims of the programme is the development of turbine rotor forgings with a minimum creep rupture strength of 100 MPa at 600°C over a time period of 10^5 hours. Initial investigations identified several promising analyses which resulted in the manufacture of prototype rotor forgings. Forgemasters Steel and Engineering Limited manufactured one such rotor in a material designated COST steel F, a 10%CrMoVNbN steel. Data obtained from this prototype concerning the control of microstructure and properties has contributed to the design of supercritical steam turbines in Europe, with projected thermal efficiencies which significantly exceed those measured for conventional power plant.

Two H.P./I.P. rotors weighing 25 tonnes each and incorporating diameters up to 1200 mm were subsequently produced utilising 10%CrMoVNbN steel, one for the Skaerbaek and one for the Nordjylland supercritical power plants in Denmark.

This paper describes the developments in manufacture and control of microstructure of advanced 9–12%Cr steels in large forgings for use in high temperature, advanced steam cycles.

INTRODUCTION

Power Generation utilities have always shown considerable interest in increasing the operating temperature and pressure of steam turbines. This enables significant improvements in the heat rate, resulting in a more economic generation of electric power. The efficiency of several different power station formats is given in Table 1. One of the material limitations to increasing the operating temperature of turbine rotors is the creep strength of 1%CrMoV and 12%Cr steels conventionally used for high temperature rotor forgings. These alloys have adequate creep strength for operation in steam temperatures in the vicinity of 540°C, but for advanced steam cycle applications which demand steam temperatures approaching 600°C, their application would require the use of complex rotor cooling systems, which limit the potential thermal efficiency.

In Europe, a collaborative project, organised within COST 501, has been working to introduce materials to overcome this limitation, through the development of new and improved steels. As well as developing steels for cas-

Table 1   Efficiency of various types of power plant

STATION	TYPE
THERMAL EFFICIENCY (%)	
38	OIL-FIRED
39	COAL-FIRED
36	NUCLEAR (AGR)
49	SUPERCRITICAL (DOUBLE REHEAT)
58	COMBINED-CYCLE

ings, bolting, blading, pipework and tubing, one of the aims of this programme was to develop rotor forgings with 10^5 hour creep strength above 100 MPa at 600°C. Other parameters such as creep ductility, hardenability, toughness, weldability and yield strength were also considered with the aim of achieving properties equivalent to or better than current materials.

The 12%CrMoV family of steels have been shown to have considerable benefits in terms of high temperature properties compared to 1%CrMoV materials (Fig. 1) and were therefore the focus of development. Extensive work within COST 501, undertaken in the 1980s utilising trial melt alloys, identified the importance of various elements including Nitrogen, Niobium, Molybdenum, Tungsten and Boron to the creep strength of the base material.

This work resulted in the manufacture of three prototype rotor forgings:

- A 10%CrMoVBNb steel designated COST steel B manufactured by Bohler Kapfenberg.
- A 10%CrMoWVNbN steel, designated COST steel E manufactured by Saarstahl Volklingen.
- A 10%CrMoVNbN steel designated COST steel F manufactured by Forgemasters Steel and Engineering Limited (FSEL), Sheffield.

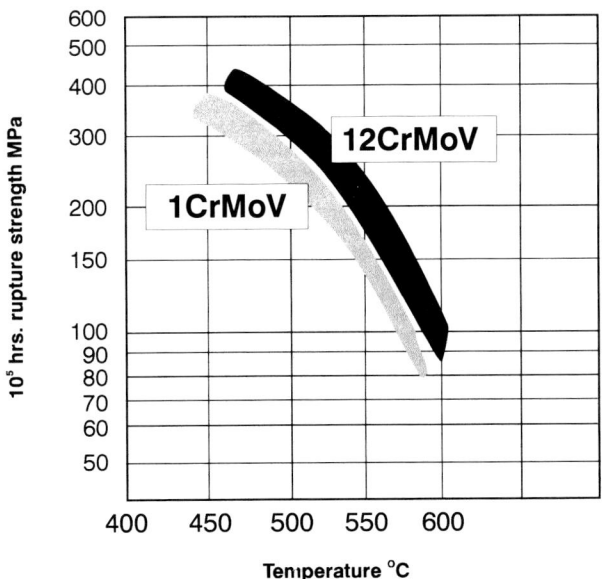

Fig. 1   Comparison of rupture strength between 1CrMoV and 12CrMoV steels (after Takenouchi[16]).

The knowledge and experience gained subsequently resulted in FSEL manufacturing two HP/IP rotors weighing 25 tonnes each, incorporating diameters up to 1200 mm, for supply to GEC ALSTHOM for the Skaerbaek and Nordjylland supercritical power plants in Denmark.

This paper describes the developments that took place and the control of microstructure and properties achieved during the production of large rotor forgings in COST steel F.

BACKGROUND TO THE DEVELOPMENT OF 9–12%CR STEELS

At conventional inlet temperatures of up to 540°C, the 1%CrMoV steels combine creep strength and thermal fatigue resistance with adequate hardenability and toughness to give a cost effective material for application in a high pressure rotor forging.

As steam temperatures become supercritical, typically 580–610°C, the properties of the 1%CrMoV steels are inadequate, particularly in terms of elevated temperature creep properties. Much of the development work has, therefore, focused upon the 12%CrMoV steels to meet the demand for higher operating temperatures.

The first 12%CrMoV steel used for turbine rotors was manufactured for KWU in 1955.[1] In the late 1950s, two 12%CrMoWV rotors were manufactured in the USA, one based on type 422 steel developed by Westinghouse[2] and the other based on a German steel, RNO-MoWV manufactured for GE.[3]

The creep rupture strength of both the 12%CrMoV and 12%CrMoWV steels were not considered to be high enough and further improvements were sought by the addition of Niobium. The initial development of Niobium containing 12%CrMoV steels was carried out in the UK by Jessop Savilles[4] and Firth Brown Limited[5] (forerunners of FSEL) and grades H46 and FV448 were extensively used in the manufacture of gas turbine discs.

GE also evaluated this material[3] but found substantially reduced levels of ductility and toughness in large rotor forgings due to the segregation of coarse NbC. It was not until the Niobium had been systematically reduced from 0.6% to 0.06% that the problems of segregated NbC were overcome.

The development of creep resistant 12%Cr steels in Japan dates back to the 1950s. Fujita[6] investigated a 12%CrMoVNbB steel designated TAF which was used in the manufacture of small rotor forgings. By the early 1960s, Toshiba had produced a 12%CrMoV steel containing Tantalum and Nitrogen with equivalent creep rupture properties to the Niobium–Nitrogen containing steels. More recent investigations initiated by Fujita at Tokyo University systematically evaluated the relationship between alloying elements in the base material and Fujita was able to optimise the chemistries in relation to creep rupture strength and defined a series of 12%CrMoV steels[7] designated TMK1, TMK2 and TR1200 which could operate at 1100°F (593°C), 1150°F (621°C) and 1200°F (648°C) respectively. TMK1 was used in the manufacture of a high temperature rotor for Wakamatsu, an experimental 50 MW supercritical steam turbine plant operating at 593°C.[8] This experimental Power Station began official operation in April 1987.

In Europe, a development programme was initiated to improve the 12%CrMoV steels under the auspices of COST 501. A series of candidate 12%Cr alloy steel variants were investigated in the form of trial melts[1,9–12] culminating in the production and evaluation of 3 prototype 12%CrMoV steel rotor forgings.

The most recent data suggests that Boron containing[12] and Cobalt containing 12%CrMoV steels[13] have the best potential for elevated temperature performance which may permit inlet steam temperatures to rise above 620°C. A summary of the various compositions developed over the last three decades is given in Table 2.

PRODUCTION AND TESTING OF PROTOTYPE COST ROTOR F

Compared to conventional low alloy steel rotor forgings, the highly alloyed 12%CrMoV rotors pose additional complexities, if an homogeneous high integrity forging, which meets the requirements of power plant designers and utilities, is to be produced. These include:

• steelmaking

Table 2. Development of 12Cr rotor forging compositions

GRADE	CHEMICAL COMPOSITION											
	C	Si	Mn	Ni	Cr	Mo	V	W	Nb	N	B	Ta
TYPE 422	0.23	0.4	0.8	0.75	13.0	1.0	0.25	1.0	–	–	–	–
RNO MoWV	0.22	–	–	0.80	11.5	1.0	0.30	0.25	–	–	–	–
H 46	0.16	0.30	0.70	–	11.67	0.60	0.30	–	0.25	0.05	–	–
FV 448	0.13	0.50	1.00	–	10.5	0.75	0.15	–	0.45	0.05	–	–
TAF	0.18	0.30	0.50	–	10.5	1.50	0.20	–	0.15	–	0.03	
SEW 555	0.23	0.25	0.50	0.50	12.0	1.00	0.30	–	–	–	–	–
12CrMoWV (WEST)	0.23	0.40	0.80	0.75	13.0	1.00	0.25	1.00	–	–	–	–
11CrMoV NbN (GE)	0.18	0.25	0.75	0.70	10.5	1.00	0.20	–	0.06	0.06	–	–
11CrMoV TaN	0.17	0.06	0.60	0.35	10.6	1.00	0.22	–	–	0.05	–	0.07
TMK1	0.14	0.05	0.50	0.60	10.2	1.50	0.17	–	0.06	0.04	–	–
TMK2	0.13	0.05	0.50	0.70	10.2	0.40	0.17	1.80	0.06	0.05	–	–
TR 1200	0.12	0.05	0.50	0.80	11.2	0.20	0.20	1.80	0.05	0.06	–	–
ROTOR B (COST 501)	0.17	0.07	0.06	0.12	9.34	1.58	0.27	–	0.059	0.02	0.008	–
ROTOR E (COST 501)	0.12	0.10	0.45	0.74	10.39	1.06	0.18	0.81	0.045	0.05	–	–
ROTOR F (COST 501)	0.11	0.03	0.52	0.58	10.22	1.42	0.18	–	0.05	0.06	–	–

- segregation
- forging
- hot workability
- heat treatment
- the production and validation of weld procedures to address the problems of 'wire wooling' on the bearings of alloys with greater than 3%Cr.

The 12%CrMoV type rotor forgings can be particularly prone to segregation[14] if careful controls are not exercised during the steelmaking operations. It is essential that segregation is minimised in order to provide uniform ambient and elevated temperature properties in the final forging and to avoid the formation of delta ferrite since many of the recently developed steels have compositions with comparatively high chromium equivalents (Fig. 2).

In general, there are two steelmaking methodologies that have been developed to minimise segregation in ingots used in 12%CrMoV rotor forgings; ESR and advanced VCD. Forgemasters Steel and Engineering refined their steelmaking techniques and ingot technologies in order to control segregation when producing the prototype COST 501 rotor F using advanced VCD (Vacuum Carbon Deoxidation). The prototype rotors B and E were produced by other manufacturers using ESR (Electro-Slag Refining) technology.

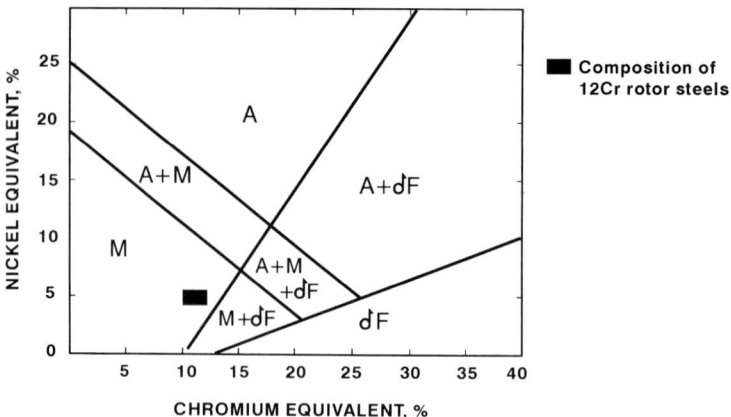

Fig. 2   Composition range of 12Cr rotor steels.

ESR for 12%CrMoV rotor forgings has been a constant source of debate over the last few years. Capellini *et al.*[15] concluded that ESR forgings were cleaner, sounder and more homogeneous than conventionally melted forgings yielding an improved tensile ductility, impact toughness and FATT. On the other hand, Takenouchi *et al.*[16] obtained excellent homogeneity with conventionally melted vacuum carbon deoxidised (VCD) steels having melted 20 rotors by this route. They concluded that segregation could be kept to a minimum by employing steep taper ingot moulds with H/D (height:diameter) ratios of approximately 1.0 or less, Ito *et al.*[17] compared the properties of 3 prototype 12%Cr rotors, two of which were manufactured using VCD and the other by ESR. The results indicated that there were no significant technical differences between the two manufacturing processes.

A schematic diagram of the steelmaking procedures used by FSEL for COST 501 rotor F, the 12%CrMoNbN prototype rotor, is outlined in Fig. 3. High grade, low residual scrap was charged into a Basic Electric Arc (BEA) furnace and dephosphorised. The 45 tonnes of liquid metal was tapped into a ladle and deoxidised using a combination of aluminium and vacuum carbon deoxidation.

The charge was then degassed and final adjustments were made to the chemistry especially with respect to Nitrogen. The melt was then uphill teemed into a steep taper 68″ (1727 mm) mould under an argon shroud.

After solidification, the ingot was stripped from the mould and transferred hot to the forge. Segregation and axial unsoundness were minimised through:

- vacuum carbon deoxidation
- a steep taper ingot mould (approximate 1.0 H/D)

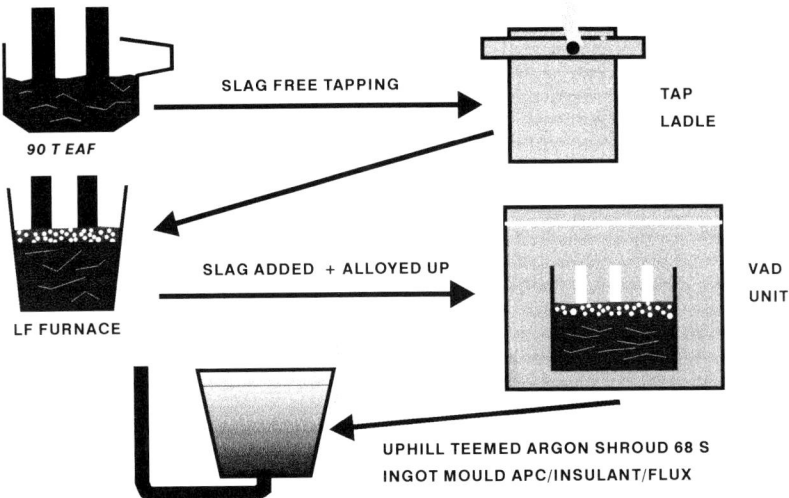

Fig. 3 Schematic diagram of melting, refining and casting of COST 501 Rotor F.

- low teeming temperature
- large head volume
- controlled teeming rate
- controlled mould preheat temperature.

The 12%CrMoV rotor forgings, as well as being significantly more creep resistant than their 1%CrMoV counterparts, are also substantially tougher.[12] However, to optimise toughness, particular attention is required during hot working operations to ensure that a fine grain size is achieved during subsequent heat treatment operations.

The 12%Cr steels are more resistant to applied pressure at forging temperatures and will generally require more reheating operations compared to low alloy steels. Increasing the maximum forging temperature is not acceptable as it could result in not only an unacceptably larger grain size, but also the possibility of overheating, a phenomenon associated with grain boundary deterioration and a loss in ductility and toughness.[18] The forging sequence used at FSEL for the COST 501 rotor F is shown in Fig. 4 and is typical of procedures used for forging 12%CrMoV rotors.

The as-forged rotor requires transformation and annealing prior to the first machining and inspection operations. Two methods of achieving transformation are feasible; slow cooling through the ferrite + carbide region or isothermal transformation at 700°C. Slow cooling is purported to promote a finer carbide distribution on completion of the preliminary heat treatment cycle which can be fully dissolved during subsequent austenitising treat-

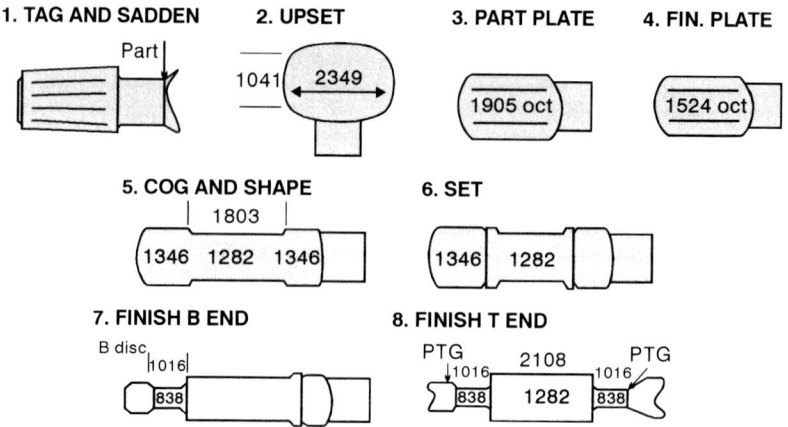

Fig. 4 Forging of the COST 501 Rotor F.

ment cycles. However, there is a risk of cracking during the initial stages of cooling due to higher transformation stresses. For this reason, the more conservative isothermal transformation is normally adopted during the preliminary heat treatment cycle and holding times at approximately 700°C are generally in excess of 150 hours to permit total transformation to ferrite + carbide.

In quality heat treatment, 12%CrMoV rotor forgings are oil quenched to achieve a fully martensitic microstructure, with the forging being equalised at a temperature of between 150°C–200°C to prevent quench cracking. This is followed by a double tempering treatment. The first tempering treatment ensures a low stressed tempered martensitic structure and, on cooling, any retained austenite is transformed to martensite. The second tempering treatment stress relieves any untempered martensite formed during the stage 1 tempering treatment.

The heat treatment cycle adopted for rotor F in the COST 501 programme is shown in Fig. 5. This complex cycle aims to produce a fully tempered martensitic structure throughout the full cross section of the rotor. This is a prerequisite to realise optimum elevated temperature properties in service. After tempering at 680°C to achieve a 700 MPa yield strength condition the rotor was comprehensively tested followed by tempering at 700°C to achieve a 600 MPa yield strength condition followed by further testing.

The trial rotor was fully ultrasonically examined prior to and after quality heat treatment. No defects were found, again confirming the validity of the steelmaking route and the consolidation achieved during forging.

The test results of the COST prototype rotor F have been extensively

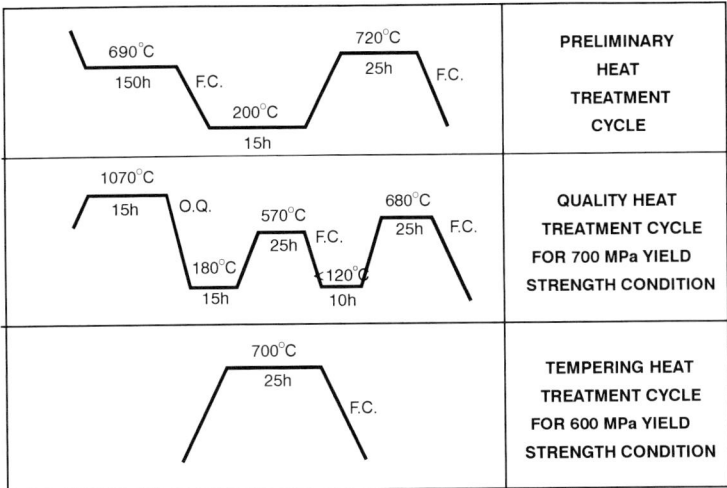

Fig. 5   Heat treatment of COST 501 Rotor F.

reported previously.[9-12] The testing positions and basic dimensions of the prototype are shown in Fig. 6.

The results of analysis showed uniform chemistry distribution throughout, and confirm the results of Takenouchi *et al.*[16] that advanced VCD steelmaking is a viable alternative to ESR.

The mechanical property results in the 700 MPa and 600 MPa yield con-

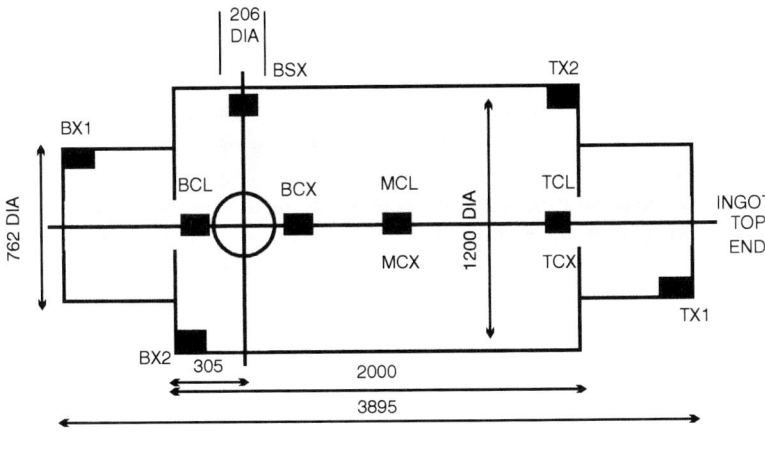

Fig. 6   Machined profile and test positions, COST 501 Rotor F.

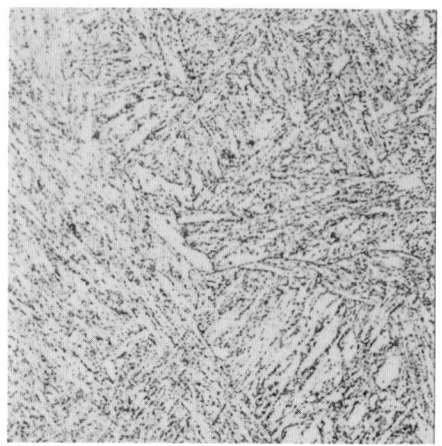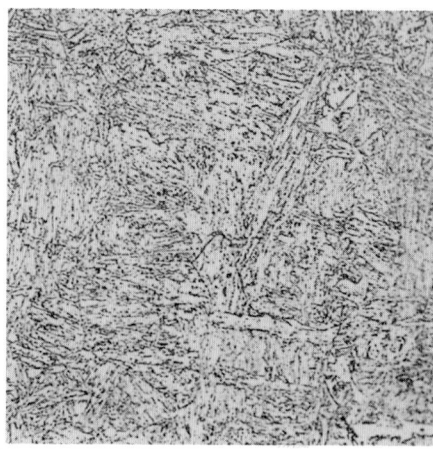

Fig. 7   Microstructure at (a) surface and (b) core of COST 501 Rotor F, × 500.

dition were generally excellent with only small variations in strength from the top to bottom of the ingot.

Compared with conventional 1%CrMoV rotors, the combination of strength and toughness was significantly improved in the 12%CrMoVNbN proto-type.[12]

Microstructural examination at the surface and core was carried out and photomicrographs are included in Fig. 7. Photomicrographs confirm that segregation can be kept to a minimum using advanced VCD steelmaking technology. A homogenous tempered martensitic microstructure is shown at both surface and core. As expected, a finer structure is shown at the surface. Measurements of grain size and delta ferrite content[1,9–12] show a uniform structure of prior austenite grain size of generally 4/6 at the surface and 2/3 at the core. Only very small amounts of delta ferrite (0.5%) were detected in the core. This is not expected to affect performance, a fact which has been confirmed during subsequent long term creep testing. The photomicrographs validate the critical aspects of the production route, including the forging, isothermal transformation and quality heat treatment processes in producing the optimum microstructure.

The creep rupture and creep strain properties have been comprehensively examined within the COST programme and have been reported extensively.[1,10–12] They are significantly better than the mean values of conventional 12%CrMoV rotor material made according to specification SEW555. The lower yield strength condition of 600 MPa shows a flatter creep rupture curve than the higher yield strength condition, and the trends reported to date appear to confirm the results of the trial melts, the material showing very good, stable performance.

PRODUCTION AND TESTING OF 10%CrMoVNbN TURBINE ROTORS FOR THE SKAERBAEK AND NORDJYLLAND SUPERCRITICAL POWER STATIONS IN DENMARK

Forgemasters Steel and Engineering Limited has manufactured two HP/IP rotors weighing 25 tonnes each and incorporating diameters up to 1200 mm utilising COST steel F (10%CrMoVNbN). These are two of the largest advanced 12%CrMoV steel forgings manufactured to date and were supplied to GEC ALSTHOM, one for the Skaerbaek and Nordjylland supercritical projects in Denmark.

The Skaerbaek and Nordjylland Power Stations incorporate a 415 MW supercritical turbine set with a steam inlet temperature of 580°C. The line diagram showing the turbine set is given in Fig. 8. The HP/IP rotors manufactured by FSEL were the largest advanced 12%CrMoV steel forgings produced for the Danish project, and although their manufacture was based on the experiences gained from the prototype rotor, further enhancements were necessary due to the increased size of the rotors, and the necessity to meet a full production specification. The BEA; LF; VAD route shown in Fig. 3 was again employed, however, the liquid metal weight was 97 tonnes compared to the 45 tonnes produced for the prototype rotor. This required the utilisation of an 80 inch (2032 mm) steep tapered ingot into which the steel was uphill teemed. The advanced steelmaking techniques employed during the manufacture of the prototype rotor, to reduce segregation and unsoundness, were again applied, close control being all the more critical with respect to the much larger mass of steel present. The forging process was also based on that for the prototype rotor (Fig. 4), however, the upsetting operation was performed in two stages, similar to the plating operation. This ensured that the working temperature remained uniform and optimum for consolidation of the centre of the forging.

Due to the larger mass, the risk of thermal gradients and transformation stresses was increased and therefore, the isothermal transformation route was

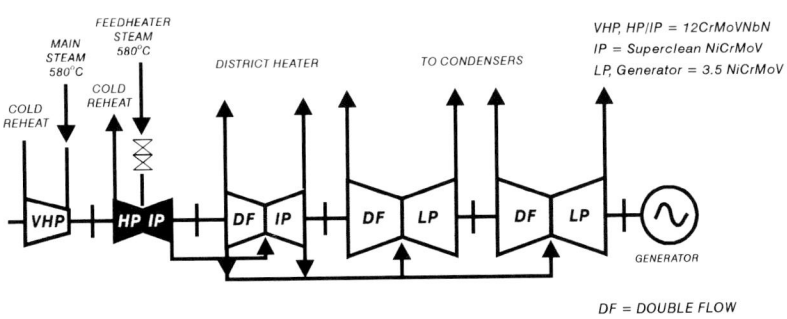

Fig. 8   Line diagram of Skaerbaek and Nordjylland Power Stations.

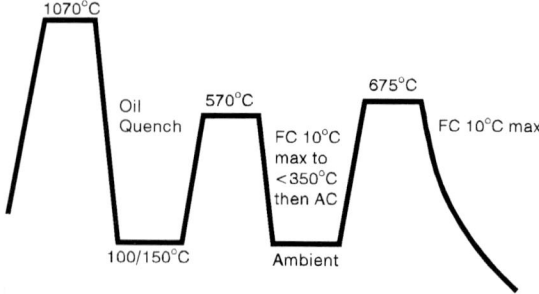

Fig. 9   Quality heat treatment cycle for Skaerbaek and Nordjylland rotors.

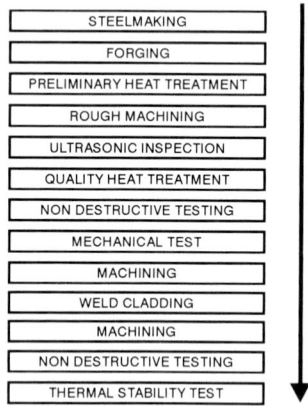

Fig. 10   Overview of manufacturing process for the Skaerbaek and Nordjylland rotors.

all the more important in order to negate the risk of cracking. The quality heat treatment cycle is given in Fig. 9. The cycle is essentially the same as that used for the prototype rotor and is aimed at producing a uniform fully tempered martensitic microstructure.

An overview of the entire production route is given in Fig. 10 which indicates the extensive manufacturing control necessary for the production rotors in terms of non-destructive testing, weld cladding (discussed later) and thermal stability testing.

Extensive testing of the two production rotors was carried out as reported previously.[19] The overall dimensions and test positions are given in Fig. 11. A full chemical analysis of the 2 rotors is given in Table 3. This shows the excel-

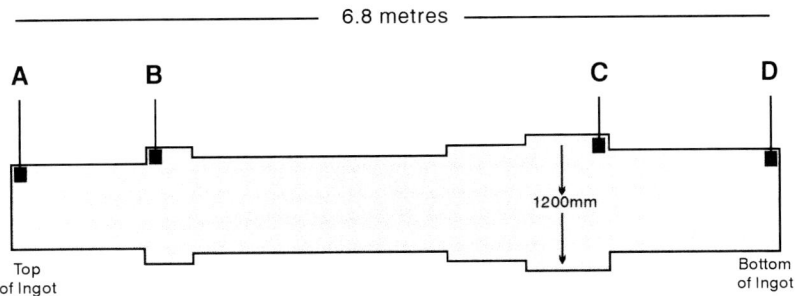

■ Location of test material

Fig. 11 Mechanical property and chemical analysis testing locations for Skaerbaek and Nordjylland rotors.

Table 3 Cast analysis of Skaerbaek and Nordjylland rotors

STATION	C	Si	Mn	P	S	Cr	Mo	Ni	V	Al	N	As	Sb	Sn	Cu	Nb
SKAERBAEK	.14	.08	.55	.011	.002	9.99	1.40	.56	.17	.006	.041	.005	.001	.004	.04	.059
NORDJYLLAND	.12	.05	.53	.010	.003	10.16	1.46	.55	.18	.003	.046	.005	.002	.004	.05	.052

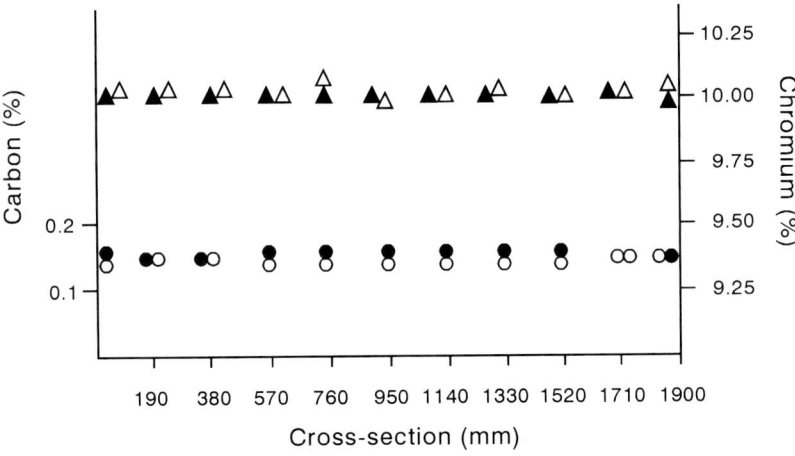

Fig. 12 Segregation survey of top and bottom discards for Skaerbaek rotor.

lent uniformity achieved with cast to cast variations limited to a few points of analysis for all the elements represented. In order to confirm this result in the cross section of the rotor, an analysis traverse of the Skaerbaek ingot was performed. The results for Carbon and Chromium, two important elements as regards properties and segregation, are plotted in Fig. 12. Discarded material was utilised representing the top and bottom of the ingot. The excellent uniformity is again evident with little variation in the analysis between top and bottom locations.

The mechanical property results from the positions specified in Fig. 11 are given in Table 4. The Skaerbaek and Nordjylland rotors were designed in the high yield condition of 700 MPa minimum. The results in Table 4 meet all customer requirements and demonstrate uniform properties throughout indicating excellent toughness.

Table 4. Mechanical properties of Skaerbaek and Nordjylland rotors

STATION	POSITION*	Rp0.2 (Mpa)	Rm (Mpa)	A (%)	Z (%)	IMPACTS AT 40°C (J)	FATT (°C)	UPPER SHELF ENERGY (J)
SKAERBAEK	A	731	872	18	54	38, 39, 41	–	–
	B	736	873	17	55	41, 41, 41	–	
	C	741	876	17	54	41, 42, 34	48	120
	D	777	915	18	53	35, 35, 39	–	–
NORDJYLLAND	A	762	892	15	48	40, 45, 37	–	–
	B	748	882	18	57	46, 42, 49	–	–
	C	763	900	16	59	40, 38, 30	54	100
	D	780	902	16	49	39, 33, 41	–	–

Microstructural examination was carried out at the body ends of both rotors and photomicrographs are included in Fig. 13. A fully tempered martensitic microstructure is evident with a prior austenite grain size of 2–4. This concurs with experience gained on the prototype rotor and reinforces the validity of the production route, particularly the forging and heat treatment processes, which are designed to produce a strong, tough and creep resistant martensitic steel.

Apart from basic chemical and mechanical property requirements, the production process for the Skaerbaek and Nordjylland rotors was designed to meet demanding inspection and validation procedures as follows:

• detailed ultrasonic examination
• creep testing

Fig. 13 Microphotographs of (a) Skaerbaerk and (b) Nordjylland rotors, body end position, × 50

- the application and non-destructive testing of a weld deposit on the journal bearings to combat the 'wire wooling' wear phenomenon.
- thermal stability testing

Both rotors were subject to intensive ultrasonic examination with 0° compression probes and a series of angle probes. No indications were reported.

Short term creep testing was also performed on samples taken from each production rotor in accordance with GEC ALSTHOM requirements. The 1000 hr creep strain was evaluated after exposure at a defined temperature and stress. All results were well within the specification requirements and comparable to data generated within COST 501.

For nearly 40 years it has been know that the surfaces of alloys with levels of Chromium above about 3% have poor tribological properties, leading to wear problems on rotor journals, which are characterised by deep scouring of the journal, commonly referred to as 'wire wooling'. Many solutions have been proposed and applied to overcome this problem, including Chromium plating, mechanical attachment of a journal sleeve and weld deposition of a protective low alloy steel cladding. A weld deposition technique which offers a robust, long term solution to this problem has been developed through collaboration between Parsons Power Generation Systems, Forgemasters Steel and Engineering and GEC ALSTHOM. The fully qualified welding programme included:

- the use of a buttering welded layer at the rotor interface to minimise dilution.
- careful control of the heat treatment parameters to minimise cracking and to optimise mechanical properties.

- a consistent structure, coherent with the rotor which can be reground during service, if necessary.
- the avoidance of residual surface tensile stresses which would reduce fatigue strength.

The thickness of the deposit applied is determined by considerations of alloy dilution, ultrasonic inspectability, rotor machining tolerances and the desire to allow for the possibility of re-machining journals should they be damaged in service. Further aspects of this important issue, that will ensure the safe operation of 12%CrMoV rotors in future supercritical steam turbine plant have been reported elsewhere.[20,21]

One of the final validation techniques, applied to confirm product quality and homogeneity, is the thermal stability test. Clearly, any local variations in chemistry or microstructure could result in rotor instability.

The thermal stability test requires a rotor to be heated to a nominated temperature (around 650°C) while under rotation. Measurements of Total Indicator Reading (T.I.R.) are then taken using dial test indications in five positions along the length of the turbine rotor. These detect any deflections which occur. Any readings greater than 0.001 mm are recorded. The Skaerbaek and Nordjylland rotors both proved to be exceptionally stable during heating, soaking and cooling confirming the homogeneity of the rotor forgings.

Considering the substantially increased mass of material in the production rotors, compared with the prototype, the chemical analysis, mechanical properties and validation procedure results, confirm both the suitability of the alloy and the choice of manufacturing route, for this size of turbine rotor.

SUMMARY

In the search for a more economic generation of electric power, the operating temperature and pressure of steam turbines is increasing. This presents a challenge to develop high temperature creep resistant alloys.

Within the COST 501 programme, in Europe several new and improved advanced 9–12%Cr steels have been developed including COST steel F, a 10%CrMoVNbN steel. The suitability of this material for high temperature, long term application in steam turbines has been assessed by the production of a prototype rotor, produced by FSEL using enhanced conventional VCD steelmaking.

The mechanical properties, creep properties and microstructural stability of the prototype rotor proved excellent and resulted in the production of two rotors for Skaerbaek and Nordjylland, which are supercritical 415 MW power plants in Denmark.

The production process for the Danish rotors, in addition to meeting basic chemistry and mechanical property requirements, was required to meet demanding inspection and validation criteria.

The turbine rotors for Skaerbaek and Nordjylland, although larger than the prototype, were shown to have comparable properties with respect to data produced for the prototype rotor. Uniform chemistry, mechanical properties and microstructure confirm that the 10%CrMoVNbN steel is suitable for application in long term, high temperature supercritical steam turbine applications.

CONCLUSIONS

A prototype 10%CrMoVNbN rotor, produced utilising advanced conventional VCD practice, has been manufactured by FSEL.

A full evaluation of the prototype showed the material and production processes suitable for manufacture of supercritical steam turbine rotors with diameters up to 1200 mm.

Development of suitably qualified welding procedures was also completed, in order to allow weld cladding of the journal bearing areas to combat the effects of 'wire wooling'.

Based on the above evaluations, FSEL have produced two production turbine rotors for GEC ALSTHOM for the Skaerbaek and Nordjylland supercritical power plants in Denmark. These are the largest 10%CrMoVNbN rotors manufactured for this project. Full production testing and inspection requirements showed:

- uniform chemical analysis and mechanical properties, including good toughness
- creep resistance comparable with data generated within the COST programme
- stable, uniform microstructure.
- satisfactory weld cladding of the journals
- successful ultrasonic examination
- excellent thermal stability

The results validate both the selection of material and production route for this size of high temperature, long term, supercritical steam turbine rotor.

REFERENCES

1. G.A. Honeyman: '12%CrMoV steels for combined cycle plant steam turbine rotor forgings', *Materials for Combined Cycle Power Plant*, Sheffield, UK, June 1991.
2. W.E. Trumpler, A.F. Le Breton, E.A. Fox and R.B. Williamson: 'Development Associated with the Superpressure Turbine for Eddystone Unit No. 1', *ASME Journal of Eng. Power*, 1960, **82** (10).

3. D.L. Newhouse, C.J. Boyle and R.M. Curran: A Modified 12% Chromium Steel for Large Temperature Steam Turbine Rotors', *ASTM 68th Annual Meeting*, June 1965.

4. G.L. Briggs, A.E. Marsh and J.W.S. Stafford: 'Properties of two highly alloyed martensitic stainless steels and a precipitation hardened austenitic stainless in the temperature range 500–700°C', *Proceedings of the Conference on High Temperature Properties of Steels ISI*, April 1966.

5. H.W. Kirkby and R.J. Truman: '12%Cr Steels; Creep Resisting and High Strength variants', *Proceedings of the Conference on High Temperature Properties of Steels'*, ISI, April 1966.

6. T. Fujita: 'Effect of MoVNb and N on creep rupture strength of TAF Steel', *Suppl. to Trans J.I.M.*, 1968, V9.

7. X. Liu, T. Fujita, A. Hizume and S. Konoshita: 'Development of High Strength Chromium Heat Resisting Steels for Turbine Rotors', *Trans. ISIJ*, 1986, 26(3).

8. T. Fujita, Y. Makabayashi, A. Hizume, Y. Takeda, T. Fujikawa, A. Takano, A. Suzuki, S. Kinoshita, M. Kahn and T. Tsuchiyama: An Advanced 12Cr Steel Rotor (TMK 1) for EPDC Wakamatsu Step 1 (593°C)', *COST – EPRI Workshop, 9–12Cr Steel for Power Generation*, October 1986.

9. C. Berger, E. Potthast, R. Bauer and G.A. Honeyman: 'Development of high strength 9–12%CrMoV Steels for high temperature rotor forgings', *11th International Forgemasters Meeting*, Terni, Italy, June 1991.

10. C. Berger, K.H. Mayer, R.B. Scarlin and D.V. Thornton: 'Improved Ferritic Rotor and Cast Steels for Advanced Steam Power Plants', *4th Int. EPRI Conf. on Improved Coal-Fired Power Plants*, Washington, USA, March 1993.

11. C. Berger, R.B. Scarlin, K.H. Mayer, D.V. Thornton and S.M. Beech: 'Steam Turbine Materials: High Temperature Forgings', *5th Int. Conf. Materials for Advanced Power Engingeering*, Liege, Belgium, Oct. 1994.

12. C. Berger, S.M. Beech, K.H. Mayer, R.B. Scarlin and D.V. Thornton: 'High Temperature Rotor Forgings of High Strength 10%CrMoV Steels', *ibid*.

13. T. Fujita: *Private Communications*.

14. D.L. Newhouse: 'Guide to 12Cr Steels for High and Intermediate Pressure Turbine Rotors for Advanced Coal-Fired Steam Plant', *EPRI Report No. C5–5277*, July 1987.

15. R.F. Cappellini, R.L. Bodnar, T.D. Nelson and K.E. Reppert: 'The production of 12Cr Rotor Forgings: A forgemasters perspective', *Proc. Int. Conf. on Improved Coal-Fired Power Plants*, Palo Alto, Not. 1986.

16. T. Takenouchi, Y. Ikeda and Y. Tanaka: 'Production of 12Cr Rotor Forgings for Steam Turbines using advanced VCD Process', *Iron and Steel Soc. Publ. on Recent Development in Rotor Forgings Steels*, 1990.

17. F. Ito, K. Kuwabara, M. Miyazaki, Y. Fukui and Y. Takeda: 'Improved 12%Cr Rotor Forgings for Advanced Steam Turbines', *Proc. of Int. Conf. Improved Coal-Fired Power Plants*, Palo Alto, Not. 1986.

18. G.E. Hale and J. Nutting: 'Overheating of low alloy steels', *Int. Metal Review*, 1984, **29** (4).

19. M.A. Walsh, S. Price and G.A. Honeyman: 'The manufacture of the large high temperature rotor forgings for Skaerbaek Supercritical Power Station', *Third International Charles Parsons Turbine Conference*, Poster 29, Newcastle upon Tyne, UK, 25–27 April 1995.

20. D.V. Thornton and R.W. Vanstone: 'Materials development for application in steam turbines for fossil fired plant, *Materials Engineering in Turbines and Compressors*, Newcastle upon Tyne, April 1995.

21. D.V. Thornton and M. Taylor: 'Experience in the manufacture of steam turbine components in advanced 9–12% chromium steels', *Advanced Steam Plant Conference*, ImechE, London, May 1997.

Alloy Design for Creep Resistant Martensitic 9–12% Chromium Steels

A. GÖCMEN, P.J. UGGOWITZER, C. SOLENTHALER,
M.O. SPEIDEL AND P. ERNST*

Institute of Metallurgy, Swiss Federal Institute of Technology ETH, Sonneggstr. 3, 8092 Zurich, Switzerland
* ABB Corporate Research Ltd., 5405 Baden – Dättwil, Switzerland

ABSTRACT

Martensitic 9–12% chromium steels are heat treated by normalising, quenching and tempering at an intermediate temperature. The resulting microstructure is a particle stabilised subgrain structure in a body centered cubic iron matrix. Strength, toughness and microstructural stability of tempered martensite depend on the grain coarsening resistance at high normalising temperature, and on the microstructural degradation during tempering. The dominant feature of microstructural degradation is the development of nonuniform precipitation states in tempered martensite. Improved combinations of strength and toughness can be achieved by proper alloying with nitrogen and vanadium. This is due to the wide solubility gap of vanadium nitrides in the temperature range between 600 and 1200°C. Uniform precipitation states of primary vanadium nitrides can be produced during forging, which can be used for effective grain growth inhibition during normalising. In the same way a uniform precipitation state of vanadium nitrides can be produced during an isothermal ausageing treatment, which can be used to decelerate the recovery processes during tempering of the martensite. Optimised alloy compositions make it possible to achieve outstanding properties in terms of yield strength (1140 MPa) and impact energy (55J) at room temperature as well as a high yield strength at 550°C (700 MPa). Ausageing enables to achieve these properties in thick section components and provides for a decelerated degradation of tempered microstructures.

1. INTRODUCTION

Martensitic 9–12% chromium steels are favoured by low costs as well as by beneficial physical and mechanical properties for various high temperature power plant plant applications. Due to the increased demands for strength, toughness, creep resistance and creep rupture strength, control of microstructural development is of fundamental interest for alloy development.

While primary properties, such as strength and toughness, depend on the microstructural development during the heat treatment, long term creep rupture strength additionally depends on the effect of microstructural degradation on creep acceleration and creep rupture. Modern alloy development aims to improve the high temperature properties by means of an intensified age hardening and increased microstructural stability of the ferrite. High temperature strength, toughness, creep resistance and creep rupture strength can be improved by alloying with nitrogen.[1-7] The beneficial effect of nitrogen on creep resistance has been mainly related to the dissolution and precipitation behaviour of vanadium nitrides during normalising, tempering and creep.

The aim of this paper is to critically review the role of vanadium nitrides in martensitic 9–12% chromium steels and to demonstrate the resulting alloy design opportunities by proper use of nitrogen and heat treatment for specific power plant applications.

2. PHYSICAL METALLURGY OF MARTENSITIC 9–12% CHROMIUM STEELS

Martensitic 9–12% chromium steels are commonly heat treated by normalising, quenching and a subsequent tempering treatment at an intermediate temperature. The main source of strength is offered by the martensitic phase transformation, which produces a strongly supersaturated microstructure with a dense distribution of dislocations and interfaces.[8] Beneficial combinations of strength and toughness are achieved by age hardening reactions coupled with dislocation recovery processes during tempering at temperatures beyond 600°C.

The microstructure in the heat treated state consists of martensite crystals (blocks) within former austenite grains, each one being split up into elongated subgrains (laths). The precipitation state in the high tempered condition is

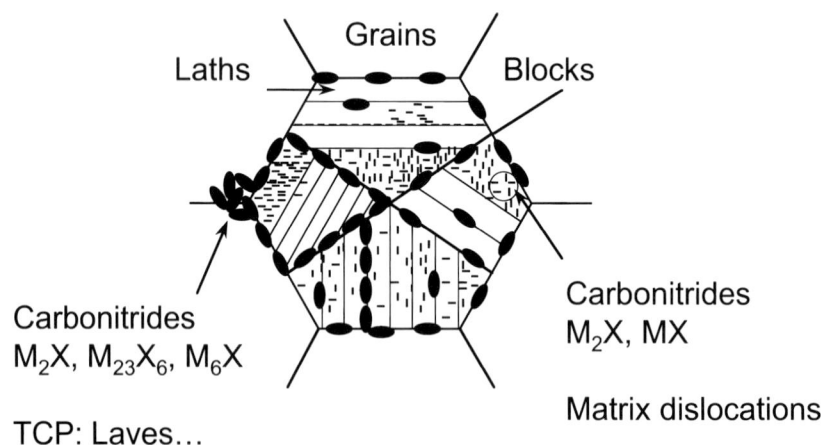

Fig. 1 Schematic representation of nonuniform precipitation states in tempered martensitic 9–12% chromium steels.

nonuniform on the scale of the interface substructure. Coarse carbonitrides are nonuniformly distributed to different interfaces and fine dispersed precipitation states are nonuniformly distributed to different subgrains (Fig. 1). This dominant feature of microstructural degradation depends on alloy composition, tempering time and temperature.

The possible range of chemical composition to improve the mechanical properties is restricted by the requirement of through- and strong hardening, offered by the martensitic phase transformation. Besides, alloy compositions must satisfy the requirement for sufficient grain coarsening resistance during normalising and a sufficiently high tempering resistance. It is well known that tempering resistance is strongly governed by ferrite forming elements such as vanadium, niobium or tungsten and molybdenum, be it either by precipitation or solid solution hardening.[9] Thus, their ferrite stabilising effect during normalising has to be compensated by austenite stabilising elements such as, nickel, manganese, cobalt, carbon or nitrogen. While some combinations, as for example alloying with vanadium, niobium, nitrogen and carbon, can be used for a twofold purpose – grain coarsening resistance and tempering resistance due to the effect of second phase particles – some other combinations, such as alloying with tungsten and cobalt in low amounts are mainly used to increase the tempering resistance by means of solid solution hardening. The fundamental question has to be addressed to the rationality of alloying, giving a maximised combination of grain coarsening resistance, martensite hardening, tempering resistance and microstructural stability during creep.

Since Zener (1948)[10] there exists a simple formula, which relates a limiting grain size D_z of the microstructure with the volume fraction f_v and radius r of second phase particles in a high temperature heat treatment. It is given by:

$$D_z \propto \frac{4r}{3f_v}$$

This relationship says, that high grain coarsening resistance is provided by a high volume fraction of small particles. Obviously, grain coarsening resistance cannot be modelled with the solubility products of carbonitrides alone, for the size of the particles will depend on the precipitation and coarsening behaviour of the particles during solidification and forging. However, a high grain coarsening resistance can be expected for an alloy composition, which exhibits a wide solubility gap of carbonitrides in the temperature range of forging and normalising. As a consequence a uniform distribution of fine particles can be produced by precipitation reactions of carbonitrides during the forging process. Two points are noteworthy. It can be expected that the tempering resistance of an alloy will increase with increasing grain coarsening resistance at high normalising temperatures, since there will be a wide solubility gap of carbonitrides with a high resistance against particle coarsening

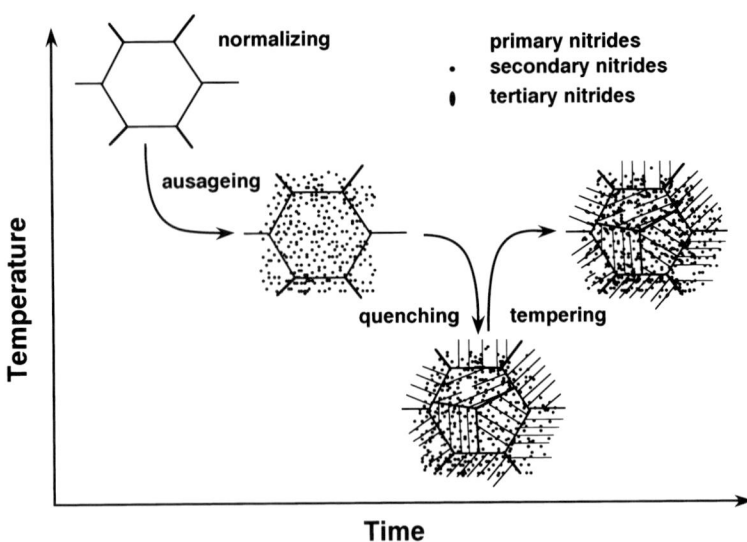

Fig. 2   Schematic representation of the development of the microstructure involving an ausageing treatment prior to the martensitic transformation.

during normalising. Besides, it has to be taken into account that the martensite hardening will then be influenced by the tendency of carbonitrides to precipitate in the austenite prior to the martensitic transformation. Thus, there will be an opportunity to influence the microstructural development by means of an ausageing treatment prior to the martensitic transformation.[12]

Figure 2 schematically represents the way in which the development of the microstructure during the heat treatment can be controlled by the choice of an alloy composition, offering a high grain coarsening resistance during normalising. The grain coarsening resistance will be controlled by the volume fraction and composition of primary nitrides or carbonitrides. The remaining amount of nitrogen and carbon can be either reprecipitated during cooling in the stable, or metastable austenite phase (secondary nitrides), or during tempering (tertiary nitrides). Morphology, dislocation density and hardness of the martensite crystals can be controlled by the volume fraction and size of uniformly distributed primary and secondary nitrides and by the activities of the dissolved carbon and nitrogen atoms. The uniformity of chemical phase evolution processes and the softening of the dislocation substructure during tempering and creep can be controlled by the volume fraction and size of secondary nitrides and by the age hardening characteristics of tertiary nitrides.

Assuming that the nonuniform development of precipitation states in the ferrite is governed by the interface substructure of the martensite or by the complex chemical phase evolution processes of the ferrite, answers to the following questions can be expected by the application of systematic variation of alloy composition as well as of the ausageing conditions temperature and time.

1) How does the development of strength and toughness depend on the effect of alloy composition by the development of (non)-uniform precipitation states during tempering?

2) How does the development of creep resistance and creep rupture strength depend on alloy composition and heat treatment by the development of (non)-uniform phase evolution processes during creep?

3) How does the development of mechanical properties in thick section components depend on alloy composition as a result of microstructural instabilities occuring during the slow cooling treatment?

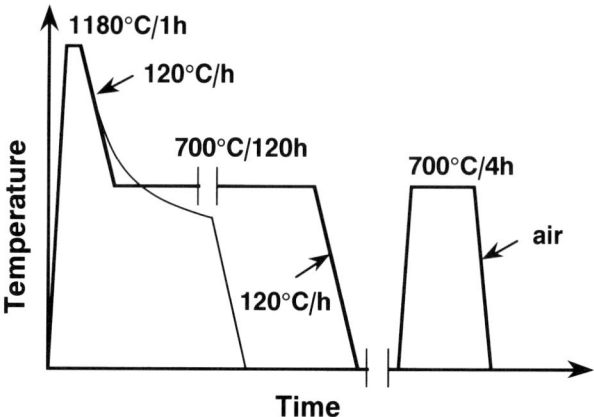

Fig. 3   Schematic representation of a heat treatment involving an ausageing treatment.

3. EXPERIMENTAL PROCEDURE

Heats of 10 kg were melted in a vacuum induction furnace at temperatures between 1500 and 1600°C with an applied nitrogen pressure of 0.9 bar during melting and solidification. A nitrogen content as much as 0.15wt.% can be dissolved without overpressure in a iron melt containing 9–12% chromium, if it additionally contains a near stochiometric amount of 0.7wt.% vanadium to nitrogen.[11] Base compositions of this kind exhibit a large solubility gap of vanadium nitrides in the temperature range of metastable or stable austenite between 600 and 1230°C. About 2–5wt.% Co or 1–3wt. % Mn are necessary to compensate the ferrite stabilising effect of vanadium at 1200°C. However,

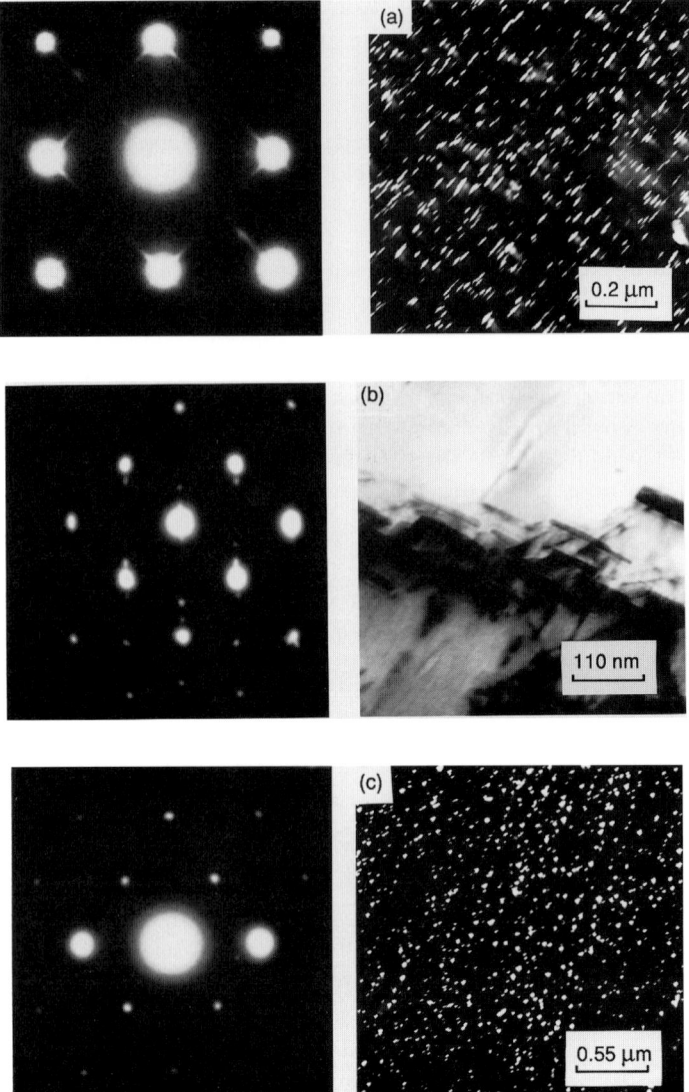

Fig. 4 Selected are diffraction pattern of f.c.c. – nitrides exhibiting Bain orientation (4a, b) and apart of Bain Orientation (4c).[12]

such compositions allow the application of flexible isothermal ausageing experiments in the metastable austenite at temperatures between 600 and 700°C, which are described by the author in.[11,12] Figure 3 schematically represents a typical heat treatment, which involves an ausageing treatment prior to the martensitic transformation.

4. EXPERIMENTAL RESULTS

4.1   The Precipitation Behaviour of Vanadium Nitrides

Vanadium nitrides with face centered cubic structure (f.c.c) commonly develop as thin plates or discs on the {100} planes of the ferrite with the Bain orientation relationship.[13]

$$^{(001)}b.c.c. \parallel {}^{(001)}f.c.c \; : {}^{[100]}b.c.c. \parallel {}^{[110]}f.c.c.$$

As a result, the age hardening behaviour of vanadium nitrides can be detected with selected area diffraction by a streaked diffraction pattern in the <001> zone of the ferrite (Fig. 4a).[11,12] The strong intensity of the streaks can be qualitatively related to hardening due to a small interparticle distance and a high volume fraction of the nitrides. Such diffraction patterns are not reported for conventional grades of 9–12% chromium steels. It must be concluded, that the local volume fractions of these nitrides are too small to be detectable or that they lose the Bain orientation relationship during overaging. Extensive metallographical studies by the author suggest, that the loss of the Bain orientation relationship results in the development of nonuniform precipitation states (Figs 5a and b) and in an intensified softening of the microstructure. This process depends on alloy composition. For one niobium free alloy it could be demonstrated that the Bain orientation was maintained during overageing of the vanadium nitrides (Fig. 4b).[11,12]

The development of nonuniform precipitation states can also be considered as an important feature of microstructural instability, because it is coupled with a more or less strong drop of strength and probably with a loss of toughness too. It can be assumed, that the driving force for the development of nonuniform precipitation states is given by the strong competitive particle

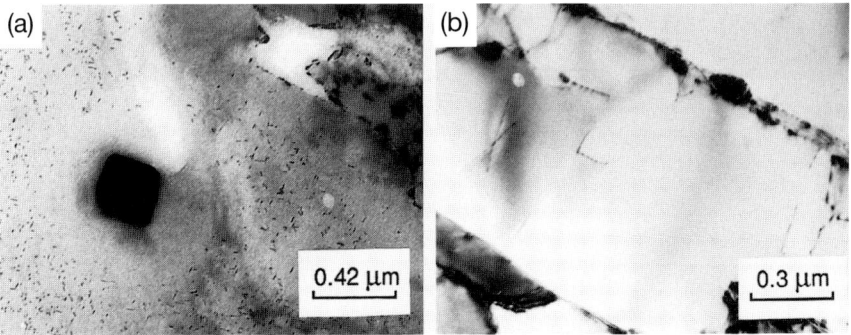

(a)   0.42 µm

(b)   0.3 µm

Fig. 5 Transformation from uniform to nonuniform precipitation states of nitrides with f.c.c. – structure during tempering from 4 to 28 h at 708°C.[12]

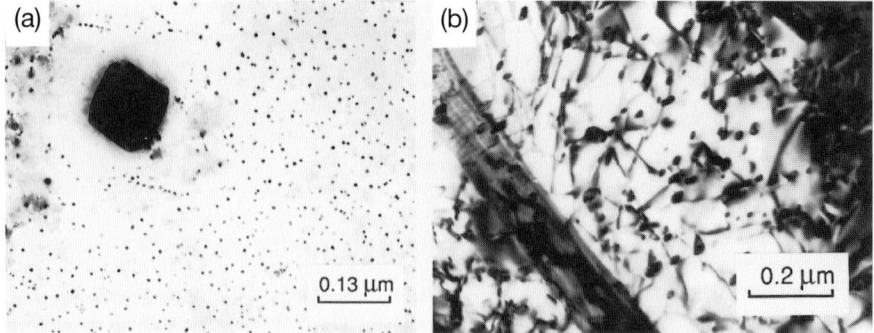

Fig. 6   Development of ausaged precipitation states of nitrides with f.c.c. – structure during tempering from 4 to 28h at 708°C.[12]

growth between interface precipitates and matrix precipitates after nucleation.[11,12]

Ausageing experiments can be used to determine the effect of particle size on the development of nonuniform precipitation states during tempering the martensite. It could be demonstrated that the uniformity of the precipitation state of vanadium nitrides maintains during tempering the martensite, if the precipitates are slightly over-aged in the austenite prior to the martensitic transformation (Figs 6a and b), even though, no well defined orientation

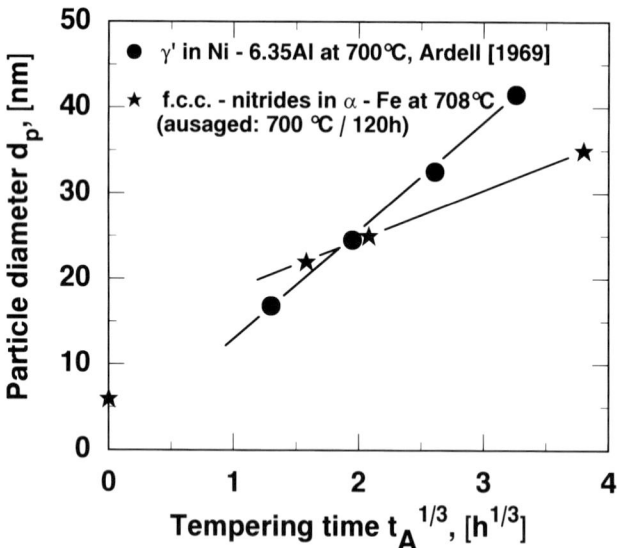

Fig. 7   Coarsening of nitrides in the normalized and ausaged condition during tempering the ferrite at 708°C in comparison with γ' – phases in a Ni – base as measured by Ardell.[14]

relationship of these precipitates to the ferrite could be identified (Fig. 4c). Thus, ausageing can be considered as an effective design tool for producing uniform precipitation states of particles with a remarkable coarsening resistance (Fig. 7) in the ferrite.

4.2 Case Study: The Effect of an Ausageing Treatment on the Mechanical Properties of Alloy AP1

The effect of an ausageing treatment on the development of the mechanical properties have been examined for an iron alloy containing in wt.% 12Cr, 0.5Ni, 2Mn, 10Co, 1.5Mo, 0.7V, 0.06Nb, 0.04Ta, 0.04Ti, 0.15N and 0.03C. 10wt.% cobalt has been added to achieve a near ferrite free solidification.

Table 1 Applied heat treatments of the examined alloy AP1

Heat treatment	Normalizing	Ausageing	Cooling	Tempering
T2	1180°C/1 h	–	>500°C/1 h	700°C/4h
T5	1180°C/1 h	700°C/120 h	120°C/h during both cooling periods	700°C/4h
T6	1180°C/1 h	–	>500°C	650°C/4h

However, only about 2–4wt.% Co would be necessary to prevent formation of ferrite during normalising at 1200°C.

The heat treatments, that have been applied for this alloy are given in Table 1. The heat treatment T2 and T6 differ from heat treatment T5 by the absence of an ausageing treatment before the martensitic transformation and by the high cooling rates. Heat treatment T5 is schematically represented in Fig. 3. The limiting grain size during normalising at 1180°C was about 35 μm. The development of the hardness of the microstructure for these heat treatments are given in Table 2.

Figure 8 represents the effect of heat treatment on yield strength and impact energy at room temperature. There is no drop of impact energy at room temperature as a result of an applied ausageing treatment prior to the martensitic transformation. However, ausageing can be used to increase the strength of tempered martensite. Thus, it can be expected, that the combination of strength and toughness in tempered martensite can be improved by proper ausageing.

Table 2 Development of hardness during the heat treatment

Heat Treatment	as quenched	as tempered
T2	450HV10	350
T5	390HV10	380
T6	450HV10	390

Figure 9 represents the effect of heat treatment on the temperature dependence of yield strength. It follows, that an extraordinary high yield strength up to 500°C can be achieved by means of a uniform precipitation hardening of nitrides in tempered martensite. However, the yield strength tends to drop to the level of conventional grades beyond a temperature of 600°C. It must be concluded, that the contribution of recovered, stable dislocation substructures to yield strength increases with increasing testing temperature.

5. SUMMARY AND DISCUSSION OF ALLOYING CONCEPTS IN MARTENSITIC 9–12% CHROMIUM STEELS

The role of vanadium nitrides and the alloy design capabilities of high nitrogen containing steels are critically reviewed with respect to the controllability of microstructural development during the heat treatment. It was shown that combinations of structural properties, such as grain coarsening resistance, martensite hardening, and tempering resistance can be optimised in a rational manner by controlled dissolution and repricipitation of vanadium nitrides during forging, normalising, cooling, and tempering. Optimised alloy compositions reflect a high grain coarsening resistance during normalising and a strong age hardening capability of metastable austenite.

Remaining fundamental questions concern the ideal development of the microstructure during tempering beyond 700°C with regard to high temperature applications beyond 550°C. It was recently suggested,[5,6] that the beneficial effect of vanadium nitrides arises from the wide solubility gap of these precipitates in the temperature range of high temperature tempering. This would allow well recovered, stable subgrain structures to be produced during tempering and the microstructure can be dynamically age hardened by vanadium nitrides during service operation exposure. However, care has to be taken on the effect of competitive phase reactions on the stability and uniformity of precipitation states of vanadium nitrides.

The alloy development approach discussed in this paper aims to increase the stability of the subgrain structure by uniform precipitation states of vanadium nitrides. It requires that uniform nucleation and coarsening of vanadium nitrides are not degraded by the precipitation of thermally unstable phases, such as $M_{23}X_6$, M_2X or intermetallic compounds. Besides, observations on the coarsening behaviour of f.c.c. – nitrides suggest, that the uniformity of nitrides will depend on the ability to maintain the Bain orientation during over-ageing. Thus, the role of further microalloying elements, such as Nb, Ti, Ta, Zr and Hf have to be evaluated not only with regard to the developing phase equilibrium, but also with regard to the coarsening behaviour of the developing equilibrium phases.

Ausageing is an effective tool to produce uniform precipitation states, and is a suitable instrument to determine the susceptibility of an alloy compostion

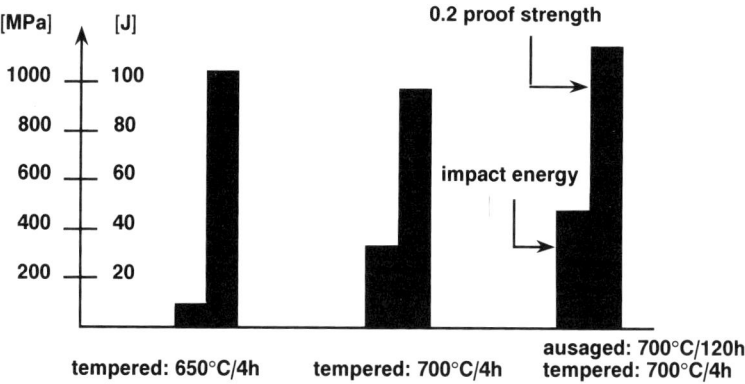

Fig. 8 Effect of an ausageing treatment on yield strength and toughness at room temperature for the experimental alloy AP1.

to develop nonuniform precipitation states during tempering or creep. The transformation kinetics from uniform to nonuniform precipitation states, which will depend on the particle size at the beginning of ageing, can now be easily measured. In addition to that, ausageing can be considered as a challenging task to control the development of local properties in thick sectioned components, such as rotors and rotor discs. The methodology allows a direct influence of local structures and properties by the application of existing simulation techniques of cooling processes.

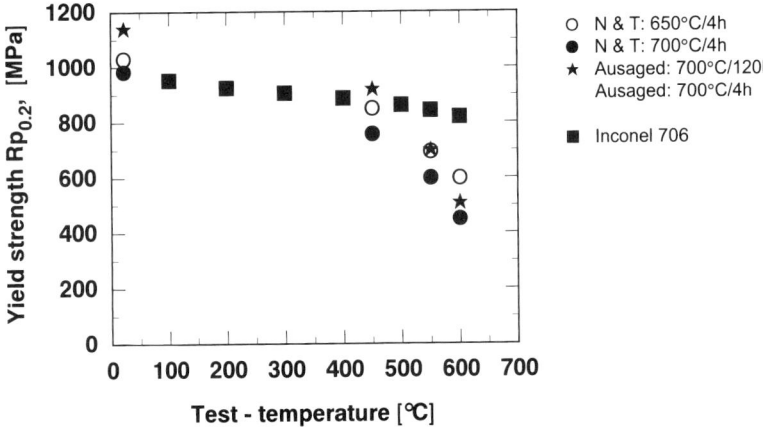

Fig. 9 Effect of an ausageing treatment on the temperature dependence of yield strength of the experimental alloy AP1 in comparison with the Ni – base alloy Inconel 706.

ACKNOWLEDGEMENTS
The financial assistance from ABB Corporate Research Ltd. and the Swiss Energy Research Fund (NEFF) is acknowledged. The authors further thank Erwin Schönfeld and Walter Brehm for technical assistance.

REFERENCES
1. B.R. Anthamatten, P.J. Uggowitzer, Ch. Sohlenthaler and M.O. Speidel: *Proc. 2nd Int. Conf. High Nitrogen Steels*, G. Stein and H. Witulski eds, Verlag Stahleisen, Düsseldorf, 1990, 436.
2. H. Berns and F. Krafft: *1st Int. Conf. High Nitrogen Steels*, J. Foct and A. Hendry eds, The Institute of Metals, London, 1989, 168.
3. P. Ernst, Y. Hasegawa, N. Tokomitsu, Y. Kawauchi and F. Masuyama: *2nd Int. Conf. High Nitrogen Steels*, G. Stein and H. Witulski eds, Verlag Stahleisen, Düsseldorf, 1990, 132.
4. R. Steins: 'Stickstofflegierte, Dispersionsgehärtete 9–12% Chromstähle für Anwendungen in Modernen Gas- und Dampfturbinen mit hohen Wirkungsgraden', *Dissertation*, ETH Zürich, 1997.
5. V. Foldyna, A. Jakobová, V. Vodárek and Z. Kubov: *Materials for Advanced Power Engineering*, D. Coutsouradis *et al.* eds, Kluwer Academic Publishers, Dordrecht, 1994, 453.
6. V. Foldyna and Z. Kubon: *Performance of Bolting Materials in High Temperature Plant Applications*, A. Strang ed, The Institute of Materials, London, 1995, 175.
7. J. Orr and D. Burton: *Materials for Advanced Power Engineering*, D. Coutsouradis *et al.* eds, Kluwer Academic Publishers, Dordrecht, 1994, 263.
8. J.W. Christian: *Strengthening Methods in Crystals*, A. Kelly and R.B. Nicholson eds, Applied Science Publishers, London, 1971.
9. K.J. Irvine and F.B. Pickering: *High Strength 12% Chromium Steels*, The Iron and Steel Institute, London, 1961.
10. C. Zener: (private communication to C.S. Smith), *Trans. Met. Soc. A.I.M.E.*, 1949, **175**, 15.
11. A. Göcmen: Grundrisse der Gefügeausbildung und der Zeitstandeigenschaften Martensitischer 9–12% Chromstähle, Diss. Nr. 12020, ETH Zürich, 1997.
12. A. Göcmen, R. Steins, C. Solenthaler, P.J. Uggowitzer and M.O. Speidel: *ISIJ Int.*, 1996, **36**, 768.
13. D.H. Jack: *Acta Metall.*, 1976, **24**, 137.
14. A.J. Ardell: *Mechanismen of Phase Transformation in Crystalline Solids*, Institute of Metals, London, 1969, 111.

The Effect of Welding on the Microstructural Development of Advanced 9–12 Cr Steels

H. CERJAK AND E. LETOFSKY

Department of Materials Science and Welding Technology, Technical University of Graz, Austria

ABSTRACT

There are strong environmental and economic pressures to increase the thermal efficiency of fossil fuel fired power stations, and this has led to a steady increase in steam temperatures and pressures resulting in world wide plans for ultra-supercritical power plants. Basic investigations on the weldability of advanced 9–12% Cr steels which are either currently in use or which are intended to fulfil this requirement were performed on pipes of P91, E911, and a tungsten containing cast steel G–X 12 CrMoWVNbN 10 1 1. Gleeble simulations representing the manual metal arc welding process were applied to produce HAZ simulated microstructures. After different post-weld heat treatments they were tested using hardness tests, metallographic investigations, constant strain rate tests, and creep tests. Particular attention was given to the softening effect in the HAZ and its influence on the creep resistance of the welded material. This decrease shown by simulated and manufacturing welded samples, seems to be less pronounced of the tungsten modified versions than observed at P91 material.

INTRODUCTION

There is a substantial and growing interest in operating thermal power plants at relatively high temperatures and/or pressures for improving thermal efficiency and reducing CO_2 emissions. Materials with ferritic/martensitic microstructures are preferred, because of their favourable physical properties, such as good thermal conductivity and low coefficient of thermal expansion, coupled with higher resistance to thermal shock.[1,2] These are some of their advantages over austenitic stainless steels. For these reasons, there has been a growing demand for high strength, high chromium ferritic steel, which has resulted in the development and application of several kinds of 9 to 12% chromium steels. The development of these materials as a function of the 100 000 h creep rupture strength at 600°C can be seen from Fig. 1.[3] The main steps in the development are different in various countries, such as in the USA, the European nations and Japan. In Europe, particularly in Germany the gap between the application range of the ferritic steel P22, German designation 10 CrMo 9 10, and the

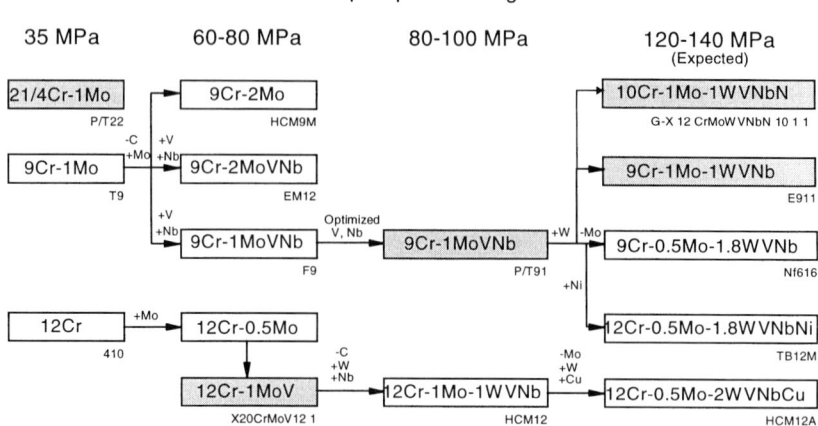

Fig. 1   Development progress of 9–12%Cr creep resistant steels.[3]

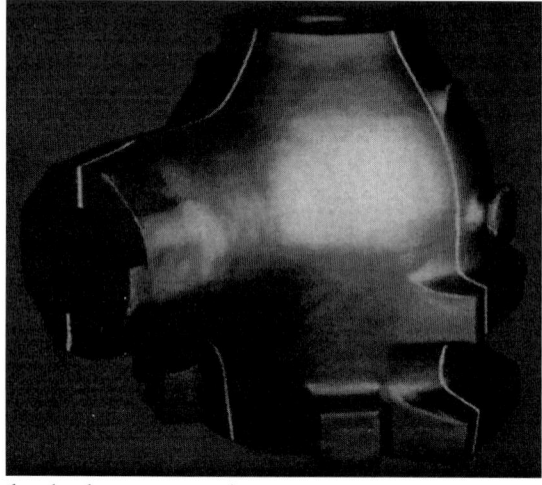

Fig. 2   Pilot valve body, German designation G–X 12 CrMoWVNbN 10 1 1, prototype casting, Georg Fischer Formtech AG.

austenitic stainless steels regarding their creep resistance was successfully bridged in the past by using the 12% chromium steel X20 CrMoV 12 1.[4] Since 1975 a new modified 9% chromium steel has been developed in the US under the leadership of ORNL and standardised i.e. as P91 in ASTM A335, (German designation X10 CrMoVNb 9 1) in the early 1980s. These developments led to the application of tungsten containing 9–12% chromium steels, which showed higher creep resistance, compared to the type P91.

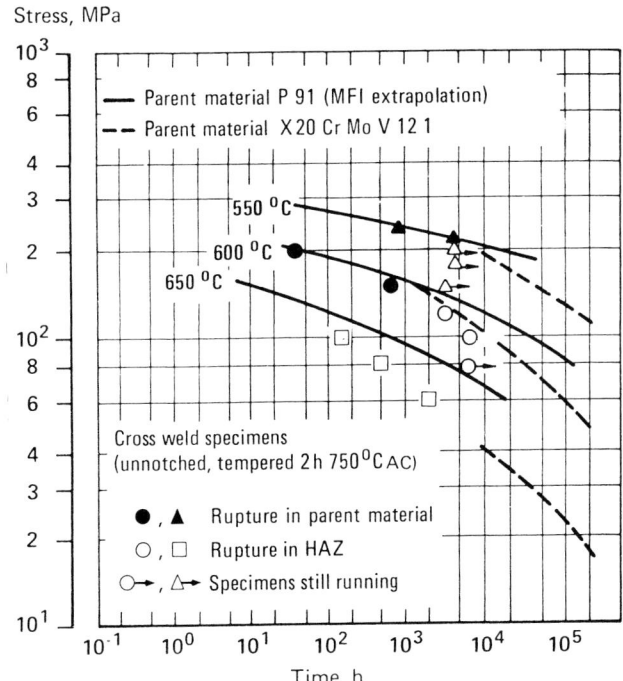

Fig. 3   Creep rupture strength of P91 as well as X20 CrMoV 12 1 of base material and crossweld specimens.[1]

Within the framework of the European COST 501 programme (Development of Materials for Advanced Steam Cycles)[5] a cast version of a tungsten modified 10%Cr-steel, see Fig. 2 and a 9%Cr version for pipes and forgings, called E911, was designed to fulfil the increased demands on the creep strength for advanced design values for fossil fired power plants.

For a successful service application and acceptance in practice, the weldability and the long time behaviour of the newly developed materials are two of the most important aspects. In Fig. 3 the creep rupture strengths of the parent materials P91 and X20 CrMoV 12 1 are compared to the creep rupture strengths of crossweld specimens.[1] It can be observed that the high temperature creep rupture strength of crossweld samples is significantly lower than that of the base material at the same temperature.

EXPERIMENTAL PROCEDURE

Materials Investigated
As for the material P91, the investigations for the study were performed on a seamless pipe produced by the pilgermill rolling process, dimension 149 mm o.d.

Table 1 Chemical composition in wt-% of materials investigated

Material	C	Si	Mn	P	S	Cr	Ni	Mo	W	V	Nb	N
X 10 CrMoVNb 9 1 (P91)	0.099	0.385	0.40	0.017	0.004	8.75	0.128	0.96	0.03	0.204	0.07	0.058
E 911	0.11	0.18	0.40	0.015	0.003	8.61	0.21	0.92	0.99	0.19	0.089	0.065
G–X 12 CrMoWVNbN 10 1 1	0.12	0.29	0.62	0.027	0.003	10.51	0.93	0.99	0.99	0.22	0.08	0.048
Cromocord 10M	0.09	0.24	1.00	0.012	0.007	9.60	0.90	1.05	1.03	0.20	0.06	0.050

and 20 mm wall thickness. The newly developed COST steel E911, heat no. 3 comes from Mannesmannröhren-Werke, Germany. The tests were realised on a seamless pipe with 336 mm o.d. and 62 mm wall thickness. For the casting material G–X 12 CrMoWVNbN 10 1 1, the material investigated comes from the valve body trail casting produced by Georg Fischer Schaffhausen, Switzerland. The manufacturing weld was carried out by Voest Alpine Stahl Linz Foundry, Austria, in the framework of COST 501 Round II. As for the welding rod, Cromocord 10M produced by Oerlikon Welding Ltd., Switzerland, was used. Table 1 shows the chemical composition of the materials investigated.

Heat Affected Zone Simulation
The difference microstructures appearing in the heat affected zone (HAZ) of fusion welds show different properties, depending on their thermal history.[6]

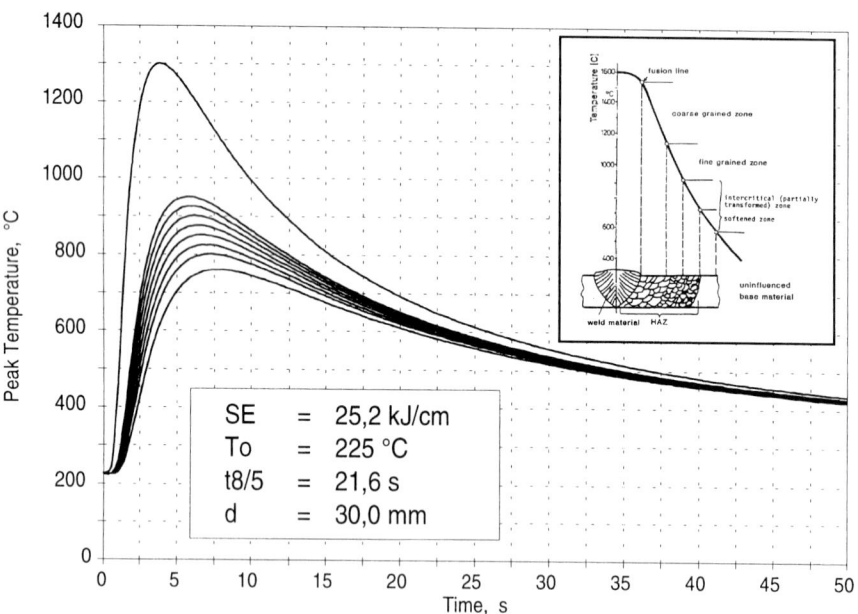

Fig. 4 Thermal cycles calculated for various peak temperatures between 1300°C and 760°C. SE, heat input; To, preheat temperature; $t_{8/5}$, cooling time between 800°C and 500°C; d, wall thickness.[8]

The simulation of weld thermal cycles is a powerful method for investigation of HAZ microstructures. Compared to the HAZ microstructure of real joints, simulated microstructures can be produced in relatively large volumes. This leads to a transparent variation of welding parameters and helps to reduce scatter in testing results which is caused by inhomogeneities in real welded microstructures. Effects of local gradients in microstructure, properties and residual stresses are not taken into account when using the HAZ simulation technique.

For a simulation a Gleeble 1500 machine was used. The weld thermal cycles which are needed as input for the simulation process were calculated from Rosenthal's solution of Fourier's heat conduction equation in a simplified version derived by Rykalin.[7] These equations allow the representation of the weld thermal cycles by the cooling times between two temperatures (respectively the cooling time between 800°C and 500°C – $t_{8/5}$), the peak temperature (T_p), the preheat temperature, and the plate thickness. For this investigation the thermal cycles applied were selected to represent a manual shielded metal arc welding process with heat input of 25.2 kJcm^{-1} and a preheat temperature of 225°C. These conditions led to a corresponding cooling time between 800°C and 500°C of 21.6 s, see Fig. 4.[8]

Investigations of HAZ Simulated and Welded Materials
Using real welds of P91, E911, and G–X 12 CrMoWVNbN 10 1 1, metallographic investigations, hardness tests, and creep rupture tests were performed. The results of P91 investigation were presented in Ref. 1, the results of cast steel investigation were summarised in Ref. 5. Concerning Gleeble HAZ simulated samples from P91, E911, and G–X 12 CrMoWVNbN 10 1 1 special attention was given to the evaluation of the soft zones in the HAZ by metallographic investigations, using light microscopy and TEM, hardness tests, and constant strain rate tests. Additionally, creep rupture tests were performed to investigate the creep strength behaviour of the soft zone at the HAZ region.

RESULTS AND DISCUSSION

Hardness
In hardness tests across a weld seam of material P91, it was noted that this type of steel shows a tendency to form a soft zone in the fine grain HAZ after post-weld heat treatment (PWHT) (Fig. 5). The hardness in the zones was found to be ≈20 HV(10 kg) lower than that of the unaffected base material. The hardness profiles across weld seams of E911 and G–X 12 CrMoWVNbN 10 1 1 show a similar curve to that of P91 (Fig. 5). To define the zone in which maximum softening occurs, specimens were subjected to HAZ simulation at

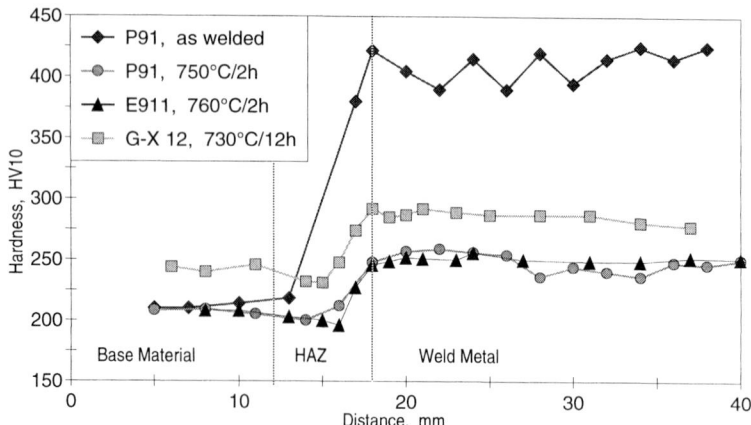

Fig. 5 Hardness of weld seam in as welded condition and after various post-weld heat treatments of P91, E911, and G–X 12 CrMoWVNbN 10 1 1.[1,5]

different peak temperatures in the range between 760 and 950°C according to Fig. 4. The hardness was tested in the as welded, simulated, and in the simulated and tempered condition. The tempering conditions after welding were 760°C/2 h air cooled for P91, 760°C/2 h for E911, and 730°C/12 h for G–X 12 CrMoWVNbN 10 1 1.

The results of the hardness tests performed on the weld simulated microstructures of the three materials are presented in Fig. 6 as a function of the peak temperature for both the tempered and as welded conditions. The hardness of the base materials tested is shown on the left side of Fig. 6. It is not influenced by thermal cycles with peak temperatures up to about 850°C. The beginning α/γ-transformation as a function of heat cycles can be observed by the increase of hardness in the as welded condition. After stress relieving, the hardening effect disappears in the materials as can be observed from Fig. 6. In the range of peak temperatures between 850 and 950°C in the materials, a minimum hardness can be observed. The hardness here is about 10 HV(10 kg) lower than the remaining values.

Constant Strain Rate Tests on HAZ Simulated Materials
Constant strain rate tests at 600°C using an constant strain rate of $\dot{\varepsilon} = 10^{-5}$ s^{-1} were applied on specimens which have been subjected to HAZ simulation and subsequent tempering to investigate the principal influence of the HAZ softening effect on the creep resistance. This test method was first applied on this type of steel in Ref. 9. The maximum stress gives on indication of the creep resistance of the microstructures tested. The results of these tests, performed at 600°C on microstructures produced by weld simulation with dif-

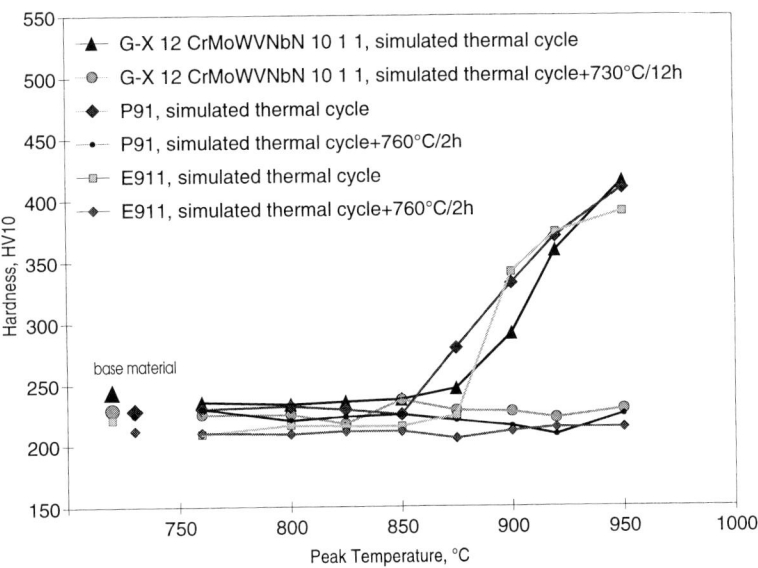

Fig. 6   Results of hardness measurements on specimens of P91, E911, and G–X 12 CrMoWVNbN 10 1 1 subjected to weld thermal cycle simulation treatments (softening behaviour of HAZ).

ferent peak temperatures, are shown in Fig. 7. For the unaffected base material P91 a maximum stress of 280 MNm^{-2}, for E911 a stress of 290 MNm^{-2}, and for G–X 12 CrMoWVNbN 10 1 1 a stress of 260 MNm^{-2} was measured. For peak temperatures in the range of 850–950°C, there is a significant decrease in the maximum stress, analogous with the decrease in hardness. The minimum in the stress versus peak temperature curve is for P91 at 250 MNm^{-2} and 920°C, for E911 at 275 MNm^{-2} and 925°C, and for G–X 12 CrMoWVNbN 10 1 1 at 240 MNm^{-2} and 875°C. The measured minimum of stress is about 10% lower than that measured for the unaffected base material.

Microstructural Aspects
As described in Ref. 10 the microstructure of the normalised and tempered P91 steel consists of tempered martensite with a large amount of $M_{23}C_6$ and MX precipitates. Microscopic examination of the specimens subjected to HAZ simulation thermal cycles revealed no significant change in the microstructures when the peak temperature did not exceed 850°C. When the peak temperature becomes higher than the AC_1 temperature, depending on the shape of the thermal cycle, more and more austenite will be formed and transformed into martensite during cooling to room or preheating temperature. When heating up to 920°C, the amount of martensite which was found

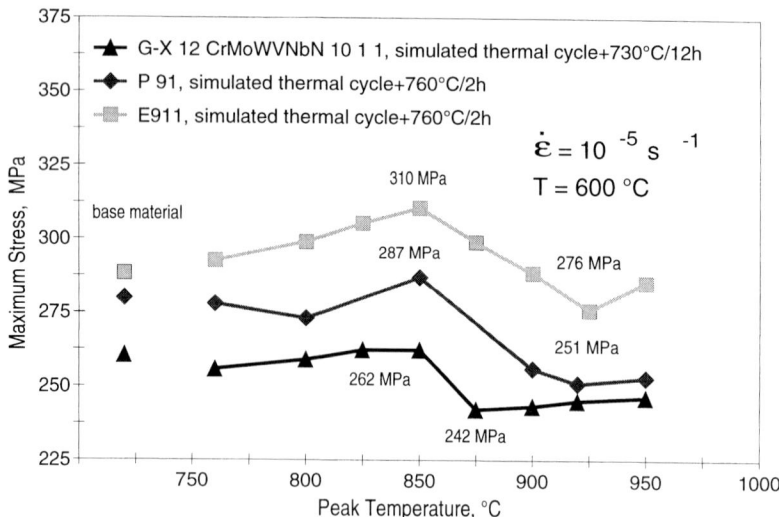

Fig. 7   Results of constant strain rate tests on specimens of P91 and G–X 12 CrMoWVNbN 10 1 1 subjected to weld thermal cycle simulation treatment followed by tempering (HAZ-softening).

under given conditions of our tests is about 70–90%. At this temperature, slightly above AC_1, and the short times caused by the weld thermal cycle, no significant amount of precipitates and hence carbon and nitrogen go into solution. Therefore, the martensite formed under these conditions lacks carbon. This is confirmed by the increase in hardness caused by higher peak temperatures (see Fig. 6). Since very small amounts of carbon go into solution at these peak temperatures, the reprecipitation of $M_{23}C_6$ carbides and MX carbonitrides during the subsequent tempering treatment is very limited. In addition, the coherent MX particles coarsen and coagulate and lose their strengthening effect to a large extent. These effects also create a higher tendency toward recrystallisation in this area which, in connection with the overaging of the precipitates, results in a significant softening of these fine grained zones of the HAZ which were heated up to 900–950°C during the weld thermal cycle after PWHT.

In order to extract the information of the distribution and size of particles and the general tendency of martensite lath wide and its development during different weld thermal cycles from the TEM microscopy a manual evaluation method was undertaken.[11] The evolution of both the particles and the martensite lath subgrains in the microstructure of the investigated conditions are separately shown in Figs 8a–8e. Figure 8a shows the TEM microstructure of the normalised and tempered G–X 12 CrMoWVNbN 10 1 1 cast steel consists of tempered martensite with a large amount of $M_{23}C_6$ and MX precipi-

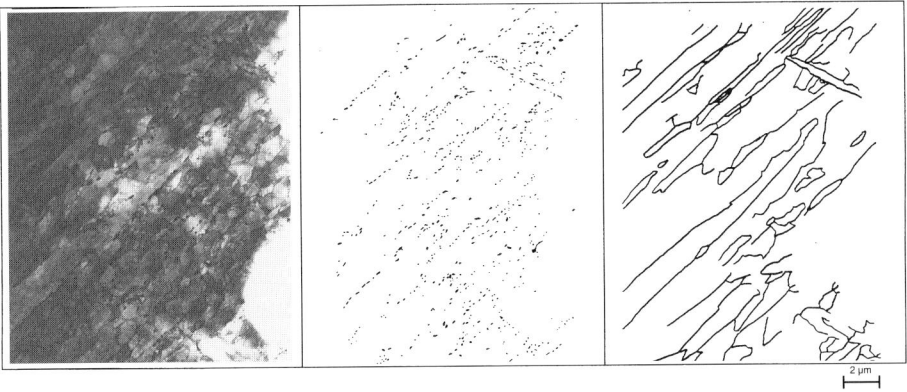

Fig. 8a   TEM microscopy, particles and subgrains of G–X 12 CrMoWVNbN 10 1 1 (as received condition – base material).

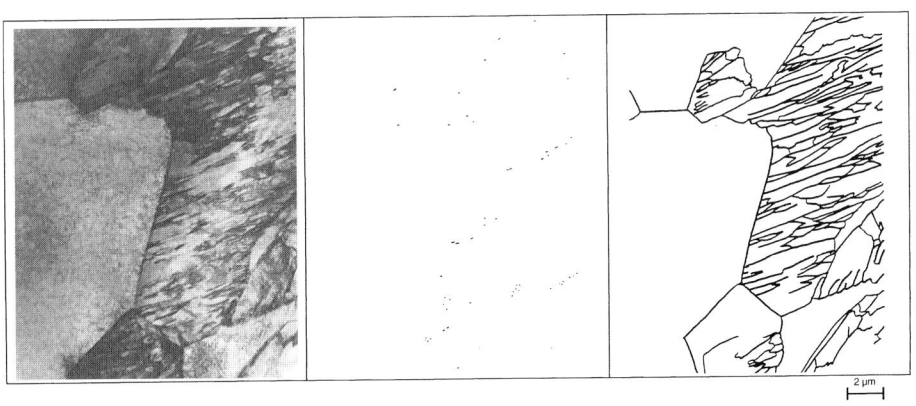

Fig. 8b   TEM microscopy, particles and subgrains of G–X 12 CrMoWVNbN 10 1 1 (coarse grained region – weld thermal cycle Tp = 1300°C, as welded).

tates. The microstructure of the coarse grained region (Tp = 1300°C) shows fine structural martensite without precipitates (see Fig. 8b). After tempering at 730°C for 12 h, the microstructure shows a large amount of fine precipitates (see Fig. 8c). At the intercritically austenitised region (Tp = 920°C) the microstructure shows an increase of the martensite lath subgrains and the precipitates (see Fig. 8d). The tempering at 730°C for 12 h bring a significant increase in the subgrain size in relation to the virgin condition (see Fig. 8e).

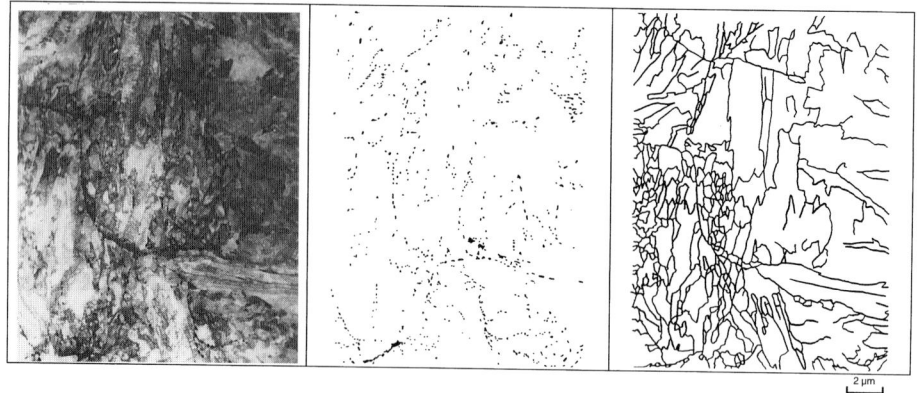

Fig. 8c TEM microscopy, particles and subgrains of G–X 12 CrMoWVNbN 10 1 1 (coarse grained region – weld thermal cycle Tp = 1300°C, and PWHT 730°C/12 h).

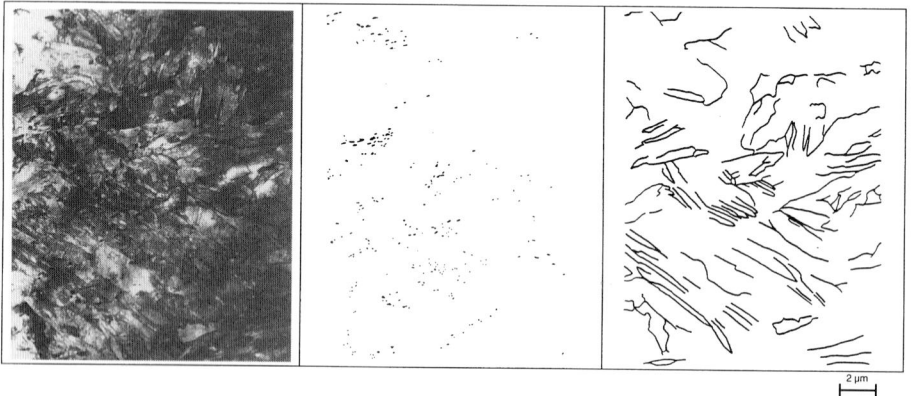

Fig. 8d TEM microscopy, particles and subgrains of G–X 12 CrMoWVNbN 10 1 1 (intercritically austenitised region – weld thermal cycle Tp = 920°C, as welded).

These effects also supports the results of the significant softening of these fine grained zones of the HAZ.

Creep Rupture Tests on HAZ Simulated Materials
Creep rupture tests on samples which were designed to represent material containing microstructures caused by a peak temperature of 920°C (HAZ simulated) were performed. The results are shown in Fig. 9 and are compared with results obtained on creep samples made from uninfluenced base material

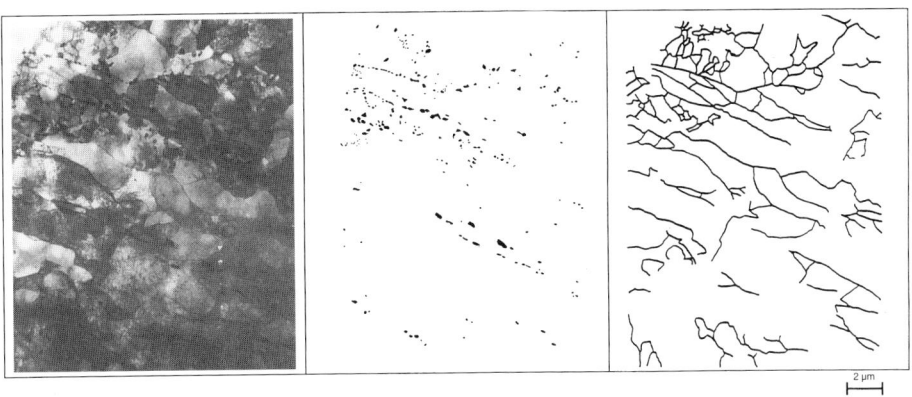

Fig. 8e TEM microscopy, particles and subgrains of G–X 12 CrMoWVNbN 10 1 1 (intercritically austenitised region – weld thermal cycle Tp = 920°C and PWHT 730°C/12 h).

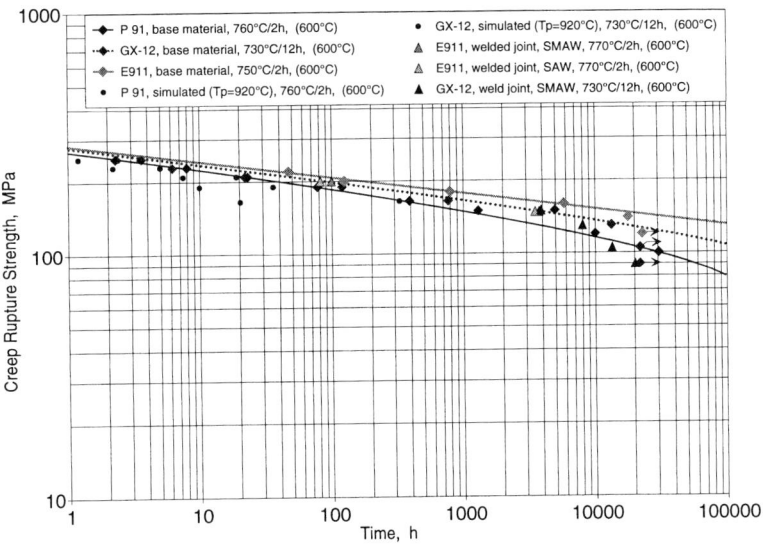

Fig. 9 Results of creep rupture tests of P91, E911, and G–X 12 CrMoWVNbN 10 1 1 base material, welded joint and soft zone HAZ simulated materials.[5]

and welded joints. As can be seen from Fig. 9 the creep resistance of plain HAZ simulated materials falls for all investigated materials at high stress levels, significantly below the creep resistance of that of the uninfluenced base material and of welded joints. Comparing the behaviour of HAZ simulated material P91 to HAZ simulated materials E911 and G–X 12 CrMoWVNbN 10 1 1 it can be seen that the softening effect in the casting tungsten modified

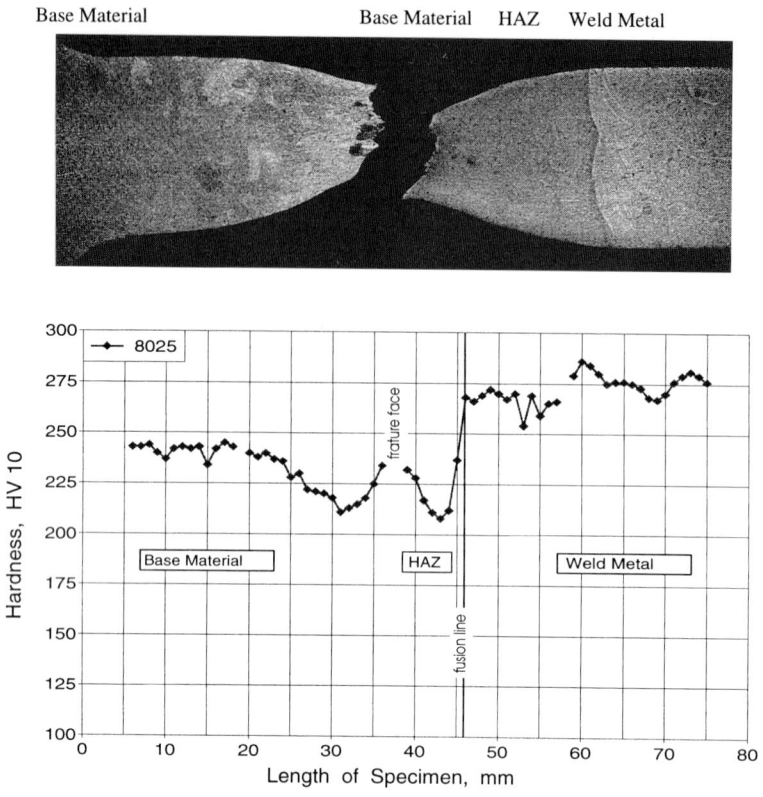

Fig. 10 Fracture location and hardness profile at crossweld sample: specimen no. 8025 G–X 12 CrMoWVNbN 10 1 1/Cromocord 10M, 600°C, 150 MNm^{-2}, 3.919 h).

material on the creep behaviour, at high stress levels, is lower than that of material P91. Creep tests performed on original welded and stress relieved samples of the cast steel show similar behaviour. At high stress levels the creep resistance of HAZ simulated samples lay below that of the base material. The creep rupture strength of welded joints lay, as a result of the restraining effect in the HAZ region, at the same level as that of the base material. At lower stress levels HAZ simulated materials and welded joints shows the same creep resistance.

Creep Rupture Tests on Crossweld Samples
The current status of the creep tests at 600°C with a testing time of 30 000 hours is shown in Fig. 9. The cast steel G–X 12 CrMoWVNbN 10 1 1 and pipe version E911 appears to have a creep strength which is at least as high as

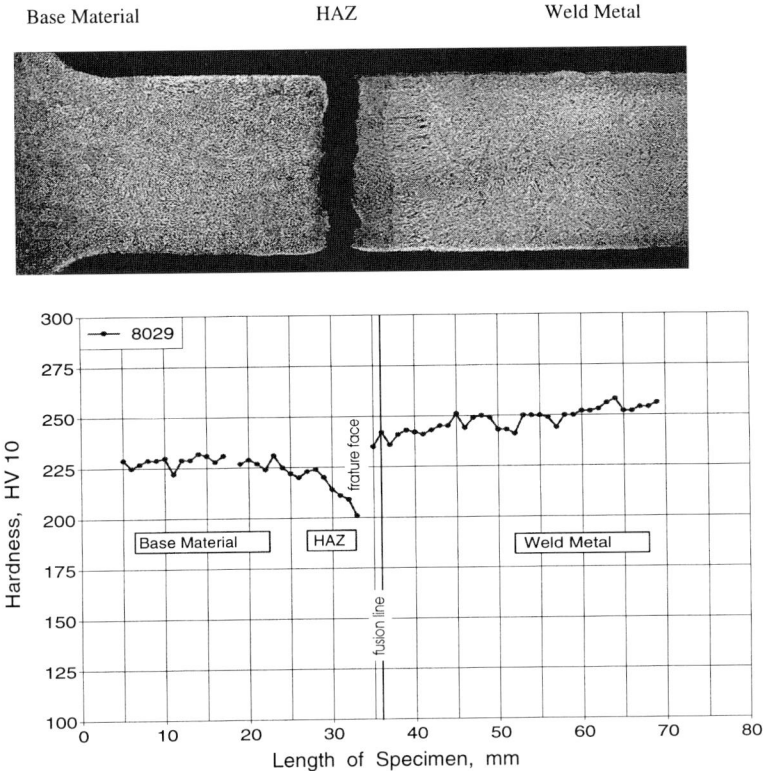

Fig. 11 Fracture location and hardness profile at crossweld sample: specimen no. 8029 G–X 12 CrMoWVNbN 10 1 1/Cromocord 10M, 600°C, 75 MNm^{-2}, 20.175 h).

that of modified 9%CrMo steel P91. Crossweld samples of the repair weld (code E) of the pilot valve body was tested.[5] At high stress levels the fracture is located in the base material. As the applied stress decreases and the rupture time increases, the fracture location shifts from the base material (Fig. 10) into the fine grain region in the HAZ. Results of microstructural investigation and hardness tests of broken creep rupture samples are shown in Figs 10 and 11.

CONCLUSIONS
Previous investigations[1,9] have shown that the weldability of the heat resistant material P91 can be described as very satisfactory. Although, in the HAZ a drop in the creep resistance can be observed. Compared to the base material a loss of creep resistance measured in crossweld samples of welded P91 material of about 20–25% have to be taken into account. Basic investigations using mainly the Gleeble HAZ simulation technique revealed that this drop

in the creep resistance occurred in the fine grained area of the HAZ, where the peak temperature reached a level of about 900–950°C. The aim of this study was to investigate whether newly developed tungsten containing materials, with higher creep resistance than P91 material, show similar behaviour in the welded condition. The results obtained revealed that the behaviour of the tungsten modified 10% chromium cast material G–X 12 CrMoWVNbN 10 1 1 shows similar behaviour regarding the creep resistance in the HAZ to P91. Hardness tests, constant strain rate tests, and short time creep tests on HAZ simulated microstructures showed that also in tungsten modified 9–10%Cr creep resistant steels a drop of the creep strengths in the welded area have to be taken into account. The first results of HAZ simulated short time creep tests showed that this drop is less than observed in P91 pipe material. At stresses lower than 150 MNm^{-2}, the fracture location shifts from the base material into the softened fine grained HAZ At 600°C the data points of the weldments are below those of the base material by more than 25%. In the design of welded components made from this type of materials, this effect must be taken into account.

ACKNOWLEDGEMENTS

This work is a part of the European Action COST 501, Round II and III and was supported by the Austrian Research Funds (FFF) which is gratefully acknowledged.

REFERENCES

1. W. Bendick, H. Cerjak, K. Niederhoff, G. Wellnitz and M. Zschau: *Proc. Int. Conf. on Trends in welding science and technology*, ASM, June 1992, 587–598.
2. R. Blum, J. Hald and E. Lund: *Proc. VGB-TB, Int. Conf. on Werkstoffe in der Schweißtechnik im Kraftwerk*, January 1991, Paper No. 9.
3. F. Masuyama: *Proc. 123rd ISIJ Annual Meeting*, April 1992, **5**, 811. Cited in: R. Blum, J. Hald, W. Bendick, A. Rosselet, and J.C. Vaillant: *Proc. VGB-TB 140, Int. Conf. on Fossil-fired power plants with advanced design parameters*, June 1993, **V2**.
4. M.K. Booker *et al.*: *Proc. Int. Conf. on Production, fabrication, properties and application of ferritic steels for high temperature applications*, October 1981, ASM, Warren.
5. D. Coutsouradis *et al.*: *Int. Conf. on Materials for advanced power engineering*, October 1994, Liège.
6. K. Easterling: *Introduction to the physical metallurgy of welding*, 1983, Butterworths Monographs in Metals.
7. N.N. Rykalin: *Berechnung der Wärmevorgänge beim Schweißen*, 1957, Berlin, VEB Verlag Technik.

8. B. Buchmayr: *Schweißen Schneiden*, 1989, **46** (2), 69–75.
9. F. Brühl: 'Verhalten des 9%-Chromstahles X10 CrMoVNb 9 1 und seiner Schweißverbindungen im Kurz- und Langzeitversuch', *Doctoral Thesis*, TU-Graz, 1989.
10. H. Cerjak and F. Schuster: *Weldability and behaviour of weldings of new developed creep resistant 9–10% Cr-Steels*, Pivisto Italiano dillo Soldtissa 4/94, 467–473.
11. P. Hofer: 'Microstructural evaluation of creep processes in a tungsten modified 9–10% chromium cast steel', *Diplomarbeit*, TU-Graz, 1994.

The Effect of Composition on Microstructural Development and Mechanical Properties of Modified 9%Cr 1%Mo Weld Metals

A.M. BARNES

TWI Abington Hall, Abington, Cambridge CB1 6AL, UK

ABSTRACT

The modified 9%Cr 1%Mo steels are finding significant application within the power generation industry, both in the life extension of existing plant and in new plant construction. These materials offer enhanced high temperature properties due to the controlled addition of Nb, V and N to a standard 9%Cr 1%Mo composition. The materials also offer good oxidation resistance and resistance to hot hydrogen attack, and are thus attracting interest from other industry sectors. However, there are uncertainties in relation to the design of consumables for these materials in order to achieve good levels of both creep strength and toughness, and these have limited the widespread exploitation of these steels. Initial work to match the base material composition often led to poor toughness, and it was acknowledged that modification of the deposit composition was required, particularly in relation to the Ni and Nb content.

This paper provides details of studies carried out at TWI to firstly evaluate selected commercial electrodes and secondly to carry out trials using a series of experimental electrodes designed to give deposits of controlled composition. The programme involved microstructural examination, including point counting to assess the delta (δ) ferrite content, hardness testing, assessment of toughness by both Charpy and CTOD testing, and creep testing.

The work has indicated that the compositional parameters of chromium equivalent and Kaltenhauser ferrite factor provide a good indication of the weld metal δ-ferrite content. It has been shown from the commercial consumable evaluation that the addition of ~1%Ni in combination with a reduction in the levels of Nb, Si and N (to 0.03, 0.19 and 0.021% respectively) led to a significant improvement in toughness after PWHT for 2 hours at 760°C, but markedly reduced the weld metal creep strength. Further, other compositional variations were studied, either in isolation or in combination; many of these did not significantly alter the transformed microstructure, although tungsten actively promoted δ-ferrite retention. For the conditions studied it was found that alloying with more than 1%Ni, or with 1–3% tungsten, markedly reduced toughness, conversely low levels of boron (up to ~20 ppm) led to a marginal improvement in Charpy toughness. High levels of tungsten (~3%) and boron (67 ppm) resulted in the development of a second phase on the prior austenite boundaries, believed to be retained δ-ferrite. These studies are on-going, and while there are a number of commercial electrodes currently in the market-place, there is clearly room for further improvement in properties, and the continued development of enhanced steel grades brings with it a need for continued consumable development.

INTRODUCTION

The modified 9Cr 1Mo steels (ASTM A387 Grade 91) have found increased usage over recent years within the power generation industry. They have been used in both the repair and upgrade of existing plant and in the construction of new plant. To date, for example, they have been used for headers in steam power plant, and in tubing for heat exchangers.

The improved high temperature creep strength of these materials is achieved by the addition of controlled levels of Nb, V, and N to a conventional 9Cr 1Mo base composition. This leads to the development of a dispersion of fine Nb carbonitride precipitates during the tempering heat treatment, and to the formation of stable precipitates containing vanadium of the form $M_{23}C_6$.[1] The modified 9Cr 1Mo steels also exhibit good oxidation resistance and resistance to hot hydrogen attack, and for this reason they are attracting interest from other industry sectors, for example the petrochemical sector.

However, the widespread exploitation of these grades has been limited by uncertainties in the design of the consumables to allow good levels of both creep strength and toughness to be achieved. In the early stages of consumable development, manufacturers aimed to match the composition of the base material, but the deposits produced exhibited poor toughness in relation to the base material. The early studies revealed that the deposit toughness was strongly influenced by changes in the levels of carbon and Nb although it was recognised that a certain level of Nb was required to ensure adequate creep strength. Whilst a number of studies have been carried out that have looked at compositional variations, and have acknowledged the need for modification of the deposit composition relative to the base material in order to ensure a deposit of adequate toughness (particularly in relation to that of the levels of Ni and Nb), they have not made specific recommendations on the levels that should be incorporated.[2,3] It is not surprising, therefore, that within current commercial consumables there is a very diverse range of compositions.

The modified 9Cr 1Mo steels typically exhibit a fully martensitic microstructure. However, their composition is at the threshold of delta (δ) ferrite retention, so in weldments where elevated temperatures and rapid cooling are encountered, it is not uncommon for a proportion of δ-ferrite to be retained within the weld metal microstructure as well as within the heat affected zone

(HAZ). It is acknowledged that an indication of the propensity for δ-ferrite retention can be obtained from a number of compositional parameters, for example the chromium equivalent (Cr_{eq}),[4] and the Kaltenhauser ferrite factor (FF).[5] Although the precise influence of δ-ferrite is not fully understood, it is believed that it can exert a detrimental effect on both toughness and ductility,[6–8] although some studies have indicated that it can reduce austenite grain growth in the HAZ and at the same time reduce creep crack propagation rates.[9] Studies of the influence of welding parameters on the tendency for δ-ferrite formation have proved inconclusive. Some studies have shown it to be independent of welding conditions, while others have indicated a decrease in δ-ferrite content at increased heat input.[6]

Recognising the potential improvement in weld metal properties, both toughness and creep strength, that could be achieved through modification of the weld metal composition, there has been a continuing programme of work at TWI to provide a systematic and quantitative study of the effect of various elements. The early work indicated poor room temperature toughness for a number of commercial consumable deposits. The effect of Nb in the range 0.02–0.09wt% following heat treatment of either 2 hours at 750°C or 8 hours at 760°C was studied.[10] An apparent insensitivity of the Charpy test to these compositional changes was observed (based on the temperature to achieve an absorbed energy of 40 J). However, CTOD testing of identical deposits indicated a detrimental influence of increased Nb following extended postweld heat treatment (PWHT).

The programme of work reported in this paper was aimed at formulating guidelines for modified 9Cr 1Mo weld metals in order to achieve good levels of both toughness and creep strength. The programme commenced with an evaluation of current commercial consumables in terms of microstructural development, hardness, creep strength and toughness. Based on the findings of this study experimental manual metal arc (MMA) electrodes of controlled composition were obtained in order to allow the effect of a number of elements including Ni, Nb, Co, W, B and N to be studied. The investigation of these deposits was restricted to microstructural examination and hardness testing although corresponding Charpy impact data were supplied by the electrode manufacturer.

EXPERIMENTAL PROCEDURES

Welding

Multipass MMA welds were produced in 25 mm thick modified 9Cr 1Mo steel (for composition see Table 1) at an arc energy of 2 kJ/mm for each of the four commercial consumables. Prior to welding all electrodes were baked for 2 hours at 350°C. A single V-preparation was used with a root face of 2 mm

Table 1   Chemical analyses of parent materials

Material	Element wt %												
	C	S	P	Si	Mn	Ni	Cr	Mo	V	Cu	Nb	Al	N
ASTM A387 Grade 91	0.08–0.12	0.010 max	0.020 max	0.20–0.50	0.30–0.60	0.4 max	8.00–9.50	0.85–1.05	0.18–0.25	–	0.06–0.10	0.40 max	0.03–0.07
Parent 1 (PM1)	0.10	<0.002	0.012	0.26	0.44	0.07	8.5	0.92	0.25	0.03	0.08	0.02	0.0393
Parent 2 (PM2)	0.10	<0.002	0.012	0.25	0.38	0.13	8.2	0.92	0.19	0.10	0.09	0.02	0.0489

TWI ref: S/92/326, S/93/173, O/N 93/73 and O/N 93/56.
B, Ca <0.0005.
Ti, Sn, As, Pb, Zr all <0.01, Co = 0.01, 0 = 0.0027.

and included angle of 55°. The panels were restrained during welding using strongbacks. Minimum preheat and maximum interpass temperatures of 200 and 300°C respectively were employed. The bead sequence was designed to produce a high degree of weld metal refinement in order to try to optimise the weld metal toughness. On completion of welding the panels were subjected to a PWHT of 2 hours at 760°C with controlled heating and cooling rates. The heating rate from ambient temperature to 700°C was controlled at a 100°C/hr and from 700 to 760°C at 20°C/hr. After completion of the hold time of 2 hours the plates were furnace cooled.

In order to facilitate study of the effect of a variety of other elements as previously mentioned, Ni, Co, W, B and N, a series of transverse sections from welds produced using experimental electrodes were supplied by the consumable producer. These deposits had been made in samples of carbon steel that had been buttered with 9Cr 1Mo weld metal, again to ensure an undiluted weld deposit. The deposits were made in accordance with ISO 2560–1973 at a nominal arc energy of 1 kJ/mm. This procedure consisted of nine layers with 3 beads per layer. Samples were supplied in both the as-welded condition and following a PWHT of 8 hours at 750°C.

Microstructural Examination & Chemical Analysis
For each of the panels produced with the commercial electrodes, a transverse section was removed for both the as-welded condition and following PWHT. One section from each of the commercial and experimental consumable deposits was used for chemical analysis. All sections were then prepared to 1 μm diamond finish using standard metallographic techniques, etched in a solution of 2.5% picric acid and 2.5% hydrochloric acid in alcohol, and exam-

ined using optical microscopy. An assessment was made for each deposit of the amount of retained δ-ferrite in the weld metal of the capping pass, using a point counting technique involving 1000 points at ×250 magnification. For each deposit a Vickers hardness survey was carried out using an indenting load of 10 kg, sampling both columnar and refined weld metal.

Toughness Testing

From each of the PWHT commercial consumable panels, standard 10 × 10 mm square cross-section Charpy V-notch specimen were removed from a location 2 mm below each of the cap and root surfaces. The samples were through-thickness notched at the weld centreline, and tested in accordance with BS EN10045–1: 1990 over a range of temperatures to allow a transition curve to be generated. Further, 23 mm square cross-section BXB CTOD specimens were machined transverse to the weld from the PWHT panels and notched at weld centreline. The notches were extended by fatigue to give a/w = 0.5, where 'a' is the crack depth and 'w' is the specimen width. Testing was carried out over a range of temperature in accordance with BS 7448 Part 1: 1991 to give a transition curve. Fracture surfaces were examined under a binocular microscope, and for those specimens exhibiting brittle fracture initiation, in a scanning electron microscope (SEM) to locate the fracture initiation site. The samples were sectioned through this site, parallel to the base of the notch, and prepared for examination under an optical microscope in order to reveal the initiation microstructure.

Creep Property Evaluation of Commercial Consumable Deposits

Cross weld creep specimens were removed from each of the PWHT panels in accordance with BS3500 Part 3 1969, with a 7 mm diameter gauge. The samples were tested in the temperature range 600 to 650°C and were subsequently sectioned transverse to the weld in order to identify the failure location.

RESULTS

Chemical Analysis

The chemical compositions of the four commercial consumable deposits and the range of compositions for the experimental consumable deposits are indicated in Tables 2 and 3. Comparison of the commercial consumable compositions with those of the base materials indicates some clear differences, particularly in the higher Mn content of the weld deposits; two of the deposits contained additions of Ni and one of these also incorporated reduced levels of Nb, Si and N relative to the other deposits. The experimental consumable deposits showed independent variations in the level of Ni, Co and W,

Table 2 Chemical analyses of multipass weld deposits made using commercial consumables

Weld No.	Element wt%														
	C	S	P	Si	Mn	Ni	Cr	Mo	V	Cu	Nb	Al	Co	O	N
W1 (PM1)*	0.09	0.006	0.012	0.41	0.97	0.08	8.7	0.99	0.17	0.02	0.06	<0.01	0.01	0.049	0.045
W2 (PM1)	0.08	0.003	0.009	0.42	1.02	0.03	9.3	0.98	0.21	0.01	0.09	<0.01	<0.01	0.046	0.048
W3 (PM1)	0.10	0.005	0.009	0.39	1.07	0.68	8.9	0.99	0.18	<0.01	0.05	<0.01	0.01	0.034	0.047
W5 (PM2)	0.08	0.004	0.010	0.19	1.12	0.70	8.9	0.91	0.19	0.01	0.03	<0.01	0.01	0.060	0.021

Ca, B <0.0005.
Ti, Sn, As, Pb, Zr all <0.01.
TWI ref: S/92/326, S/93/69, S/93/176, O/N 93/39, 93/73.
* Coding in brackets indicates the parent material used.

Table 3 Range of experimental consumable deposit compositions

Element wt%															
C	S	P	Si	Mn	Ni	Cr	Mo	V	Cu	Nb	B	Co	W	O	N
~0.09	~0.006	~0.011	~0.18	~0.98–2.94	0.04+	~9.2	~0.97	0.20	~0.01	0.048 –0.0067	<0.0005 –2.88	<0.01	~0–3.28	~0.063 0.049	0.016–

Ti, Al, Sn, As, Pb and Zr all <0.01.
Ca all <0.0005.

combined additions of W with Ni or Co, and a varied boron level at two differing levels of nitrogen.

Microstructural Examination

The commercial electrode deposits all exhibited a predominantly martensitic microstructure in the as-welded condition, with varied levels of retained δ-ferrite. Following PWHT, a tempered martensitic microstructure remained, with a fine dispersion of spheroidised carbides. There was some variation in the delineation of the prior austenite grain boundaries, this being most pronounced in the Ni-free high Nb deposit, W2. This deposit also contained the highest level of retained δ-ferrite, as reflected in the higher levels of Cr_{eq} and FF.

For the experimental consumable deposits the microstructure was once again predominantly martensitic with some retained δ-ferrite. It was apparent that the retention of δ-ferrite was reduced or eliminated by the addition of Ni. However, at the high Ni levels there was considerable microstructural modification after PWHT, and localised carbide free laths were observed (Fig. 1a). Tungsten quite clearly had a very marked effect on the amount of retained δ-ferrite (Fig. 1b); this is reflected in the considerable increase in the Cr_{eq} parameter. However, tungsten does not feature in the equation for FF and therefore, based on FF, the increase in δ-ferrite content would not be predicted. It was apparent from the 1%Ni 1%W containing deposit that the amount of δ-ferrite was essentially zero. The addition of Co appeared, like Ni, to reduce the propensity for δ-ferrite retention, although the effect was less marked than that of Ni. Cobalt did not, however, appear to noticeably influence the response to PWHT. The effect of combined additions of W and Co once again indicated the weaker influence of Co in relation to Ni; with the addition of 1% Co in combination with W a small amount of δ-ferrite was still retained, and a level of 2% Co was required in order to achieve a fully martensitic deposit. This is again reflected in the Cr_{eq} equation but not for the FF as the latter takes no account of Co. For the boron and nitrogen containing deposits the microstructure was once again predominantly martensitic, although there was a small increase in the amount of retained δ-ferrite with increased boron content. However, the more notable effect of boron was in the enhancement of the prior austenite grain boundaries with increased boron content, and the apparent development of a light etching phase on the grain boundaries, somewhat atypical of δ-ferrite (Fig. 1c).

Hardness Testing

The hardness data for the commercial consumable deposits (Table 4) indicated a softening of the order of 160 HV10 on PWHT for 2 hours at 760°C and did not suggest any significant hardness difference between the columnar

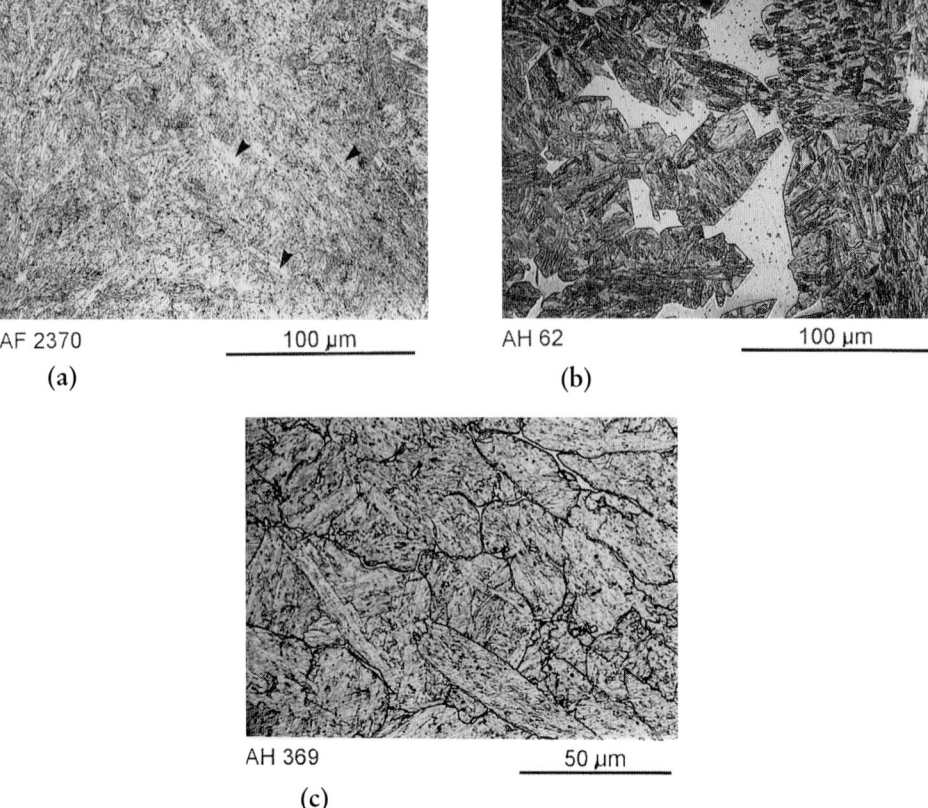

AF 2370   100 μm   AH 62   100 μm

(a)   (b)

AH 369   50 μm

(c)

Fig. 1 Optical Micrographs showing:
(a) Localised carbide free laths in high (~2.5%) Ni deposit following PWHT;
(b) Large amount of retained δ-ferrite in high (~3%) W deposit;
(c) Light etching grain boundary phase in 67 ppm B, 155 ppm N deposit.

and refined weld metal regions. Commercial consumable deposits, W1, W2 and W5, all exhibited similar levels of hardness after PWHT (~225 HV10), but W3 exhibited a hardness approximately 30 HV10 above the other deposits attributed largely to the increased levels of alloying and in particular its higher carbon content. The two Ni-containing deposits, namely W3 and W5, also exhibited marginally higher hardness (approximately 20 HV10) in the grain coarsened HAZ.

The hardness data for the experimental consumable deposits (Table 5) indicated a systematic increase in hardness with increased Ni content for both the as-welded and PWHT conditions and, particularly at the higher levels of Ni, a reduction in the degree of tempering was observed. Cobalt, however, appeared to have little effect on hardness. The addition of approximately 3% tungsten to

Table 4 Summary of Vickers hardness data for the commercial consumable deposits

Weld No.	Condition	Hardness (HV10)		$\frac{\text{max–min}}{\textit{average}}$
		Weld metal	GCHAZ	SC/HAZ Parent Interface
W1 (PM1)	As-welded	437–373 405	380–359 367	206–200 202
W2 (PM1)	As-welded	427–360 392	387–350 372	206–182 192
W5 (PM2)	As-welded	401–286 355	399–380 391	197–185 191
W1 (PM1)	PWHT	239–201 225	226–218 223	178–173 176
W2 (PM1)	PWHT	245–225 235	222–189 210	180–176 178
W3 (PM1)	PWHT	263–244 253	242–232 238	182–178 180
W5 (PM2)	PWHT	240–216 225	260–245 252	192–185 189

the deposit led to a reduction in the as-deposited weld metal hardness, but on PWHT a lesser degree of softening was recorded bringing the PWHT hardness in-line with those of other deposits. A reduction in nitrogen content appeared to reduce the as-deposited hardness, but did not significantly alter the hardness after PWHT. The addition of boron at high levels of nitrogen did not alter the as-deposited or mean PWHT hardness, but there was in general a systematic increase in the maximum hardness level recorded. The same was also observed following PWHT for the lower nitrogen containing deposits, but no trend was observed for these deposits in the as-welded condition.

Charpy Testing of the Commercial Consumable Deposits
The 40 J transition temperatures, obtained from the 'by-eye' lower bound transition curves presented in Fig. 2, are given in Table 6. Deposits W1 and W5 showed a marked difference in the performance of the cap and root sub-surface positions, with typically the root requiring a higher temperature to obtain 40 J. Welds W1 to W3 exhibited similar toughness, but in general superior toughness was exhibited by W5. The form of the transition curves for the four deposits were similar, although the upper shelf position of W5 was approximately 20 J below those of the other deposits.

Table 5 Summary of Vickers hardness data on experimental consumable deposits

Composition	Hardness HV10 (max–min/average)	
	As-welded	PWHT
Baseline	468–347/408	218–198/210
(1%Ni)	433–368/410	235–219/230
(3%Ni)	478–417/446	317–309/315
(1%Co)	433–383/414	221–207/216
(3%Co)	442–375/412	221–210/217
(1%W)	446–359/402	242–218/232
(3%W)	390–333/360	245–222/231
(1%W 1%Ni)	446–389/423	232–220/227
(1%W 1%Co)	425–357/405	216–201/209
(1%W 2%Co)	450–377/424	217–202/210
<5 ppm B 462 ppm N	423–407/415	222–202/211
12 ppm B 437 ppm N	431–322/399	231–218/223
29 ppm B 447 ppm N	457–380/417	238–222/234
63 ppm B 447 ppm N	423–365/400	251–234/243
<5 ppm B 187 ppm N	413–356/391	225–213/219
12 ppm B 168 ppm N	413–327/370	240–226/235
31 ppm B 155 ppm N	411–383/397	248–232/244
67 ppm B 155 ppm N	397–341/375	253–236/245

CTOD Testing of the Commercial Consumable Deposits

The CTOD data represented by the 0.1 mm transition temperature are summarised in Table 6 and were once again taken from 'by-eye' lower bound transition curves (Fig. 3). Weld W5 exhibited the highest toughness, while W3 exhibited the poorest toughness reflecting an inferior resistance to cleavage fracture initiation. The high Nb deposit W2 exhibited a higher upper shelf position and a steeper transition than the other three weld metals. In the majority of cases the post-test assessment of the CTOD specimens indicated that fracture had initiated in the columnar weld metal close to the weld root. For W3, however, the lowest CTOD values were associated with fracture initiation in refined weld metal.

Charpy Data for Experimental Consumable Deposits

The toughness testing of the experimental consumable deposits was not carried out at TWI, but the Charpy impact data supplied by the manufacturer

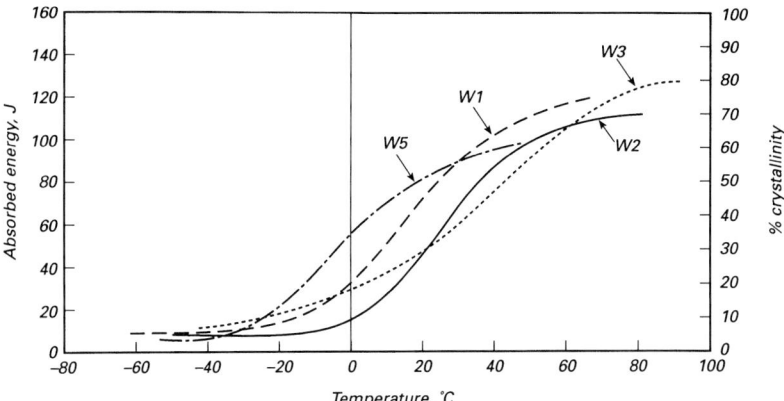

Fig. 2(a)   Charpy transition data for the cap sub-surface location of the commercial consumable deposits notched at weld centreline.

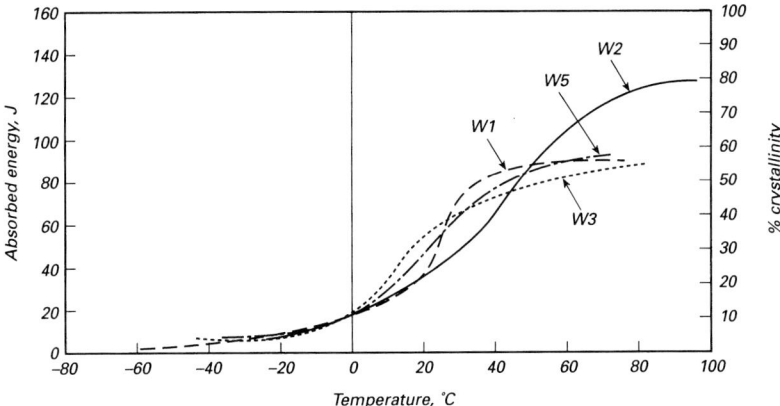

Fig. 2(b)   Charpy transition data for the root sub-surface location of the commercial consumable deposits notched at weld centreline.

are given in Fig. 4. These indicate that increased Ni (>1%) and the addition of tungsten, led to a marked deterioration in toughness. Conversely, the addition of Co (1–3%) had little effect. The combined additions of Ni or Co with tungsten gave marginal improvement in toughness relative to the addition of tungsten in isolation.

There was a small improvement in toughness with the addition of up to 20 ppm boron but thereafter toughness deteriorated, particularly at the 'normal' nitrogen level of ~460 ppm.

Creep Rupture Testing of Commercial Consumable Deposits
The results of the cross-weld creep tests are presented in Fig. 5 and the test

Table 6 Summary of mechanical testing and compositional data for welds made using commercial consumables

Weld	Composition	Ferrite Factor	Cr_{eq}	Ferrite Content**		Temperature for absorbed energy of 40 J (°C)*		Temperature for 0.1 mm CTOD (°C)*
				As-welded	PWHT	Cap	Root	
W1	Low Ni Medium Nb	10.0	7.7	0	0.2	3	21	32
W2	Low Ni High Nb	11.7	8.8	4.5	2.5	15	22	27
W3	~1%Ni Medium Nb	7.1	4.6	−+	0	10	12	56
W5	~1%Ni, Low Si, Low N, High O	7.0	4.7	0.3	0	−9	14	8

* Following PWHT 2 hrs at 760°C.
** Measured in capping pass.
+ Sample not available.

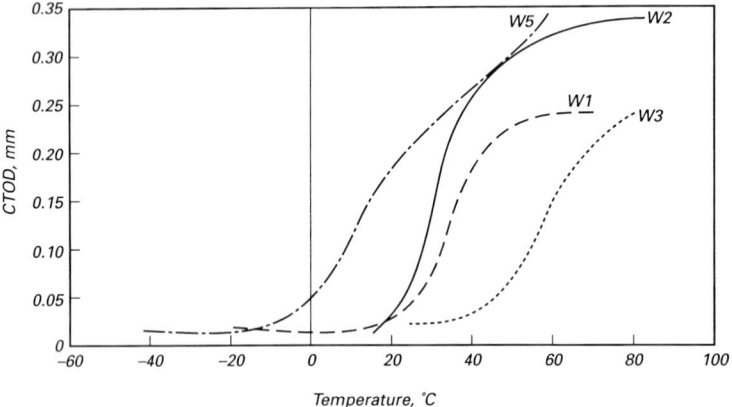

Fig. 3 CTOD transition curves for the commercial consumable deposits notched at weld centreline.

data summarised in Table 7; these indicate that at 600°C, 110 MPa the initial behaviour of deposits W1 to W3 was similar, but that W2 exhibited a higher secondary creep rate. Overall however, the performance of W3 was superior. At the higher stress of 130 MPa, the primary creep behaviour of W2, W3 and W5 was similar, but the secondary creep rate increased from W3 to W2 to W5. Weld W1 showed a relatively low creep rate but exhibited a short time to failure. Overall the behaviour of W3 was superior to that of the other welds, and that of W5 was markedly inferior. Examination of the tested specimens indicated that for W1, W2 and W3, failure had occurred remote from the weld metal in the outer HAZ/parent material region, generally referred to as Type

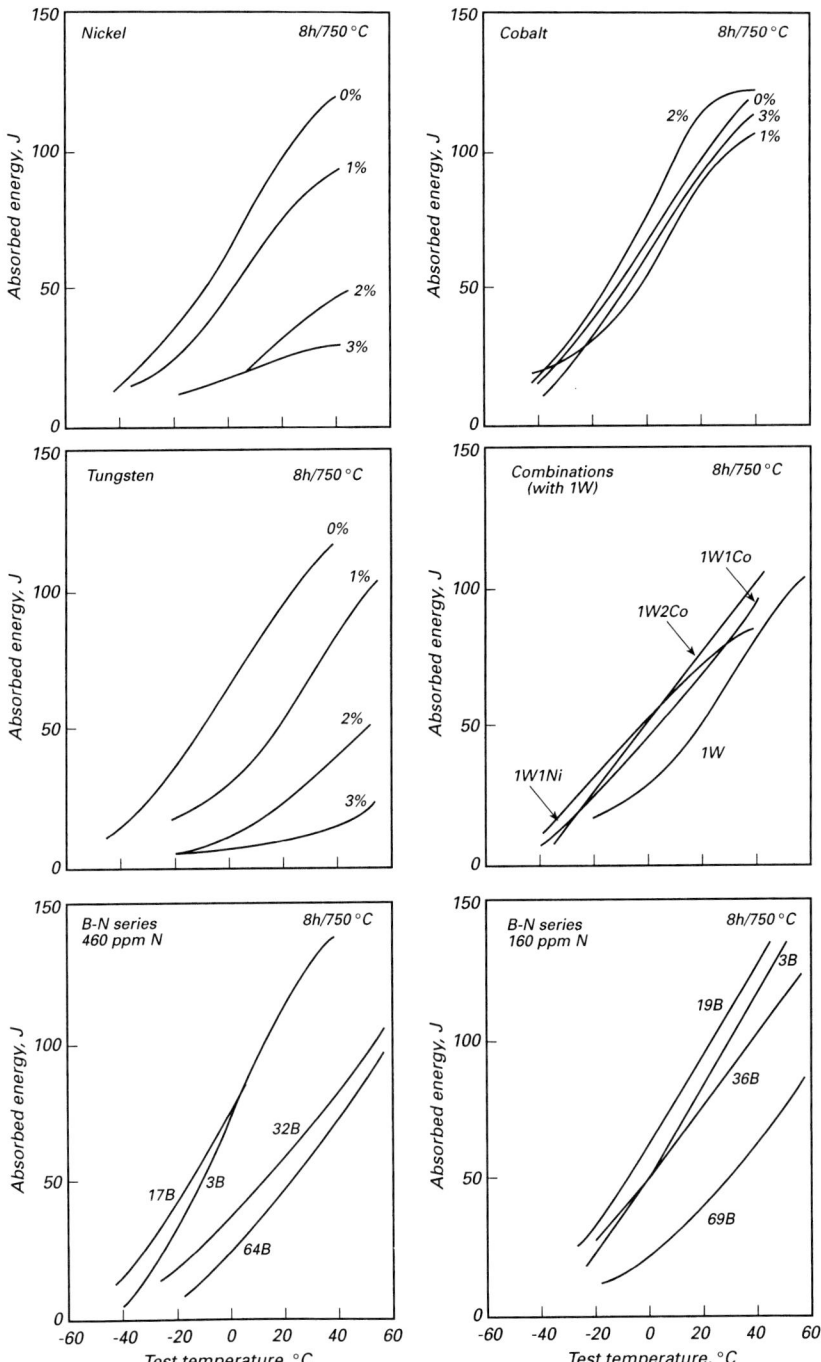

Fig. 4   Charpy impact curves for experimental consumable deposits (supplied by consumable manufacturer).

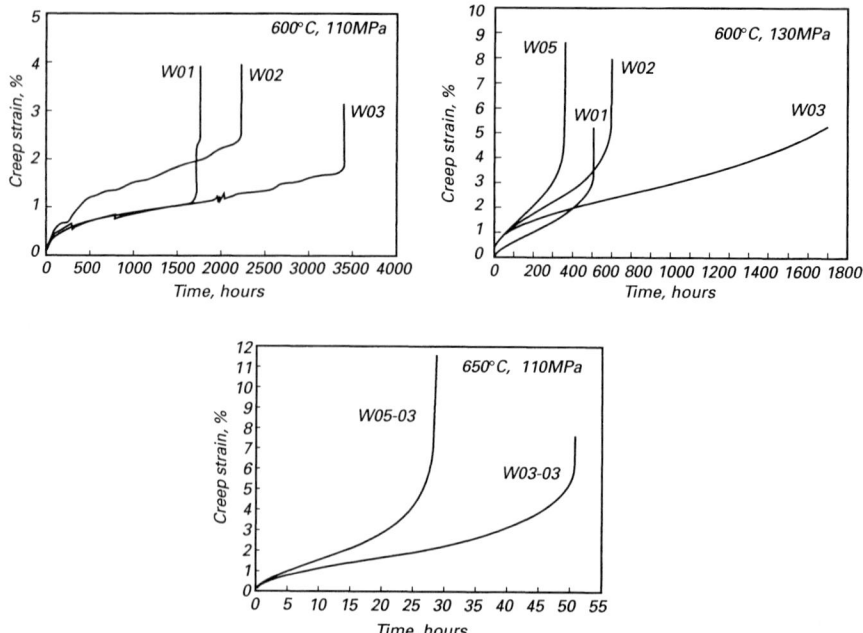

Fig. 5 Graphs showing cross-weld creep data generated at:
(a) 600°C, 110 MPa (b) 600°C, 130 MPa; (c) 650°C, 110 MPa.

Table 7 Summary of cross-weld creep data

Specimen number	Temp °C	Stress MPa	Time to failure (hours)	Elongation (%)	Reduction in area (%)	Failure location
W01–04	600	130	515	5.2	58	Type IV zone
W01–05	600	110	1780	3.8	41	Type IV zone
W02–01	600	130	607	8.0	63	Type IV zone
W02–02	600	110	2271	3.9	39	Type IV zone
W03–01	600	110	3429	3.0	31	Type IV zone
W03–02	600	130	1730	5.3	37	Type IV zone
W03–03	650	110	51	7.7	70	Type IV zone
W05–02	600	130	367	8.7	57	Weld metal
W05–03	650	110	29.1	11.7	76	Weld metal

IV zone. The samples from W5, which contained the addition of approximately 1% Ni with reduced levels of Si, Nb and N all failed in the weld metal. Only welds W3 and W5, the two Ni containing deposits, were tested at 650°C

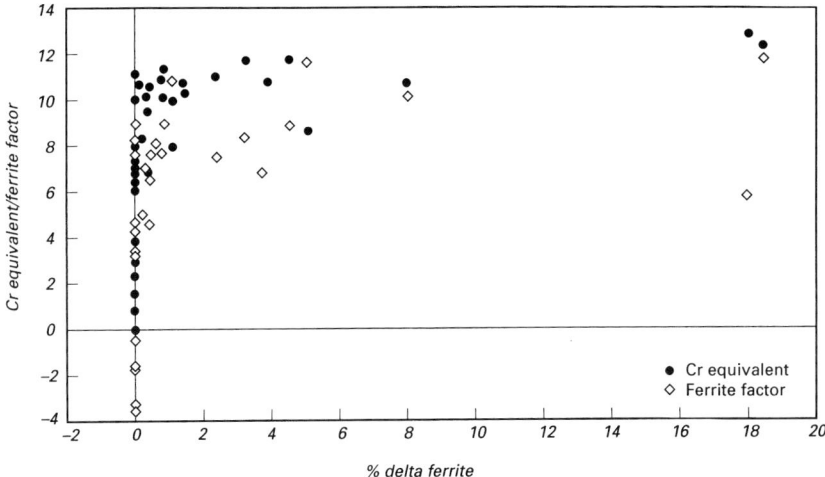

Fig. 6 Graph of chromium equivalent/ferrite factor against δ-ferrite content taken from various TWI studies.

at a stress of 110 MPa. The secondary creep rate of W5 was higher than that of W3, but in both cases an extremely short time to failure was observed.

DISCUSSION

The Effect of Composition on Microstructural Development
A diverse range of weld metal compositions has been studied in this programme, and whilst in all cases the microstructure was predominantly martensitic there were some differences in the amount of retained δ-ferrite. Considering the relationship between volume fraction of δ-ferrite and the compositional parameters of Cr_{eq} and FF for the data from various TWI studies (Fig. 6) it is apparent that, for the range of compositions that have been studied, Cr_{eq} provides the best overall correlation with the measured δ-ferrite content, and in general a fully martensitic deposit was obtained for Cr_{eq} <8. Reasonable agreement was obtained with FF for the non-tungsten or cobalt containing deposits, and in this case the threshold level was FF <6 to achieve a fully martensitic deposit. Whilst the compositional variations studied did affect the amount of retained δ-ferrite within the weld deposit, they did not lead to any marked modification of the as-transformed microstructure. Some difference was observed in the definition of the prior austenite grain boundaries for both the as-welded and PWHT conditions. This was not studied in any detail but it is postulated that it may be due, in part, to differing amounts of grain boundary segregation during solidification, but may also be influenced by the transformation behaviour of the

deposit. Indeed it is conceivable that the boundary delineation may result from the retention of fine colonies of δ-ferrite, and may also explain the light-etching phase observed in the deposits containing varied boron and nitrogen levels.

The Effect of Composition on Toughness

In light of work carried out elsewhere involving MMA welds at 1 kJ/mm subjected to PWHT of 2 hours 760°C, the superior toughness exhibited by commercial consumable deposit W5 is not unexpected;[11] it has been reported that there is a beneficial effect of reduced Nb and N and of the addition of up to 1%Ni. A reduction in the level of Si is also considered to be beneficial provided that sufficient is present to achieve adequate deoxidation. The commercial consumable deposits generally showed rather poor ambient temperature toughness particularly in the root region. This is believed to have arisen principally because of dilution effects and the higher strain occurring in the root region. The lower upper shelf level observed in W5, approximately 20 J below the other deposits, is consistent with the expected higher inclusion content of this deposit, associated with the higher oxygen level probably resulting from the reduction in the Si level. As has been observed in earlier TWI studies,[10] Charpy testing is relatively insensitive to the compositional variations studied, and a greater disparity is observed from the CTOD data. In this instance, W5, the Ni containing deposit in combination with reduced Nb, N and Si levels, exhibited notably better toughness than the other three deposits. Deposit W3, on the other hand, exhibited significantly poorer toughness than the other three deposits; it was also found that in W3 fracture tended to initiate in refined weld metal, whereas in the other three deposits fracture typically initiated in columnar regions. It did, however, appear that W3, principally by virtue of the higher Si, Mn and N levels and more specifically the C level, contained a slightly higher carbide/carbonitride density, and a higher hardness level, ~20–30 HV10 above the other deposits. Thus it is perhaps not surprising that, despite the higher Ni content relative to W1 and W2, W3 exhibited poorer CTOD behaviour i.e. the poorest fracture initiation resistance.

Developments in modified 9Cr1Mo are looking to improve the base material toughness and creep strength, and are considering the introduction of tungsten, an element believed to markedly enhance creep performance. This modification is considered equally viable for the weld metal. However, the present data show a dramatic reduction in weld metal toughness for the addition of 1–3% tungsten, due, it is believed, to the retention of a significant quantity of δ-ferrite, even at the 1%W level. It is clear that if this approach is to be pursued to give enhanced creep properties, its powerful effect on δ-ferrite retention, and hence toughness, must be counteracted. The microstructural study indicated that this could be achieved through the alloying combi-nation of 1%W and either 1%Ni or 2%Co. The consumable producers' data show an improvement in toughness (relative to the addition of 1%W in iso-lation) for these combinations, and also for the combined addition of 1%W and 1%Co, although a small amount of δ-ferrite remained in this instance. It should, however be noted that the use of Co is both economically and eco-logically less favourable than the use of Ni.

The data for the boron containing deposits indicated a marginal improve-ment in toughness through the addition of up to 20 ppm boron, but thereafter a reduction toughness, particularly at the 'normal' level of ~460 ppm nitro-gen. This toughness drop is believed to arise, in part, from the precipitation of boronitrides, but, in view of the microstructural observations, may also be due to the formation of a light etching phase, thought to be δ-ferrite, on the prior austenite boundaries. However, the deposit analyses also indicate an increase in Nb content at the higher levels of boron and, in view of the com-ments made earlier, the effect of this in reducing toughness is also likely to be a contributory factor.

The Effect of Composition on Creep Properties
Considering the creep data for the four commercial consumable deposits, it is clear that there were some significant differences; comparison involving W5 is perhaps less valid than the comparison of the other three deposits as this weldment employed a different base material. Whilst the three deposits W1, W2 and W3 all exhibited creep failure in the Type IV zone i.e. at the parent material/HAZ interface, there were significant differences in the test dura-tion. It appears from the data that the Type IV life is strongly influenced by the weld metal employed. From the strain redistribution during testing this result is perhaps not surprising, but what is surprising is that despite con-siderable work by other researchers this effect has not been revealed.[12] It is predicted from creep theory that during cross-weld creep testing strain will redistribute between the parent material, weld metal and HAZ regions, and will be influenced by the relative creep strengths and deformation rates of these constituent regions. Thus considering a butt weld in pipe, the weld metal employed is generally weaker than the parent material and cannot sus-tain the applied load, it will therefore deform preferentially to the parent and in so doing will try to shed load onto the region adjacent to the weld metal, thereby inducing a Type IV failure.[13] If the weld metal creep strength is reduced still further, this effect will be enhanced until eventually the weld metal becomes so creep weak that it is unable to tolerate the applied load and a weld metal failure will be recorded. A similar redistribution effect would perhaps therefore be expected if the weld metal significantly overmatches the base material in creep strength. Considering the overall total strain on the

weldment, as indicated by the elongation values, this is made up of the component of strain in each of the microstructural regions. Similarly the creep strain rate that is measured is the total rate from the three component regions. Considering the test data for W1–W3, (at 600°C, 130 and 110 MPa) in which Type IV failures were consistently achieved, a higher reduction of area value (RA) indicates more localised deformation in the HAZ. Assuming that the contribution from the parent material to the total strain is essentially constant, the strain component of the weld is likely to be less, and this indicates that a differential in creep strength must exist between the weld metal and the parent material. Looking at the shapes of the graphs for W3, and note that the W3 specimens exhibited reduced RA compared to W1 and W2, it is apparent the majority of test-time was spent in accumulating strain in the weld metal while little strain was actually transferred to the HAZ. Thus, although essentially identical secondary creep rates were recorded for W3 and W1, the more uniform ductility and the absence of significant strain concentration in W3 has given rise to a considerable extension in the creep life. It can therefore be concluded from these data that a greater differential in creep strength must have been evident in welds W1 and W2 relative to the base metal, thus giving rise to considerable strain concentration and hence reduce creep life for the Type IV zone. The data also suggests that the creep properties of W3 are not dissimilar from the parent material and thus the ultimate creep life is dictated by the inherent creep strength of the Type IV region.

As anticipated, when the test temperature was increased from 600°C to 650°C there was a marked reduction in the creep life for W3 in the Type IV zone, and this was accompanied by a significant increase in the RA and elongation indicating greater total strain to which the principle contribution was made by the HAZ.

Weldment W5, in contrast to W1 to W3, consistently gave rise to failure in the weld metal, although employing a different base material to the other three deposits it would not be expected from comparison of composition and hardness that the two materials would behave particularly differently under creep conditions. It is evident, therefore, that the creep strength of weld metal W5 must be significantly inferior to the parent material and Type IV region, as well as the other three deposits. Clearly weld metal W5 was unable to sustain the applied load, failing before strain concentration, creep cavitation, and ultimately Type IV failure were able to take place. This significant reduction in the weld metal creep strength is believed to have arisen from the reduction in the Nb and N contents of the deposit as well as the introduction of Ni.

It is suggested from the data generated in this study that the differing performance of weldments W1–W3 was caused by the differential between the weld metal and base metal creep strength, rather than from the absolute creep strength of the deposit. The absolute creep strength variations can be explained from the deposit analyses, where differences were detected in the levels of C, Nb, N, Ni and V.

SUMMARY AND CONCLUSIONS

Multipass butt welds have been produced in modified 9Cr1Mo steel using commercial MMA consumables, the effect of the inherent compositional variations on microstructural development, toughness and hardness has been investigated. In addition, experimental consumable deposits containing variations in Ni, Co, W, B and N have been studied in order to further elucidate the effect of compositional variations on microstructural development, hardness, and Charpy toughness. From the results of this study the following conclusions can be drawn for the welding conditions employed.

1. Within the compositional ranges exploited in the commercial consumables, little change in microstructure was observed and only small variations in Charpy toughness (40 J transition temperature) achieved after PWHT for 2 hours at 760°C.
2. Addition of ~1%Ni, with an associated reduction in Nb, Si, and N to levels of 0.03, 0.19 and 0.021% respectively, significantly improved the 0.1 mm CTOD behaviour after PWHT at 760°C for 2 hours. This composition, however, also led to a significant reduction in weld creep strength following PWHT for 2 hours at 760°C.
3. The compositional variations present within the commercial consumable deposits gave rise to significant differences in creep rupture behaviour. Marked variation in the creep life of the Type IV region was recorded with changes in weld metal composition.
4. The study suggested that optimum creep rupture life of a weldment was achieved when the creep rupture strength of the weld metal closely matched that of the parent material.
5. A good indication of the amount of retained δ-ferrite was provided by the chromium equivalent (Cr_{eq}) and Kaltenhauser ferrite factor (FF), with fully martensitic deposits developing for Cr_{eq} value <8 and FF values <6. The FF is not suitable for use with tungsten or cobalt containing deposits.
6. The compositional variables studied did not significantly alter the transformed microstructure, although tungsten actively promoted δ-ferrite retention, and the addition of high levels of tungsten (3%) and boron (67 ppm) appeared to encourage the formation of a second phase, possibly δ-ferrite, on the prior austenite grain boundaries.
7. The addition of ~1%Ni had no effect on toughness under standard PWHT conditions. In combination with ~1%W, 1%Ni or 2%Co was sufficient to prevent δ-ferrite retention and improve the Charpy toughness relative to the addition of 1%W in isolation.

8. Increased Ni content above ~1%, or alloying with tungsten in the range 1–3% had a marked detrimental effect on toughness. Low levels of boron, up to ~20 ppm gave rise to a marginal improvement in Charpy toughness.

ACKNOWLEDGEMENTS

This work was funded jointly by Industrial Members of TWI and the information and manufacturing technology division of the UK Department of Trade and Industry. The author would like to acknowledge assistance of various colleagues with this work and thanks are also due to those Industrial Members who kindly supplied material and consumables and contributed by useful discussion. In particular thanks are extended to Dr G.M. Evans formerly of Oerlikon.

REFERENCES
1. V.K. Sikka: 'Development of modified 9%Cr 1%Mo steel', *Seventh Annual Conf. Material for coal conversion and utilisation*, National Bureau of Standards, Gaithersburg, Maryland, USA, Nov. 1982, 411–437.
2. T. Honda, T. Kusano, K. Motoki, S. Kihara, M. Kuribayashi and I. Kajigaya: 'Development of modified 9%Cr 1%Mo cast steel for advanced steam cycle values', *Proc. Conf. Properties of high strength steels for high pressure containments*, Chicago, Illinois, July 1986. Publ. ASME.
3. J.F. King, V.K. Sikka, M.L. Santella, J.F. Turner and E.W. Pickering: 'Weldability of modified 9Cr 1Mo steel', *Oak Ridge National Laboratory Report ORNL 6299*, September 1986.
4. P. Patriarca: 'US advanced materials development program for steam generators', *Nuclear Tech.* 1976, **28**(3), 516–536.
5. R.H. Kaltenhauser: 'Improving the engineering properties of ferritic stainless steels', *Met. Eng. Quarterly*, May 1971, **11**(2), 41–47.
6. N. Abe, H. Tagawa, H. Suzuki, S. Shimada, S. Sugiyama and T. Nagamine: 'Heavy section 9%Cr 1Mo steel plate with improved weldability and creep rupture strength of welded joints', *Pressure Vessel Technology Vol. 2*, Proc. 6th International Conference, Beijing, China, Publ. Pergamon Press 11–15 Sept. 1988, 1005–1012.
7. N.P. Haworth and C.A. Hippesly: 'The influence of heat affected zone micro-structures on the ductility and hydrogen embrittlement of 9%Cr 1%Mo steel', *UKAEA Report AERE R11473*, Harwell, June 1985.
8. N. Abe, T. Ikoma and M. Tamura: 'Weldability of low C−9Cr 1MoNbV steel', *Transactions ISIJ*, 1984, **24**.
9. R.S. Fidler and D.J. Gooch: 'The hot tensile properties of simulated heat affected zone structures in 9%Cr Mo and 12Cr Mo V steels', *Proc. Conf. Ferritic steels for fast reactor steam generators*, BNES London 1978, 128–135.

10. R. Panton Kent: 'Weld metal toughness of MMA and electron beam welded modified 9%Cr 1%Mo steel', *TWI Members Report 429/1990*.
11. S. Dittrich and H. Heuser: *SMAW of P91 piping with optimised filler metals*, Presented 73rd Annual AWS Convention, Chicago, 1992.
12. B.S. Greenwell and J.W. Taylor: 'The properties of candidate welds in 9CrMo–NbV steel', *Proc. Conf. Steam plant for the 90's*, I Mech E London, 4–6 April 1990, Paper C386/041, 283–296.
13. J.A. Williams: 'Methodology for high temperature failure analysis', *Proc. European Symposium on Behaviour of joints in high temperature materials*, (Eur 8021), 14–15 May, 1981 Peten, The Netherlands, 187–212.

Properties of Weldments and Aged 9–12 Cr Steels

I. ARTINGER

Technical University of Budapest, H 1521 Budapest, Müegyetem rkp. 1–3, Hungary

INTRODUCTION

New modified hardenable ferritic steels with 9–12% Cr were developed in the COST 501 project for components in power engineering structures. A part of planned collaborative work was carried out by Institute for Mechanical Technology and Materials Science at Technical University of Budapest. The major impetus for this development has been provided by the needs of power engineering industry to improve the stability of the structure and creep rupture properties of new advanced steels. They replace the traditional X20 CrMoV 12.1 steel because of enhanced high temperature rupture strength and corrosion resistance, and toughness and good cyclic deformation behaviour in low cycle fatigue regime.

In order to achieve an acceptable high longterm stability of the structure for use at high temperatures (590–610°C) these steels are alloyed with W, Mo, sometimes Co, B, and tempered above the operation temperature, near but still below Ac_1 temperature. The microstructure of these steels consists of tempered martensite matrix containing a high volume fraction of carbide particles. The mean size of original austenitic grains are about 50–350 μm.

In spite of the stabilisation treatment however, the structure of these steels and volume fraction and morphology of non-equilibrium carbide, nitride particles change more during welding and all the operation and ageing times which affect unfavourably the stability of virgin mechanical properties. The type of welding technology and the long term ageing has a great effect on the properties of these steels. Sometimes significant reductions in impact energy were observed.

The goal of this work was to investigate the changes of behaviour of forged and casted steels with different alloying contents after different welding processes (SAW, SMAW, GMAW) and after long term (1000, 3000, 6000, 10 000 hours) ageing at temperatures 575, 600 and 650°C.

This paper reports the changes in the behaviour of new steels after different welding processes, of pipes and ageing of samples with different chemical composition.

EXPERIMENTAL RESULTS

Table 1 shows the chemical composition of the investigated pipe, forged and

Table 1   Chemical Composition

Code	Cr	Mo	W	Co	Nb	N	B	S
	%				ppm			
D	8.9	0.94	0.95	–	690	840	–	20
FN4	11.5	0.5	1.8	1.9	650	600	49	50
R*	8.9	0.46	1.7	–	640	460	30	20
F	8.7	0.97	0.96	–	620	655	–	70
CF2	9.7	1.4	–	0.9	660	380	–	110
CB1	9.2	1.5	–	1.1	800	190	59	40

Table 2   Base Materials Impact Energy at 100°C/RT/0°C/−20°C Joules

D	190	170	143	110
F	143	136	122	100
FN4 a	175	120	100	77
FN4 b		60	40	35
R*	154	118	82	60
CF2$_{top}$	100	50	30	
CF2$_{bottom}$	80	34	18	
CB1$_{top}$	62	30	16	
CB1$_{bottom}$	59	36	18	

cast materials. Five modified types of steel were investigated first in the as received virgin material condition: two pipes from 1Mo–1W (911), pipe from 1.8W–B (P92), forged 1.8W–B, casted 1.4Mo–0.9Co and 1.5Mo–B heats.

The toughness of the base pipe materials (D, F, R,*) and forged FN4 are excellent Table 2. The impact energy is at a very high level in the wide temperature range. The FATT transition temperature is approximately at temperature −20°C.

It was found the pipe materials with thin (20 mm) wall and the lowest S content without B have the highest impact energy and resistance the best L.C.F. data, while, the cast material with Co and B content has the lowest impact energy, and L.C.F. failure cycles.

The forged material with Co and B content may have high toughness depending on heat treatment.

WELDING OF PIPES

The large-diameter (∅ 490 mm) and small-diameter (∅ 159 mm) thick-walled (70 and 20 mm) pipes of E 911 materials and different filler materials and

Table 3 Welding process for Shielded Metal Arc Welding (SMAW), Code: Db

Welding process		Runs number	Consumable	Travel speed mm/min	Current* A	Voltage V	Preheat °C	Interpass °C	PWHT
S	GTAW $R_w = 3,2$	Root 1	C 9 MV-IG Ø 2.4 × 1000		DCSP 200	24	200–220	220	750°C/2 hrs
M									
A									
W	SMAW	Filler 2–9	FOX C 9 MV Ø 4.0	150	DCRP 150	22–23	200–220	220–270	

* REHM INVERTIG – 460.

Table 4 Welding process for dissimilar welding, Code: Dd

D = Ø 159 × 20 mm + X8 CrNi 18.9 Ø 159 × 20 mm

Welding process	Runs number	Consumable	Gas	Current A	Voltage V	Preheat °C	Travel speed mm/min	PWHT
GMAW	Surfacing 1–3	NIBAS 70/20–IG Ø 1.2		Pulsed Arc* 25–275 (85)	15–30 (2.8–53 Hz)	250–300	350	700°C/2 hrs
GTAW	Root 1	FOX NIBAS 70/20 Ø 1.2		DCSP** 140	22–23		160	
GMAW	Filler 1	NIBAS 70/20–IG Ø 1.2	Argon	Pulsed Arc* 80–290 (160)	15–30 (100 Hz)		400	
	2			70–275 (125)	15–30 (93 Hz)			
	3			50–275 (110)	14–30 (89 Hz)			

* FRONIUS TA 500 LIMAT Rt 860 – 6 Robot.
** REHM INVERTIG 160 GW.

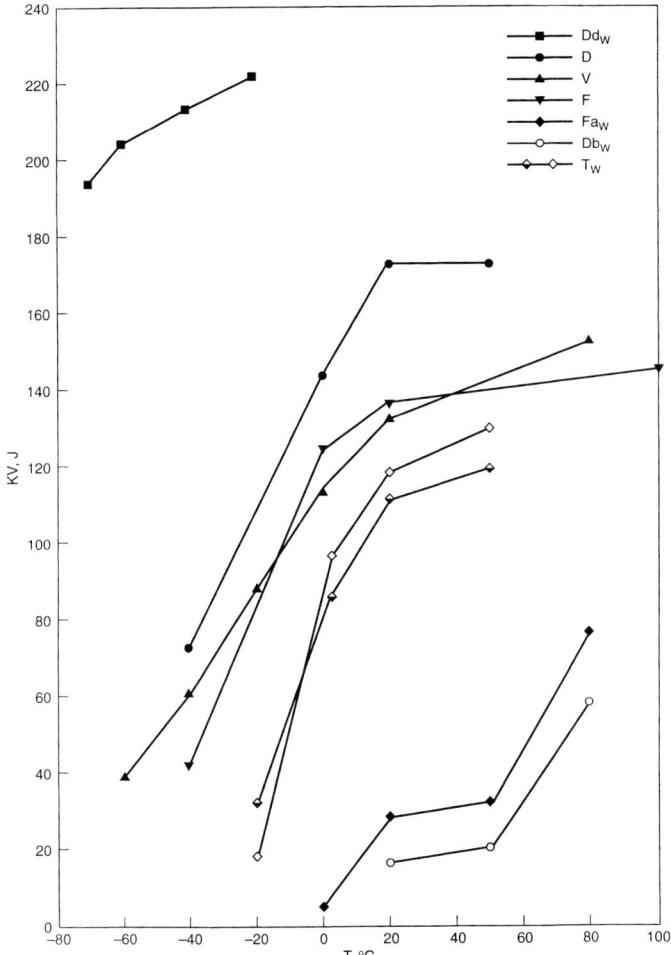

Fig. 1   Charpy impact test results for base materials and weldments.

welding processes were chosen for investigations. We have investigated different welding procedures: Submerged Arc Welding (SAW) for thick wall (70 mm) pipe, Shielded Metal Arc Welding (SMAW) for 20 mm wall-thickness and Gas Metal Arc Welding (GMAW) for dissimilar welding.

Suitable welding parameters, preheating temperatures and post weld heat treatments (PWHT) were planned and used. The welding process conditions used are summarised in Tables 3–4.

The new martensitic steels have a fully martensitic structure after cooling and must be tempered at relatively high temperature. Therefore crosswelds from these steels must be preheated before welding and heat-treated (stress relieved) after welding (PWHT).

The crossweld sections after SAW and SMAW welding were exposed to PWHT at temperature 750°C/2 hours. Figure 1 shows the Charpy impact tests results of the base materials and weldments. Positions of notch were located at the centre of the weld metal, in the heat affected zone and in the base metal.

The toughness of the heat affected zone was as high as the base metal. Sometimes the values of impact energy obtained were even higher than the values of the base metal. Much lower impact energy values were reached in the weld metal with the C 9 MV type filler material. Impact energy of both welded metals (SAW, SMAW) showed sufficiently low levels of 15–25 Joule at 20°C. The transition temperatures compared with the base metal were shifted by 80°C towards higher temperatures. This effect could be attributed to the relatively high carbon content (~0.2%) in the used type of filler metal and to the high value of heat input per unit length (2.1 kJ/mm) at SAW welding process.

The impact energy of the dissimilar welded joint (GMAW), where we used NIBAS 70/20 type filler material showed a much higher level (220 Joule) at −20°C than the base material (Fig. 1).

Recently new filler materials have been developed. We have investigated the effect of heat input and PWHT on the impact energy of the weld metal from the low carbon (0.10–0.11%) and sulphur (0.006–0.008%) content filler material type: MTS 9.11 (0.9 Mo–1 W) and MTS 6.16 (0.5 Mo–1.6 W). The welding parameters and PWHTs are shown in the Table 5.

Table 5 Welding processes for Shielded Metal Arc Welding (SMAW), weld metal tests

Consumable	Runs number	Current (DCRP) A	Travel speed mm/min	q/v kJ/mm	Interpass °C	PWHT
Thermanit MTS 9.11 Ø 3.25	1–6	100–125	138–172	0.5–0.85		a.) 750°C 2 hours
Ø 4	7–17	120–150	154–205	0.535–0.986	250	b.) 730°C 12 hours
Thermanit MTS 6.16 Ø 3.25	1–6	100–125	138–172	0.5–0.85		a.) 750°C 2 hours
Ø 4	7–17	120–150	154–205	0.535–0.986	250	b.) 730°C 12 hours
Thermanit MTS 9.11 Ø 3.25	1–6	100–140	118–127	0.8–1.1		a.) 750°C 2 hours
Ø 4	7–10	140–160	106–119	1.1–1.44	250	b) 730°C 12 hours

Table 6 Impact energy at 100°C/RT/0°C temperatures, Joules

Code	Base material	OVERAGEING						
		at 600°C				at 575°C		at 650°C
		1000 h	3000 h	6000 h	10 000 h	1000 h	10 000 h	3000 h
D	190/170/143		124/88/52	100/65/48				
F	147/136/122	120/85/57				95/68/41	115/85/62	100/55/30
CF2top	100/50/30			70/22/19				
CF2 bottom	80/34/18		54/19/11	70/24/17				61/23/14
CB1 top	62/30/16		37/15/8	42/17/10				39/19/10
CB1 bottom	59/36/18		43/15/8	55/18/12				44/22/13

There was no large difference in the toughness of base and weld metal from filler material MTS 6.16 with small value of heat input (0.5–0.98 kJ/mm) and different (750°C/2 hours, 730°C/12 hours) PWHT. Both heat input and PWHT have beneficial effects on the toughness of weld metal from filler material MTS 9.11. The decreased heat input (0.5–0.98 kJ/mm) at lower PWHT temperature (730°C/12 hours) is more suitable for the higher toughness.

The creep rupture lives of welded F material at a temperature of 600°C were similar to the base material. Sometimes short-term failures occurred in weld metal.

EFFECT OF AGEING

Pieces from different pipe and cast materials were aged at temperatures 575, 600 and 650°C for 1000, 3000, 6000 and 10 000 hours, then tested for Charpy impact energy, low cycle fatigue (L.C.F.) and long term creep rupture.

The impact energy values after different types of ageing are shown in Table 6.

The absorbed energy of aged pipe specimens tends to decrease continuously with ageing during and temperature. The toughness values have remained high enough after long term ageing. The impact values of pipe materials after ageing at 600°C up to 10 000 hours are larger than 65 and 30 Joule at RT and −20°C respectively.

The impact energy of the cast materials has decreased similarly at temperatures 600°C and 650°C for 3000 hours and has increased a little between 3000–6000 hours of ageing at 600°C.

The number of cycles to failure at L.C.F. of different steels is shown in Fig. 2. The cycles to failure of long term aged specimens have decreased significantly (Fig. 3).

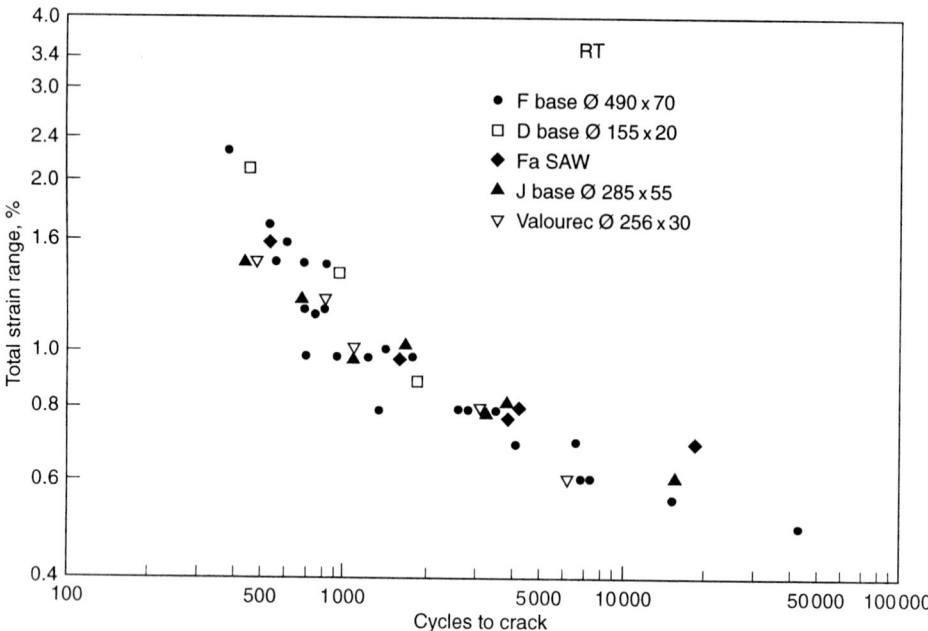

Fig. 2 L.C.F. test results for base and SAW welded materials.

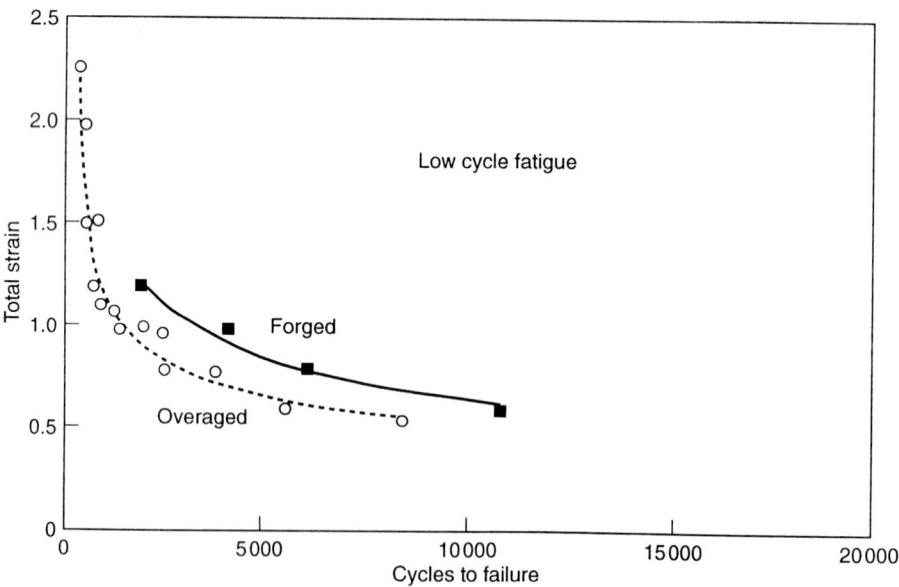

Fig. 3 Effect of ageing on L.C.F. test results.

The L.C.F. data of welded F material were similar to that of the base material.

As significant drop in creep rupture time was observed with overageing. The cycles to failure of long term overaged specimens also decreased significantly.

CONCLUSION
– The new modified pipe and forged materials have excellent impact toughness and L.C.F. data.
– It is difficult to achieve the standard required toughness level in weldments.
– The impact energy of welded pipe may remain high enough with suitable welding parameters and using new filler materials.
– Weld metal toughness strongly depends on:
    – chemical composition of consumables
    – welding process parameters
    – post weld heat treatment temperature.
– Ageing effects on the L.C.F. and the creep rupture time: significantly.

ACKNOWLEDGEMENTS
This work has been supported by the Hungarian National Fund for Supporting Scientific Research, Grant No OTKA T 017021 and by the European Commission (COST 501 ERBCIPECT 926007), which is gratefully acknowledged.

Microstructural Evolution and its Effects on the Creep Performance of High Temperature Alloys

B.F. DYSON AND M. McLEAN

Department of Materials, Imperial College of Science, Technology and Medicine, Prince Consort Road, London SW7 2BP, England

ABSTRACT

Most structural materials are developed by optimising microstructure to deliver a specified short-term mechanical performance. During high temperature service, this carefully tailored microstructure may progressively degrade with often unexpected consequences for long-term creep performance. The potential benefits of taking microstructural degradation into account when predicting creep lifetime has long been recognised but in the main, empirical procedures using a single universal equation have prevailed and with some success. Yet the fact remains that the same superficial shape of creep curve can arise from the operation of quite different mechanisms of material degradation (damage), each leading to contrary conclusions when extrapolations are made to either long term service or, more importantly, complex loading conditions. The thrust of this paper is that an effective life-assessment strategy not only requires a robust creep mechanics framework but also a quantitative description of the various microstructural damage mechanisms. There are three key aspects: to identify those features of the microstructure that have a significant influence on creep behaviour; to quantify their evolution during creep life; and to incorporate each type of damage into a quantitative description of creep rate. Physically-based Continuum Creep Damage Mechanics (CDM) provides a suitable framework and examples have been given of its usage in quantifying the shapes of creep curves caused by several microstructural damage mechanisms.

INTRODUCTION

Great care is taken when alloys are being developed to withstand stresses at high temperatures but the process is still mainly evolutionary and iterative. Effects on creep performance induced by changes in chemical composition and heat-treatment are usually monitored by combining microstructural observations with relatively short-term mechanical tests which are themselves often constrained by requirements imposed in some national or international code of practice. The commercial benefits from such time-consuming and expensive procedures for optimising the initial microstructure may be blighted when a *thermally-induced* change of the microstructure (such as

371

particle-coarsening or matrix-solute depletion) occurs during service at times much longer than those experienced in laboratory testing. Thermally-induced microstructural changes are not the only mechanisms that can degrade the alloy-strength: two further engineering categories of 'damage' have been conceived as a result of analysing laboratory creep tests or service-exposed material. Within the *strain-induced* category, microstructures evolve at rates that are proportional to the current state of deformation – no strain, no damage; while damage kinetics in the *environmentally-induced* category are dependent on testpiece or component dimensions and the chemical composition and reactivity of the surrounding fluid environment.

The term 'damage' is often used in a totally empirical way in engineering practice, with neither a definition nor indeed any physical concept given for the cause of any degradation in material performance. That is not the case in the present paper: various mechanisms of damage have been reviewed; placed within the three categories described above; and for each, their evolution and effect on creep response has been quantified. A specific major change in microstructure may or may not have a significant influence on alloy performance and it is often difficult to discount what may have been high-quality and painstaking-acquired metallographic measurements that claim to support such a relationship without an accompanying kinetic theory of creep and damage evolution. Research into damage caused by grain boundary creep cavitation serves to illustrate this point very well. In alloys where the level of cavitation is low (but still easily measurable), its effect on creep performance is negligible, influencing at best the local reduction in area at failure in a necking testpiece without significantly changing lifetime or elongation. At the other extreme of high levels of cavitation, not only is elongation reduced to a fraction of that found in the absence of cavitation, but its presence can also accelerate strain rates and shorten lives appreciably. These differences are now well-understood theoretically and their consequences are modelled quantitatively in this paper. An interesting outcome of the understanding of cavitation is that the maximum acceleration of creep rates and the biggest reductions in ductility are achieved with cavity distributions that *cannot* be resolved using the conventional observational tool of optical microscopy; only SEM and TEM have the necessary resolving power.

The objective of this paper is to review and categorise the microstructural damage mechanisms that are believed to be important in alloys used for critical components in power engineering and to illustrate how physically-based continuum creep damage mechanics (CDM) can be used for design- and remanent-lifetime predictions.

MICROSTRUCTURE AND THE CONCEPT OF DAMAGE

An effective way to improve creep resistance by manipulating microstructure

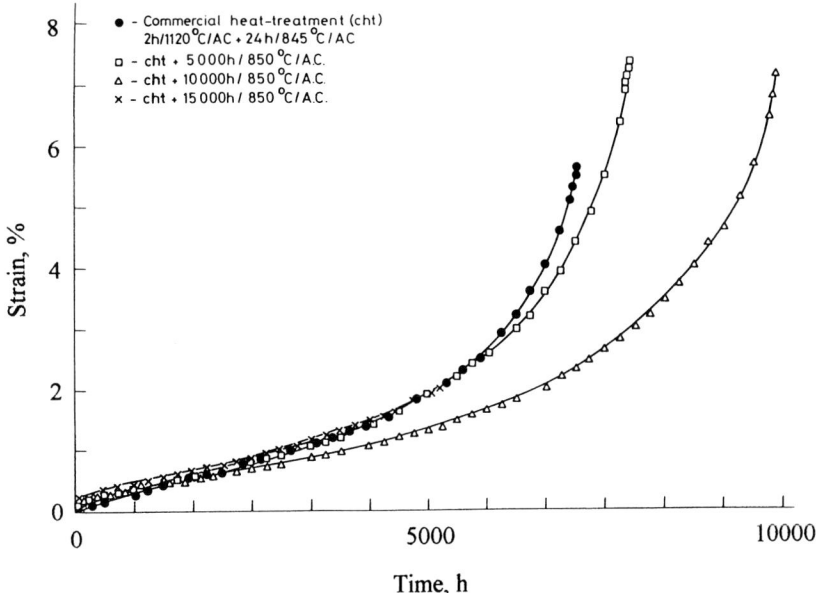

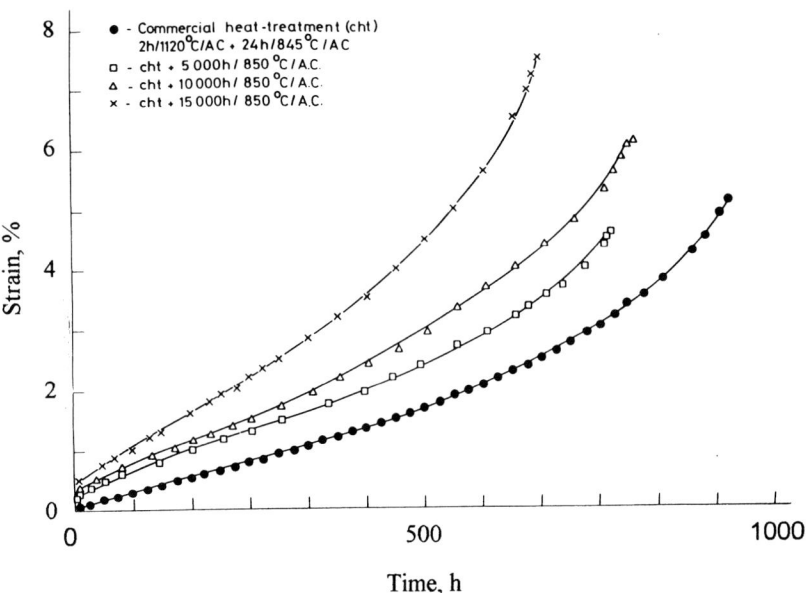

Fig. 1   Effects of prior thermal exposures up to 15 000 h on creep behaviour of the conventionally-cast nickel base superalloy In738LC (1): (a) 170 MPa/850°C; (b) 250 MPa/850°C. Taken from Tipler and Peck.[1]

is to introduce a dispersion of particles within the alloy matrix grains. These can take a variety of forms and be introduced by a range of processing routes. For example, in nickel-based superalloys, high volume fractions (20–70%) of ordered coherent particles of the phase Ni_3Al (γ') can be precipitated by means of an optimised set of thermal treatments. The finite solubility of γ' guarantees that there will be significant thermally-induced particle-coarsening and morphological changes in service, even though alloy design successfully minimises their kinetics. However, it has been suggested that particle coarsening has relatively little effect on the long-term creep performance of superalloys and consequently does not constitute a significant *damage*. This has been clearly demonstrated at service levels of stress by Tipler and Peck,[1] who found no effect on creep behaviour of prior ageing the conventionally cast alloy IN738LC, Fig. 1(a). Although the ageing treatment radically altered the γ' size and distribution, it had relatively little effect on grain size and on the contribution from grain boundary cavitation, as manifested in the almost constant ductility. This information suggests strain-induced cavitation as the likely damage mechanism. But other data on the directionally-solidified version of the same alloy show similar strain/time trajectories and yet little cavitation is exhibited (and consequently ductility is much higher). This casts doubt on cavitation being the major intrinsic damage mechanism and the suggestion has been made that another strain-softening mechanism, the gradual accumulation of mobile dislocations, is responsible.[2] In contrast to the data at low stresses, Fig. 1(b) shows that at stresses greater than about half the yield stress, there are systematic reductions in creep resistance after prior thermal exposure in much the same way as particle-ageing reduces ambient temperature yield stress. The essential features of this complex behaviour will be modelled in the section on Shapes of Creep Curves by incorporating thermally- and strain-induced mechanisms of damage within the strain rate equation.

The nickel-base superalloys used in power propulsion are the most creep resistant metallic alloys currently available but the option of introducing a large volume fraction of precipitate into the matrix is not available in the cheaper iron-based alloys used in electricity generation plant. Alternative strategies are therefore being employed to improve the inherent creep resistance given by the small ($\approx$1%) volume fraction of carbide/nitride precipitates. Advanced martensitic alloys appear to rely on two additional means of increasing creep strength: very fine subgrains within prior austenite grains and additions of W and Mo into the ferrite matrix of the normalised alloy. Damage in these materials can be caused by both thermal- and strain-induced changes to these three components of the microstructure, acting either alone or in combination. Figure 2 is a schematic illustration of the microstructures and potential damage mechanisms in the gauge and head sections of 12% Cr

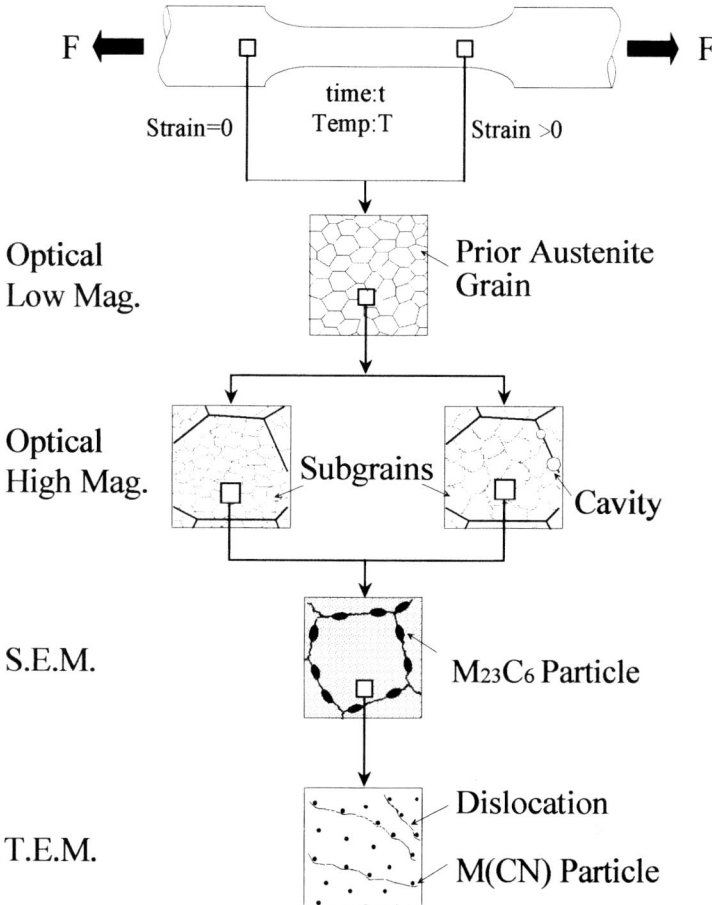

Fig. 2   Illustrating the microstructures and potential damage mechanisms in the gauge and head sections of 12% Cr steel, observed at different magnifications.

steels observed at different magnifications. At the lowest optical level, both gauge and head sections display an identical prior austenite grain size. In contrast, at the highest optical (or lowest SEM) magnification, it is found that the subgrains are larger in the gauge section where the only difference is the strain accumulated.[3] Sometimes isolated grain boundary cavities are also observed at this level of magnification but these martensitic steels rarely display a significant cavity density and failure under uniaxial tension is usually due to a necking instability. At higher magnifications, the subgrain boundaries are decorated with $M_{23}C_6$ particles which could in principle, control subgrain growth by a thermally-induced mechanism (driven by surface energy alone);

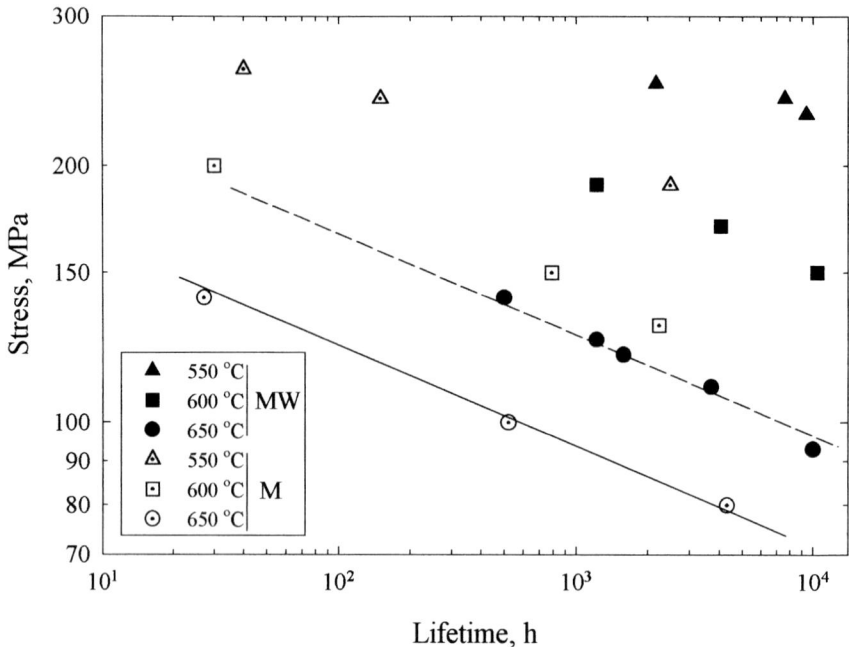

Fig. 3 Demonstrating the potent and stress-independent influence of tungsten (alloy MW) in improving creep strength of a 12% Cr steel, (alloy M).[8]

the consequence would be that subgrain growth rates would be identical in the gauge and head sections. The observation that subgrains are larger in the gauge section provides unambiguous evidence of the dominance of strain-induced (or dynamic) subgrain growth over thermally-induced growth and is an important experimental finding. Blum and coworkers[4,5] have quantified dynamic subgrain growth kinetics and found agreement with data obtained by Eggeler *et al.*[6,7] At still higher magnifications (TEM), particles of M(CN) can be seen in the subgrain interiors. According to the creep model described in the next section, the volume fraction and state of dispersion of these particles play an important role in determining the intrinsic (or base-line) creep resistance of these alloys and so their coarsening rate is potentially an important thermally-induced damage mechanism.

Additions of W and Mo are made to martensitic 9–12% Cr steels to improve lifetimes; the mechanism is usually assumed to be due to some form of 'solid solution strengthening' of the ferrite grains. Figure 3 illustrates very well that the 'solid solution strengthening' of W/Mo is in fact a kinetic effect since, to a first approximation, any strengthening is independent of stress. This observation has influenced subsequent modelling of the effect of these elements on creep resistance.[8] Damage can occur during thermal exposure because W and Mo are removed from solid solution as precipitation of the Laves phases, Fe_2Mo and Fe_2W, occurs. Thus Fig. 2 is incomplete. Matrix depletion of W and Mo and the accompanying precipitation of Laves phases have now been quantified by Kadoya et al.[8] by relating measured decreases of W/Mo in solid solution to concurrent increases in creep rate. The kinetics come within the thermally-induced category. This is another example of how metallographic observations alone can be misleading since precipitation of Laves phase might be expected to increase creep resistance or even reduce creep ductility. However, the data of Kadoya et al.[8] demonstrate that the depletion of W and Mo in solid solution that necessarily accompanies this – and which is much more difficult to monitor – has a far greater creep weakening effect.

PHYSICALLY-BASED CDM

Physically-based continuum creep damage mechanics (CDM) is a multi-state variable formulation for creep rate that has its origins in the single-state variable description of tertiary creep introduced by Kachanov[9] and Rabotnov.[10] However, CDM differs considerably in detail from the original empirical concept of Kachanov/Rabotnov and additionally considers primary creep, with secondary creep being merely an inflection in the strain/time trajectory. The CDM formulation therefore contrasts with all but two of the treatments of creep in the metallurgical literature (the exceptions being Graham-Walles[11] and the θ Projection Method[12]) that purport to provide an equation for steady-state creep without reference to the possibility of distortions due to tertiary damage processes. Whether the resultant secondary creep rate is close or not in magnitude to a (theoretically achievable) steady-state rate depends on specific circumstances and will be assessed quantitatively later in the section on Shapes of Creep Curves.

Formally, CDM in uniaxial tension can be represented by a set of linear differential equations:

$$\dot{\epsilon} = \dot{\epsilon}\,(\sigma,T,H,D_i)$$
$$H = H(\sigma,T,H,D) \qquad (1)$$
$$\dot{D}_i = \dot{D}_i(\sigma,T,H,D_i)$$

where the strain rate $\dot{\epsilon}$ is a function of the applied stress σ, temperature T, an evolving hardening parameter H to represent primary creep and the set of evolving damage parameters D_i.

Creep Rate Equation
Lifetime predictions require explicit functional forms for $\dot{\varepsilon}$, $\dot{H}$ and $\dot{D}_i$. Initial efforts (Dyson and McLean,[2] Ashby and Dyson,[13] Ion et al.[14]) were directed mainly towards quantifying the latter two functions which were then incorporated into the empirical Norton power-law function, modified to include a threshold stress, σ_{th}:

$$\dot{\varepsilon} = A(\sigma - \sigma_{th})^n \tag{2}$$

Primary hardening and thermally-induced tertiary damage can easily be incorporated into eqn (2), through the terms A and σ_{th}, but problems do arise. The threshold stress was introduced historically into eqn (2) in order to facilitate using a constant value of n ($\cong 4$) and an activation energy close to that for matrix diffusion. A serious criticism of eqn (2) is that threshold behaviour has never been reported in nickel-base superalloys or ferritic steels and creep occurs at stresses where conventionally defined 'n' values approaching unity have been found without any evidence of diffusion creep.[15] Equation (2) therefore becomes just another empirical (and complicated) expression for creep.

An alternative model of creep in precipitation-strengthened alloys has recently been developed which overcomes these deficiencies:[16,17]

$$\dot{\varepsilon} = \dot{\varepsilon}_0 \sinh\left(\frac{\sigma(1 - H)}{\sigma_0}\right) \tag{3}$$

where

$$\dot{\varepsilon}_0 = k' \rho_m \frac{1 - \phi_p}{\phi_p} c_j D_m \tag{4}$$

and

$$\sigma_0 = \frac{kT}{\alpha G b^3} \sigma_{or} \tag{5}$$

k' and α are constants of the order unity, ρ_m the mobile dislocation density, ϕ_p the particle volume fraction, c_j the jog density, D_m the volume diffusivity, G the Shear Modulus and σ_{or} the Orowan Stress for particle by-pass.

Primary creep in the alloys of interest to power engineering is believed to be kinematic in character and arises from stress redistribution around hard regions of the microstructure – particles, subgrains, pearlitic regions etc. It should not be confused with the ideas of dislocation-hardening and -recovery that apply to pure or lightly alloyed metals which in the main, lead to isotropic hardening. The model used in this paper is a slight modification of the one proposed by Ion et al.:[14] an internal back stress is defined by $\sigma_i = \phi_p \sigma_p$, where σ_p is the traction acting across the hard-region/matrix interface and having a hard-region volume fraction ϕ_p. The dimensionless hardening parameter $H = \sigma_i/\sigma$ has an evolution rate at constant stress given by

$$\dot{H} = \frac{h'}{\sigma} \left(1 - \frac{H}{H^*} \right) \dot{\epsilon} \qquad (6)$$

where $h' = \phi_p E$, E is Young's Modulus and the superscript star denotes saturation of the redistribution. Whether or not the exhaustion of primary creep coincides with the time to achieve a minimum creep rate (and whether that rate is sensibly equivalent to steady state creep) depends as noted above, on the damage accumulation rate and is quantified below in the section on Shapes of Creep Curves.

Incorporation of Creep Damage
Damage is introduced into eqn (3) by defining additional dimensionless parameters which are then incorporated in one of two ways. In the first, there is a progressive augmentation of the stress applied to the alloy due to a decrease in load-bearing area that is either *real* (when damage is caused by reduction in external section or loss of internal section due to the volume fraction of cavities) or *effective* (when damage is due to creep-constrained cavitation or coarsening of strengthening particles). The *real* reduction of stress due to cavities is trivial (because of its low volume fraction) but the *effective* reduction may become very large through the mechanism of constrained cavity growth. This type of damage provides a physical basis for the seminal ideas of Kachanov/Robotnov.[9,10] But damage can also be introduced in a second way by increasing the intrinsic rate of deformation though the dislocation flux term; a method not considered by Kachanov/Robotnov. Two mechanisms have been identified: (i) by a progressive depletion of solid solution strengthening elements caused by precipitating a mechanically innocuous phase; and (ii) by an increase in the density of mobile dislocations. This type of damage is disconnected from the applied stress with the consequence that a fixed amount of damage imparts a stress-independent strain rate acceleration-ratio. It is not possible, therefore, to devise a universal equation to account for the evolution of 'damage', rather the nature of the damage mechanism must be reflected in a specific form of equation. The proponents of universal equations imply the reverse: that somehow the underlying kinetics of hardening and softening must obey certain restrictive laws. For example, the form of the θ equations have led its proponents to conclude that all relevant hardening and softening mechanisms must obey first order reaction rate kinetics.[18]

Although there is some arbitrariness in the precise mathematical definition of each type of damage, the important point to note is that any change in defi-

Table 1 Creep Damage Categories, Mechanisms and Incorporation into CDM

CREEP DAMAGE CATEGORY	DAMAGE MECHANISM		DAMAGE PARAMETER D	DAMAGE RATE $\dot{D}$	STRAIN RATE $\dot{\epsilon}$
Strain-Induced	Cavity Nucleation Control; Growth Constrained		$D_N = \dfrac{\pi d^2 N}{4} = \omega$	$\dot{\omega} = \dfrac{k_N}{\epsilon_u}\dot{\epsilon}$	$\dot{\epsilon} = \dot{\epsilon}_0 \sinh\left[\dfrac{\sigma(1-H)}{\sigma_0(1-\omega)}\right]$
	Cavity Growth Controlled by Creep-Constraint		$D_N = \dfrac{\pi d^2 N}{4} = \omega$ $D_G = \left(\dfrac{r}{l}\right)^2$	$\dot{\omega} = 0$ $\dot{D}_G = \dfrac{d}{2lD_G}\dot{\epsilon}$	$\dot{\epsilon} = \dot{\epsilon}_0 \sinh\left[\dfrac{\sigma(1-H)}{\sigma_0(1-\omega)}\right]$
	Dynamic Subgrain Coarsening		$D_{sg} = 1 - \left(\dfrac{r_{sg,i}}{r_{sg}}\right)$	$\dot{D}_{sg} = (1-D_{sg})\dfrac{\dot{r}_{sg}}{r_{sg}}$	?
	Multiplication of Mobile Dislocations		$D_d = 1 - \dfrac{\rho_i}{\rho}$	$\dot{D}_d = C(1-D_d)^2\dot{\epsilon}$	$\dot{\epsilon} = \dfrac{\dot{\epsilon}_0}{(1-D_d)}\sinh\left[\dfrac{\sigma(1-H)}{\sigma_0}\right]$
Thermally-Induced	Particle-Coarsening		$D_p = 1 - \dfrac{P_i}{P}$	$\dot{D}_p = \dfrac{K_p}{3}(1-D_p)^4$	$\dot{\epsilon} = \dot{\epsilon}_0 \sinh\left[\dfrac{\sigma(1-H)}{\sigma_0(1-D_p)}\right]$
	Depletion of solid-solution elements		$D_s = \beta\phi_p$	$\dot{D}_s = K_s D_s^{1/3}(1-D_s)$	$\dot{\epsilon} = \dfrac{\dot{\epsilon}_0}{(1-\beta D_s)}\sinh\left[\dfrac{\sigma(1-H)}{\sigma 0}\right]$
Environmentally-Induced	Fracture of Surface Corrosion Product		$D_{cor} = \dfrac{2x}{R}$	$\dot{D}_{cor} = \dfrac{1}{R}\left(\dfrac{K_c \dot{\epsilon}}{\dot{\epsilon}^*}\right)^{1/2}$	$\dot{\epsilon} = \dot{\epsilon}_0 \sinh\left[\dfrac{\sigma(1-H)}{\sigma_0(1-D_{cor})}\right]$
	Internal Oxidation		$D_{ox} = \dfrac{2x}{R}$	$\dot{D}_{ox} = \dfrac{K_{ox}}{R^2 D_{ox}}$	$\dot{\epsilon} = \dot{\epsilon}_0 \sinh\left[\dfrac{\sigma(1-H)}{\sigma_0(1-D_{ox})}\right]$

nition must be accompanied by a corresponding mathematical transformation in the equation describing its evolution rate. In other words, the definition of damage and its evolution are inextricably coupled. Computational advantages are gained by defining damage parameters so that they fall within the range zero to unity. Engineers deal in numbers and equations while showing little interest in underlying mechanisms and so to reflect this, creep damage has been classified into three engineering categories: strain-induced; thermally-induced; and environmentally-induced, but with each category containing more than one physically distinct mechanism. The damage parameters for the range of mechanisms that have been identified to date, together with their effects on creep rate are listed in Table 1. The mechanisms and classification scheme in Table 1 is built upon previous work[13] and only a brief description follows.

Strain-induced damage   The kinetic law is $\dot{D} = \dot{D}(\sigma,T,H,D)$, signifying that creep has to occur before any damage can accumulate. Four damage mechanisms are shown in Table 1: two cavitation; dynamic subgrain coarsening and multiplication of mobile dislocations. In each case the damage parameter has been defined to lie between zero and unity and each damage rate has been derived from either theory or experiment. Cavity growth that is controlled by creep-constraint is unique in having two damage parameters since the idea is that a certain fraction of grain facets ω are cavitated at $t = 0$ but fracture only occurs when individual holes, characterised by D_G, have grown by strain to coalescence. In both cavitation mechanisms, strain rate is powerfully dependent on the magnitude of ω; the difference between the two being that ω evolves with strain in the first but takes on its maximum value at $t = 0$ in the second. Clearly the latter can have a profound effect on the initial creep rate, as shown in the section on Shapes of Creep Curves.

Dynamic subgrain growth has been extensively researched by Blum and co-workers[4,5] but has yet to be put within the CDM framework. It seems to be well established that subgrains grow at a rate that is proportional to strain rate, but theoretical work is required before it can be incorporated within the strain rate equation.

Multiplication of mobile dislocations is characterised by a single parameter 'C' and enters the strain rate equation by modifying the rate function, $\dot{\epsilon}_0$ rather than the stress.

Thermally-induced damage   The kinetic law is $\dot{D} = \dot{D}(T,D)$, signifying that thermal exposure alone can cause damage. This can be a potent source of damage during long service times or after excursions at higher temperatures that either occur unexpectedly during service or deliberately prior to service. Particle-coarsening has long been recognised and the damage rate equation is

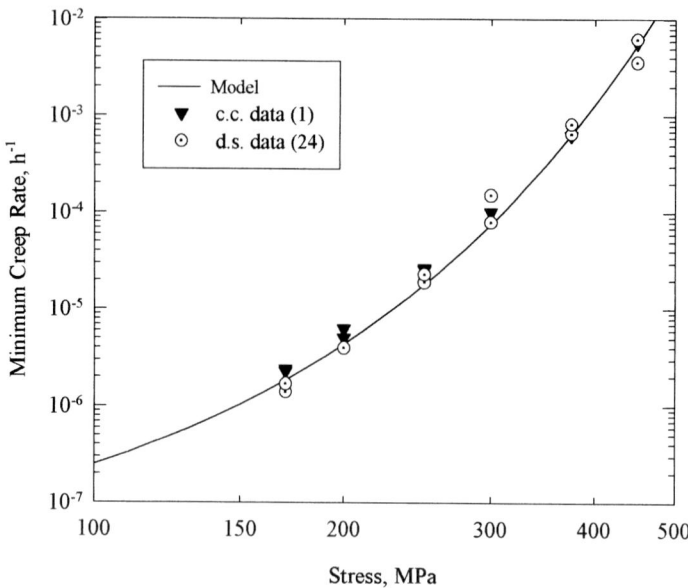

Fig. 4   Comparison of model predictions of minimum creep rate as a function of stress in IN738LC at 850°C using the parameter-set in Table 2 with datasets from conventionally-cast (1) and directionally-solidified (24) materials.

the matrix diffusion-controlled Ostwald Ripening equation due to Liftshitz[19] and Wagner.[20] The new theoretical feature[16,17] is that coarsening enters into the denominator of the argument of the strain rate equation rather than the previous method of reducing the magnitude of the threshold stress in eqn (2).[21,22] Depletion of solute elements to precipitate widely spaced particles of second phase (aimed specifically to date at the technologically important precipitation of Laves phase in 12% Cr martensitic steels) has been modelled[8] using the precipitation and growth equation of Wert and Zener.[23] The depleting atoms (in this case W and Mo) are assumed to affect only the rate constant $\dot{\epsilon}_0$, to be consistent with the findings of Fig. 3.

Environmentally-induced damage   There are two kinetic laws, reflecting the fact that the testpiece or component radius, R, can influence both a strain-induced and a thermally-induced category: $\dot{D} = \dot{D}(\sigma,T,H,D,R)$ and $\dot{D} = \dot{D}(\sigma,T,D,R)$. If surface-oxidation is periodically fractured by creep strain to expose new alloy surface, then the oxidation kinetics are accelerated and this, coupled with a smaller radius will increase the effective stress and the faster the testpiece or component will fail. When damage is by internal oxidation (to produce for example, gas bubbles that are not load-bearing) then it is long thermal exposures and small testpiece radii that cause the most damage.

SHAPES OF CREEP CURVES

Model Predictions

In order to illustrate the effect of different types of damage on the shapes of creep curves, a model parameter-set for undamaged material was derived using minimum creep data from the directionally-solidified nickel-base superalloy In738LC at 850°C and is given in Table 2. These parameter-values are part of a more extensive analysis covering several temperatures.[24] Each value lies within a range dictated by limits set by the theory discussed above. To simulate the conditions of a creep test, the parameters were input into eqns (3) & (6) and integrated isothermally under constant load to generate the solid line in Fig. 4. There is also a very good fit with the experimental dataset taken from the work of Tipler and Peck[1] on the conventionally-cast version of In738LC, although the tendency is for these data to lie towards the upper bound of the dataset for directionally-solidified material.

Table 2 Model parameter-set for non-damaged IN738LC used in the generation of the curves in Fig. 4

$\dot{\epsilon}_0$ h^{-1}	σ_{or} MPa	h' MPa	H^*
2.9×10^{-8}	430	4×10^4	0.4

Damage mechanisms acting alone

The effect of introducing a damage mechanism singly on the non-damage strain trajectory was computed for four types of damage and the resultant creep curves are illustrated quantitatively in Figs 5 (a)–(j). Each damage mechanism was studied at two levels of stress (150 and 300 MPa) using the parameter-set in Table 2 and the appropriate strain rate and damage evolution equations given in Table 1. To illustrate base-line behaviour, Figs 5 (a) and (f) present the strain/time trajectories under constant stress and constant load conditions. Although a damage parameter could have been defined for a constant-load test in the manner of Table 1, it is computationally more efficient to impose the constraint $\dot{\sigma} = \sigma\dot{\epsilon}$ to reflect the increasing stress with creep strain. The values for minimum creep rate are nearly the same under constant stress and constant load and identical to those plotted during the generation of Fig. 4. The convergence is a consequence of the stress redistribution process modelled during primary creep being essentially complete at small strains. It is worth noting that the strain at which primary creep is finished is predicted to increase with applied stress, as observed experimentally. This is a consequence of the normalised internal stress H being in the argument of the

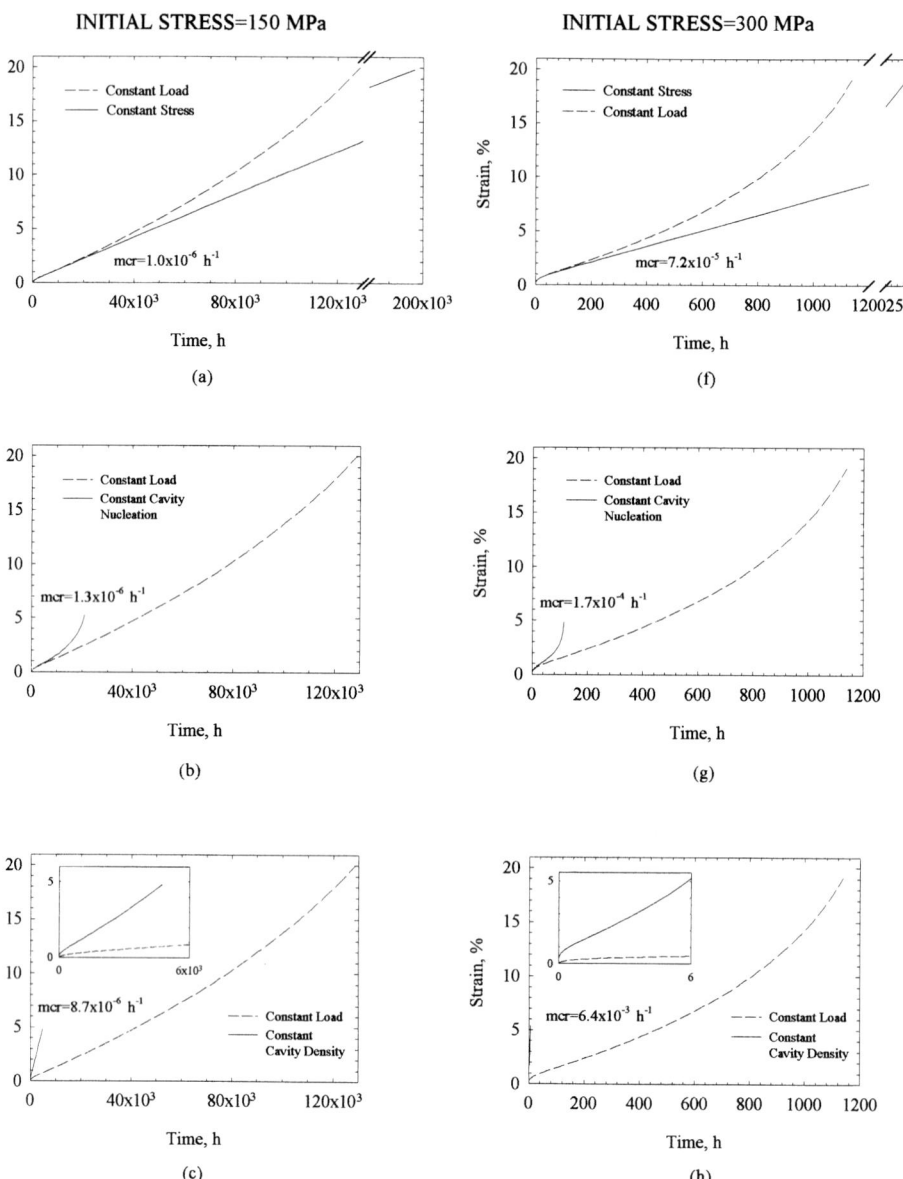

Fig. 5 Comparison of strain/time trajectories based on the model parameter-set in Table 2 and the various damage mechanisms taken individually: (a)–(e) 150 MPa; (f)–(j) 300 MPa.

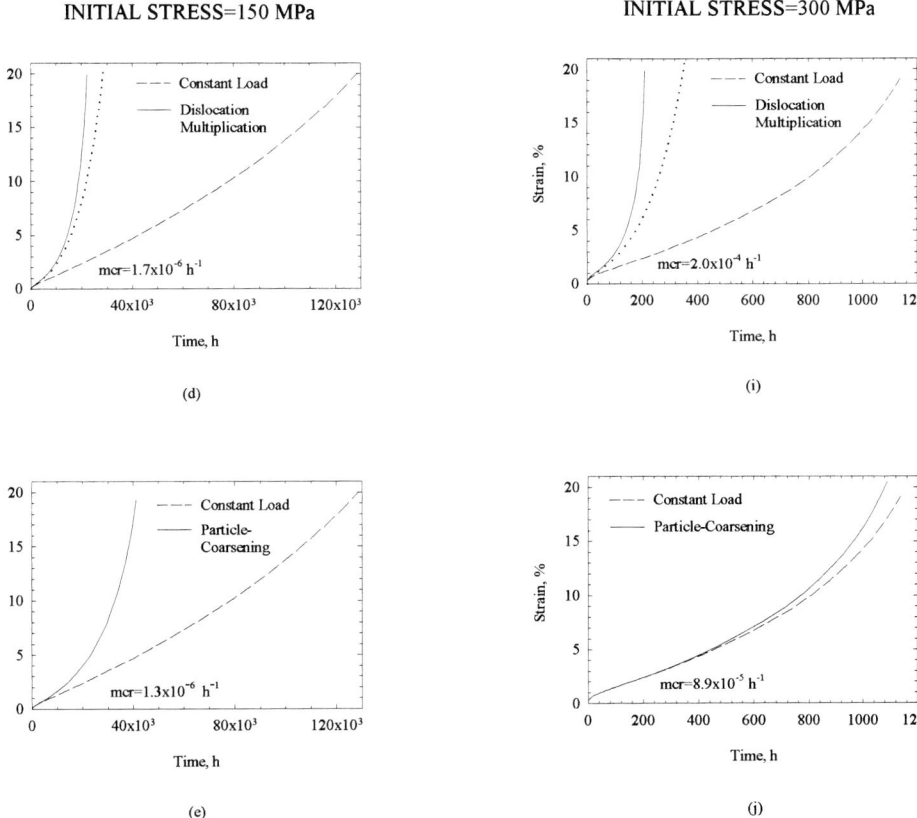

Fig. 5 (continued)   Comparison of strain/time trajectories based on the model parameter-set in Table 2 and the various damage mechanisms taken individually: (a)–(e) 150 MPa; (f)–(j) 300 MPa.

sin*h* function rather than any arbitrary stress-dependent change in model parameters – it is yet another example of how an experimental observation can be misleading without guidance from theory. Deviations in strain trajectory become apparent more at the higher load because the effective '*n*' value is greater. The two figures illustrate very clearly that differences in behaviour become increasingly apparent at higher strains and stresses. At 300 MPa, the time to reach 20% strain is approximately 2½ times longer at constant stress than at constant load. It will be shown below that such differences between constant load and constant stress become very much reduced when the material suffers some intrinsic form of damage evolution.

Grain boundary creep cavitation is probably the most quoted form of damage and often erroneously thought of as being the only important manifestation of creep damage. It is likely however, to be the only damage mechanism having an important influence on ductility in engineering alloys and it can also accelerate creep rates dramatically in certain conditions. Creep cavitation can only occur in polycrystalline materials (wrought or conventional cast) and the detailed kinetics are complicated: Figs 5 (b), (c), (g) and (h) are therefore simply an illustration of its influence on creep behaviour and use the damage equations in Table 1. We consider two possible situations: (i) where cavities are continuously nucleated with increasing strain, Figs 5 (b) and (g); and (ii) where cavity nucleation is complete on loading, Figs 5 (c) and (h). The calculations require creep ductility as the sole input parameter and result in creep rates accelerating much more rapidly than for the case of changes in external specimen section (only constant load curves are shown for comparison); the ductility of 5% has seen chosen because it is approximately the lower bound acceptable in engineering practice. Notice that minimum creep rates are changed only slightly when nucleation is continuous whereas large increases in creep rate are apparent even at the start of creep when a constant density of cavities is present throughout the whole lifetime. Clearly, the type of material behaviour represented by Figs 5 (c) & (h) should be avoided if possible, but it may be characteristic of welds and has been shown to be operating after plastic pre-straining in at least two instances in widely different material bases – nickel base superalloys[25] and an aluminium alloy.[26]

Tertiary creep resulting from the multiplication of mobile dislocations is shown in Figs 5 (d) and (i) and again compared with the much milder effect of constant load deformation. Notice that when this damage is dominant, the lifetime reduction factor is independent of stress level. The only additional model parameter is C (Table 1) and a value of C = 100 was chosen for illustration purposes only. In this case there is no influence on the failure strain; rather there will be a gradual (but so far unquantified) development of a geometrical instability at higher strains in the strain-softening material. The computed strain trajectory at constant stress is shown as a dotted curve. In sharp contrast to Figs 5 (a) & (f), there is now very little difference in creep lifetimes between the two forms of loading, particularly at the lower stress.

When tertiary creep is the result of coarsening of the particulate microstructure, lifetime reductions at constant temperature become dependent upon time of test and so its influence becomes greater at lower stress levels when tests are performed under the usual constant conditions. Figures 5 (e) & (j) illustrate this well. The curves were computed using the parameter K_p in the coarsening equation (see Table 1):

$$K_p = K_{p,0} \exp\left(-\frac{Q}{RT}\right) \tag{7}$$

where Q is the sum of the activation energy for diffusion and the solution enthalpy of the particles. Values of $K_{p,0} = 10^6 \text{ s}^{-1}$ and $Q = 300 \text{ kJ/mol}$ were estimated by a combination of physical reasoning (Q) and the arbitrary constraint of ensuring that lifetime was similar to that given by the dislocation strain softening mechanism at the lower stress. The creep curves produced by dislocation strain-softening and particle-coarsening are then superficially similar at 150 MPa. Yet at 300 MPa, comparison of Figs 5 (e) & (j) with Figs 5 (d) and (i) demonstrate a quite different behaviour which could lead to large inaccuracies in long-term extrapolations if the wrong damage mechanism were to be identified.

Figures 5 (a) to (j) demonstrate that minimum creep rates (m.c.r.) depend upon the operating damage mechanism. This is the physical explanation of the point often made by Wilshire *et al.* when using the θ Projection Method that the m.c.r. is merely a point of inflection between primary and tertiary creep and does not represent a steady state. If the damage mechanism changes an m.c.r. significantly from the one predicted by Table 2, then it may create difficulties when a set of creep curves are being analysed to extract their full parameter-set, because of difficulties associated with parameter-correlation. In the current example, a sophisticated numerical procedure would need to be used to extract a physically sensible parameter-set if cavitation were a dominant damage mechanism, particularly at higher stresses (Fig. 5(g)) or when cavitation is present at zero time (Figs. 5(c)&(h)). However for high ductility alloys, the computations show that the elevation of m.c.r. is small compared with the material and testing uncertainties involved in creep data.

Multiple damage mechanisms   The comparisons given above indicate the relative magnitudes of effects due to each damage mechanism acting alone. In practice all could operate simultaneously, although it is likely that within any restricted stress/temperature regime, there will be a dominant mechanism controlling the rate of accumulation of creep strain and thus lifetime. There are many situations in service when it is important to know whether there is more than one damage mechanism operating. For example, when stress/temperatures are varying the response of each mechanism will be different. To illustrate the consequences, let us suppose that damage is due to the simultaneous operation of dislocation multiplication and particle-coarsening. Then, using the nomenclature given in Table 1, the two-damage model is

$$\dot{\epsilon} = \frac{\dot{\epsilon}_0}{1 - D_d} \sinh\left(\frac{\sigma(1 - H)}{\sigma_0(1 - D_p)}\right)$$

$$\dot{H} = \frac{h'}{\sigma}\left(1 - \frac{H}{H^*}\right)\dot{\epsilon}$$

$$\dot{D}_d = C\,(1 - D_d)^2\,\dot{\epsilon} \tag{8}$$

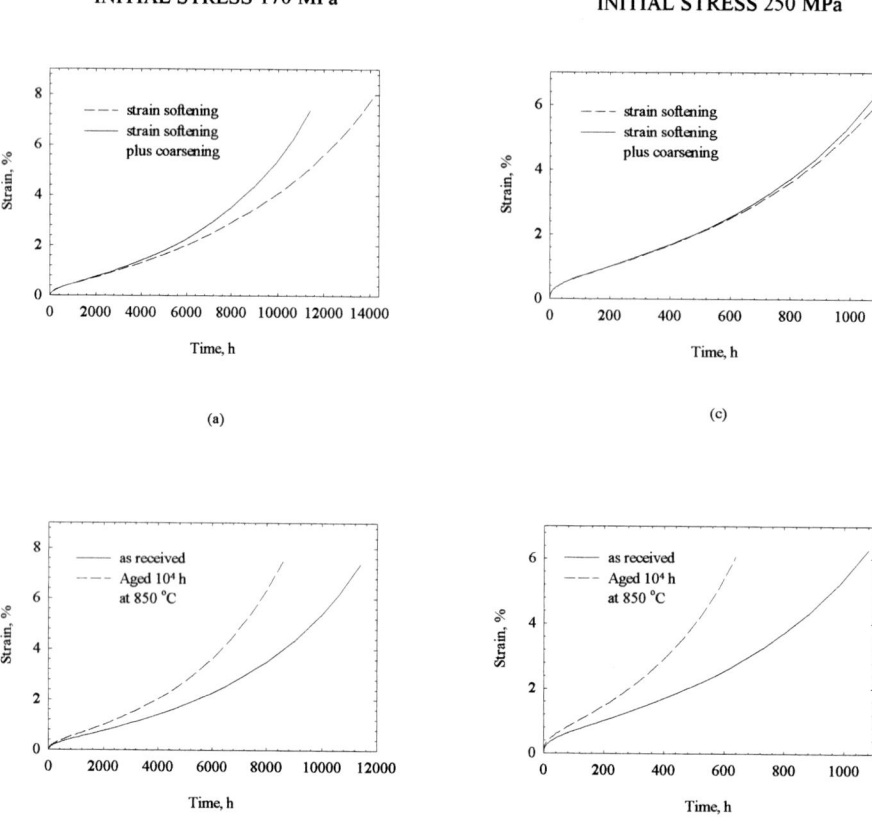

Fig. 6 Two-damage modelling of the effect of prior thermal exposure on strain/time trajectories to be compared with In738LC data in Fig. 1.

$$\dot{D}_p = \frac{K_p}{3}(1 - D_p)^4$$

Equation-set (8) has been used to try to simulate the results on conventionally-cast In738LC reported by Tipler and Peck[1] and given in Figs 1 (a) & (b). Integrating under constant load and using the model parameters in Table 2, with C = 50, $K_{p,0}$ = 2× 10^5 s^{-1} and the same strains to failure reported by Tipler and Peck,[1] result in the curves shown in Fig. 6. Figures 6 (a) & (c) are the usual steady-load, isothermal tests and demonstrate the expected result that the effect of particle-coarsening is to reduce lifetime (solid curve) at the lower stress but to have a negligible effect at 250 MPa. This is equivalent to the parameter determined in a one-damage representation increasing with decreasing stress, as reported many times by Wilshire *et al.* in terms of the

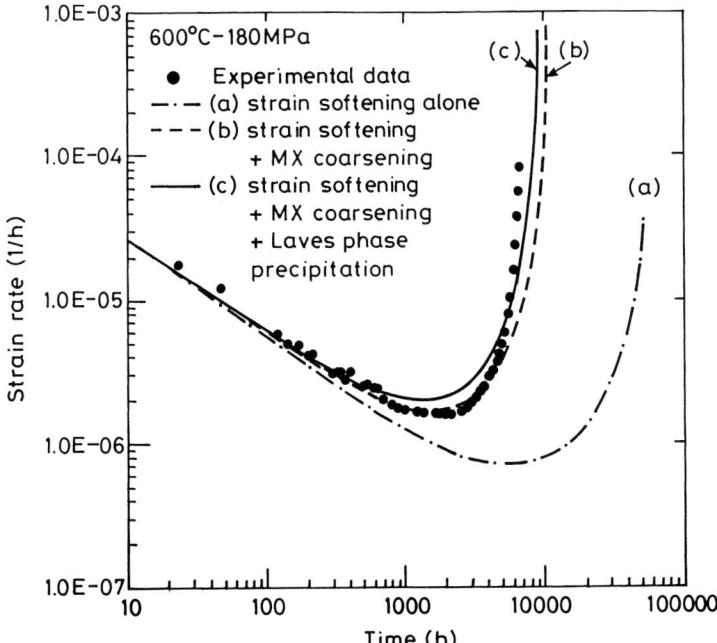

Fig. 7   Comparison of experimental strain rate versus time curve for a 12%Cr martensitic steel tested at 600°C/180 MPa with model predictions incorporating single and multiple damage mechanisms.[8]

stress dependency of θ_3 and by Ghosh *et al.*[27] for SRR99 in terms of the C parameter. Figure 6 (b) & (c) compute the effect of ageing the material prior to creep – in this case, for 10 000 h to simulate the experiments of Tipler and Peck[1] given in Fig. 1. In sharp contrast to the previous figure, pre-ageing weakens the material more during creep at 250 MPa than at 170 MPa and is qualitatively similar to the data reported by Tipler and Peck. The zero effect of ageing reported by Tipler and Peck at 170 MPa cannot be reproduced with this model (except near zero stress) and needs to be investigated further. Nevertheless, the conclusion can be drawn from these computations that although strain-softening appears to be the dominant cause of tertiary creep in IN738C, there is an additional thermally-induced contribution from particle-coarsening that will assume more importance under service conditions.

Theory and Fig. 6 suggests that there can be a strong synergy between the various damage mechanisms. This is further illustrated in the work of Kadoya *et al.*,[8] who consider the relative contributions of particle coarsening, dislocation multiplication and solid-solution depletion as a consequence of Laves phase precipitation. Figure 7 shows a relatively long-term creep curve displayed as strain rate as a function of time; the CDM predictions for different combinations of damage are also shown. Clearly, in this case, particle-coarsening has a major synergistic influence on the already large 'base-line' damaging effect of dislocation multiplication, while solid solution depletion is relatively unimportant. However, in extrapolating to long service times, the Laves phase formation increases to the extent that solid solution depletion can reduce the service life to half that found from extrapolations of short term data that ignore the effects of Laves phase formation.[8]

DISCUSSION

In the United States during the 1950s, great strides were made in developing parametric equations for predicting service creep lifetimes from laboratory data obtained at much higher stresses and/or higher temperatures (for example, Larson–Miller). These methods can be used in conjunction with Robinson's Life-Fraction Rule to predict behaviour under variable stresses and temperatures. They are in fact the mainstay in the armoury of techniques available to design engineers in spite of periodic criticisms over the years of their failure to predict lifetimes of certain materials. These equations have indeed stood the test of time but cannot be developed further to meet the needs of computer-based predictions of design and remanent-lifetimes. Not only are they in the main empirical, although some have a good physical foundation, but all are based on the concept of steady state with no hint that the material may damage during service. But of even more importance is that parametric equations were not conceived for coping with the type of conditions found in a turbine blade for example: spatially varying, complex and time-dependent stresses and temperatures.

Creep is a kinetic phenomenon and can only be approached with any generality by computing behaviour from a set of rate equations that reflect the evolution of microstructure as well as inelastic strain. We have chosen CDM, which is a specific state variable description of creep damage, as the vehicle to achieve this. The CDM equation-sets can be integrated under various external constraints without adding further assumptions: isothermally, under constant load/stress to produce the usual creep curve; under constant total strain, to produce a stress relaxation curve; under constant total strain rate to produce a stress/strain curve. Thermal/mechanical cycles that involve through-zero stress cycling invariably require more information concerning alloy behaviour (to reflect properly the kinematic hardening parameter, H) or sometimes further equations to cope with additional complexities. An apparent disadvantage of the CDM approach is that it may seem overly complicated. Indeed, if the aim is just to extrapolate to service stress levels from a limited laboratory database then, with several equations and perhaps 6 parameters needing evaluation, the complexity is hardly warranted. Particularly when a simple Larson–Miller extrapolation (two equations and a minimum of

3 easily computed parameters) is probably good enough anyway, considering the uncertainties in engineering uniaxial creep data. It is only when more complex extrapolations are required that the CDM approach starts to come into its own. Because the equations are used in a modular manner, then once the parameters have been established for a base-line condition (for example, a standard heat-treatment), the creep behaviour after subjecting the material to say a prior plastic straining can be predicted using most of the previous parameters with the addition of either further parameters or damage evolution equations (for example, cavitation).[25] This modular approach of CDM can be contrasted with the Theta Projection Method of Wilshire and co-workers,[12] who are the only other group to have successfully broken out of the conventional mould. At first glance, the method appears easy to implement and understand, since it is couched in strain/time terms, with only 4 'θ' parameters to consider when describing creep response at a single stress and temperature. But the number of parameters escalates to 16 when strain predictions are required over a range of temperatures and stresses and therefore CDM with a minimum of 8 parameters (rising to perhaps 10 depending on the number of important damage parameters operating) compares very favourably. This 'parameter-efficient' advantage of CDM does not stop here. In predicting the effect of changing stress and/or temperature, the necessary information for choosing the correct path is already contained within the damage-rate equations of the CDM approach, in contrast to θ-Projection, where empirical rules ('strain-hardening/softening' or 'time-hardening/softening') must be devised. An illustration of the power of these benefits of CDM is given by Maldini and Lupinc.[28]

Enhanced competitiveness in the power generation and propulsion industries accrues from improved capability in component-design and remanent-lifetime assessments and finite element implementation of mathematical descriptions of material behaviour is increasingly of paramount importance. The CDM approach is uniquely placed in this regard: computers prefer rate equations and the dimensionless form of the parameters in CDM make the approach extremely computer-friendly.

CONCLUSIONS

- the creep response of engineering alloys can be profoundly influenced by the evolution rate of different aspects of alloy-microstructure (damage);
- creep curves of similar shape can arise from quite different mechanisms of damage;
- CDM provides a powerful formalism for representing quantitatively the relative contributions of different types of microstructural damage and for accounting for their interactions as demonstrated in Figs 5 (a)–(j);
- the CDM equation-sets can be integrated to account for arbitrary types of

loading (stress relaxation, cyclic creep, etc.) without recourse to the usual associated empirical assumptions of 'strain-hardening/softening' or 'time-hardening/softening';

- material response to specific temperature/load histories are predicted by CDM to diverge significantly for different damage mechanism as demonstrated in Fig. 6;
- the CDM formalism is 'parameter-efficient', allowing the potential effects of different types of damage to be assessed quantitatively with only minimal additions of parameters to the base-line equation describing creep in the undamaged material.

REFERENCES

1. H.R. Tipler and M.S. Peck: *NPL Report DMA A33*, 1981.
2. B.F. Dyson and M. McLean: 'Particle coarsening σ_0 and tertiary creep', *Acta Metall.*, 1983, **31**, 17–27.
3. A. Strang: private communication, 1997.
4. W. Blum: *Scripta Metall.*, 1984, **18**, 1383–1388.
5. W. Blum and S. Straub: 1991, *Materials Technology, steel research 62 No. 2*, 72–74.
6. G. Eggeler, N. Nilsvang and B. Ilschner: 1987, *Materials Technology, steel research 58 No. 2*, 97–103.
7. G. Eggeler: *Acta Metall.*, 1989, **37**, 3225–3234.
8. Y. Kadoya, N. Nishimura, B.F. Dyson and M. McLean: *Creep & Fracture of Engineering Materials & Structures*, J.C. Earthman and F.A. Mohammed eds, TMS, Warrendale, USA, 1997, 343–352.
9. L.M. Kachanov: *Izv. Ak. Nauk SSSR Otdel. Tekh. Nauk*, 1958, **8**, 26–31.
10. Y.N. Rabotnov: *Proc. XII IUTAM Congress*, Stamford, Hetenyi and Vincenti eds, Springer, 1969, 137.
11. A. Graham and K.E.A. Walles: *J Iron & Steel Inst.*, 1955, 179.
12. R.W. Evans, J.D. Parker and B. Wilshire: *Recent Advances in Creep and Fracture of Engineering Materials and Structures*, B. Wilshire and D.R.J. Owen eds, Pineridge Press, Swansea, 1982, 135–184.
13. Ashby and Dyson, *Advances in Fracture Research*, S.R. Valluri *et al.* eds, Pergamon Press, 1984, 1, 3–30.
14. J.C. Ion, A. Barbosa, M.F. Ashby, B.F. Dyson and M. McLean: *NPL Report DMA A115*, 1986.
15. Wilshire: *Creep & Fracture of Engineering Materials & Structures*, B. Wilshire and R.W. Evans eds, IoM, London, 1990, 1–9.
16. B.F. Dyson and S. Osgerby: *NPL Report DMA A116*, 1993.
17. B.F. Dyson: work to be published.
18. B. Wilshire: *Engineering Application Through Scientific Insight*, E.D. Hondros and M. McLean eds, The Institute of Materials, London, 1996, 155–172.

19. I.M. Liftshitz and I. Slyozov: *J Phys Chem Solids*, 1961, **19**, 35.
20. C. Wagner: *Z. Electrochem.*, 1961, **65**, 581.
21. R. Lagneborg: *J Materials Science*, 1968, **3**, 596.
22. P.L. Threadgill and B. Wilshire: *Creep Strength in Steel and High Temperature Alloys*, ISI, Sheffield, UK, 1972, **8**.
23. C. Wert and C. Zener: *J Appl. Phys.*, 1950, **21**, 5.
24. NPL databank.
25. B.F. Dyson: *Engineering Application Through Scientific Insight*, E.D. Hondros and M. McLean eds, The Institute of Materials, London, 1996, 213–227.
26. J. Przydatek, B.A. Shollock and M. McLean: *Creep & Fractures of Engineering Materials & Structures*, J.C. Earthman and F.A. Mohamed eds, TMS, Warrendale, USA, 1997, 29–38.
27. R. Ghosh, R. Curtis and M. McLean: *Acta Metall. Mater.*, 1990, **38**, 1997–1992.
28. M. Maldini and V. Lupinc: *Creep & Fracture of Engineering Materials & Structures*, B. Wilshire and R.W. Evans eds, IoM, London, 1990, 951–960.

Kinetics of Precipitation Reactions in Ferritic Power Plant Steels

J.D. ROBSON AND H.K.D.H. BHADESHIA

Department of Materials Science and Metallurgy, Pembroke Street, Cambridge CB2 3QZ

ABSTRACT

A model has been developed to enable the kinetics of precipitation reactions in power plant steels to be predicted. The model allows for simultaneous precipitation or dissolution of up to six phases and has been verified using steels ranging from 2.25Cr1Mo to 10CrMoV. Predictions have been made to investigate how changes in the concentration of carbon and nitrogen are expected to influence the precipitation kinetics in a 10wt% chromium steel. An increase in nitrogen is shown to intensify and stabilise M_2X precipitates whilst having little effect on the other phases. A decrease in carbon also results in an increase in M_2X precipitation whilst reducing the volume fraction of $M_{23}C_6$ and delaying the onset of its formation. Apart from volume fractions, the model also gives other useful information, such as the growth rate of each phase as a function of time and temperature. The growth rate of Laves phase is compared with that of the other phases, and the influence of the molybdenum and tungsten level on its rate is calculated. These predictions suggest that, provided Laves phase is stable, addition of both of these elements increases the growth rate by a comparable amount.

INTRODUCTION

A critical part of the microstructure of a steel designed for prolonged service at elevated temperatures is the stability, size and distribution of a variety of

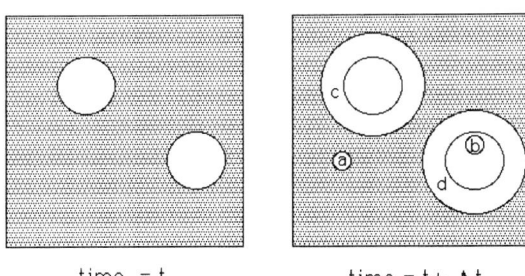

Fig. 1 An illustration of the concept of extended volume. Two precipitate particles have nucleated together and grown to a finite size in the time *t*. New regions *c* and *d* are formed as the original particles grow, but *a* & *b* are new particles, of which *b* has formed in a region which is already ß.

precipitates. There are a number of steps involved in formulating a kinetic theory which enables the estimation of precipitation reactions as a function of the heat treatment and chemical composition. A model for a single trans-formation would begin with the calculation of the nucleation and growth rates using classical theory, but an estimation of the volume fraction requires impingement between particles to be taken into account. This is generally done using the extended volume concept of Johnson, Mehl and Avrami, illus-trated in Fig. 1. Suppose that two particles exist at time t; a small interval δt later, new regions marked a, b, c & d are formed assuming that they are able to grow unrestricted in extended space whether or not the region into which they grow is already transformed. However, only those components of a, b, c & d which lie in previously untransformed matrix can contribute to a change in the real volume of the product phase (which we shall call ß):

$$dV_\beta = (1 - \frac{V_\beta}{V}) \, dV_\beta^e$$

where the superscript e refers to extended volume, V_β is the volume of ß and V is the total volume. Multiplying the change in extended volume by the probability of finding untransformed regions has the effect of excluding regions such as b, which clearly cannot contribute to the real change in volume of ß. This equation can easily be integrated to obtain the real volume fraction as a function of time.

There is, however, an important complication in the context of power plant steels, that reactions rarely, if ever, occur in isolation. The different reactions interfere with each other in a way which is seminal to the development of power plant microstructures. Therefore, considerable effort has recently been devoted to the development of an Avrami model for simultaneous reac-tions.[1,2] A simple explanation for two precipitates (ß and α) is that the above equation becomes two equations,

$$dV_\beta = (1 - \frac{V_\beta + V_\alpha}{V}) \, dV_\beta^e$$

$$dV_\alpha = (1 - \frac{V_\beta + V_\alpha}{V}) \, dV_\alpha^e$$

which must now be solved numerically and can be extended to incorporate an indefinite number of reactions happening together. The model takes the chemical composition and free-energy changes associated with formation of each phase as inputs, and gives as outputs the volume fraction of each phase as a function of time and temperature. The phase compositions and driving forces are calculated from a knowledge of the alloy composition and the tem-perature using MTDATA,[3] a computer package for the calculation of ther-

Table 1 Concentration (in weight %) of the major alloying elements in the steels used to demonstrate the model

	C	N	Mn	Cr	Mo	Ni	V	Nb	Fe
2.25Cr1Mo	0.15	–	0.50	2.12	0.9	0.17	–	–	bal
3Cr1.5Mo	0.1	–	1.0	3.0	1.5	0.1	0.1	–	bal
10CrMoV	0.11	0.056	0.50	10.22	1.42	0.55	0.20	0.50	bal

Table 2 Values for interfacial energy, σ, and site density, N_0, for nuclei of each phase which give the best fit between prediction and experiment

Phase	σ/Jm^{-2}	N_0/m^{-3}
M_2X	0.248	2.7×10^{16}
M_7C_3	0.249	2.7×10^{23}
MX	0.260	5.4×10^{13}
$M_{23}C_6$	0.269	2.7×10^{15}
Laves	0.331	2.7×10^8

modynamic parameters. In this way it has been possible to estimate the precipitation kinetics for power plant steels with a wide range of compositions, as well as investigate the effects of individual alloying elements on these kinetics.

Further details are described fully in previous papers;[1,2] the purpose here is to show results from calculations which have included more phases than ever before: M_2X, MX, M_7C_3, $M_{23}C_6$ and Laves phase. The change in the chemical composition of cementite following its initial paraequilibrium precipitation is also taken into account.

There are two unknown parameters for each phase, the nucleation site density and the interfacial energy per unit area. These must be found by fitting the predicted results to measured data in such a way that the fitting parameters are consistent with the thermodynamically predicted sequence of precipitation, bearing in mind that the work can be of most use when the parameters are the same for a wide range of steels. For M_2X and $M_{23}C_6$ predictions were compared with measurements made on heat treated samples of 10CrMoV steel (composition given in Table 1).[2] For Laves phase, data from a 9CrMoWV steel due to Hald,[4] was used for comparison. The values used for MX and M_7C_3 were estimated from a knowledge of the positions of these phases in the precipitation sequence. The complete set of fitted parameters is shown in Table 2. It may be that a different set of parameters might equally describe the experimental data but it is noteworthy that the interfacial energy correlates with the known position of the phase concerned in the precipitation sequence. This set of parameters has indeed been shown to be applicable to a wide range of power plant alloys, and is used throughout the work presented here.

RESULTS
The compositions of three power plant alloys are shown in Table 2. These three alloys, whilst of quite different chemical compositions, show similar

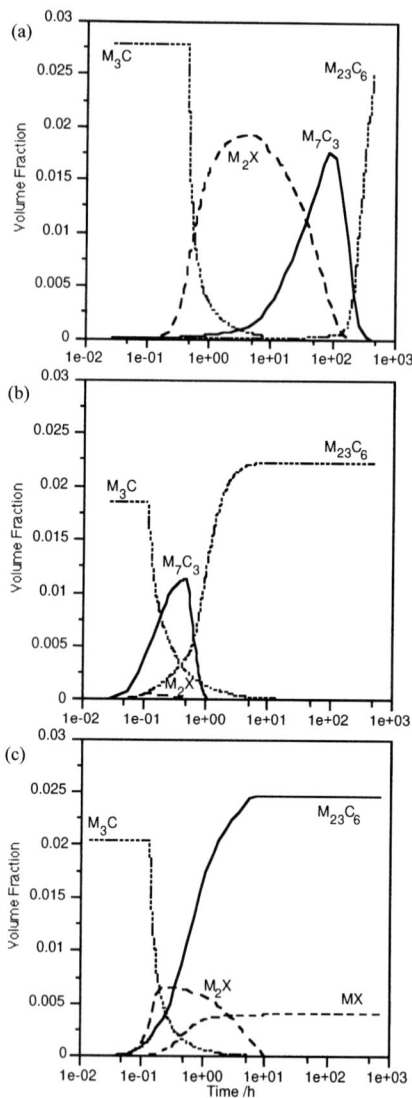

Fig. 2   The predicted evolution of precipitate volume fractions at 600°C for three power plant materials (a) 2.25Cr1Mo (b) 3Cr1.5Mo and (c) 10CrMoV.

precipitation *sequences*[2,5,6] but with vastly different rates. For example, at 600°C the time taken before $M_{23}C_6$ is observed is 1 h in the 10CrMoV steel,[2] 10 h in the 3Cr1.5Mo alloy[5] and in excess of 1000 h in the 2.25 Cr1Mo steel.[6] The alloys listed in Table 1 cover a good proportion of creep-resistant power plant steels and thus are representative for the purposes of model calibration. It was also hoped that the model might reveal why the precipitation kinetics differ so much between the alloys.

A plot of the predicted variation in the volume fraction of each precipitate as a function of time at 600°C is shown in Fig. 2. Consistent with experiments, the precipitation kinetics of $M_{23}C_6$ are predicted to be much slower in the 2.25Cr1Mo steel compared to the 10CrMoV and 3Cr1.5Mo alloys. One contributing factor is that in the 2.25Cr1Mo steel a relatively large volume fraction of M_2X and M_7C_3 form prior to $M_{23}C_6$. These deplete the matrix of chromium and molybdenum and therefore suppress $M_{23}C_6$ precipitation. The volume fraction of M_2X which forms in the 10CrMoV steel is relatively small, and there remains considerable excess of solute in the matrix, allowing $M_{23}C_6$ to precipitate rapidly. Similarly, in the 3Cr1.5Mo steel the volume fractions of M_2X and M_7C_3 are insufficient to suppress $M_{23}C_6$ precipitation to the same extent as in the 2.25Cr1Mo steel.

$M_{23}C_6$ is frequently observed in the form of coarse particles which do not contribute directly to the creep strength of the alloy.[7] Delaying its precipitation would have the effect of stabilising the finer dispersions of M_2X and MX to longer times with a possible enhancement of creep strength. The model was used investigate composition variations which lead to a change in the kinetics of $M_{23}C_6$ precipitation, using the 10CrMoV steel in Table 1 as a base.

Effect of Carbon
A reduction in carbon concentration in the 10CrMoV steel would be expected to suppress the formation of $M_{23}C_6$ relative to M_2X and MX, both of which are nitrogen-rich. The predicted evolution of volume fraction of the precipitates at 600°C for a 10CrMoV steel with a reduced carbon concentration of 0.05 wt% C) is shown in Fig. 3. Whereas this may be undesirable in practice (e.g. lower carbon encourages δ ferrite formation) the calculation is useful in revealing the role of carbon. As expected, for any given time, the volume fraction of $M_{23}C_6$ is reduced, and that of M_2X increased. The reduced driving force for $M_{23}C_6$ precipitation, combined with the increased volume fraction of M_2X suppresses $M_{23}C_6$ so that the time taken it to reach a volume fraction of 1% is an order of magnitude greater in the reduced-carbon alloy. Naturally, the final volume fraction of $M_{23}C_6$ is halved.

Effect of Nitrogen
The nitrogen content of the 9–12 wt% Cr steels is often adjusted by alloy

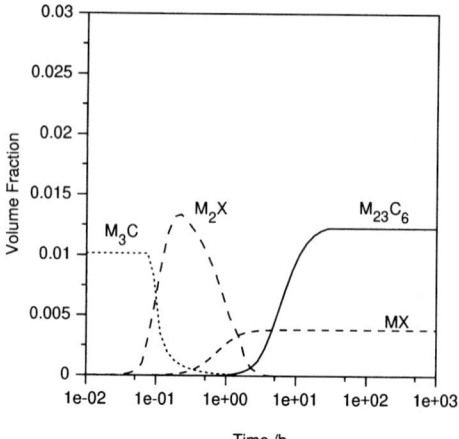

Fig. 3   The predicted evolution of precipitate volume fraction at 600°C for a 10CrMoV steel with a carbon content half that of the standard 10CrMoV alloy (0.05 wt% C).

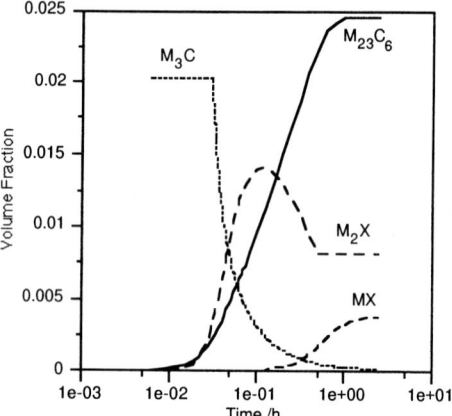

Fig. 4   The predicted evolution of precipitate volume fraction at 600°C for a 10CrMoV steel with twice the nitrogen content of the standard 10CrMoV alloy (0.112 wt% N).

designers to optimise performance. The predicted effect of doubling the nitrogen level in the standard 10CrMoV steel is shown in Fig. 4. Additional nitrogen stabilises M_2X which becomes an equilibrium phase at 600°C. Therefore, unlike in the standard 10CrMoV steel, M_2X does not completely redissolve as $M_{23}C_6$ precipitates. Furthermore, the maximum volume fraction of M_2X achieved prior to dissolution is found to be greater with the larger nitrogen concentration. However, it is not predicted to be sufficient to retard

$M_{23}C_6$ precipitation which occurs at approximately the same time in the standard and high-nitrogen steels. The addition of nitrogen does not appear to change the volume fraction of MX. This is because MTDATA predicts the maximum fraction of this phase is limited in this alloy by the availability of suitable 'M' atoms (i.e. V and Nb). In practice, if other alloying elements also partition into MX then the nitrogen content may limit the MX volume fraction, in which case additional nitrogen would result in an increase in the amount of MX precipitate.

Growth Rate of Laves Phase

Laves phase formation is generally believed to be undesirable and associated with a reduction in toughness.[8] The precipitation of this phase also can remove molybdenum and tungsten from solid solution, leading to a reduction in the long-term creep strength since Laves phase itself is typically ineffective as a pinning dispersion.

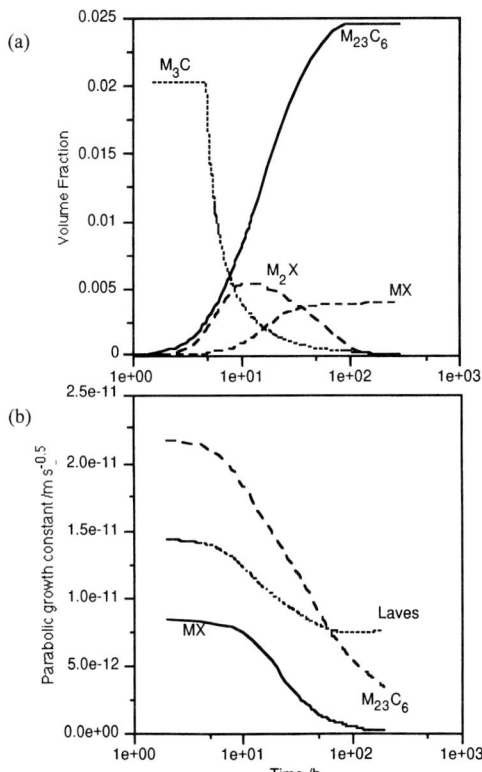

Fig. 5 (a) Predicted evolution of precipitate volume fraction at 500°C for a 10CrMoV steel (b) Variation in parabolic growth constant for MX, $M_{23}C_6$ and Laves phase at 500°C.

The variation of the parabolic growth constant for the three phases which are modelled assuming a spherical particle morphology is shown in Fig. 5 for the 10CrMoV steel at 500°C. It can be seen that the growth constant for MX is considerably less than that of either $M_{23}C_6$ or Laves phase, consistent with observations.[9] It is predicted that $M_{23}C_6$ initially grows faster than Laves phase, but as the chromium in the matrix is depleted the growth rate of $M_{23}C_6$ falls more rapidly than that of Laves phase. Eventually $M_{23}C_6$ will stop growing, Laves phase will continue to grow since there remains excess molybdenum in the matrix. Since there are relatively few Laves particles (the nucleation rate is low), growth will continue for much longer compared to the other phases before all the excess solute is removed from the matrix. This leads to the observed distribution of widely spaced, large particles at long times. Note that coarsening, which is not accounted for in this analysis, will also lead to an increase in the size of the larger particles at the expense of the

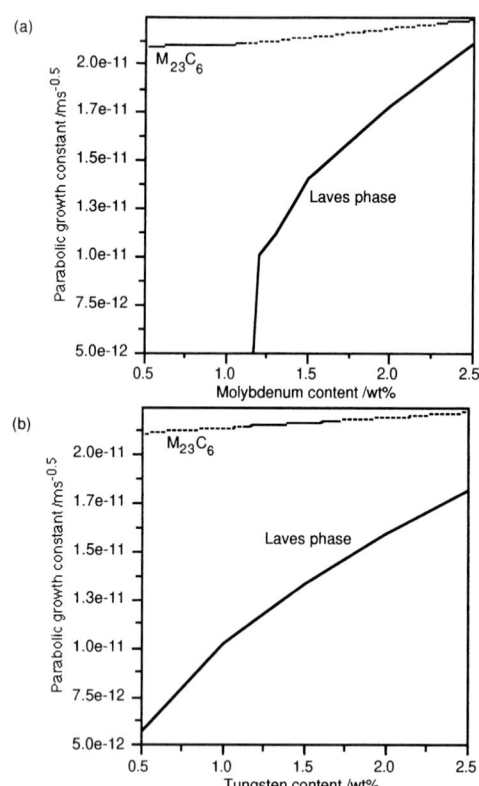

Fig. 6   Predicted variation of the initial parabolic growth constant at 600°C (a) for a 10CrMoV alloy with a varying molybdenum level (b) a 10CrWV alloy with a varying tungsten level.

smaller ones. The particle sizes calculated on the basis of the growth constant alone will therefore be an underestimate of the true particle sizes.

The Effect of W and Mo on Laves Growth

Both tungsten and molybdenum additions encourage the formation of Laves phase. Increasing levels of either element would therefore be expected to enhance Laves phase growth. Figure 6 shows the predicted variation in the initial parabolic growth constant of both $M_{23}C_6$ and Laves phase at 500°C in an alloy based on the standard 10CrMoV composition for two cases: (a) when the molybdenum level is varied between 0.5 and 2.5 wt% with zero tungsten and (b) when the molybdenum level is set to zero and the tungsten varied between 0.5 and 2.5 wt%. Note that in Fig. 6(a) for molybdenum levels of 1.2 wt% and below it is predicted that Laves phase will not form at all at 500°C (zero growth rate). At higher levels of addition the growth rate of Laves phase is similar for the same mass of tungsten or molybdenum. Per added mass, molybdenum enhances the growth rate slightly more than tungsten. This is a consequence of the lower atomic mass of molybdenum, since per added mass, the increase in the atomic fraction of molybdenum, which governs the growth rate, is greater. Note that additions of either tungsten or molybdenum have little effect on the growth rate of $M_{23}C_6$.

CONCLUSIONS

A model has been produced to predict the simultaneous precipitation reactions which occur in power plant steels. For each phase a pair of fitting parameters is required, namely the nucleation site density and interfacial energy. It has been shown that, once obtained, a single set of parameters is applicable for a wide range of power plant steels.

The model has been used to make a number of predictions for a 10CrMoV steel. It is predicted that reducing carbon delays the precipitation of $M_{23}C_6$ and increases the maximum volume fraction of M_2X. Nitrogen additions intensify and stabilise M_2X.

The growth rate of Laves phase is predicted to lie between that of MX and $M_{23}C_6$ in the very early stages of transformation. $M_{23}C_6$ reaches its equilibrium fraction at a relatively short time and the growth rate becomes zero. Laves phase continues to grow for a much longer time, leading to the large particles which are observed experimentally.

Addition of excess tungsten or molybdenum to the standard 10CrMoV steel increases the growth rate of Laves phase significantly, with both elements showing a similar potency. These additions have little effect on the growth rate of $M_{23}C_6$.

ACKNOWLEDGEMENTS

The authors are grateful to National Power plc. for financial support via Dr. David Gooch. We would also like to thank David Gooch, Andrew Strang, Rod Vanstone and Rachel Thomson for their help during the course of some of this work. The help of Hugh Davies and Susan Hodgson with the provision and support of MTDATA is greatly appreciated. HKDHB is grateful to the Royal Society for a Leverhulme Trust Senior Research Fellowship.

REFERENCES

1. J.D. Robson and H.K.D.H. Bhadeshia: *Mat. Sci. Tech.*, 1997, **13**, 631–639.
2. J.D. Robson and H.K.D.H. Bhadeshia: *Mat. Sci. Tech.*, 1997, **13**, 640–644.
3. MTDATA: Metallurgical and Thermochemical Databank, National Physical Laboratory, UK, 1995.
4. J. Hald: COST 501, WP 11 Report, 1994.
5. N. Fujita: Private Communication, University of Cambridge, 1996.
6. R.G. Baker and J. Nutting: Journal of the Iron and Steel Institute, 1959, 192, 257–268.
7. A. Bjärbo: 'Microstructural Changes in a 12% Chromium Steel During Creep', Report, Royal Institute of Technology, Stockholm, Sweden, 1994, 11.
8. Y. Hosoi, N. Wade and T. Urita: Trans. Iron and Steel Inst. Japan, 1986, 26, 30.
9. R.W. Vanstone: 'Microstructure and Creep Mechanisms in Advanced 9–12% Cr Creep Resisting Steels', Materials for Advanced Power Engineering, Liège, Belgium, 1994.

Simulation of the Creep Behaviour of 9–12% CrMoV-Steels on the Basis of Microstructural Data

P. POLCIK, S. STRAUB, D. HENES AND W. BLUM

Institut für Werkstoffwissenschaften, Lehrstuhl I, Universität Erlangen-Nürnberg, Martensstraße 5, D–91058 Erlangen, Germany

ABSTRACT

Martensitic 12% CrMoV-steels are in common use in power plants. There are continuous efforts to develop this group of steels in order to enhance the efficiency of power plants. The initial microstructure after heat treatment (austenitisation and annealing) consists of subgrains with free dislocations inside and carbides preferentially located at the subgrain boundaries. In previous works the microstructure of different heats of 9–12% CrMo(W)VNb(N, B)-steels was quantitatively characterised for the initial and various long-term annealed and crept conditions. On the basis of these data it was possible to formulate self-consistent evolution laws for the microstructural parameters, which are of relevance for creep deformation. During creep deformation the microstructure coarsens. The evolution of the dislocation structure was found to be strain-dependent, the carbide structure was found to coarsen in proportion to the time. The strength characteristics used for the design and the service of a plant are often extrapolated from short-term creep tests. In order to check the validity of the extrapolation, a simulation of the deformation behaviour on a microstructural basis appears highly desirable. In the present paper we use the composite model of plastic deformation to model the deformation behaviour of 9–12% CrMo(W)VNb(N, B)-steels. The composite model treats the material as consisting of a (plastically) 'hard' phase with high dislocation density and a 'soft' phase with low dislocation density, which deform in parallel under the condition of equal total (elastic plus inelastic) strain. The model is applied to simulate strain time–, strain rate–strain-curves and creep limit diagrams. The comparison of experimental data with modelled results shows that the simulation describes the experiment quite well. This clearly demonstrates the potential and efficiency of a microstructure based simulation.

1. INTRODUCTION

In the past 15 years new martensitic 9–12% CrMo(W)VNb(N, B)-steels were developed for use in power plants with the aim of enhancing the thermal efficiency.[1-4] The new steels have superior creep strength compared to the conventional steels of the type X 20(22) CrMoV 12 1. The strength charac-

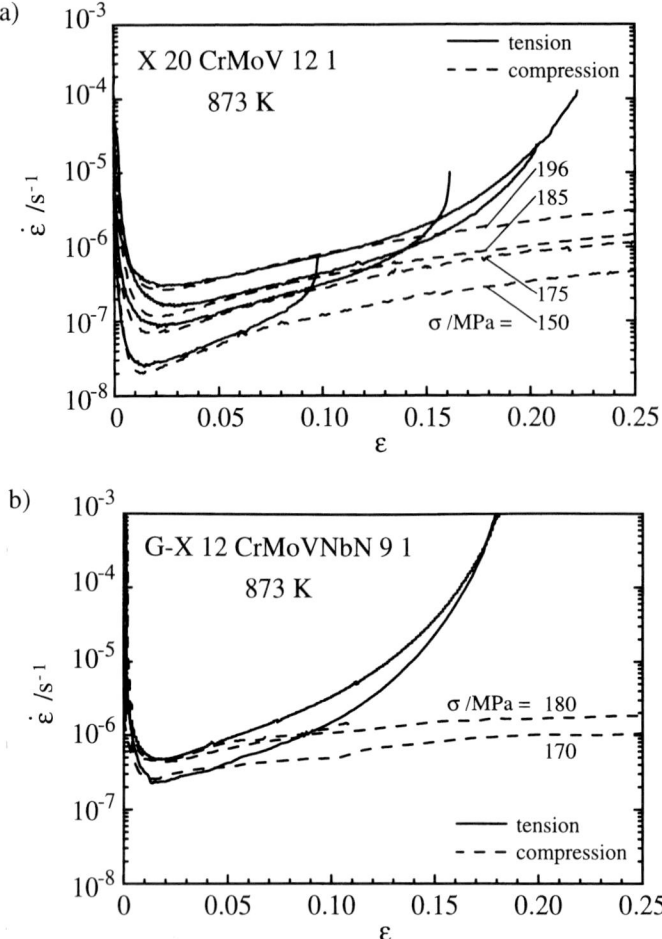

Fig. 1   Strain rate($\dot{\epsilon}$)-strain(ϵ)-curves for tensile- and compression-creep tests, (a) X 20 CrMoV 12 1 and (b) G-X 12 CrMoVNbN 9 1.

teristics which are necessary for design and service of power plants are often determined by extrapolation of short-term data. Experience shows however, that the creep rupture strength may be overestimated by this method.[5–7] Therefore the simulation of strength properties on the basis of the relevant microstructural parameters will gain in importance as a complementary method.

The creep strength of the 9–12% CrMo(W)VNb(N, B)-steels is mainly determined by the hardening due to precipitated carbides and to the subgrain structure. Detailed microstructural investigations have shown that the dislocation and particle structure, which is the result of the initial heat treatment of the steels, is changing significantly during creep.[6,8–22] In the present work a model is presented which allows simulation of the high temperature creep behaviour of a range of conventional (X 20(22) CrMoV 12 1) and modern 9–12% CrMoV-steels (X 10 CrMoVNb 9 1, G-X 12 CrMoVNbN 9 1, X 16 CrWMoVNbN 11 2, X 12 CrMoWVNbN 10 1 1 and X 18 CrMoVNbB 9 1). Experimental details concerning the heat treatment, the chemical composition of the steels and the quantitative characterisation by electron microscopy will not be treated in the following as they have been reported in previous publications.[10–13,23–27]

2. MICROSTRUCTURAL PARAMETERS

In the following, the microstructural parameters which are considered to be relevant for the creep strength and their evolution during creep are briefly presented. A simple result is that the hardness measured at room temperature decreases significantly with increasing duration of high temperature creep.[9–13,18,19,22–27] The reasons for this drop in hardness may lie in the formation of pores or the coarsening of carbides and subgrains.

Pores are in fact formed in uniaxial creep tests of the conventional steels X 20(22) CrMoV 12 1.[9–11,13,19] The modern 9–12% CrMo(W) VNb(N, B)-steels rupture after pronounced necking.[12,19,23–25] The comparison of the creep curves in tension and compression revealed that damage by pores as well as necking influences the creep rate $\dot{\epsilon}$ only at strains beyond half of fracture strain ϵ_{fr}, where $\approx$80% of the creep rupture life has been consumed (Fig. 1). The minimum of the creep rate is observed at strains much below $\epsilon_{fr}/2$. This means that the softening during creep, which is responsible for the minimum creep rate $\dot{\epsilon}_{min}$, cannot be explained by pores and/or necking, but must be due to processes in the dislocation and particle structure. Similarly, the loss in room temperature hardness could be related to the changes in the subgrain structure rather than the formation of pores.[10,15]

The dislocations form a subgrain structure. The subgrain boundaries are essentially planar networks of dislocations. In addition to the subgrain boundary dislocations there are free dislocations in the subgrain interior. The particles are precipitated as carbides and carbonitrides partly in the subgrain interior and preferentially at the subgrain boundaries. Quantitative electron microscopy showed that the structure coarsens during high temperature creep.[9–19,21–29] The subgrain size w and the mean spacing of free dislocations $\rho_f^{-0.5}$ (ρ_f: density of free dislocations) develops strain dependent towards its steady state value (subscript '∞') at the rate $d\log x/d\epsilon$:[10–13,19,22–28]

$$d\log x/d\epsilon = -(\log x - \log x_\infty)/k_{\log x} \text{ with } x = w, \rho_f^{-0.5} \qquad (1)$$

The slightly stress dependent constant $k_{\log w}$ was experimentally determined for X 20 CrMoV 12 1 as ≈ 0.12.[10,11] Both w_∞ and $\rho_{f,\infty}^{-0.5}$ can be expressed in good approximation as a unique function of the stress σ normalised by the shear modulus G :[10,27,28]

$$w_\infty = 10.0 \cdot bG/\sigma, \qquad \rho_{f,\infty}^{-0.5} = 3.9 \cdot bG/\sigma \qquad (2)$$

using the values for $G(873\ K) = 63\,550$ MPa and the length of the Burgers vector $b = 0.248$ nm for X 20(22) CrMoV 12 1.[10]

Figure 2 displays the evolution of w for three experimental melts of modern 9–12% CrMo(W)VNb(N, B)-steels, developed in COST 501 for large steam turbines.[2,3] The quantified subgrain coarsening (symbols) is well represented by the evolution (lines) calculated with eqns (1) and (2). The fact that the observed w-data for X 16 CrWMoVNbN 11 2 and X 18 CrMoVNbB 9 1 lie somewhat below the calculated curves can partly be explained by necking: w has been measured outside the necked area in the region of uniform elongation, where the strain is smaller than the average values of ϵ. This means that the measured points must be shifted to the left.

The particles in the conventional steels X 20(22) CrMoV 12 1 consist mainly of Cr-rich carbides of type $M_{23}C_6$.[8–10,20,29] The alloying of the modern steels with Nb and N leads to an additional precipitation of MX-carbonitrides (VN, Nb(C, N)), which due to their small size and slow coarsening cause a second hardening contribution in addition to that caused by the $M_{23}C_6$-carbides (Fig. 3).[6,7,12,19,21,23–25,30,31] W results in formation of relatively

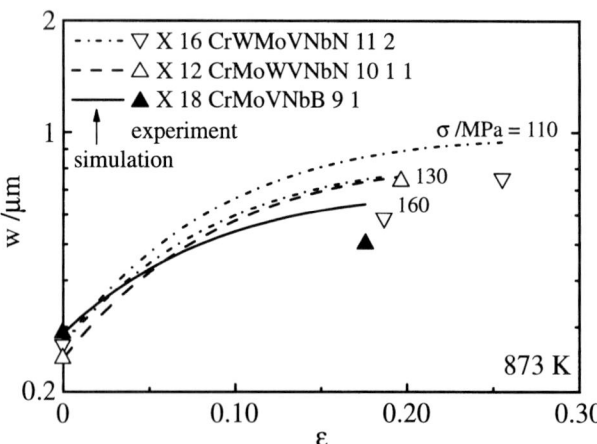

Fig. 2   Subgrain size w as a function of strain ϵ. The evolution of w (lines) was calculated with eqn (1).

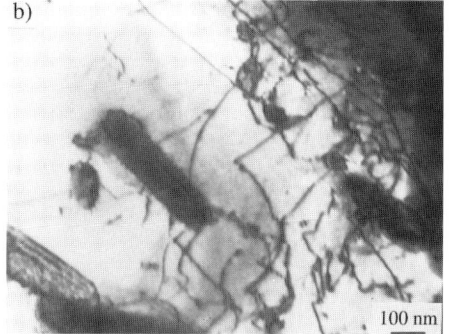
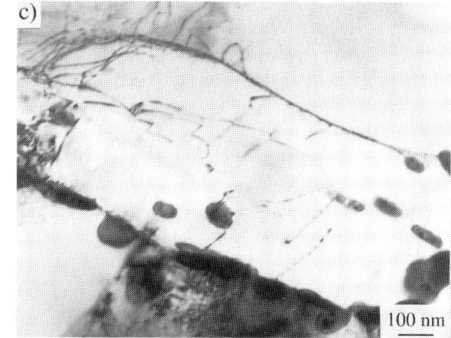

Fig. 3 Interactions between carbides and dislocations after long-term creep tests, (a) X 16 CrWMoVNbN 11 2, (σ_0 = 110 MPa, ϵ_{fr} = 0.255, t_{fr} = 28589 h), (b) X 12 CrMoWVNbN 10 1 1 (σ_0 = 130 MPa, ϵ_{fr} = 0.196, t_{fr} = 19626 h) and (c) X 18 CrMoVNbB 9 1 (σ_0 = 160 MPa, ϵ_{fr} = 0.173, t_{fr} = 23855 h) at 873 K (bright field TEM images).

large platelike precipitates of Laves-phase (($Fe,Cr)_2$ (Mo,W)) with low hardening efficiency.[8,19,21,25,29,31,32]

Due to the MX-precipitation the particle size distribution curves $F_{cum}(d_{p,i})$ for X 12 CrMoWVNbN 10 1 1 (Fig. 4), X 16 CrWMoVNbN 11 2 and X 18 CrMoVNbB 9 1 exhibit a bimodal shape.[25] As a consequence of the MX-precipitation the cumulative frequency F_{cum} at which small particles are observed increases. The increase may be so strong that the average size of all particles decreases relative to the initial state.[12,18,19,23,25] The MX-precipitation is more pronounced during creep than during annealing indicating that the MX-precipitation is of dynamic nature, i.e. enhanced by concurrent creep. Using the average particle size to calculate the particle hardening term is not sufficiently exact. Therefore the distribution curves have been separated[12,19,23–25] into the distribution curves $F_{cum,i}$ for the $M_{23}C_6$-carbides (i = 1) and the MX-carbonitrides (i = 2):

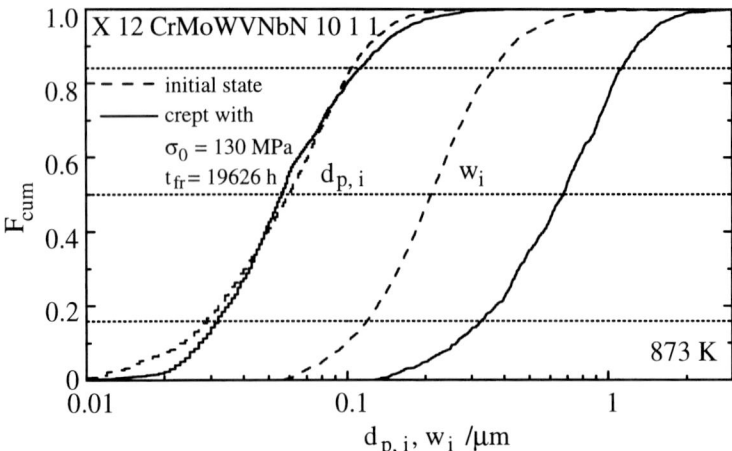

Fig. 4   Cumulative frequencies F_{cum} curves of individual carbide sizes $d_{p,i}$ and subgrain sizes w_i for the initial and a long-term deformed state of X 12 CrMoWVNbN 10 1 1.

$$F_{cum} = \phi_1 \cdot F_{cum,1} + \phi_2 \cdot F_{cum,2} \qquad \phi_1 + \phi_2 = 1 \qquad (3)$$

with the relative amount ϕ_i of precipitations of the type i. The method is described in.[12] In this manner the average sizes $d_{p,i}$ and volume fractions $f_{p,i}$ of $M_{23}C_6$ and the MX can be determined.

The phenomenological description of the particle growth starts from:

$$d d_{p,i}^3 / dt = k_{p,i}(T) \qquad (4)$$

$k_{p,i}$ is proportional to the rate at which the average volume of the particles grows. Assuming that $k_{p,i}$ is a constant (this assumption is violated during dynamic precipitation) depending on temperature T one obtains:

$$d_{p,i}^3 + d^3_{p,i,0} = k_{p,i}(T) \cdot t = k_{p,i}(T) \cdot t' \qquad (5)$$

Figure 5 shows measured data of $d_{p,i}$ as a function of t'. The constants $k_{p,1} = 5.0 \cdot 10^{-29} m^3 s^{-1}$ and $k_{p,2} = 1.5 \cdot 10^{-30} m^3 s^{-1}$ have been chosen in accordance with the existing data of 9–12% CrMo(W)VNb(N, B)-steels. In steels containing W and B the $d_{p,1}$-values of $M_{23}C_6$-carbides grow more slowly than in X 20(22) CrMoV 12 1 (X 12 CrMoWVNbN 10 1 1: $k_{p,1} = 3.5 \cdot 10^{-29} m^3 s^{-1}$, X 16 CrWMoVNbN 11 2: $k_{p,1} = 2.0 \cdot 10^{-29} \ m^3 s^{-1}$ and X 18 CrMoVNbB 9 1: $k_{p,1} = 1.5 \cdot 10^{-29} \ m^3 s^{-1}$);[19,25] this may be due to the necessity for diffusion of W and

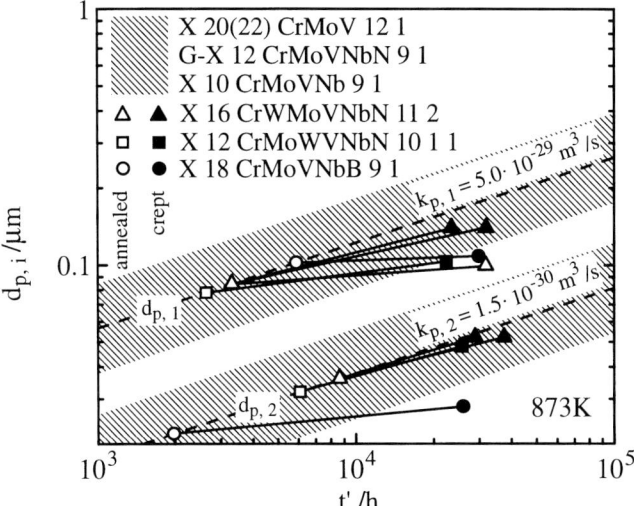

Fig. 5   Carbide sizes $d_{p,1}$ ($M_{23}C_6$) and $d_{p,2}$ (MX) as a function of the time t' at 873 K corresponding to eqn (5). Data from references 10, 12, 13, 22–25.

B during particle growth, as both elements are found in the particles.[24,25,33–35] In the B-containing steel the growth of $d_{p,2}$ is decelerated ($k_{p,2}$ = $5.0 \cdot 10^{-31}$ $m^3 s^{-1}$), too; part of this deceleration must be due to dynamic precipitation. The volume fractions $f_{p,1}$ of $M_{23}C_6$ remains fairly constant for the different 9–12% CrMo(W)VNb(N, B)-steels.[10,19,23–25] This is consistent with the report in[36] that the equilibrium value of $f_{p,1}$ is reached already after 1 h in a 9% CrMoWV-steel at a relatively low annealing temperature of 600°C. For MX-carbides[19,24,25] and the Laves phase[29,31,36] the situation is different. $f_{p,2}$ increases in the 9–12% CrMo(W)VNb(N, B)-steels during creep. This increase is more pronounced than that in the undeformed specimen heads. MX-precipitation has also been reported by Foldyna[7,20,37] and Fujita *et al.*[30] The growth rate of $f_{p,2}$ during precipitation of MX from the initial value $f_{p,2,0}$ after heat treatment to the final value $f_{p,2,\infty}$ is phenomenologically described by an equation of the Johnson–Mehl–Avrami–type:[38]

$$df_{p,2}/dt = n \cdot k \cdot (f_{p,2,\infty} - f_{p,2}) t^{n-1} \text{ or } f_{p,2} = (f_{p,2,\infty} - f_{p,2,0}) \cdot$$

$$(1 - \exp(-k \cdot t^n)) + f_{p,2,0} \tag{6}$$

A similar equation has been used by Hald[39] for calculating the precipitation of Laves-phase in NF 616 (9% Cr–0.5% Mo–1.8% W–V–Nb-steel). In the present case, however, the constants k and n must be made depen-

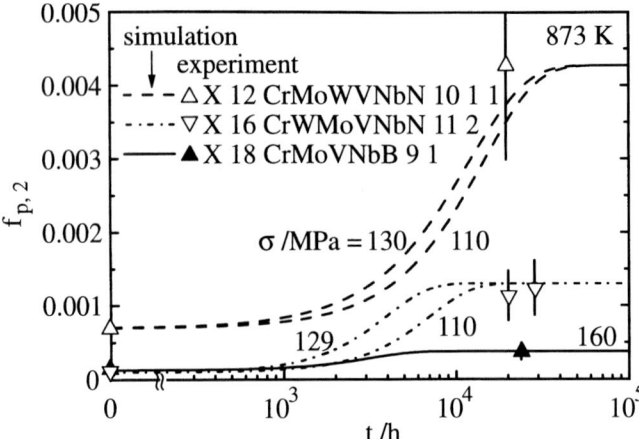

Fig. 6   Volume fraction $f_{p,2}$ as a function of time t at 873 K. $f_{p,2}$ (t) was calculated with eqn (6). Data from reference 25.

dent on stress in order to capture the dynamic nature of the process. Figure 6 shows the $f_{p,2}(t)$-curves used for the simulation of creep of X 16 CrWMoVNbN 11 2, X 12 CrMoWVNbN 10 1 1 and X 18 CrMoVNbB 9 1 (see Figs 8 and 9).

3. THE MODEL
The composite model of plastic deformation was used for modelling the progress of plastic deformation in the course of creep. The model goes back to Mughrabi[40,41] who proposed to treat the material as a composite of a (plastically) 'hard' (subscript h) phase with high dislocation density and a 'soft' (subscript s) phase with low dislocation density, which deform in parallel under the condition of equal total (elastic plus inelastic) strain. The basic idea is that the heterogeneous dislocation distribution (Figs 7a and b) causes a heterogeneous stress distribution (Fig. 7c). The model was applied by Blum [42–46] to calculate the kinetics of deformation. It follows from the model that internal forward stresses σ_f are formed at the (hard) subgrain boundaries corresponding to internal back stresses σ_b in the subgrain interior. The existence of long-range internal stresses was verified in high temperature creep deformed cooper.[47,48] The internal stresses average to zero so that:

$$\sigma = f_s \cdot \sigma + f_h \cdot \sigma_h, \qquad f_s + f_h = 1 \qquad (7)$$

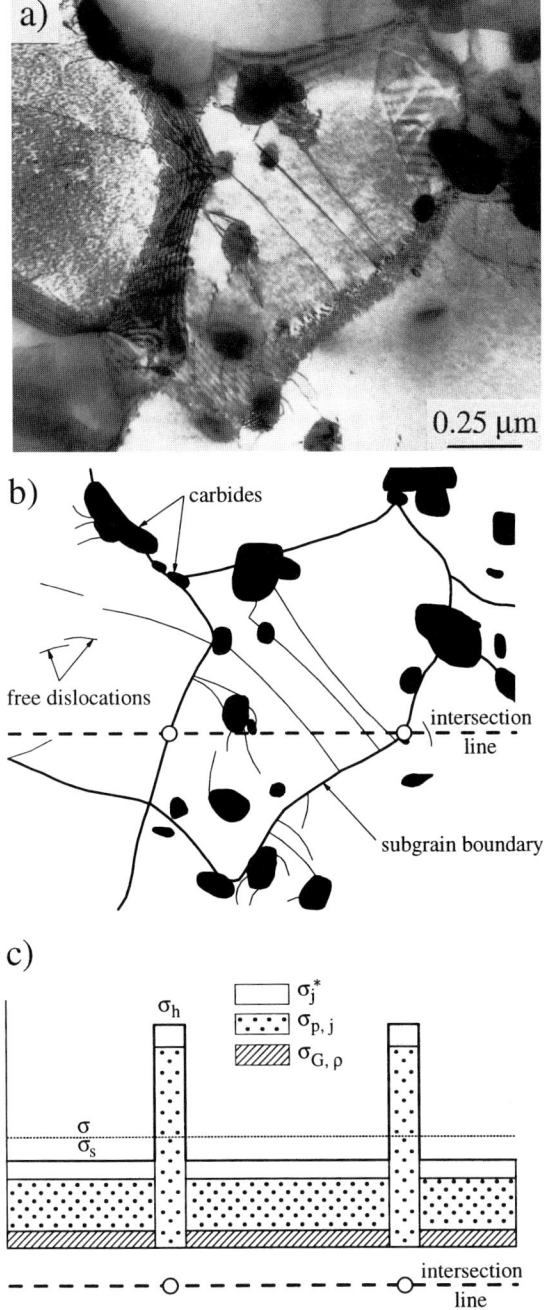

Fig. 7   Carbide and dislocation structure in X 10 CrMoVNb 9 1 after a long-term creep test at 873 K ($\sigma_0 = 100$ MPa, $\epsilon_{fr} = 0.138$, $t_{fr} = 38944$ h), (a) bright field TEM image and (b) schematic drawing. In (c) are the stress components in the soft (j = s) and hard (j = h) region displayed resulting from the composite model for $\epsilon = 0.03$.

where $f_h = 2a/w$ is the volume fraction of the hard region (a: width of hard region, $2/w$: subgrain boundary area per volume) and f_s that of the soft region. The inelastic deformation rate follows from the model as:[44]

$$\dot{\epsilon} = (1 - f_h)\cdot\dot{\epsilon}_s + f_h\cdot\dot{\epsilon}_h + \dot{f}_h(\epsilon_h - \epsilon_s) \tag{8}$$

Usually the last term in eqn (8) is negligibly small and the local (inelastic) deformation rates $\dot{\epsilon}_s$ and $\dot{\epsilon}_h$ are nearly identical. $\dot{\epsilon}_s$ and $\dot{\epsilon}_h$ are formulated as:

$$\dot{\epsilon}_s = \frac{b}{M}\cdot\rho_f v_s \quad \text{and} \quad \dot{\epsilon}_h = \frac{b}{M}\cdot\rho_f v_h \tag{9}$$

with the Taylor factor M. The velocity v_s of the free dislocations was phenomenologically expressed as:

$$v_s = v_{s,0}\cdot\sinh\left(\frac{b\cdot\Delta a_s\cdot\sigma^*_s}{M\cdot k_B\cdot T}\right) \tag{10}$$

with the activation area Δa_s in the soft region ($160/b^2 < \Delta a_s < 650/b^2$) and the Boltzmann constant k_B. The effective stress σ^*_s is the difference between the applied stress σ and the sum of the athermal stress components:

$$\sigma^*_s = \sigma - \sigma_{G,\rho} - \sigma_{p,s} - \sigma_b \tag{11}$$

$\sigma_{G,\rho} = \alpha M b G \sqrt{\rho_f}$ results from the interaction between free dislocations (α: interaction constant), σ_b is the long-range internal back stress resulting from the difference between the local strain rates $\dot{\epsilon}_s$ and $\dot{\epsilon}_h$. The particle hardening terms were calculated from the Orowan stresses for passing the particles in the soft ($j = s$) and the hard ($j = h$) regions:

$$\sigma_{p,j} = C(\sigma_j)\cdot\sigma_{Orowan,j} \quad \text{with} \quad \sigma_{Orowan,j} = 3.32\cdot Gb\cdot\sum_{i=1}^{2}\left(\sqrt{f_{p,j,i}}/d_{p,i}\right) \tag{12}$$

The factor $C(\sigma_j)<1$ takes care of the fact that σ^*_j must remain positive even when σ_j becomes smaller than the Orowan stress. The local Orowan stress is the sum of the contributions from $M_{23}C_6$ ($i = 1$) and MX-particles ($i = 2$). The dislocation velocity v_h in the hard region was similar expressed as in the soft region as:

$$v_h = v_{h,0}\cdot\sinh\left(\frac{b\cdot\Delta a_h\cdot\sigma^*_h}{M\cdot k_B\cdot T}\right) \tag{13}$$

with the activation area Δa_h ($25/b^2 < \Delta a_h < 65/b^2$). The effective stress σ^*_h in the hard region is given by:

$$\sigma_h^* = \sigma + \sigma_f - \sigma_{p,h} \tag{14}$$

$\sigma_{p,h}$ is calculated by eqn (12) as described above. Further details of the model are described elsewhere, e.g. ref. 49. The above system of equations has been successfully used in the past to model the deformation behaviour of X 20(22) CrMoV 12 1,[10,13,49] G-X 12 CrMoVNbN 9 1[12, 49] and X 10 CrMoVNb 9 1.[23] It should be noted that the present simulations were done without changing these equations so that the differences between the different steels result only from the individual microstructure.

So far rupture has not been considered in the model. Rupture occurs in tension creep tests by accumulation of damage in the form of pore formation and necking. Both processes lead to stress concentrations on different (microscopic or macroscopic) scales which enhance the rate of deformation due to its non-linear stress dependence, in addition to the small direct increase in the observed relative rate of elongation due to increasing porosity. According to the experimental results reported above, the increase of $\dot{\epsilon}$ with ϵ due to damage becomes more significant after half of fracture strain $\epsilon_{fr}/2$[10,11,22,26,27] (see Fig. 1). This result is accounted for by multiplying the simulated creep rate $\dot{\epsilon}$ at a given nominal stress σ and microstructure with the factor $\exp(D) > 1$ in order to arrive at the experimentally observed average tensile strain rate $\dot{\epsilon}_{tensile}$.[50]

$$\dot{\epsilon}_{tensile} = \dot{\epsilon} \cdot \exp(D) \quad \text{with} \quad D = C_D \cdot (\epsilon/\epsilon_{fr}) \cdot \epsilon^{k_D} \tag{15}$$

with the constant $k_D \approx 4.5$. The parameter c_D is a function of σ ($5.0 \cdot 10^2 < c_D < 1.5 \cdot 10^4$).

4.   RESULTS OF THE MICROSTRUCTURAL MODELLING

One essential advantage of microstructural modelling is that the model can be applied to new materials simply by changing the microstructural parameters (w, $\rho^{-0.5}_p$, $d_{p,i}$ and $f_{p,i}$). Figure 8 shows the $\dot{\epsilon}$– ϵ–curves at 873 K and constant stress $\sigma = 130$ MPa calculated for the three steels X 22 CrMoV 12 1, X 10 CrMoVNb 9 1 and X 12 CrMoWVNbN 10 1 1 with the model described above. The bottom part of the same figure presents the microstructural evolution starting from the experimentally determined initial values as predicted by eqns (1), (5) and (6). The variations of the different σ-components with ϵ are displayed in the centre part. The initial build-up of the athermal σ-components $\sigma_{G,\rho}$ and $\sigma_{p,s}$ at small ϵ (for simplicity the build-up was modelled by an exponential function so that the build-up is finished at $\epsilon \approx 5 \cdot 10^{-3}$) leads to a significant decrease of the effective stress σ_s^* and consequently of $\dot{\epsilon}$ in the primary range of creep. The growth of particles which is particularly pronounced at small $\dot{\epsilon}$, where the specimens spend most of their lifetime, causes

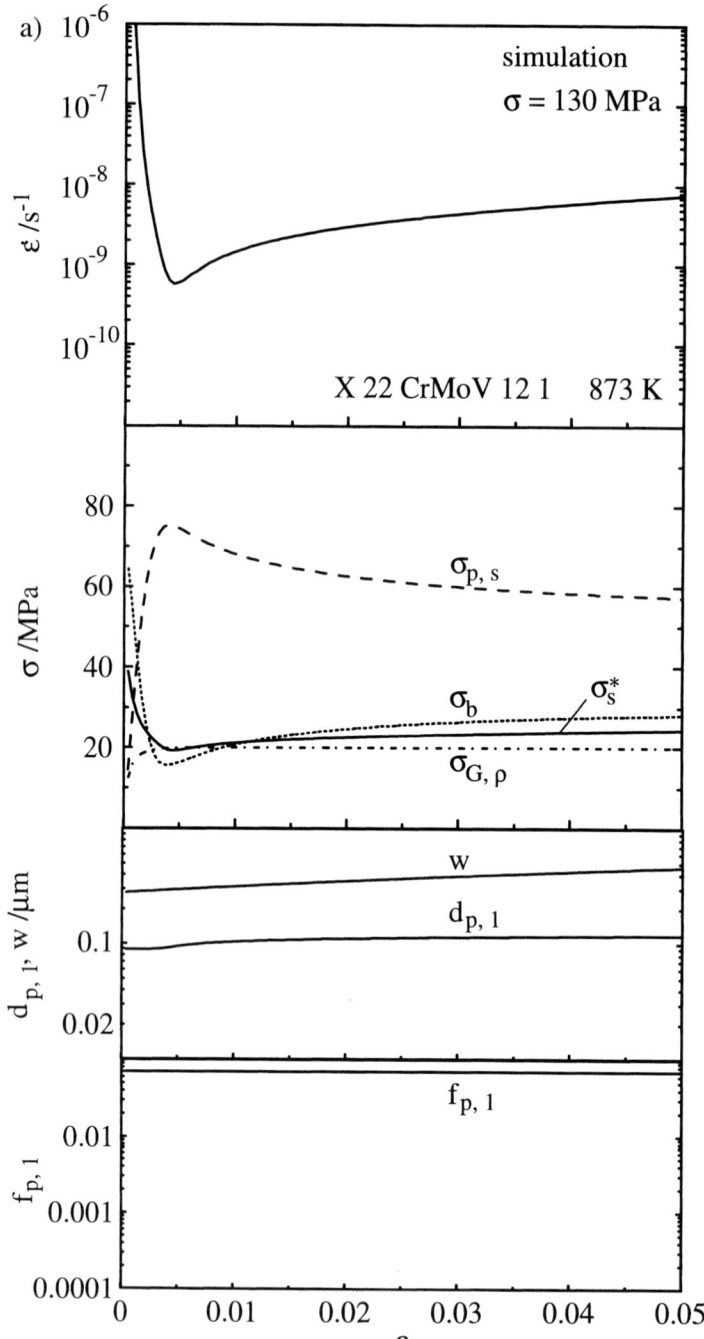

Fig. 8 Simulation of long-term creep tests with $\sigma = 130$ MPa at 873 K for (a) X 22 CrMoV 12 1. Upper part: $\dot{\epsilon}-\epsilon$-curves, centre part: development of the stress components and lower part: evolution of the microstructure.

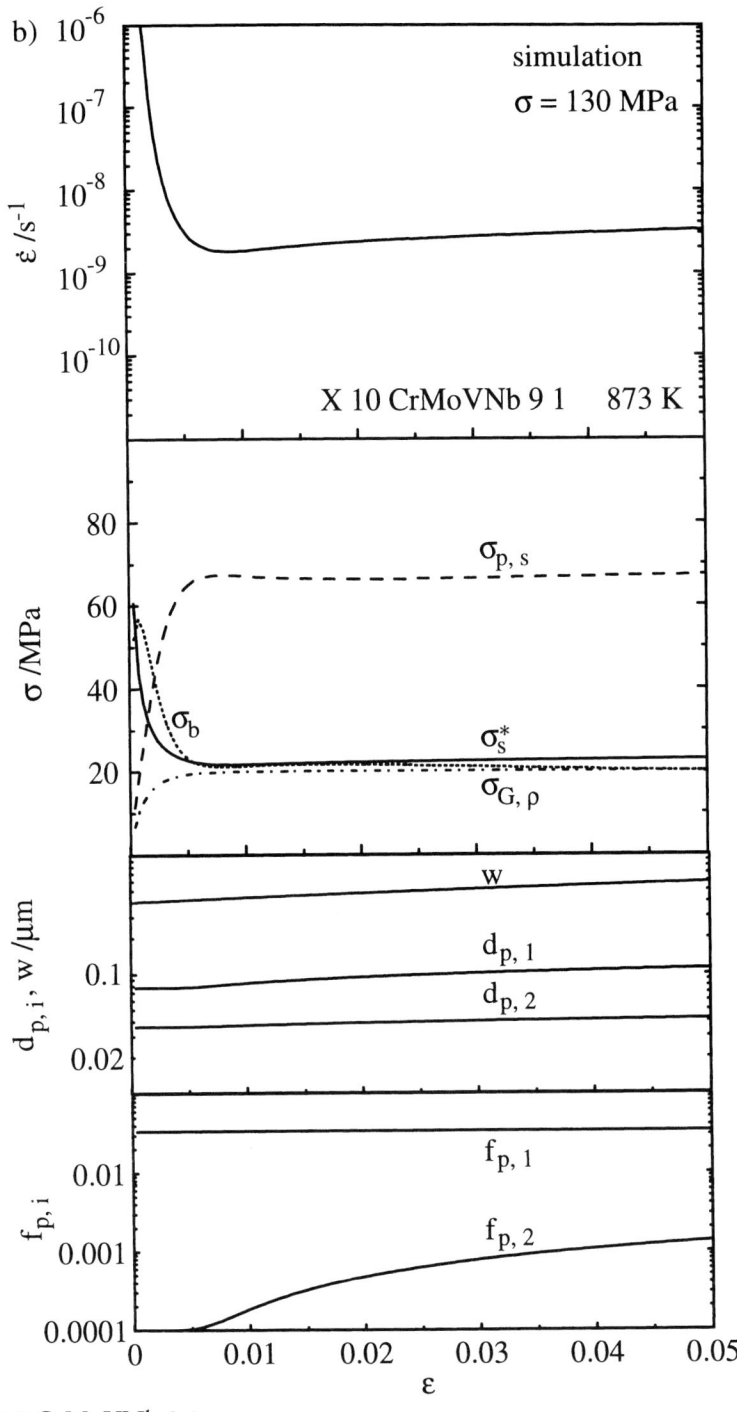

(b) X 10 CrMoVNb 9 1.

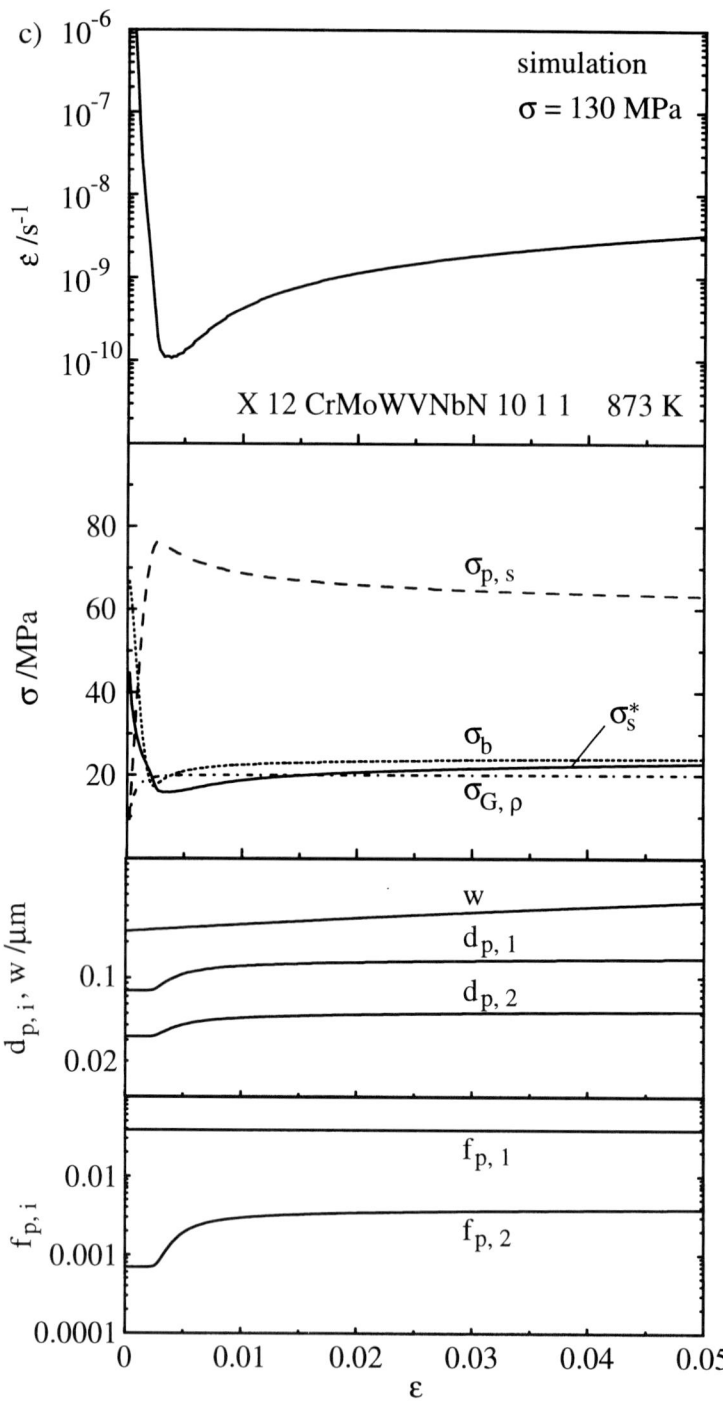

(c) X 12 CrMoWVNbN 10 1 1.

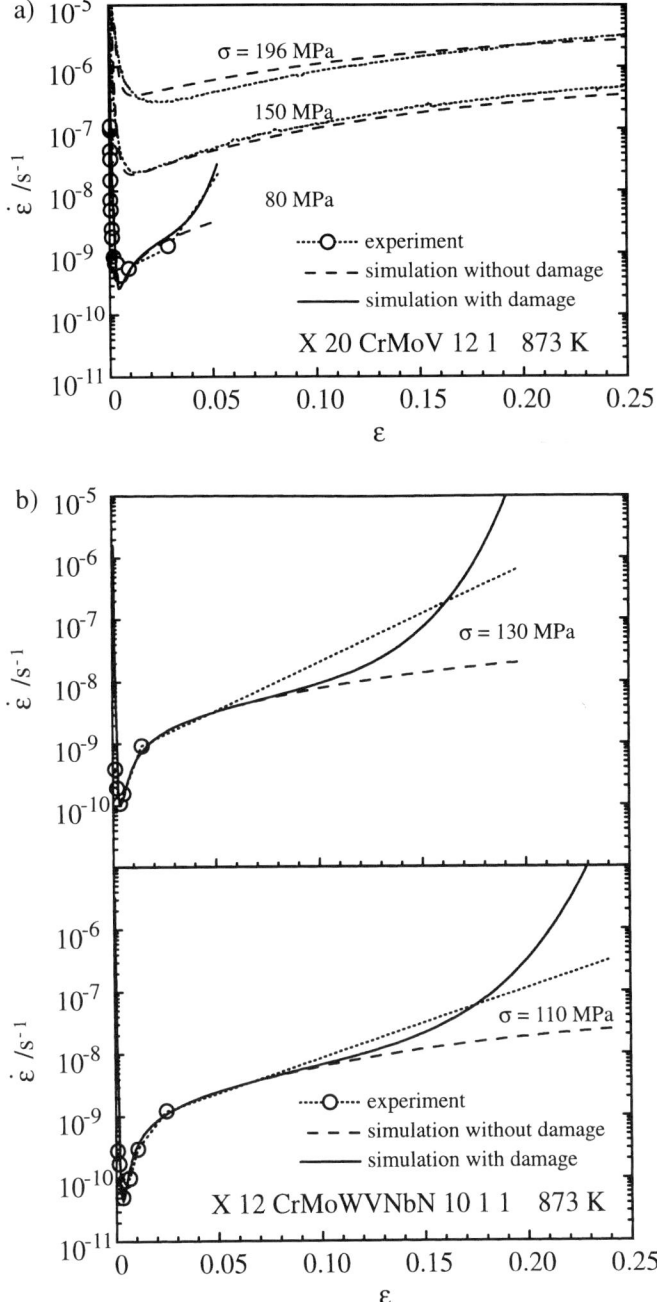

Fig. 9 Simulation of long-term creep tests for different 9–12% CrMo (W)VNb(N, B)-steels at 873 K, for (a) X 20 CrMoV 12 1, (b) X 12 CrMoWVNbN 10 1 1. Damage was taken into account after eqn (15).

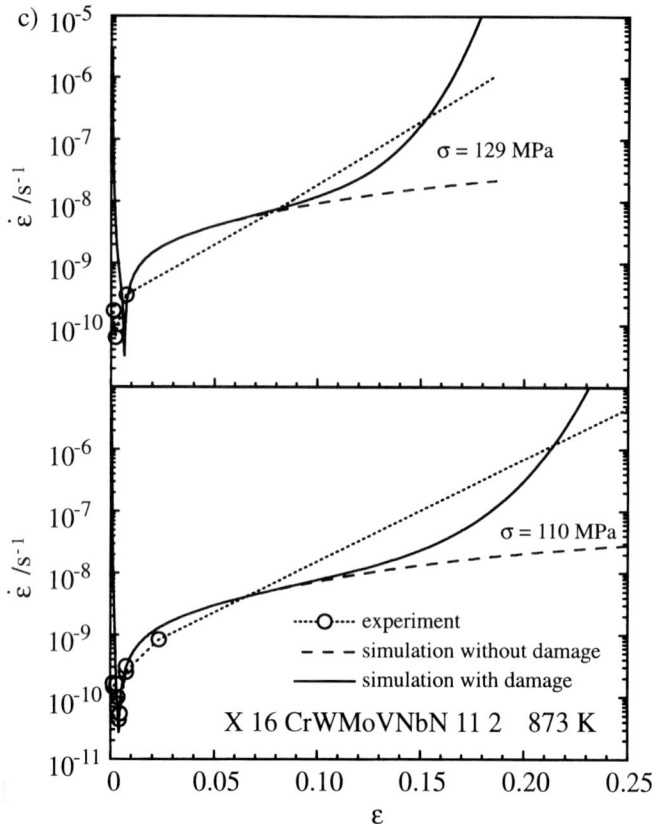

(c) X 16 CrWMoVNbN 11 2.

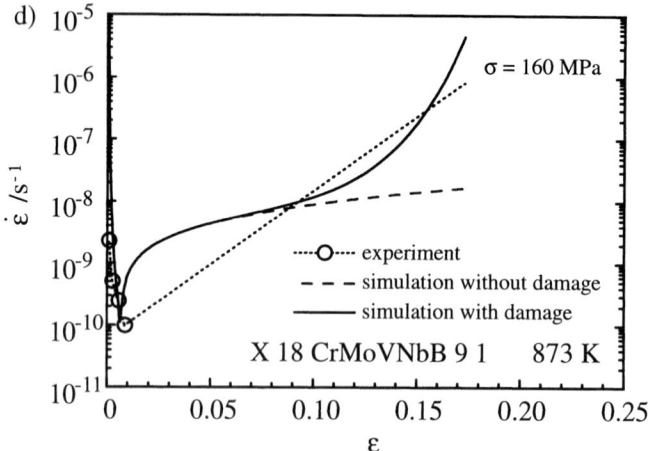

(d) X 18 Cr-MoVNbB 9 1.

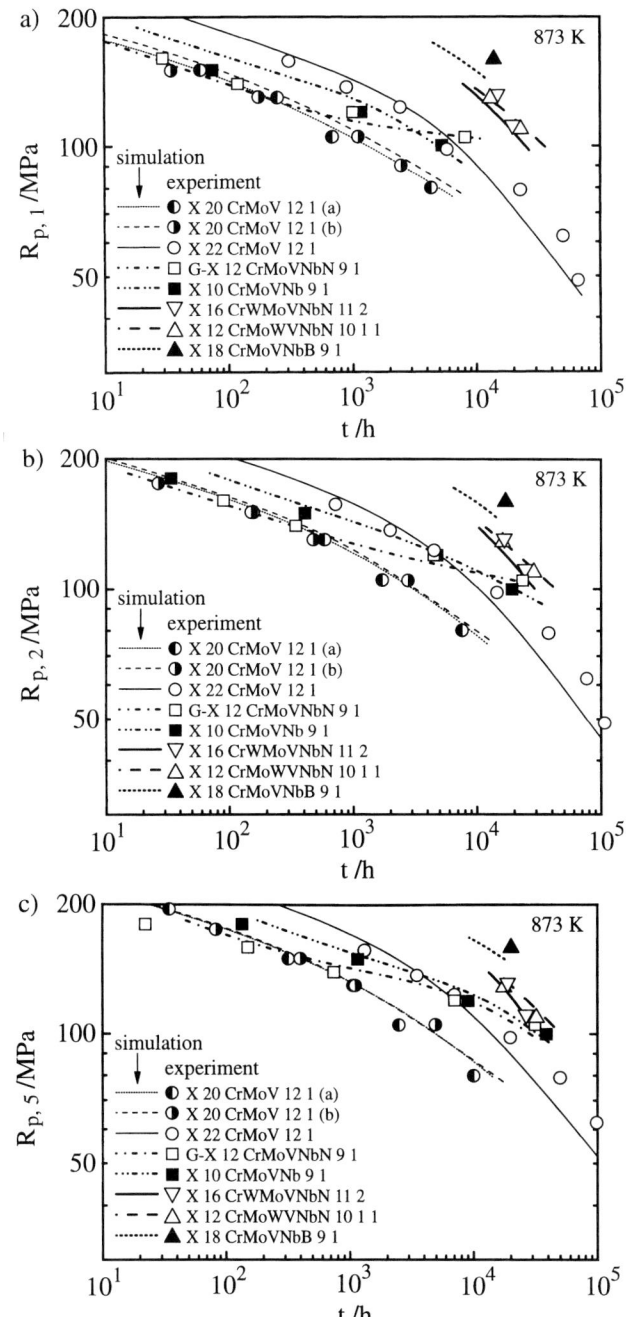

Fig. 10   Creep limit diagrams (stress $R_{p,\epsilon}$ to reach a certain plastic strain ϵ as function of t) for different 9–12% CrMo(W)VNb(N, B)-steels at 873 K for (a) $R_{p,1}$ for $\epsilon = 0.01$, (b) $R_{p,2}$ for $\epsilon = 0.02$ and (c) $R_{p,5}$ for $\epsilon = 0.05$. Data from references 10, 12, 23, 25, 51, 52.

a loss of particle hardening. This leads to increase in σ_s^* and $\dot{\epsilon}$. The increase of $\dot{\epsilon}$ with ϵ is less pronounced in the cases where dynamic precipitation of MX (increase of $f_{p,2}$) occurs (X 10 CrMoVNb 9 1 and X 12 CrMoWVNbN 10 1 1). The kinetics of MX-precipitation is also of significant importance for the minimum creep rate and thus for the creep life. It should be noted that the relative contribution of the particle hardening terms $\sigma_{p,i}$ to the local stresses σ_i is quite high at low stresses which means that the strain rate $\dot{\epsilon}$ reacts sensitive to changes in $\sigma_{p,i}$.

Figure 9b–d compares the result of simulations of tensile creep tests with experimental data for the case of the new turbine steels X 12 CrMoWVNbN 10 1 1, X 16 CrWMoVNbN 11 2 and X 18 CrMoVNbB 9 1. For comparison are results for X 20 CrMoV 12 1 included in Fig. 9a. It is seen that the shapes of the simulated $\dot{\epsilon}$–ϵ-curves and the simulated positions of the $\dot{\epsilon}_{min}$-values are in reasonable agreement with experiment, taking into account that only the microstructural data entering the model have been changed and that these data are unavoidably subject to a certain degree of experimental inaccuracy. For the B-containing steel X 18 CrMoVNbB 9 1 the simulation is uncertain in the tertiary range of creep beyond $\dot{\epsilon}_{min}$ due to the fact that not enough experimental ϵ–t-data are available. So there may be an worst agreement between experimental and simulated $\dot{\epsilon}$–ϵ-curve, even though the particle growth constants $k_{p,1}$ and $k_{p,2}$ are by a factor of 3 smaller than for the other 9–12% CrMoV-steels.[25]

In spite of the inaccuracies mentioned above the time to reach a certain strain under long-term creep conditions is well predicted by the simulation. This is seen from Fig. 10. The measured points are well met by the calculated curves over a large stress range and for a number of different steels. The differences between the steels result from microstructural differences. The creep strength of the conventional 12% CrMoV-steels results from its fine subgrain structure and the $M_{23}C_6$-carbides. Due to subgrain and particle growth the two simulated melts of X 20 CrMoV 12 1 as well as X 22 CrMoV 12 1 exhibit a decrease in creep strength with time. This decrease is significantly less pronounced in the modern 9-12% CrMo(W)VNb(N, B)-steels. The reason for this lies in the dynamic precipitation of MX-carbonitrides and the slower growth of $M_{23}C_6$- and MX-particles. When the dynamic MX-precipitation is complete, the growth of particles dominates again, as the generation of new small particles ceases. Consistent with that all the curves in Fig. 10 are relatively steep in spite of the low growth constants $k_{p,i}$ in the steels alloyed with W and B. The curvature of the lines makes extrapolation to long times uncertain.

Figure 11 shows the influence of different heat treatments on the creep behaviour of X 16 CrWMoVNbN 11 2. If the heat treatment results in an increase of the initial values of subgrain size w_0 and of the size $d_{p,1,0}$ of $M_{23}C_6$-carbides the model predicts an shift of the ϵ–t and the $\dot{\epsilon}$–ϵ-curve to higher

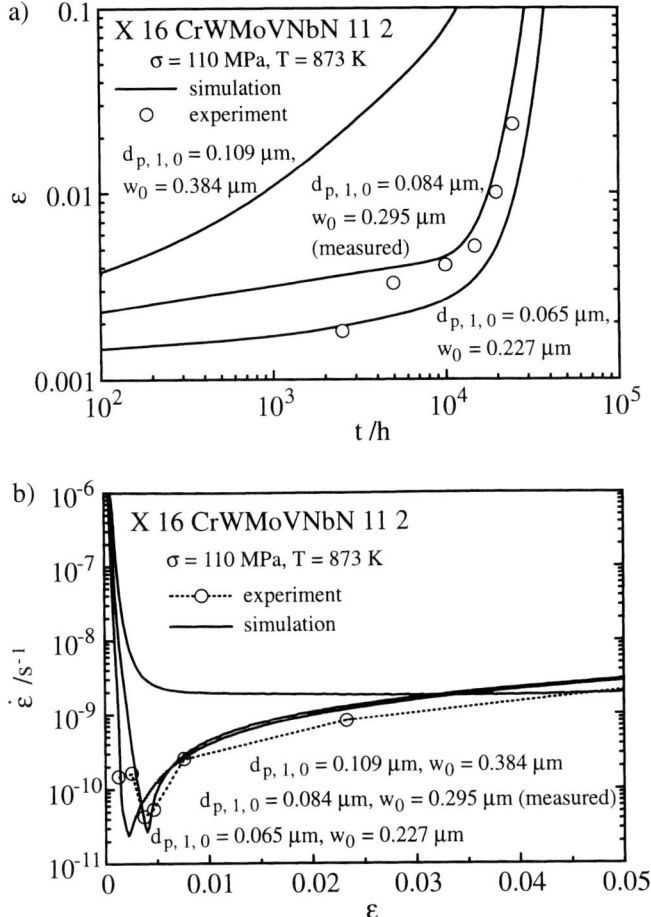

Fig. 11   Influence of varying initial values of the subgrain size w_0 and the size $d_{p,1,0}$ of $M_{23}C_6$-carbides on the (a) ϵ–t– and (b) $\dot{\epsilon}$–ϵ– behaviour.

values of ϵ and $\dot{\epsilon}$, respectively. Smaller initial values of w_0 and of $d_{p,1,0}$ are increasing the creep strength. In this context it would be efficient to combine the composite model of plastic deformation with an model, which is able to predict the initial values of w_0 and of $d_{p,1,0}$ from the chemical composition and the heat treatment.

If the microstructural simulation presented above is correct, it has import-ant implications for the understanding of creep resistance and the further development of the 9–12% CrMo(W)VNb(N, B)-steels. The simulation uses only the differences in the hardening by dislocations (including subgrains) and particles. Differences in solid solution hardening do not exist in the simulation, because the expression (eqns (10) and (13)) for the relation between the dislocation velocity and the effective stress was the same for all steels. This is in contradiction to frequent statements of solid solution hardening effects (e.g. by Mo or W) in the literature.[21,29,31] In our model the effects of alloying elements like W and B exert only an indirect effect on the creep strength through their influence on the particle structure. In view of the dominance of particle hardening found at low stresses in the course of the present simulation, it seems reasonable that the indirect influence of alloying elements through the particles is larger than their direct influence on the velocity of free dislocations in the matrix between the particles.

A general remark seems appropriate regarding the limitations of the microstructural simulation. Principally, the simulation is not exact. The reasons are manifold. One reason is that the expression for the dislocation velocity is not exactly known. For instance, the validity of the definition of the effective stress becomes dubious once the effective stress is a small fraction of the applied stress which means that the phenomenological parameter C in eqn (12) becomes very important. Given these basic limitations the success of the microstructural simulation performed in the present work is rather astonishing. It may mean that the relations (10) to (14) capture some physical truth. This may result from the sequence of fitting procedures which have been performed in the development of the model and its application to many different cases (different steels, heat treatments, loading conditions). Nevertheless one must be aware that the intelligence of the simulation does not rest primarily on the quantification of the microstructure, although this alone gives important insights, but mainly on the way how the microstructural information is used in the equations of the model. This means that the reliability of the model in future applications is uncertain and that the model will not be able to replace long-term tests in the foreseeable future. However, the model is in fact an extremely valuable means to make and test predictions within a short time and comparatively little effort once it is available, raise questions and deepen the microstructural understanding of the creep.

5. SUMMARY

In the present work results of simulations on a microstructural basis are presented. The microstructural data, which are used as 'input' values for the simulation, can be determined by quantitative transmission electron microscopy.

The evolution of the microstructural parameters (w, $\rho^{-0.5}_f$, $d_{p,i}$ and $f_{p,i}$), which are of relevance for the long-term creep behaviour, can be calculated by relative simple equations. Uncertain is the dynamic precipitation of MX-carbonitrides. In the present version of the composite model this process is phenomenological described by an Johnson–Mehl–Avrami–equation, fitted on the experimental values of $f_{p,2}$. Also phenomenological formulated is the influence of damage in form of external and internal necking.

The composite model allows to simulate the creep behaviour of different heats of 9–12% CrMo(W)VNb(N, B)-steels with one set of kinetic equations. The individuality of a single heat is taken into account by its own set of microstructural parameters, which are for example characteristic for the initial state. The initial state itself is a result of the chemical composition and the heat treatment. One advantage of modelling on a microstructural basis is an easy application to other steels, which was demonstrated.

The presented composite model explains the differences in the long-term creep behaviour in terms of the observed microstructure. The three characteristic parts of creep are explained by the model in the following manner: In the primary range, where $\dot{\epsilon}$ decreases with ϵ, are the athermal stress components $\sigma_{G,p}$ and $\sigma_{p,j}$ build up, which are related to the dislocation and particle structure. The secondary range of creep, the minimum creep rate $\dot{\epsilon}_{min}$, results from the coarsening of carbides and subgrains (decrease of $\sigma_{p,j}$). In the tertiary range, where $\dot{\epsilon}$ increases with ϵ, dominates the softening the creep behaviour. For $\epsilon > \epsilon_{fr}/2$ influences internal and external necking the deformation behaviour.

ACKNOWLEDGEMENT

The authors want to thank the German Federal Minister of Economics, the Arbeitsgemeinschaft industrieller Forschungsvereinigungen e.V. (AiF) and the Verein Deutscher Eisenhüttenleute (VDEh) for the financial support of part of the work within the AiF-project no. 9732. Thanks also due to Mr. K.H. Mayer from GEC ALSTHOM Energie GmbH, Nürnberg, Germany, for having the possibility to investigate trial heats, developed in the European Research Program COST 501.

REFERENCES

1. R. Blum: *Materials for Advanced Power Engineering 1994, Part I*, D. Coutsouradis *et al.* eds, Liège, Belgium, 1994, Kluwer Academic Publishers, Dordrecht, 15.
2. Ch. Berger, R.B. Scarlin, K.H. Mayer, D.V. Thornton and S. M. Beech: *Materials for Advanced Power Engineering 1994, Part I*, D. Coutsouradis *et al.* eds, Liège, Belgium, 1994, Kluwer Academic Publishers, Dordrecht, 47.
3. R.B. Scarlin, Ch. Berger, K.H. Mayer, D.V. Thornton and S.M. Beech: *Materials for Advanced Power Engineering 1994, Part I*, D. Coutsouradis *et al.* eds, Liège, Belgium, 1994, Kluwer Academic Publishers, Dordrecht, 73.
4. J. Orr and D. Burton: Commission of the European Communities,

ECSC Information Day, The Manufacture and Properties of Steel 91 for the Power Plant and Process Industries, VDEh, Düsseldorf, J. Orr *et al.* eds, Swinden, England 1992, British Steel Technical, contribution 2.1.

5. W. Bendick and M. Ring: *Steel Research*, 1996, **67**, 382.

6. A. Zielińska-Lipiec, A. Czyrska-Filemonowicz, P.J. Ennis and O. Wachter: *Proc. of the 9th International Symposium on Creep Resistant Metallic Materials*, J. Purmensky *et al.* eds, Hradec nad Moravicí, Czech Republic, 1996, VÍTKOVICE a.s., Ostrava, 254.

7. V. Foldyna, Z. Kuboň, M. Filip, K.H. Mayer and Ch. Berger: *Steel Research*, 1996, **67**, 375.

8. G. Eggeler, N. Nilsvang and B. Ilschner: *Steel Research*, 1987, **58**, 97.

9. N. Nilsvang: *PhD thesis*, Ecole Polytechnique Federale de Lausanne (EPFL), Lausanne, 1989.

10. S. Straub: VDI-Fortschritt Berichte, Dissertation Universität Erlangen-Nürnberg 1994, VDI-Verlag, Reihe **5**, Nr. **405**, Düsseldorf, 1995.

11. S. Straub, M. Meier, J. Ostermann and W. Blum: *VGB Kraftwerkstechnik*, 1993, **73**, 646.

12. S. Straub, P. Polcik, W. Besigk, W. Blum, H. König and K.H. Mayer: *Steel Research*, 1995, **66**, 402.

13. S. Straub, T. Hennige, P. Polcik and W. Blum: *Steel Research*, 1995, **66**, 394.

14. F. Abe and S. Nakazawa: *Metall. Trans.*, 1992, **23**A, 3025.

15. F. Abe, S. Nakazawa, H. Araki and T. Noda: *Metall. Trans.*, 1992, **23**A, 469.

16. V. Sklenička, K. Kuchařová, A. Dlouhý and J. Krejčí: *Materials for Advanced Power Engineering 1994*, Part I, D. Coutsouradis *et al.* eds, Liège, Belgium, 1994, Kluwer Academic Publishers, Dordrecht, 435.

17. G. Eggeler, J. Hald, M. Cans and J. Phillips: *Proc. of the 5th Int. Conf. on Creep and Fracture of Engineering Materials and Structures*, Swansea, England, 1993, Pineridge Press, London, 527.

18. A. Buršik, A. Orlová, K. Kuchařová and V. Sklenička: *Proc. of the 9th International Symposium on Creep Resistant Metallic Materials*, J. Purmensky *et al.* eds, Hradec nad Moravicí, Czech Republic, 1996, VÍTKOVICE a.s., Ostrava, 234.

19. S. Straub, P. Polcik, D. Henes and W. Blum: VGB-Konferenz *Werkstoffe und Schweißtechnik im Kraftwerk 1996*, Cottbus, Germany, 1996, VGB Technische Vereinigung der Großkraftwerksbetreiber e.V., Essen, Band VGB-TB 514, 6.1.

20. V. Foldyna, A. Jakobová, R. Řiman and A. Gemperle: *Steel Research*, 1991, **62**, 453.

21. V. Foldyna, Z. Kuboň, A. Jakobová and V. Vodárek: *Proc. of the 9th*

International Symposium on Creep Resistant Metallic Materials, J. Purmensky *et al.* eds, Hradec nad Moravicí, Czech Republic, 1996, VÍTKOVICE a.s., Ostrava, 203.

22. S. Straub, P. Polcik and W. Blum: *Proc. of the 10th Int. Conf. on the Strength of Materials (ICSMA 10)*, H. Oikawa *et al.* eds, Sendai, Japan, 1994, The Japan Institute of Metals, Sendai, 623.

23. P. Polcik, S. Straub, M. Kiesbauer, W. Blum and W. Bendick: Arbeitsgemeinschaft für warmfeste Stähle und Arbeitsgemeinschaft für Hochtemperaturwerkstoffe, 18. Vortragsveranstaltung *Langzeitverhalten warmfester Stähle und Hochtemperaturwerkstoffe*, Verein Deutscher Eisenhüttenleute (VDEh), Düsseldorf, 1995, 93.

24. S. Straub, P. Polcik, M. Kiesbauer, W. Blum and K.H. Mayer: Arbeitsgemeinschaft für warmfeste Stähle und Arbeitsgemeinschaft für Hochtemperaturwerkstoffe, 18. Vortragsveranstaltung *Langzeitverhalten warmfester Stähle und Hochtemperaturwerkstoffe*, Verein Deutscher Eisenhüttenleute (VDEh), Düsseldorf, 1995, 106.

25. S. Straub, D. Henes, P. Polcik, W. Blum, K.H. Mayer and J. Hald: Arbeitsgemeinschaft für warmfeste Stähle und Arbeitsgemeinschaft für Hochtemperaturwerkstoffe, 19. Vortragsveranstaltung *Langzeitverhalten warmfester Stähle und Hochtemperaturwerkstoffe*, Verein Deutscher Eisenhüttenleute (VDEh), Düsseldorf, 1996, 112.

26. S. Straub and W. Blum: Arbeitsgemeinschaft für warmfeste Stähle und Arbeitsgemeinschaft für Hochtemperaturwerkstoffe, 15. Vortragsveranstaltung *Langzeitverhalten warmfester Stähle und Hochtemperaturwerkstoffe*, Verein Deutscher Eisenhüttenleute (VDEh), Düsseldorf, 1992, 81.

27. S. Straub, P. Polcik, P. Weidinger and W. Blum: Arbeitsgemeinschaft für warmfeste Stähle und Arbeitsgemeinschaft für Hochtemperaturwerkstoffe, 16. Vortragsveranstaltung *Langzeitverhalten warmfester Stähle und Hochtemperaturwerkstoffe*, Verein Deutscher Eisenhüttenleute (VDEh), Düsseldorf, 1993, 21.

28. S. Straub and W. Blum: *Proc. of the Int. Symp. Hot Workability of Steels and Light Alloys-Composites*, H.J. McQueen *et al.* eds, Montreal, Quebec, Canada, 1996, The Canadian Institut of Mining, Metallurgy and Petroleum, Montreal, 189.

29. J. Hald and Z. Kuboň: *Proc. of the 9th International Symposium on Creep Resistant Metallic Materials*, J. Purmenský *et al.* eds, Hradec nad Moravicí, Czech Republic, 1996, VÍTKOVICE a.s., Ostrava, 53.

30. N. Fujita, K. Ohmura, M. Kikuchi, T. Suzuki, S. Funaki and I. Hiroshige: *Scripta Materialia* , 1996, **35**, 705.

31. J. Hald: *Steel Research*, 1996, **67**, 369.

32. K. Spiradek, R. Bauer and G. Zeiler: *Materials for Advanced Power*

Engineering 1994, Part I, D. Coutsouradis *et al.* eds, Liège, Belgium, 1994, Kluwer Academic Publishers, Dordrecht, 251.

33. L. Lundin: *Dissertation*, Chalmers University of Technology and Göteborg University, Göteborg, Schweden, 1995.
34. L. Lundin and H.-O.Andréen: *Surface Science*, 1991, **266**, 397.
35. L. Lundin and B. Richarz: Applied Surface Science, 1995, **87/88**, 194.
36. J.D. Robson and H.K.D.H. Bhadeshia: *Proc. of the 9th International Symposium on Creep Resistant Metallic Materials*, J. Purmenský *et al.* eds, Hradec nad Moravicí, Czech Republic, 1996, VÍTKOVICE a.s., Ostrava, 83.
37. A. Jakobová, V. Foldyna and V. Vodárek: *Proc. of the 9th International Symposium on Creep Resistant Metallic Materials*, J. Purmenský *et al.* eds, Hradec nad Moravicí, Czech Republic, 1996, VÍTKOVICE a.s., Ostrava, 306.
38. J.W. Christian: *The Theory of Transformations in Metals and Alloys*, 2nd edition, Part I, Equilibrium and General Kinetic Theory, Pergamon Press, Oxford, 1975.
39. J. Hald: *New Steels for Advanced Plant up to 620°C*, The EPRI/National Power Conference, E. Metcalfe ed., London, England, 1995, Pica Publishing Services, London, 152.
40. H. Mughrabi: *Proc. 5th Conf. on the Strength of Metals and Alloys (ICSMA 5)*, P. Haasen *et al.* eds, Aachen, Germany, 1980, Pergamon Press, London, 1615.
41. H. Mughrabi: *Acta Metall.*, 1983, **31**, 1367.
42. W. Blum and H. Schmidt: *Res Mechanica*, 1983, **9**, 105.
43. W. Blum: *Scripta Metall.*, 1984, **18**, 1383.
44. W. Blum, A. Rosen, A. Cegielska and J.L. Martin: *Acta Metall.*, 1989, **37**, 2439.
45. S. Vogler and W. Blum: *Proc. of the 4th Int. Conf. on Creep and Fracture of Engineering Materials and Structures*, B. Wilshire *et al.* eds, Swansea, England, 1990, The Institute of Metals, London, 65.
46. U. Hofmann and W. Blum: *Proc. of the 7th JIM International Symposium on Aspects of High Temperature Deformation and Fracture in Crystalline Materials*, Y. Hosoi *et al.* eds, Sendai, Japan, 1993, The Japan Institute of Metals, Sendai, 625.
47. A. Borbély, H.J. Maier, H. Renner, S. Straub, T. Ungár and W. Blum: *Scripta Metall. mater.* 1993, **29**, 7.
48. S. Straub, W. Blum, H.J. Maier, T. Ungár, A. Borbély and H. Renner: *Acta Mater.* 1996, **44**, 4337.
49. P. Polcik, S. Straub and W. Blum: *Proc. of the 4th European Conference on Advanced Materials and Processes (Euromat '95)*, Symposium D – Structural Metallic Materials, Padua/Venice, Italy, 1995, Associazione Italiana di Metallurgia, Milano, 313.

50. H. Riedel: *Fracture at High Temperatures, Materials Research and Engineering Series*, B. Ilschner *et al.* eds, Springer Verlag, Berlin, 1987.
51. S. Straub, P Polcik, W. Blum, R. Mohrmann and H. Riedel: Ergebnisse im Rahmen einer Zusammen arbeit zwischen dem Fraunhofer Institut für Werkstoffmechanik, Freiburg i.Br., und dem Lehrstuhl I für Allgemeine Werkstoffeigenschaften der Universität Erlangen-Nürnberg, Erlangen, 1995.
52. M. Oehl: VDI-Fortschritt Berichte, Dissertation Technische Hochschule Darmstadt 1993, VDI-Verlag, Reihe 5, Nr. 335, Düsseldorf, 1994.

Forecasting Grain Boundary Precipitation Kinetics in Austenitic Steels

R.G. FAULKNER,* D. MEADE,* C.C. GOODWIN* AND
M. SPINDLER†

* Power and Aerospace Materials Group, IPTME, Loughborough University,
Loughborough, Leicestershire, LE11 3TU, UK
† Nuclear Electric Ltd, Barnett Way, Barnwood, Gloucester, Gloucestershire, GL4
3RS, UK

1. INTRODUCTION

Earlier models for forecasting the kinetics of grain boundary precipitate growth have been modified and extended. The precipitate system being investigated is inter- and intra-granular $Cr_{23}C_6$ in heats of type 316 austenitic stainless steels.

The original grain boundary precipitation models of Aaron and Aaronson,[1] Caisley and Faulkner[2] and Faulkner and Carolan[3] assumed a lenticular shaped precipitate morphology and a constant collector plate size with regard the solute supply. Some consideration of a variable collector plate size was made by Carolan and Faulkner, who concluded that this model was applicable only at temperatures close to the precipitate solvus. Mass transfer to assist growth was assumed to take place by a combined lattice and grain boundary diffusion controlled mechanism. Nucleation was also included in the kinetics appraisal and for the case of $M_{23}C_6$ in an austenitic steel matrix the nucleation kinetics were lattice diffusion controlled. Other work on aluminium alloys by Aaronson's group highlighted the importance of grain boundary diffusion in this process, but this was probably because the precipitate matrix interfaces (θ-prime in Al–Cu) were semicoherent in their work.

The model developed here has been applied to the alloys whose respective compositions are listed in Table 1. The prime difference between the alloys are the carbon contents, which will obviously affect the rates of carbide nucleation and growth within the systems being investigated.

Table 1 Material compositions

Alloy	C	Si	Mn	P	S	Cr	Mo	Ni	B	Nb	Ti	V	N	Co
A	.083	.35	.77	.033	.016	16.7	2.52	11.7	.0001	.083	.035	.77	.0285	.09
B	.034	.18	1.87	.021	.013	16.62	2.3	10.45	.0043	.01	.006	.02	.033	.035
C	.05	.61	1.75	.031	.01	17.2	2.25	12.2	.0002	.035	.005	.025	.04	.295

2. MODEL DETAILS

2.1 Introduction

More recently Jiang and Faulkner[4,5] have taken the Carolan models further and applied them to high strength 7 000 series aluminium alloys. Specifically the improvements to the model include the following:

1. Abandonment of analytical equations to describe the kinetics in favour of an iterative approach, dividing the precipitate growth into a series of small steps and changing the driving force in accordance with the precise conditions applicable at each step.
2. Inclusion of grain boundary segregation, using models of Seah and Hondros[6] for equilibrium segregation and of Faulkner[7] for non-equilibrium segregation. The effect of the segregation behaviour modifies the solute supply rate.
3. Application of classical nucleation equations to define the activation energy to form the critical nucleus at the ageing temperature; this is used to calculate the nucleation site density, the reciprocal of which is the collector plate area at the onset of growth.
4. Assumption of coalescence: this is quantified by calculating the change in collector plate area as a function of time assuming a normal statistical distribution of precipitates. Increasing the collector plate size increases the solute supply rate accordingly.
5. Inclusion of the Gibbs–Thompson effect, which allows for the effects of interface curvature on solute solubility in very small precipitates.
6. Zener modelling of intra-granular precipitation.

We have incorporated this approach into a new model dedicated to describing $Cr_{23}C_6$ inter- and intra-granular precipitation in heats of 316 austenitic stainless steel.

2.2 Segregation During Quench

The model begins by describing the amount of non-equilibrium segregation of solute (in this case Chromium) created around the grain boundary during the quench. The effective quench time is calculated assuming that the quench is performed at a constant temperature. The quench conditions set the solute concentration expected on the boundary as a result of non-equilibrium segregation, which can be given by:[8]

$$C_b = C_g \exp\left(\frac{E_b - E_f}{kT_i} - \frac{E_b - E_f}{k0.55T_{mp}}\right)\frac{E_b}{E_f} \tag{1}$$

where C_g is the bulk solute concentration, E_b is the solute-vacancy binding

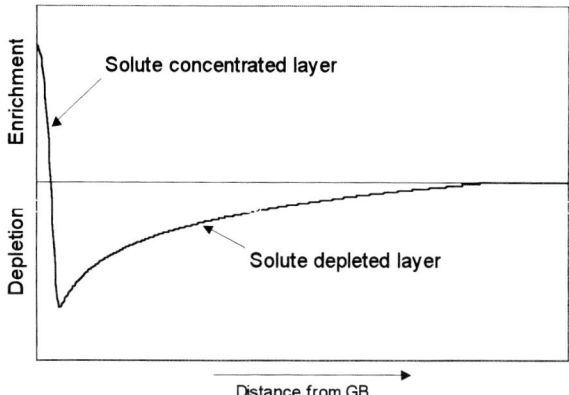

Fig. 1   Segregation profile due to quench

energy, E_f is the vacancy formation energy, k is Boltzmanns constant, T_i is the solution treatment temperature and $0.55T_{mp}$ is the temperature below which no more segregation is assumed to take place since complexes are insufficiently mobile, taken as 0.55 times the melting point of the alloy.

Two regions, the solute concentrated layer (SCL) and the solute depleted layer (SDL) are created near to the boundary (see Fig. 1). These originate from solute being taken into the SCL by solute-vacancy complex diffusion during the quench time.

During quenching from solution treatment temperature, or during high temperature ageing where the ageing temperature, T_{age}, is greater than $0.55T_{mp}$, equations (2a)–(2c) govern the segregation behaviour.

$$\text{If } y \leq w_1 \qquad C\left(y,t_{eff}\right) = C_b \exp\left(-\alpha y^2\right) \tag{2a}$$

$$\text{If } w_1 < y \leq w \quad C\left(y,t_{eff}\right) = C_b \exp\left(-\alpha y^2\right) + \frac{y - w_1}{w} C_g \tag{2b}$$

$$\text{If } y > w \qquad C\left(y,t_{eff}\right) = C_b \exp\left(-\alpha y^2\right) + C_g \tag{2c}$$

where y is the distance from the grain boundary,

$$w_1 = \frac{C_g\sqrt{2D_{ci}t_{eff}}}{C_b}, \ w = \frac{2w_1\left(C_b - C_g\right)}{C_g}, \ \alpha = \frac{\pi C^2_b}{8D_{ci}t_{eff}C^2_g}$$

t_{eff} is the effective quench time (determined later) and D_{ci} is the diffusion coefficient of the vacancy-solute complexes at the solution treatment temperature.

2.2.1 Determination of the effective quench time, t_{eff} The cooling time $t_{T_0}^{Ti}$, from the solution treatment temperature, T_i, to room temperature, T_0, can simply be determined from:

$$t_{T_0}^{Ti} = \frac{\left(T_i - T_0\right)}{\theta} \tag{3}$$

where θ is the cooling rate between temperatures T_i and T_0. The effective cooling time at the solution treatment temperature for this temperature change can thus be expressed as:

$$t_{eff} = t_{T_0}^{Ti} e\left(\frac{-Qc\left(T_i-T_0\right)}{kT_iT_0}\right) \tag{4}$$

where Q_c is the activation energy term for the calculation of the diffusion coefficient of the vacancy-solute complexes.

However, quenching processes generally have different cooling rates at different temperatures, so this approach will only give an approximate value. In order to incorporate the effect of varying cooling rates, it would be necessary to alter the equation so that it incorporates different cooling rates over different temperature ranges.

2.3 Precipitate Nucleation

Nucleation is important for two reasons. First, it defines the nucleation site density on the grain boundary which is crucial to predicting the interparticle spacing and hence the collector plate area at the onset of growth. Second, it is important for determining the nucleation time, which is needed to describe the total time necessary to grow a precipitate to a given size.

As mentioned earlier, the initial collector plate size, A_{m0}, is given by the inverse of the number of precipitates per unit area of grain boundary, thus:

$$A_{m0} = \frac{1}{N_0} \tag{5}$$

where the nucleation site density function, N_0, given by Russell[9] is determined from:

$$N_0 = \frac{N}{x_0} \exp\left(\frac{-\Delta G^*}{kT_{age}}\right) \tag{6}$$

where x_0 is the solute concentration in the precipitate phase, ΔG^* is the Gibb's free energy of critical grain boundary nucleus formation and N is the number of atom sites at the grain boundary, which can be evaluated from:

$$N = d_0 \rho_\alpha N_A \tag{7}$$

where d_0 is the grain boundary width, ρ_α is the molar density of the matrix and N_A is Avagadro's number.

The driving force, ΔG^*, to produce the critical nucleus size can be determined thus:

$$\Delta G^* = \frac{4\pi\sigma^3_{\alpha\theta}}{3\Delta G_V^2} \left(2 - 3\cos\left(\psi\right) + \cos^3\left(\psi\right)\right) \tag{8}$$

where $\sigma_{\alpha\theta}$ is the matrix/precipitate interfacial energy and ψ is the angle between the grain boundary and the tangent plane to a precipitate at their intersection.

The driving force for the solid-state transformation, ΔG_v, is given by:

$$\Delta G_v = \frac{RT_{age}}{2V_\theta} \ln\left(\frac{x_b}{x^{\alpha\theta}_{\alpha Tage}}\right) \tag{9}$$

where R is the gas constant, V_θ is the molar volume of the precipitate phase and x_b is grain boundary solute concentration modified to take into account the segregation behaviour of the material, which is taken to be equal to C_b.

The only unknown factor in eqn 9 is $x^{\alpha\theta}_{\alpha Tage}$, and this is the equilibrium concentration existing between the precipitate phase and the matrix at the nucleation temperature, which we are assuming to be the ageing temperature. This can be calculated from the solubility product equation found, for example, in Faulkner and Caisley.[3]

$$x^{\alpha\theta}_{\alpha Tage} = \left[\frac{1}{C_c} \exp\left(\frac{-28800}{RT_{age}} - 0.9\right)\right]^{\frac{6}{23}} \tag{10}$$

where C_c is the carbon content of the alloy being considered.

The only method by which the collector plate size can increase is through particle coalescence and coarsening. The times for the onset of coarsening are inappropriately long in this study and so this aspect will not be considered further here. Coalescence can be important and we have considered it by assuming that there exists a normal distribution of collector plate sizes about the mean that we have calculated from eqn 5. This given by eqn

11, where A_v is the variable collector plate size defined by the normal distribution:

$$n_0 = \frac{N_0}{2\sqrt{\pi\sigma}} \exp\left[-\frac{(A_v - A_{m0})^2}{2\sigma^2} \right] \tag{11}$$

where σ is the standard deviation of the collector plate areas and is taken as:

$$\sigma = \frac{(A_{m0} - A_{min})}{3} \tag{12}$$

where A_{min} is the minimum collector plate area, given by:

$$A_{min} = 16D_{bnuc}\tau \tag{13}$$

where D_{bnuc} is the grain boundary diffusion coefficient of the solute atoms at the nucleation, or ageing, temperature and τ is the nucleation time, derived later. Equation (11) is used later in the calculation of precipitation growth.

The nucleation times are generally very short at typical ageing temperatures, i.e., <600°C. They are calculated from the Van der Velde et al.[10] equation for cap shaped nuclei which are those present in the material:

$$\tau = \frac{32kT_{age}a^4\sigma^2{}_{\alpha\theta}N_A^2}{D_{bnuc}d_0x_bV_\theta^2\Delta G_v^3 \sin\psi} \tag{14}$$

where a is the lattice parameter of the matrix. This time is added to the calculated growth time for the specified heat treatment which will be discussed in the next section.

2.4 Segregation During Ageing

The width of the SCL increases with time until the temperature drops below the temperature at which non-equilibrium segregation is expected to cease, usually about $0.55T_{mp}$, where T_{mp} is the absolute melting point of the material. After this time no more non-equilibrium segregation is expected to occur and the SDL is gradually filled up by solute diffusion down the concentration gradient formed by the non-equilibrium segregation process (Fig. 2). During ageing where the ageing temperature, T_{age}, is less than $0.55T_{mp}$:

$$C(y,t) = C_b \exp(-\alpha'y^2) \quad \text{if } y < w_1 \tag{15}$$

$$C(y,t) = C_b \exp(-\alpha'y^2) + c_I(y,t) + c_{II}(y,t)$$

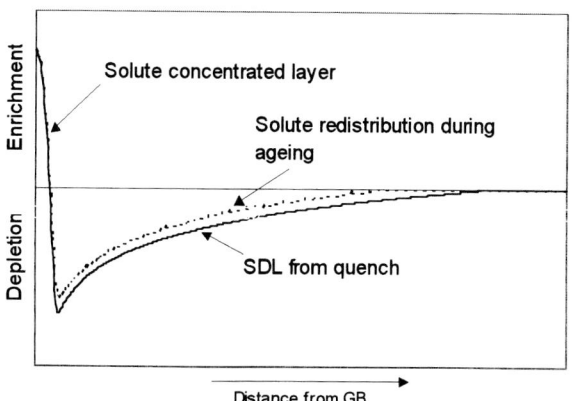

Fig. 2   Segregation during ageing at temperatures below 0.55T_{mp}

$$+ c_{III}(y,t) + c_{IV}(y,t) \ \text{if} \ y \geq w_1 \tag{16}$$

where:

$$\alpha' = \frac{\pi C^2_{\ b}}{8 D_{cl} t_{eff} C^2_{\ g}} \tag{17}$$

$$c_I(y,t) = C_g + \frac{(y - w_1)C_g}{w} erf \frac{y - w_1}{2\sqrt{D_c t}} + \frac{2\sqrt{D_c t} C_g}{\sqrt{\pi} w} \exp \frac{-(y - w_1)^2}{4 D_c t} \tag{18}$$

$$c_{II}(y,t) = -\frac{C_g}{2} \left(erf \frac{w + y - w_1}{2\sqrt{D_c t}} + erf \frac{w - y + w_1}{2\sqrt{D_c t}} \right) \tag{19}$$

$$c_{III}(y,t) = -\frac{\sqrt{D_c t} C_g}{\sqrt{\pi} w} \left[\exp \frac{-(w + y - w_1)^2}{4 D_c t} + \exp \frac{-(w - y + w_1)^2}{4 D_c t} \right] \tag{20}$$

$$c_{IV}(y,t) = -\frac{y C_g}{2w} \left(erf \frac{w + y - w_1}{2\sqrt{D_c t}} - erf \frac{w - y + w_1}{2\sqrt{D_c t}} \right) \tag{21}$$

where t is the ageing time, $w_1 = \dfrac{C_g \sqrt{2 D_{cl} t_{eff}}}{C_b}$ and D_c is the diffusion coeffi-

cient of the vacancy-solute complexes at the ageing temperature.

The key requirement of this part of the model is to provide information at any stage in the heat treatment about the amount of segregated solute on the subsequent calculations of precipitate growth kinetics. The amount of extra solute, $x_{\alpha i}$, at any time during the ageing process, t', is given by:

$$x_{\alpha i} = \frac{1}{\delta l} \int_0^{\delta l} C(y,t') \, dy \qquad (22)$$

where δl is the diffusion distance of the solute in the matrix within time t', given by:

$$\delta l = \sqrt{2 D_I t'} \qquad (23)$$

where D_I is the volume diffusion coefficient of the solute at the ageing temperature. Equations 15–21 above are used for this part of the calculation of $C(y,t')$.

2.5 Precipitate Growth

The precipitate growth is again treated in a series of steps, δt, with each step adding to the size calculated from the sum of the preceding steps. The representative equation for the precipitate radius, L_i, is given by:

$$L_i = \left\{ \frac{2A_{mi-1} D_I^{1/2}(x_{\alpha i} - x_{\alpha Tage}^{\alpha\theta})[t'^{1/2} - (t' - \delta t)^{1/2}]\rho_\alpha}{\rho_\theta \pi^{1/2} f(\psi)(x_\theta - x_{\alpha Tage}^{\alpha\theta})} + L_{i-1}^3 \right\}^{1/3} \qquad (24)$$

where x_θ is the solute concentration in the precipitate, $f(\psi)$ is a geometrical parameter dependent on the particle shape, ρ_θ is the molar density of the precipitate phase and L_{i-1} is the precipitate size determined for the previous time.

The A_v term in eqn 11 becomes the variable collector plate size as a function of time. This is determined by assessing the reduction in number of precipitates per unit area as a result of coalescence. The probability of coalescence occurring is given by n_0/N_0 taken from eqn 11 and so at each time step A_v can be recalculated to give A_{mn} as in

$$A_{mn} = \frac{1}{N_0 \prod\limits_{i=1}^{i=n} \left\{ 1 - 0.25 \left[erf\left(\frac{A_{mi-1} - A_{min}}{\sqrt{2}\sigma} \right) - erf\left(\frac{A_{mi-1} - 4L_i^2}{\sqrt{2}\sigma} \right) \right]^2 \right\}} \qquad (25)$$

where A_{mi-1} is the previous collector plate area and L_i is the precipitate size at the time being considered. In principle, the model can be used not only to calculate precipitate size as a function of quenching and ageing treatments, but also to calculate interparticle spacing, through $1/(A_i)^{1/2}$ and any precipitate-free zone width from an assessment of the shape of the curve given in Fig. 2, which is described by eqns 15–21. In this case we have only used the model to forecast precipitate size.

2.6 Intragranular Precipitation

The simple Zener model based on ref. 11 is used for this type of growth. It simply applies a spherical diffusion field around the precipitate, controlled by the lattice diffusion coefficient for the solute. In one dimension the amount of solute supplied from a right angled triangle of composition-distance space whose height is $(x_\alpha - x^{\alpha\theta}_{\alpha\text{Tage}})$ and length is $2(Dt)^{1/2}$ is used to grow a particle whose area in composition-distance space is $(x_\theta - x_\alpha) \times$ (*particle radius*). Thus, with certain approximations, the particle grows with a power law relationship with time of $t^{1/2}$ using the following equation

$$x = \sqrt{\frac{2D_I(x_\alpha - x^{\alpha\theta}_{\alpha\text{Tage}})t}{x_\theta}} \tag{26}$$

3. RESULTS

Material data used in the calculations are given in Table 2. The variable parameters used in the model are the carbon content, the solute content, and the

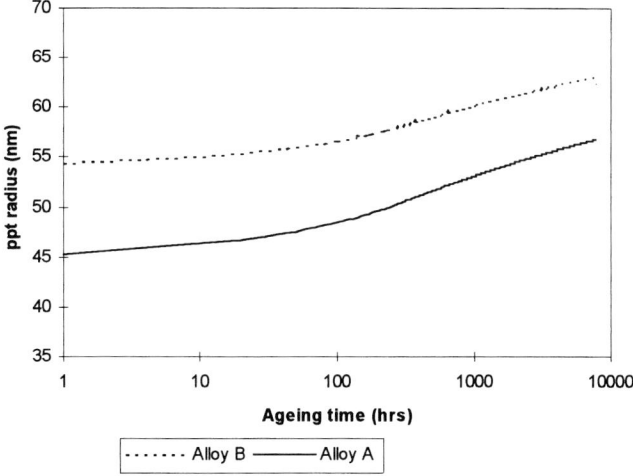

Fig. 3   Predictions for alloys A and B at 923K

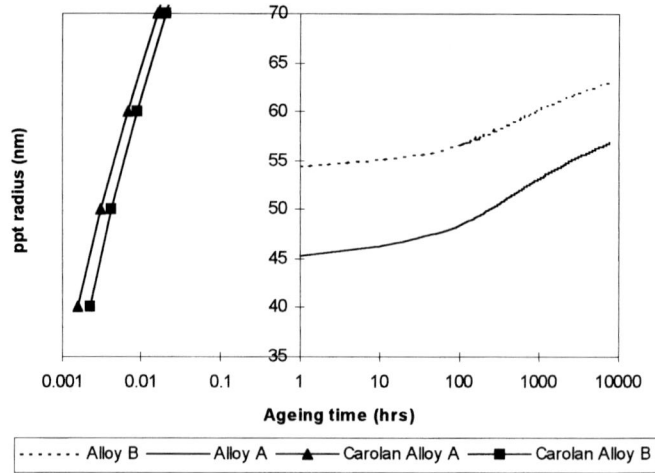

Fig. 4   Comparison of Carolan model and present model for alloys A and B

ageing temperature, time and calculation time interval. The carbon and solute contents can be found from Table 1. The ageing processes modelled are temperature ranging from 773K to 923K, with time up to 100,000 hours. The quench was assumed to be uniform at 50K s^{-1} through the three temperature ranges, from the solution treatment temperature of 1323K.

The intergranular $Cr_{23}C_6$ precipitate growth curve at 650°C is shown for alloy A and alloy B in Fig. 3. The predictions show the effect of carbon content, alloy B has a 0.034% C content whilst alloy A contains 0.083% carbon. Comparisons of the Carolan model with the new model for alloy A and alloy B are shown in Fig. 4. Isothermal precipitation curves for 100 nm radius pre-

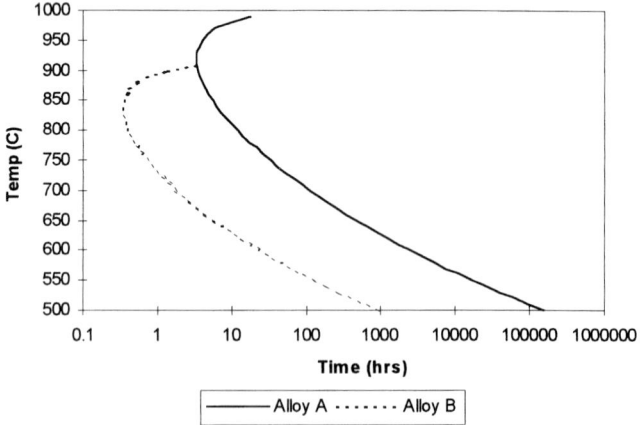

Fig. 5   Carolan predictions for 100nm precipitate in alloys A and B

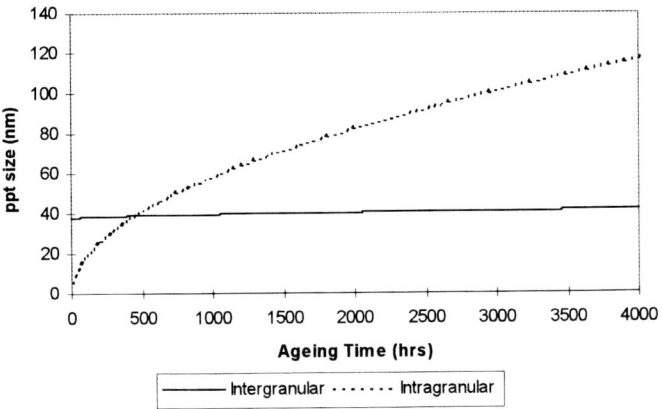

Fig. 6 Inter- and intra-granular precipitation predictions for alloy A at 823K

Table 2 Material data

Parameter	Units	Value
T_s	K	1323
$0.55T_{mp}$	K	1023
d	m	0.000005
ψ	°	57
gbw	m	0.0000000001
a	m	3.649e–10
av	m²	0.0000063
Qv	Jk⁻¹mol⁻¹	243 000
ab $\}D_b$	m² s⁻¹	0.003
Qb	Jk⁻¹mol⁻¹	191 000
Eb	eV.atom	0.0361
Efv	eV.atom	1.6
Ea $\}D_{ci}$	eV.atom	1.96825
DoC	m² s⁻¹	0.00083
$\rho\alpha$	mol m⁻³	140 920
$\rho\theta$	mol m⁻³	5496
V_θ	mol⁻¹	0.000181932
x_θ	mol.fr	0.78
$\sigma_{\alpha\theta}$	Jm⁻²	0.668
r'	m	0.0000001

$\}D_{ci}$

cipitates predicted by the Carolan model are given for alloy A and alloy B in Fig. 5. Comparisons of the inter- and intra-granular precipitation rates for alloy A are shown in Fig. 6.

4. DISCUSSION

At lower temperatures, 500–700°C, lower growth rates observed here with higher C material reflect the fact that the nucleation is more difficult with lower C contents. This leads to larger collector plates at the early stages of growth and this results in a higher solute supply to individual grain boundary precipitates.

The solvus temperature is higher for higher carbon contents. This means that at temperatures near to the solvus temperature, 900–1000°C, growth rates will be faster in higher carbon content materials because the available solute for precipitation according to the solubility product equation is lower. Therefore temperatures must be higher for the equilibrium solute concentration to reach the alloy solute concentration, which is what occurs at the solvus temperature.

The Carolan model predicts faster rates of growth; thus indicating that, although the Cr grain boundary segregation is playing a small but important role in the precipitation kinetics, the smaller collector plate size being calculated in the new model is not allowing such high levels of solute supply.

Figure 6 shows that intergranular precipitation rates are only faster than intragranular rates for times of up to 1,000 hours at 823K. This could be due to that fact that Ostwald ripening has not been included in the intergranular precipitate modelling as yet, or that the classical Zener models do not work well after long times.

5. CONCLUSIONS

The introduction of a new iterative modelling approach to prediction of grain boundary precipitation kinetics has been applied to Al alloys and austenitic steels. The results for $M_{23}C_6$ in 316 steel are shown here for the first time. Results indicate that, because of the revised method for estimating the collector plate size in the modelling results in a smaller collector plate size, precipitation rates are reduced, when compared with results of earlier models. Carbon content is shown to be a sensitive parameter when considering the overall isothermal precipitation curve behaviour. Intragranular precipitation rates are seen to exceed those of intergranular precipitation when ageing times exceed 1,000 hours at 650°C.

REFERENCES

1. H.B. Aaron and H.I. Aaronson: *Acta Metall.*, 1968, **16**, 789.
2. R.A. Carolan and R.G. Faulkner: *Acta Metall.*, 1988, **36**(2), 257.

3. R.G. Faulkner and J. Caisley: *Met. Sci.*, 1977, **11**, 200.
4. H. Jiang and R.G. Faulkner: *Acta Mater.*, 1996, **44**(5), 1857.
5. H. Jiang and R.G. Faulkner: *Acta Mater.*, 1996, **44**(5), 1865.
6. M.P. Seah and E.D. Hondros: *Int. Met. Rev.*, 1977, **222**, 262.
7. R.G. Faulkner: *Acta Metall.*, 1987, **35**(12), 2905.
8. R.G. Faulkner, S. Song and P.E.J. Flewitt: *Met. Trans.*, 1996, **27A**, 3381.
9. K.C. Russell: *Acta Metall.*, 1969, **17**, 1123.
10. G. Van der Velde, J.A. Velasco, K.C. Russell and H.I. Aaronson: *Metall. Trans.*, 1976, **7A**, 1472.
11. H.B. Aaron, D. Fainstein and G.R. Kotler: *J. Appl. Phys.*, 1970, **41**, 4404.

Power-Law and Viscous Creep in Advanced 9%Cr Steel

L. KLOC, V. SKLENIČKA, A. DLOUHÝ AND
K. KUCHAŘOVÁ

Institute of Physics of Materials, Academy of Sciences of the Czech Republic, Žižkova 22, CZ–61662 Brno, Czech Republic

ABSTRACT

Short-term creep tests were performed on a 9%Cr (P-91 type) steel at temperatures from 823 K to 923 K and wide range of stresses from 1 MPa to 300 MPa. The conventional constant stress tensile creep tests were performed for stresses above 100 MPa, while the helicoid spring specimens technique was used for low strain rates at lower stresses. Current knowledge of the creep mechanisms is based mainly on short-term laboratory experiments in power-law creep regime with creep rates higher than $10^{-10}\,\text{s}^{-1}$. However, service loading of real high temperature components may lead to rates lower than $10^{-12}\,\text{s}^{-1}$, which are typical for viscous creep. Unfortunately, experimental data describing viscous creep regime in structural materials are very rare, because their limited microstructural stability does not enable experiments to be run at very high temperatures, where the viscous creep can also be observed at higher strain rates.

The steady state creep rates correspond to viscous behaviour under stresses below approximately 100 MPa at 600°C , characterised by the apparent stress exponent close to 1. Since the stress exponent at higher stresses is around 10, the change in the deformation mechanism at lower stresses is proved. The deformation mechanisms map resulting from the presented data shows that the service loading conditions respond to the viscous creep. Extrapolation from power-law creep regime to low stresses can cause serious underestimation of predicted deformation rates.

The service loading conditions for the steel under consideration lay close to the transition boundary from viscous to power-law creep regime. None of both deformation mechanisms is then negligible at the service conditions. Thus, any realistic model describing creep properties of the steel must take into account both power-law and viscous mechanisms.

The effect of the viscous deformation mechanism on the creep life remains questionable. Since the materials behaves like Newtonian viscous fluid, it is possible that the viscous deformation mechanism has only little effect on the creep damage development. Unfortunately, only long term creep experiments (tens or hundreds of years) can shed light into this problem. Nevertheless, viscous creep should be taken into consideration, mainly for the parts where the dimension stability is critical.

1. INTRODUCTION

An increasing demand for power plants with higher efficiencies finds the solution in the use of plants operating at higher temperatures and higher pressures. It can be assumed that with the improved chromium ferritic-martensitic steels the present limit for steam admission can be considerably raised. However, more fundamental understanding of mechanisms responsible for high temperature creep and its microstructural aspects is needed for further development of these steels.

The present investigation contributes to a systematical study of high temperature creep behaviour and the role of microstructure in a tempered martensite ferritic steel (P91). Since the viscous creep had been found for several steels at very low stresses,[1-4] the main effort of present study was focused to the transition from power-law to viscous creep regime.

2. EXPERIMENTAL PROCEDURES

The material used in this study was a trial melt of P-91 type steel, supplied by

Table 1 Chemical composition of the material (in mass percent)

Cr	Mo	Si	Mn	V	Ni	Nb	C	N	Al	P	S
8.5	0.88	0.43	0.4	0.23	0.1	0.1	0.1	0.045	0.018	0.015	0.006

Vítkovice Steel a.s. The initial state was formed by the two stage heat treatment consisting of (i) normalisation at 1060°C for 1 hour followed by air cooling and (ii) tempering at 750°C for 2 hours followed by air cooling. Chemical composition of the material is summarised in the Table 1. Two different creep testing techniques were used to cover wide range of applied stresses and creep rates. The conventional uniaxial tensile creep tests with constant stress were used for stresses above 100 MPa, while the helicoid spring specimen technique[5-7] was used for lower stresses down to 1 MPa. All creep tests were performed in a protective atmosphere of purified argon at the temperatures stabilised to ±1°C.

2.1 Tensile Tests

Flat creep specimens having a 50 × 5 × 3.2 mm gauge section were machined from the heat treated blank. The changes in specimen length were measured using a linear variable differential transducer, the output of which was recorded. All the tests were conducted to the final fracture.

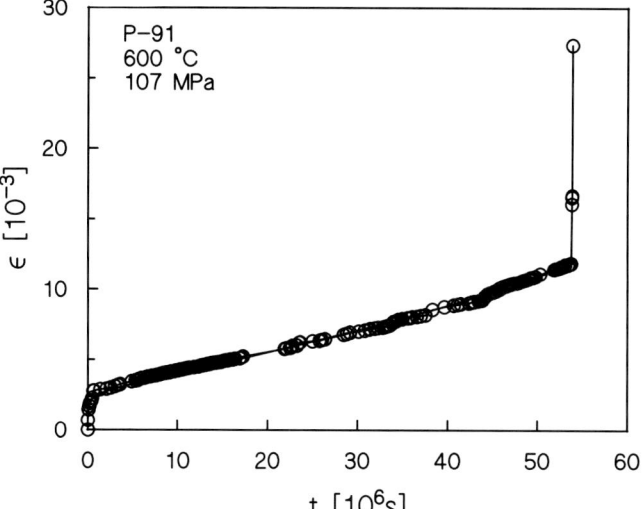

Fig. 1   Typical creep curve obtained by the conventional tensile test.

2.2   Helicoid Spring Tests

Helicoid spring specimens were machined from the tube. Most specimens were used for creep testing without any additional treatment, but some of them were annealed at 600°C for 5000 hours. Optical measurements of individual coil spacings were performed periodically and the creep strains were derived. Since the stress and strain in helicoid spring are essentially shear ones, they were recalculated to the equivalent tensile quantities using a well-known relations $\sigma = \tau\sqrt{3}$ and $\epsilon = \gamma/\sqrt{3}$ where σ is the tensile stress, τ is the shear stress, ϵ is the tensile strain and γ is the shear strain. Self-loaded specimens provided results at stresses below 12 MPa, and additional weight was used for the tests at higher stresses up to 100 MPa.

3.   RESULTS

3.1   Creep Curves

Under given conditions and stresses above 100 MPa, the material exhibits normal three-stage creep behaviour. The primary stage of decreasing creep rate is well pronounced, but relatively short, followed by short secondary stage. The long tertiary stage represent the major part of the test duration and the major part of total creep strain reached in the test (Fig. 1). The creep curves were fitted by the semiempirical equation[8]

$$\epsilon = \dot{\epsilon}_s t_p \ln\left(1 + k_p\left(1 - \exp\left(\frac{-t}{t_p}\right)\right)\right) + \dot{\epsilon}_s t + k_t\left(\exp\left(\frac{t}{t_t}\right) - 1\right) \tag{1}$$

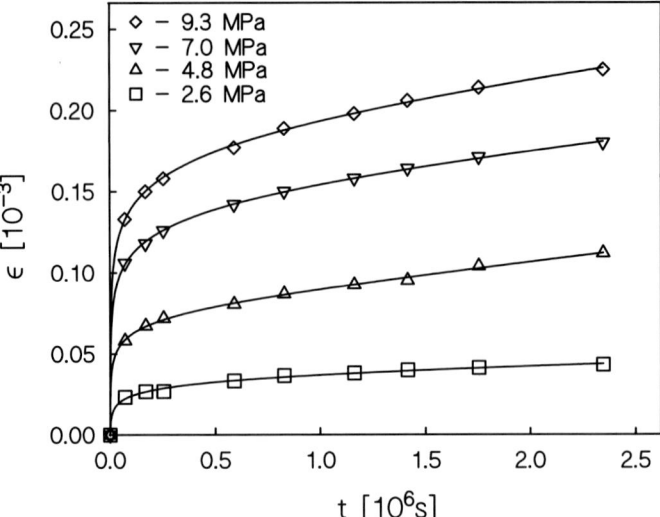

Fig. 2 Typical creep curves obtained by the helicoid spring specimen technique at low stresses.

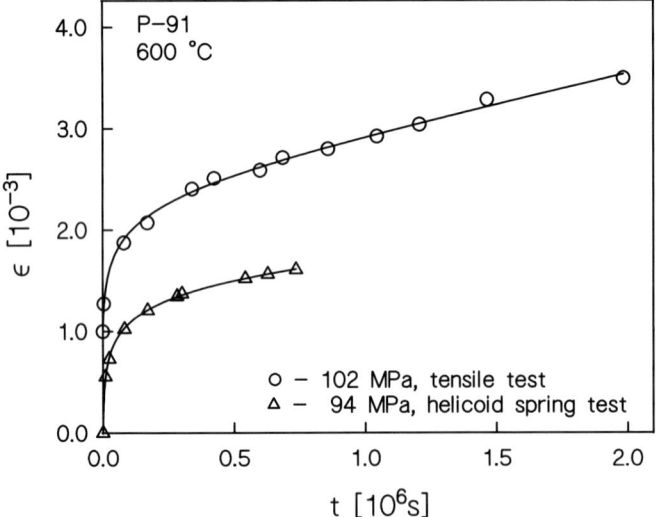

Fig. 3 Comparison of the creep curves obtained by the both conventional and helicoid spring specimen techniques. The upper curve is shifted up by $\epsilon = 0.001$ to avoid overlapping of the curves.

where t is time and $\dot{\epsilon}_s$, t_p, k_p, k_t and t_t are parameters. The last part of creep curve was cut off at 0.9 t_f (time to fracture) before fitting.

For the technique of helicoid spring specimen at stresses below 100 MPa, the tests were conducted to reach secondary state. Creep curves were fitted

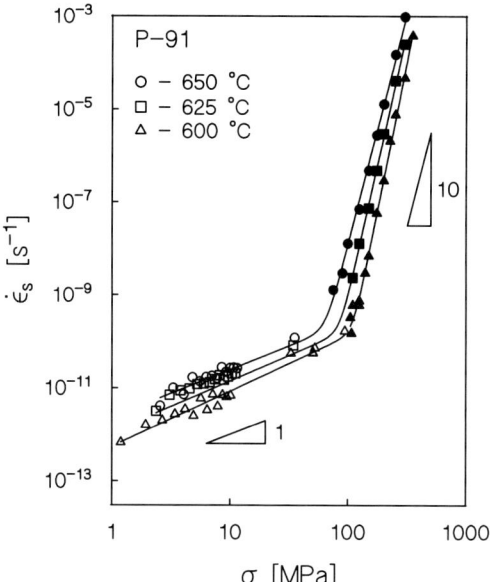

Fig. 4   Stress dependence of the steady state creep rate in the both power-law and viscous regimes. The data obtained by the helicoid spring specimen technique are marked by the open symbols.

by the eqn (1) with the tertiary term neglected. The condition $t \geq 6\,t_p$ was used to prove that the secondary state was reached. A set of typical creep curves obtained at low stresses is displayed in the Fig. 2. Figure 3 compares the creep curves obtained at similar conditions by uniaxial tensile test and helicoid spring specimen test. It is clear that both curves are well comparable.

3.2   Secondary Stage Creep Rates

The secondary stage creep rate $\dot{\epsilon}_s$ was derived as a parameter from the eqn (1). Its dependence on the applied stress is shown in the Fig. 4. The slope of the dependence, i.e. the apparent stress exponent, is close to unity at low stresses up to about 100 MPa at 600°C, and then increases drastically to the values higher than 10. This behaviour clearly indicates the change in the controlling mechanism. The viscous creep at lower stresses is replaced by the power-law creep at higher stresses. The curves in the Fig. 4 are fitted to the experimental creep rates under an assumption, that both mechanisms act in parallel, that is

$$\dot{\epsilon}_s = \dot{\epsilon}_v + \dot{\epsilon}_p \tag{2}$$

where $\dot{\epsilon}_v$ and $\dot{\epsilon}_p$ are creep rates of viscous and power-law mechanisms, respectively. It can be seen from the figure, that the assumption is in a good agreement with the measured data.

The secondary stage creep rate dependence on the temperature can be described using an apparent activation energy, defined as

$$Q_a = - \frac{\partial \ln(\dot{\epsilon}_s)}{\partial(1/RT)} \tag{3}$$

where T is the absolute temperature and R has the obvious meaning. The dramatic increase of apparent activation energy at stresses around 100 MPa provides another evidence for the change in the creep controlling mechanism (Fig. 5).

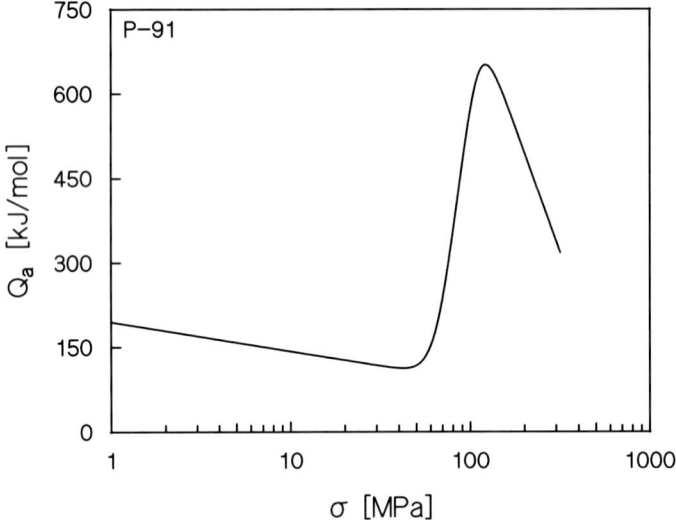

Fig. 5   Dependence of the apparent activation energy on the applied stress. The curve is derived from the fitted lines in the Fig. 4.

3.3   Time to Fracture
The time to fracture was measured with the conventional tensile creep tests only. There are two main reasons for it: (i) the time to fracture is too long for the tests at low stresses and (ii) the helicoid spring specimen technique is convenient only for very low strains. The times to fracture for power-law creep regime fulfil the Monkman–Grant relation, as can be seen from the Fig. 6.

4.   DISCUSSION

4.1   Power-Law Creep Regime
The results in the power law creep regime are consistent with that obtained by other authors[9] on similar material. The apparent activation energy is about

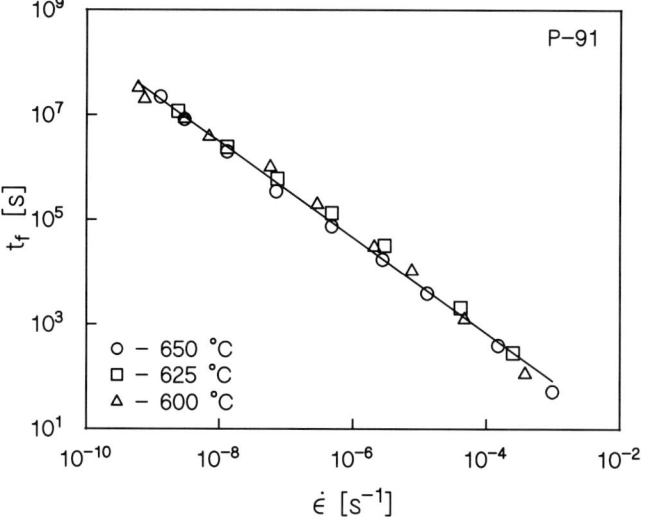

Fig. 6   Monkman–Grant relation in the power-law creep regime.

three times higher than activation enthalpy of diffusion.[10–12] This high value increasing with decreasing stress, together with the relatively high value of stress exponent suggests a temperature dependent threshold stress. Nevertheless, this study was not focused on deep analysis of creep behaviour in power-law regime.

4.2   Viscous Creep Regime

Two different mechanisms have been proposed for viscous creep. The first is Nabarro–Herring[13,14] and/or Coble[15] diffusional creep. The second is Harper–Dorn dislocation creep.[16] The main difference between both mechanisms consists in grain size dependence. Since the material of only one grain size was used, the results do not provide any basis for the creep mechanism identification. The measured value of apparent activation energy for viscous creep regime is comparable or even lower than activation enthalpy of grain boundary diffusion.[9–12] On the other hand, the dislocation Harper–Dorn creep mechanism should be controlled by the pipe diffusion under the applied conditions.[17] The low value of apparent activation energy then provides no additional information about the possible mechanism.

4.3   Transition from Viscous to Power-Law Creep

The data obtained in this work show relatively sharp transition between the power-law and viscous creep regime. Since the difference in stress sensitivity is very large for both regimes, even the assumption of two independent processes leads to such a sharp transition, as was shown in section 3.2.

The serious difficulty arises from the fact, that the different creep regimes were revealed by the different testing technique. Recently, Kowalewski[18] published results showing that the uniaxial tensile stress is not equivalent to the shear stress at special conditions. Since most theories of creep deformation mechanisms are based on this assumption, such results can influence the understanding of creep deformation mechanisms very seriously. Nevertheless, creep curves obtained by both techniques at similar conditions do not justify hesitations about the influence of testing technique on the presented results.

4.4 Creep Life

In the power-law creep regime, the Monkman–Grant relation for time to fracture t_f is fulfilled. The equation representing the Monkman–Grant relation,

$$t_f = k(\dot{\epsilon}_s)^{-g}, \qquad (4)$$

where k and g are parameters, seems to be a good tool for prediction of time

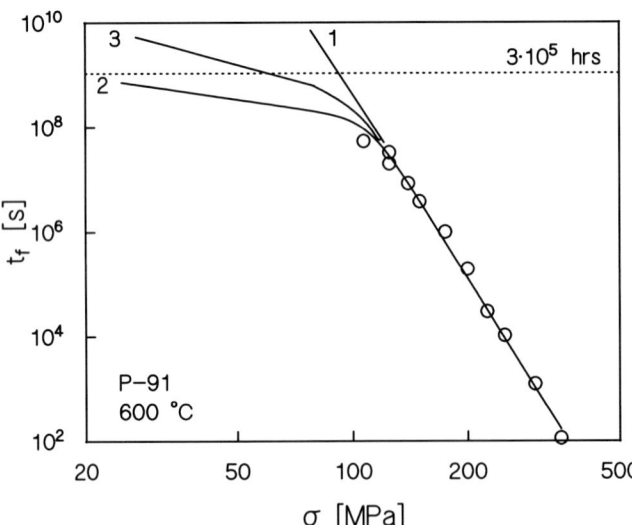

Fig. 7 The time to fracture dependence on the applied stress. Possible extrapolations to lower stresses are marked as follows:
1 – Simple extrapolation from power-law regime. Viscous regime completely neglected.
2 – Validity of the eqn (4) assumed for viscous creep regime, too.
3 – Limited contribution of viscous creep to damage development.

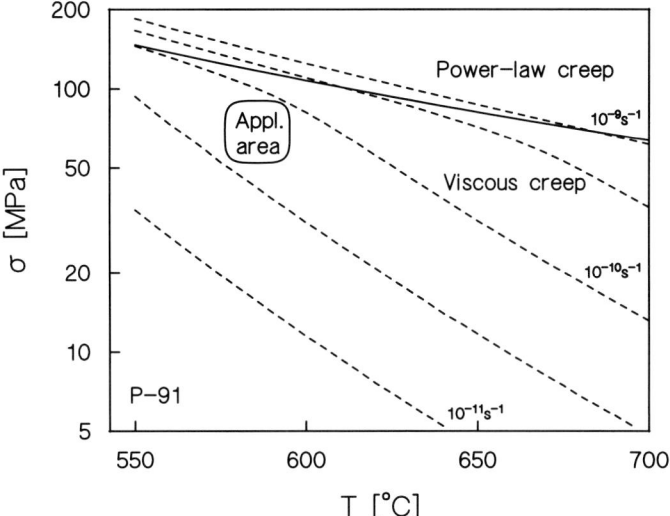

Fig. 8 Creep deformation mechanism map derived from the results on the Fig. 4.

to fracture, since the $\dot{\epsilon}_s$ can be measured in relatively short term tests. Regression analysis for the present material gives $g = 0.906$ and $k = 0.168$. But the validity of the eqn (4) may break down when the creep deformation mechanism changes. Under the viscous creep regime, the material behaves as a Newtonian fluid. Since the development of creep damage can be treated as a plastic instability in a micro-scale and the Newtonian fluid is resistant against plastic instabilities, it is possible that the viscous creep mechanism does not contribute to the creep damage development. Possible dependencies of time to fracture on applied stress are depicted in the Fig. 7. The creep life longer than that predicted by the line 2 must be balanced by the higher total creep strain. In the authors' opinion, the most probable behaviour is described by the line 3.

4.5   A Map of Creep Mechanisms

When the material may deform by several different mechanisms, it is convenient to present these mechanisms in the form of a deformation mechanism map.[19] The creep deformation mechanisms map in Fig. 8 was derived from the results in the co-ordinates temperature vs. stress. The potential service loading conditions for the steel under consideration, i.e. temperatures 575–600°C and stresses 50–100 MPa are fully involved in the viscous creep regime, though it is not very distant from the transition to power-law creep. Therefore, viscous creep should be taken into consideration, mainly for the applications where the dimension stability is critical.

5. CONCLUSIONS

The creep behaviour of a 9% chromium steel (P-91 type) was investigated at temperatures from 600°C to 650°C and at wide range of stresses from 1 MPa to 300 MPa, using the helicoid spring specimens technique and the conventional uniaxial tensile technique. An analysis of creep data showed that the viscous creep regime is inherent to the creep conditions, which correspond to the practical use of the steel. Mutual comparison of the creep results obtained by the helicoid spring specimen technique with the results of standard uniaxial tensile creep tests on the steel investigated has shown very good coherency of the both creep testing techniques. The transition from power-law creep with the stress exponent about 10 to the viscous creep was found at stresses around 100 MPa at 600°C. Any extrapolation from the power-law creep regime to stresses below 100 MPa may lead to serious and potentially dangerous underestimation of the creep rate predicted.

ACKNOWLEDGEMENTS

Financial support for this work was provided in part by the Grant Agency of the Academy of Sciences of the Czech Republic under Contract No. A2041702, and in part by the Ministry of Education of the Czech Republic under Contract No. OC 501.20

REFERENCES

1. T. Sritharan and H. Jones: *Acta Metall*, 1980, **28**, 1980, 1633.
2. T. Sritharan and H. Jones: *Metal Sci.*, 1981, **15**, 365.
3. S.J. Dobson, H. Jones and G.W. Greenwood: *Perspectives in Metallurgical Development*, J. Beech ed., The Metals Society, London 1984, 291.
4. R.S. Mishra, H. Jones and G.W. Greenwood: *Acta Metall. Mater.*, 1990, **38**, 461.
5. B. Burton and G.W. Greenwood: *Metal. Sci. J.*, 1970, **4**, 21.
6. J. Fiala and J. Čadek: *Metall. Mater.* **20**, 1982, 27 (in Czech).
7. J. Novotný, J. Fiala and J. Čadek: *Metall. Mater.*, 1985, **23**, 32 (in Czech).
8. J.C.M. Li: *Acta Metall.*, 1963, **11**, 126.
9. S. Spigarelli, E. Evangelista, E. Cerri and C. Guardamagna: in *Creep Resistant Metallic Materials*, Vítkovice steel R & D, Ostrava 1996, 335.
10. A.M. Huntz, P. Guiraldenq, M. Aucouturier and P. Lacombe: *Mem. Sci. Rev. Metall.*, 1969, **66**, 8.
11. F. Chaix and A.M. Huntz: *Mem. Sci. Rev. Metall.*, 1974, **71**, 11.
12. J. Čermák, J. Růžičková and A. Pokorná: *Scripta Mater.*, 1996, **35**, 41.
13. F.R.N. Nabarro: *Report of Conference on Strength of Solids*, Phys. Soc., London 1948, 7.
14. C. Herring: *J. Appl. Phys.*, 1950, **21**, 43.

15. R.L. Coble: *J. Appl. Phys.*, 1963, **34**, 167.
16. J.G. Harper and J.E. Dorn: *Acta Metall.*, 1957, **5**, 65.
17. A.H. Chokshi: *Scripta Metall.*, 1985, **19**, 52.
18. Z.L. Kowalewski: *Arch. Mech.*, 1995, **47**, 13.
19. M.F. Ashby: *Acta Metall.*, 1972, **20**, 88.

The Contribution of Microstructural Investigations to the Development of Improved Materials for Future High Temperature Steam Turbines

R.W. VANSTONE

GEC ALSTHOM, Large Steam Turbines, Newbold Road, Rugby, CV21 2NH, UK

ABSTRACT

The development of advanced creep resistant 9–12%Cr steels is outlined. The understanding of microstructural evolution and its role in determining long term behaviour, gained from microstructural investigations and material modelling, is discussed. The potential significance of such tools to continuing and future development, both of steels and nickel-based alloys is described.

THE HISTORICAL DEVELOPMENT OF ADVANCED 9–12%CR STEELS

Creep resistant 9–12%Cr alloys were first developed for gas turbine applications in the 1950s. Steels like the 12CrMoVNb alloys developed in the UK have high alloy content giving them high hardenability so that on cooling after solution treatment martensite is formed even in thick section and at relatively low cooling rates. This martensitic matrix contains a dense network of dislocations which is stabilised by a dispersion of carbides formed during tempering. Dislocation movement is inhibited by interaction with the particle dispersion and with the high density of other dislocations, resulting in high creep strength. These steels contain high levels of C which combines with large amounts of carbide forming elements like Mo and V but especially Nb. These steels were not generally considered appropriate for the large forgings and castings required for steam turbines, being susceptible to segregation and having poor forgeability, poor weldability and limited fracture toughness. Although these steels have very high creep strength for durations of 10 000 hours, appropriate to their jet engine application, at temperatures above about 550°C their creep strength falls rapidly as durations are extended towards the 100 000 hours which is the typical basis of steam turbine design, This reduction in creep strength occurs as a result of microstructural instability; the carbide dispersion becomes much coarser after long durations and is then less effective in inhibiting the movement of dislocations.

457

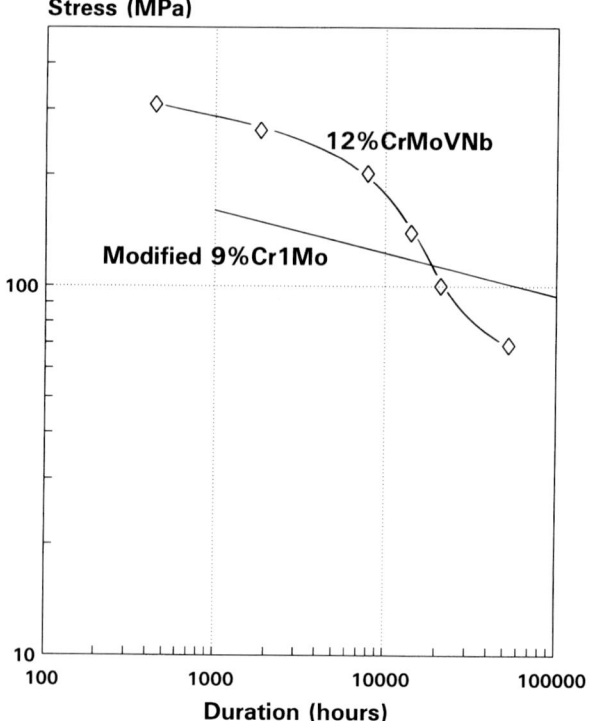

Fig. 1 In the short term at 600°C 12CrMoVNb has higher rupture strength but in the longer term modified 9%Cr1Mo is stronger.

In the mid 1970s ORNL were charged by the US Department of Energy with the development of a ferritic steel with high creep strength and good weldability for application in fast breeder reactors. They took as their basis the 9%Cr1%Mo alloy used for tubing in the UK and modified it through additions of V, Nb and N. The resultant alloy, known as modified 9%Cr1Mo, bore a superficial resemblance to the earlier generation 12CrMoVNb steels but it differed in certain key respects. Nb content was reduced drastically, from levels of 0.3% and above to less than 0.1%. This eliminated the occurrence of large particles of primary NbC, not taken into solution during heat treatment, which contributed to the poor toughness of the earlier 12CrMoVNb alloys. Carbon content was also reduced, to about 0.1%. The optimisation of alloy content resulted in more stable long term creep properties.[1] At 600°C the 100 000 hour creep strength of the new alloy proved to be around 50% greater than that of the previous generation of 12CrMoVNb steels (Fig. 1).

Around the same time that modified 9%Cr1Mo was being developed in the USA, work began in Japan on materials suitable for rotor forgings. This work, which involved tests on a wide range of materials to find the best chemical analysis, led to the development of an alloy designated TMK1.[2] This steel is similar to modified 9%Cr1Mo but contains slightly more Cr (about 10%) and C (about 0.14%) and Mo is increased from 1.0 to 1.5%. The increase in C content, together with the adoption of lower tempering temperatures, enables yield strengths greater than 700 MPa to be achieved while at the same time retaining a high level of fracture toughness. At 600°C the 100 000 hour creep strength of this alloy is similar to, or slightly greater than, that of modified 9%Cr1Mo.

In Europe a collaborative project between all the main turbine makers under the auspices of COST 501 began in the early 1980s. A large number of small scale melts in different alloys and different heat treatment conditions were assessed and two of the most promising materials were applied to the manufacture of full scale rotor forgings which were sectioned for short and long term characterisation.[3] These alloys were an optimised 10%CrMo VNbN alloy (Steel F) very similar to TMK1 and an alloy with an addition of 1% tungsten (Steel E).

The new steels are currently being applied to rotors in machines operating at high temperature and pressure. In Japan TMK1 has been applied at a number of power stations, as has an alloy containing additions of tungsten. In Europe, the COST developments have been used as the basis for selection of 10%CrMoVNbN (Steel F) for the VHP and HP/IP rotors of the steam turbines for the Skaerbaek and Nordjylland power stations, currently under construction in Denmark. The application of these steels in place of the low alloy steels conventionally used has enabled an increase in operating temperatures resulting in increased power generation efficiency, lower fuel consumption and lower emissions.[4]

MICROSTRUCTURAL CHARACTERISATION

During the last decade, microstructural investigations have been carried out in an attempt to understand the improved properties of these materials. With such understanding comes the ability to improve properties still further. Studies have been carried out in the USA and in Japan but probably the greatest effort has been made in Europe where 13 bodies from 7 countries have worked together to investigate the alloys developed within COST 501[5] (Table 1). These studies have led to a much better description of microstructural evolution during heat treatment and service: of dislocation structure, of secondary phase dissolution and precipitation, and of particle size and the influence of these factors on matrix composition.

Nearly all of the newly developed steels are designed to be fully martensitic in order to give the optimum combination of creep strength and toughness. After heat treatment of the steels thin film electron metallography reveals elongated dislocation cells or subgrains containing a dense dislocation net-

Table 1   Participants in COST 501 microstructural characterisation and modelling activities.

Participant	Location
Austrian Research Centre	Seibersdorf, Austria
Technical University of Graz, Institute of Materials	Graz, Austria
Institute of Physics of Materials (Czech Academy of Sciences)	Brno, Czech Republic
Skoda Research	Plzen, Czech Republic
Vitkovice j.s.c. Research Institute	Ostrava, Czech Republic
ELSAM/ELKRAFT	Lyngby, Denmark
Tampere University of Technology, Institute of Materials Science	Tampere, Finland
VTT Metals Laboratory	Espoo, Finland
ENEL Thermal and Nuclear Research Centre	Milano, Italy
University of Ancona Dept. of Mechanics	Ancona, Italy
Swedish Institute for Metals Research	Stockholm, Sweden
Cambridge University, Dept. of Materials Science and Metallurgy	Cambridge, United Kingdom
GEC ALSTHOM Turbine Generators (co-ordinator)	Rugby, United Kingdom

work (Fig. 2). Interaction between dislocations and precipitates is important, especially in pinning the subgrain walls. Carbide extraction replicas clearly reveal the alignment of the larger carbides, mainly $M_{23}C_6$ with the subgrain walls but also reveal the presence of smaller VN or M_2X particles inside the subgrains (Fig. 3).

Fig. 2   COST 501 Steel B: Carbides pin the elongated subgrain boundaries. (Courtesy of K. Spiradek, OFZS, Austria)

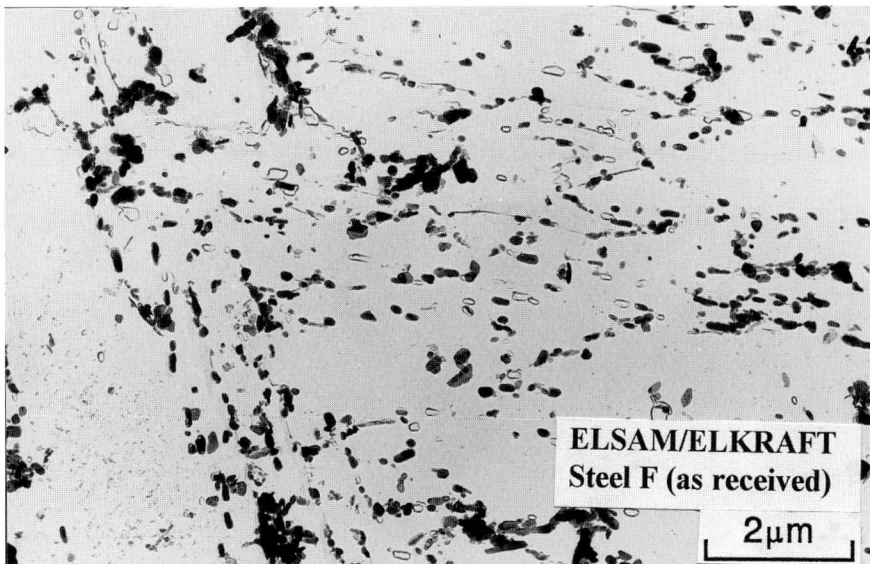

ELSAM/ELKRAFT
Steel F (as received)
2μm

Fig. 3   COST 501 Steel F: Larger carbides generally coincide with subgrain boundaries. Finer carbides appear within the subgrains. (Courtesy of J. Hald, ELSAMPROJEKT, Denmark)

Creep testing results in marked changes to this microstructure. This is evident from reductions in hardness. Some softening occurs due to thermal exposure in the heads of creep test pieces but the reduction of hardness in the gauge length is much more marked. This softening is accounted for by a number of microstructural changes.

Particles coarsen and changes in the mean particle size approaching an order of magnitude are not uncommon. Much of this particle coarsening is accounted for by the growth of existing particles through Ostwald ripening but some is due to the precipitation of new particles which coarsen rapidly. Significant quantities of Laves phase precipitate in all steels containing more than about 0.5% Mo or W. The precipitation of M_6C particles is encouraged in steels with high levels of Ni[6] and formation of Z-phase has been observed in Nb bearing steels.[7] The precipitation of Laves phase is particularly significant from a number of aspects, especially in the most modern steels which contain high levels of Mo and W and therefore form large quantities of this phase. Firstly, Laves phase is observed not to form at high temperatures. It is therefore invalid to use the results of creep tests at these temperatures to predict behaviour at lower temperatures where Laves phase does form and influences creep behaviour. Secondly the influence of Laves phase on creep strength is still a matter of debate. By withdrawing Mo and W from solid solution Laves phase would be expected to reduce creep strength but the precipitates formed might result in additional precipitation hardening. To date the metallographic evidence suggests that when Laves phase forms it coarsens rapidly and is thus observed only as large particles which have no significant strengthening effect.

Significant changes in dislocation structure occur in parallel with these changes in particle content. Dislocation density falls so that few dislocations are observed inside the subgrains. In the very early stages of creep this can result in a reduction in mean subgrain size. Free dislocations migrate and combine to form new subgrain walls. However these subgrains then grow larger and become more equiaxed.

These changes in particle distribution and dislocation structure are observed to occur more rapidly in the gauge length of creep test pieces than in the head, explaining the greater reductions in hardness seen in the gauge length.

Resistance to increments in creep strain in these steels depend on barriers to dislocation movement through climb and glide. A dense dislocation network and a dispersion of fine particles together with solid solution hardening of the matrix represent effective barriers to dislocation movement. As these barriers become less effective through particle coarsening, sub-grain growth and the loss of elements from solid solution through precipitation creep strain rates increase. Microstructures providing high creep strength in the short term condition can be achieved in a wide range of steels and are most simply produced by tempering at lower temperatures, treatments which also give high proof and tensile strength. However materials tempered at lower temperatures are less stable, their microstructures coarsen more rapidly during creep so that any advantage they hold over more highly tempered structures is lost in the longer term. The key to long term creep strength lies with those steels which undergo these microstructural changes at the lowest rates. The strongest steel tested in the COST 501 programme contained about 100 ppm boron and was designated Steel B. This steel was remarkable for maintaining a high dislocation density for significantly longer than the other steels investigated in COST 501.[5,8]

MODELLING MICROSTRUCTURAL EVOLUTION AND CREEP STRENGTH

Models have been developed which go some way to explaining these microstructural changes and predicting their effect on creep strength. During creep the steels evolve towards thermodynamically more stable conditions. Computer models have been developed which use available thermochemical data to predict the most thermodynamically stable combination of phases in the steel, the equilibrium condition where free energy is at a minimum. Equilibrium phase diagrams are produced which predict the phases present and their fraction as a function of temperature. The chemical composition of

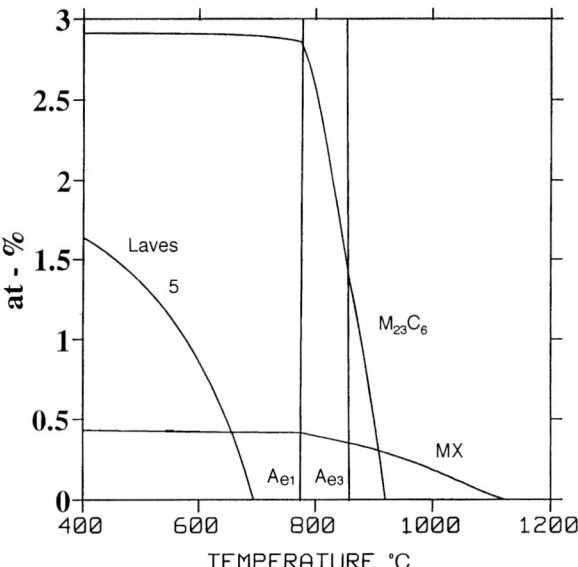

Fig. 4 Equilibrium phase diagram for COST Steel E generated by use of Thermocalc. (Courtesy of H. Cerjak, Technical University of Graz, Austria)

each phase can also be predicted. An example is shown for COST steel E in Fig. 4 and it can be seen that at typical tempering temperatures just above 700°C the presence of only $M_{23}C_6$ and MX is predicted. At temperatures representing likely service temperatures for this steel, around 600°C, the phase diagram predicts the presence of Laves phase in addition to these phases. This is consistent with metallographic observations. Laves phase is not present after heat treatment but after long term exposure it does precipitate, in line with the predicted equilibrium. Thus many of the microstructural observations can be explained on the basis of evolution towards a state of minimum free energy.

Microstructure evolves only slowly towards the state of equilibrium because of the slow kinetics governing many of the processes of this evolution. In order to understand and predict the rate of evolution, models describing microstructural kinetics are required. Major efforts in modelling the kinetics of precipitate nucleation and growth have already been made. The model developed by Robson and Bhadeshia[9] uses thermochemical data in combination with models for nucleation and growth to predict the sequence in which phases precipitate, their fraction as a function of time and the distribution of particle size for each phase.

Understanding and predicting microstructural evolution is not a goal in itself. The link between microstructure and creep strength must be modelled.

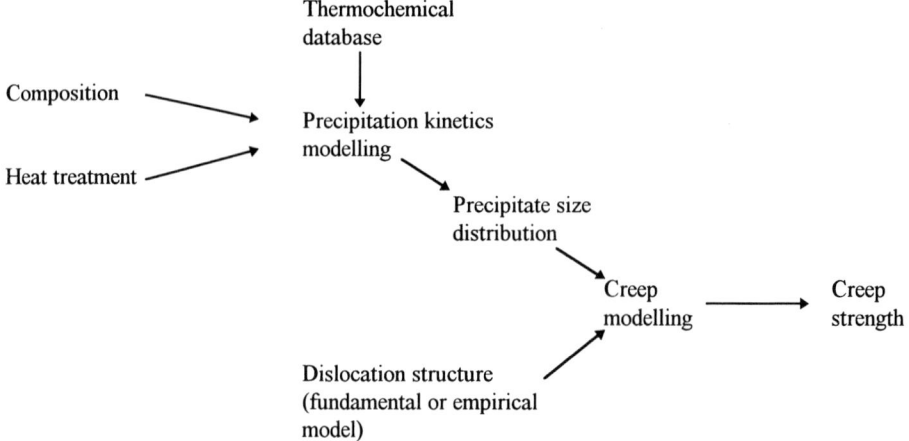

Fig. 5 The potential for combination of thermodynamic, kinetic and creep modelling to predict creep strength.

Only then can modelling be used to predict creep strength and thus provide a powerful tool for alloy development. Models describing the accumulation of creep strain as a function of microstructural parameters are also being developed. For example, the composite model of Straub *et al.*[10] predicts creep strain as a function of particle size, dislocation density and subgrain size. Therefore many of the elements are in place for a combined model which will predict creep strength from a knowledge of chemical composition and heat treatment (Fig. 5). However no models are yet available which describe the evolution of dislocation density and subgrain size. This input to a combined model must currently be made through empirical observation.

FURTHER DEVELOPMENT IN MODELLING MICROSTRUCTURAL EVOLUTION AND CREEP STRENGTH

To date, a good qualitative description of microstructural development has been achieved and some progress has been made in development of models explaining this evolution and the link between microstructure and creep strength. However validation of the models requires more quantitative descriptions of microstructure to be available. For a full description of the microstructure, quantitative measurements of a great many parameters are required: chemical composition of the matrix; dislocation density and sub-grain size distribution; and size distribution, shape, and chemical analysis for each precipitate species at each of the characteristic locations within the microstructure (grain boundaries, subgrain boundaries, subgrain interiors). These parameters must be known at the beginning of service life and their evolution must be known throughout life up to the point of rupture. This is clearly a massive task and most investigations to date have investigated only a few of these parameters, have simplified some of the parameters by, for example, not discriminating between different species when measuring particle size distribution, and have typically focused only on the as-received and end of life conditions.

The size of the task necessary for establishing a database by which models can be validated puts it beyond the capability of a single metallographer and even beyond the capability of a single laboratory. Therefore collaborative action by many laboratories is necessary and this requires that the results generated by different laboratories are comparable. Such collaboration has been one of the strengths of the COST 501 project but experience has revealed that significantly different values can be reported for measurements of the same parameter. For example, variations of a factor of 2 have been reported when different laboratories measured mean subgrain width in the same material. Investigation into the source of such variation has suggested that much of it arises from the use of different image analysis procedures and packages. There is a clear need for standard procedures to be developed for quantitative metallography.

Even if quantitative databases on microstructural evolution are generated to provide a basis for validation of models, further development of the models existing today will be required. The models for prediction of equilibrium microstructure depend on thermochemical data. In principle the dependence of the free energy of every phase present in the steel on every element present in the steel must be known to allow definition of the minimum free energy condition. Since there are typically more than five phases to be considered and the number of elements significantly affecting microstructure is generally in excess of twelve, it can be seen that a very large thermochemical database is required. Potentially important interactions, such as the influence of nickel on certain carbides, are currently unknown. Further development of the existing thermochemical database is required.

Other influences are also not incorporated in available models. Strain is observed to be very significant in accelerating microstructural changes but current models do not address this influence. Models linking microstructure to creep behaviour will also require further development so as to take into account the influence of solid solution hardening in addition to the influence of particle dispersions and dislocation structure. It has already been mentioned that no models describing the evolution of dislocation structures are currently available.

In summary, physically based models have the potential to provide powerful tools for alloy development. However further work is necessary to extend and combine existing models; to provide the necessary thermochemical data; and to develop a microstructural database for validation of the models, this in turn requiring the standardisation of metallographic procedures.

CONTINUED EMPIRICAL ALLOY DEVELOPMENT IN ADVANCED 9–12%CR STEELS

The current status of these physically based models is sufficient to provide only peripheral support to the continuing development of 9–12%Cr steels. This development continues, as in the past, on a largely empirical or semi-empirical basis. Currently attempts to raise creep strength are based principally around the addition of tungsten and boron.

The Use of Tungsten to Elevate Creep Strength

The intellectual basis for this approach was laid by Fujita. He reported that creep strength was optimised when the Mo + 0.5 W content was around 1.5%. TMK1 obeys this principle with a Mo content of 1.5% and no tungsten. However he also reported improved creep strength when W rather than Mo was used.[11]

Some of the rotor materials already developed contain up to 1%W but such additions have not resulted in creep strength discernibly greater than that of the W-free compositions. Similarly, in an attempt to raise creep strength of cast materials, an alloy containing 1% tungsten was successfully applied to the manufacture of a valve chest casting within the COST 501 project.[12] Although the chemistry and heat treatment of the alloy give it enhanced tensile strength in comparison with modified 9%Cr1Mo and hence an advantage in short-term creep strength,[13] its advantage in the long term has still to be established.

More generous additions of tungsten have sometimes been more successful. One of the most mature developments is that of NF616, containing about 1.8%W and 0.5%Mo. This alloy has 100 000 hour strength at 600°C estimated to be about 30% greater than that of the tungsten free alloys and has already gained acceptance under ASME codes for pipework applications.[14] On the other hand an alloy with similar W level investigated within COST 501 showed poor long term strength. Although a rotor forging with 1.8%W has been produced in Japan[15] it seems to offer at best only a very marginal advantage in creep strength over the W free alloys like COST 501 Steel F and TMK1.

The introduction of large amounts of W encourages the formation of delta ferrite. This is avoided in NF616 by the limitation of Cr content to around 9%. However if alloys like these are to be applied at temperatures above 600°C then oxidation resistance becomes a concern and it becomes desirable to increase Cr content to nearer 12%. The combination of such Cr levels with high W content would certainly result in significant quantities of delta ferrite so elements which increase the austenite stability field and hence suppress delta ferrite, such as cobalt or copper, are added. In Japan a 12%Cr alloy containing 2.5%W balanced by 2.5%Co, designated HR1200, has been developed for rotor forgings.[16] A very similar alloy is also being developed for tubing applications under the designation NF12.[17]

The Use of Boron to Elevate Creep Strength
Work on steels containing B dates back to the origins of the COST 501 project in the early 1980s. It soon became apparent that large additions of this element led to severe segregation and an optimum level of around 100 ppm was established. A steel containing this level of B was further developed in later rounds of COST 501 and showed promise of creep strength slightly greater than that of the B free steels.[3] A VGB funded project has recently applied this steel to the manufacture of a full scale rotor forging which will be sectioned for long term characterisation.

The use of boron has also been adopted in Japan. In addition to 2.5%W, the HR1200 alloy contains significant levels of boron. In line with the European experience, early work on this alloy with levels of B around 180 ppm resulted in segregation and later development has focused on reduced B levels, around 100 ppm.[16] NF616 also contains some B, typically about 30 ppm, and it is uncertain what contribution this element, rather than the high W content, makes to its high creep strength.

In Japan and Europe development is continuing on alloys with 12%Cr and high levels of W balanced by elements like Co or Cu and with additions of B. In COST 501 a series of trial melts is being tested and metallographic investigations are continuing to throw light on the mechanisms of creep resistance. The success of these developments would enable the design and construction of steam turbines with inlet temperatures approaching 650°C.

EXPLOITATION OF NICKEL-BASED ALLOYS

Even the most optimistic view does not suggest that the 12%Cr alloys will enable steam turbines to operate at temperatures greater than 650°C. Nonetheless advances in steam turbine operating temperature to around 700°C are planned through the exploitation of nickel-based alloys. It will be important to understand the essential features of the microstructure of these alloys and its evolution in service. Where alloys dependent on precipitation strengthening are employed it will be important to have confidence in the stability of these precipitate dispersions over long term at the operating temperature. Another important feature will be the tendency of these alloys to undergo long range ordering bringing about changes in lattice parameter and hence changes in component dimensions. This effect is well known in Nimonic 80A which has nonetheless been successfully used as a bolting alloy for many years. The tendency of bolts to contract after long term exposure is potentially serious since it can lead to overloading of the bolt. However given knowledge of this phenomenon the bolted joint can be designed to accommodate such contractions.

Much of the knowledge necessary for the application of nickel base alloys in steam turbines is available through technology transfer from the gas turbine industry and its supply chain. However there will be significant differences in the components to which these alloys will be applied and the conditions under which they will operate. Components for both static, pressure containing purposes and for rotating parts will be required in sizes up to, and possibly slightly beyond, the limits of current experience and capability. These components will then be required to remain in service for periods of up to 30–40 years. The steam turbine operating regime is typically a more lowly stressed but longer term regime than is typical for gas turbines. The application of nickel-based alloys, even where they are existing alloys, is likely to require additional long term characterisation to account for these changes in product form and service conditions. The sensitivity of microstructural stability to these new circumstances must also be established.

CONCLUSION
Microstructural investigations, and models arising from these investigations, have provided a good qualitative understanding of microstructural evolution in advanced 9–12%Cr steels. Refinement and combination of the models has the potential to provide a powerful tool for alloy development and optimisation. Such a tool would greatly reduce the time and expense involved in the development of new steels and the exploitation of new, stronger steels in the power generation industry would bring widespread economic and environmental benefits. In the meantime most alloy development continues now, as it has done in the past, on a largely empirical basis.

New initiatives to apply nickel based alloys to steam turbines are just beginning. It is to be hoped that advances made in microstructural investigation and planned applications of material modelling to alloy design, processing and property prediction will prove to be sufficient to make a significant contribution to these initiatives in parallel with more empirical techniques.

REFERENCES
1. V.K. Sikka, M.G. Cowgill and B.W. Roberts: 'Creep properties of modified 9Cr-1Mo steel', *Proc. Conf. Ferritic alloys for use in nuclear energy technologies*, Snowbird, Utah, June 1983.
2. Y. Nakabayashi *et al.*: 'Advanced 12Cr steel rotor (TMK1) for EPDC's 50MW high temperature turbine step 1 (593°C/593°C)', *Proc. of 1st EPRI Int. Conf. on Improved Coal-Fired Power Plants (ICPP)*, Palo Alto, November 1986.
3. C. Berger, R.B. Scarlin, K.H. Mayer, D.V. Thornton and S.M. Beech: 'Steam turbine materials: high temperature forgings', *Proc. of COST 501*

Conf. Materials for Advanced Power Engineering 1994, Liège, October 1994.

4. R.W. Vanstone and D.V. Thornton: 'New materials for advanced steam turbines', *Proc. of I Mech E conference on Advanced Steam Plant*, London, May 1997.

5. R.W. Vanstone, H. Cerjak, V. Foldyna, J. Hald and K. Spiradek: 'Microstructural development in advanced 9–12%Cr creep resisting steels – a collaborative investigation in COST 501/3 WP11', *Microstructural development and stability in high chromium ferritic power plant steels*, The Institute of Materials, 1997, 93.

6. A. Strang and V. Vodarek: 'The effects of Ni on the precipitation characteristics and microstructural stability of creep resistant martensitic 12CrMoV and 12CrMoVNb turbine steels', *this publication*.

7. A. Strang, and V. Vodarek: 'Z phase formation in martensitic 12CrMoVNb steel', M. S. Technol., 1996, **12** (7), 552–556.

8. K. Spiradek, R. Bauer and G. Zeiler: 'Microstructural changes during the creep deformation of 9%Cr steel', *Proc. of COST 501 Conf. Materials for Advanced Power Engineering 1994*, Liège, October 1994.

9. J. Robson and H.K.D. Bhadeshia: 'Kinetics of carbide precipitation in ferritic power plant steels', *this publication*.

10. S. Straub, P. Polick and W. Blum: 'Simulation of the creep behaviour of 9–12%CrMoV steels on the basis of microstructural data', *this publication*.

11. T. Fujita, K. Asakura and T. Sawada: 'Creep rupture strength of low C, 10Cr-2Mo heat resisting steels', *Metall Trans.*, 1981, **2A**, 1071.

12. R.B. Scarlin, C. Berger, K.H. Mayer, D.V. Thornton and S.M. Beech: 'Steam turbine materials: high temperature castings', *Proc. of COST 501 Conf. Materials for Advanced Power Engineering 1994*, Liège, October 1994.

13. D.V. Thornton and R.W. Vanstone: 'Materials developments for application in steam turbines for fossil fired plant', *3rd Int Charles Parsons Conference 'Materials engineering in turbines and compressors'*, Newcastle upon Tyne, April 1995.

14. New steels for advanced plant up to 620°C, *EPRI/National Power conference*, London, May 1995.

15. M. Kamada *et al.*: 'An advanced turbine rotor of 12Cr steel (TMK2) applicable to 600°C steam temperature', *3rd Int Charles Parsons Conference 'Materials engineering in turbines and compressors'*, Newcastle upon Tyne, April 1995.

16. K. Hidaka *et al.*: 'Development of large scale 12Cr ferritic steel turbine rotor aiming the application of ultra supercritical steam power plant at 650°C', *ibid*.

17. M. Ohgami, Y. Hasegawa, H. Naoi and T. Fujita: 'Development of 11CrMoWCo heat resistant steel for fossil thermal plants' *Proc. of I Mech E conference on Advanced Steam Plant*, London, May 1997.

List of Delegates

Allen, D.;
  Powergen, UK
Artinger, I.; (361–70)
  TU Budapest, Hungary
Barnes, A.M.; (339–60)
  TWI, UK
Berglin, L.;
  ABB Stahl AB, Sweden
Bjarbo, A.O.;
  Royal Inst. Technology, Sweden
Bralsford, P.;
  Forgemasters Steel & Engineering, UK
Brett, S.;
  National Power plc, UK
Buršík, J.; (89–106)
  Academy of Sciences, Czech Republic
Cawley, J.;
  Sheffield Hallam University, UK
Cerjak, H.; (323–38)
  TU Graz, Austria
Dulieu, D.; (1–26)
  Avesta Sheffield Ltd, UK
Dyson, B.; (371–94)
  Imperial College, UK
Ennis, P.J.; (135–44, 215–22)
  Research Centre Juelich, Germany
Ernst, P.; (311–22)
  ABB Corporate Research Ltd, Sweden
Foldyna, V.; (257–70)
  Vitkovice/GECA, Czech Republic
Forster C.;
  Rolls-Royce, UK
George, R.;
  British Steel Engineering Steels, UK
Gladh, M.;
  CITU, Sweden

Gladman, T.; (49–68)
  Leeds University, UK
Göcmen, A.; (311–22)
  ABB Corporate Research Ltd, Switzerland
Gravill, N.;
  Ross & Catherall Ltd, UK
Greenwood, G.W.;
  University of Sheffield, UK
Green, J.;
  Parsons, UK
Hald, J.; (173–84)
  ELSAM/Elkraft, Denmark
Hammersley G.;
  Metal Improvement Company inc, UK
Hättesrand M.; (197–214)
  Chalmers University, Sweden
Hellsing, M.;
  CITU, Sweden
Henderson, P.;
  Vattenfall Energisystem AB, Sweden
Holmes, P.;
  ABB Corporate Research Ltd, Switzerland
Honeyman, G.;
  Forgeed Rolls, UK
Kallqvist, J.;
  Chalmers University, Sweden
Kastensson, P.;
  ABB Stahl AB, Sweden
Kern, T.-I.;
  Siemens KWU, Germany
Kimura, K.; (185–96)
  National Research Institute for Metals, Japan
King, S.;
  Special Melted Products Ltd, UK

Authors in these proceedings have the corresponding page numbers listed after their names.

Index

473